NATO ASI Series

Advanced Science Institutes Series

A series presenting the results of activities sponsored by the NATO Science Committee, which aims at the dissemination of advanced scientific and technological knowledge, with a view to strengthening links between scientific communities.

The Series is published by an international board of publishers in conjunction with the NATO Scientific Affairs Division

A Life Sciences B Physics	Plenum Publishing Corporation London and New York
C Mathematical and Physical Sciences D Behavioural and Social Sciences E Applied Sciences	Kluwer Academic Publishers Dordrecht, Boston and London
F Computer and Systems Sciences G Ecological Sciences H Cell Biology I Global Environmental Change	Springer-Verlag Berlin Heidelberg New York London Paris Tokyo Hong Kong Barcelona Budapest

PARTNERSHIP SUB-SERIES

1. Disarmament Technologies	Kluwer Academic Publishers
2. Environment	Springer-Verlag/Kluwer Academic Publishers
3. High Technology	Kluwer Academic Publishers
4. Science and Technology Policy	Kluwer Academic Publishers
5. Computer Networking	Kluwer Academic Publishers

The Partnership Sub-Series incorporates activities undertaken in collaboration with NATO's Cooperation Partners, the countries of the CIS and Central and Eastern Europe, in Priority Areas of concern to those countries.

NATO-PCO DATABASE

The electronic index to the NATO ASI Series provides full bibliographical references (with keywords and/or abstracts) to about 50000 contributions from international scientists published in all sections of the NATO ASI Series. Access to the NATO-PCO DATABASE compiled by the NATO Publication Coordination Office is possible in two ways:

- via online FILE 128 (NATO-PCO DATABASE) hosted by ESRIN, Via Galileo Galilei, I-00044 Frascati, Italy.
- via CD-ROM "NATO Science & Technology Disk" with user-friendly retrieval software in English, French and German (© WTV GmbH and DATAWARE Technologies Inc. 1992).

The CD-ROM can be ordered through any member of the Board of Publishers or through NATO-PCO, Overijse, Belgium.

2. Environment – Vol. 15

Springer
Berlin
Heidelberg
New York
Barcelona
Budapest
Hong Kong
London
Milan
Paris
Santa Clara
Singapore
Tokyo

Water Supply Systems

New Technologies

Edited by

Čedo Maksimović

IRTCUW, Institute of Hydraulic Engineering
P.O. Box 895
11000 Belgrade, Yugoslavia

Francesco Calomino

Dipartimento di Difesa del Suolo
Università della Calabria
I-87040 Montalto Uffugo (Cs), Italy

John Snoxell

Wessex Water, Wessex House
Passage street, B22 OJQ Bristol, UK

Springer

Published in cooperation with NATO Scientific Affairs Division

Proceedings of the NATO Advanced Study Institute "New Technologies for Large Water Supply Projects", held in Varna, Bulgaria, October 24 – November 4, 1994

Library of Congress Cataloging-in-Publication Data

Water supply systems : new technologies / edited by Čedo Maksimović,
Francesco Calomino, John Snoxell.
p. cm. -- (NATO ASI series. Partnership sub-series 2,
Environment ; vol. 15)
Includes bibliographical references and index.
ISBN 3-540-61533-4
1. Water-supply--Management--Technological innovations.
I. Maksimović, Č. II. Calomino, Francesco, 1949- . III. Snoxell,
John, 1948- . IV. Series.
TD353.W39 1996
628.1--dc20 96-32374
CIP

ISBN 3-540-61533-4 Springer-Verlag Berlin Heidelberg New York

Printed in Germany

Typesetting: Camera-ready by authors/editors
SPIN: 10470207 31/3137 – 5 4 3 2 1 0 – Printed on acid-free paper

New Technologies for Water Supply Systems

Preface

Technology now affects almost every aspect of Water Supply Management, Operation, Planning and Design; the speed of development means that assessing what is "new" is sometimes difficult. Old ideas can now be applied because of new technology; technology is now revealing problems that were unnoticed 10 years ago. Some emerging technologies promise much but are still underdeveloped for use in real world conditions, while we should always remember that "new" technology depends upon the state of development in respective countries, a point which is particularly relevant to the NATO Advanced Study Institute, for which this book has been produced.

Thus our objective in producing the book has been to highlight, in a wide range of technical areas, where and how technology is being applied, what is "new" and what the limitations of these technologies are in the real world. We have also tried to provide an European and American perspective where possible to illustrate how problems are tackled in different cultural environments. It is probably true that "technology" is also somewhat dependent upon the political, economic and organisational climate in different countries and we have included a chapter covering these aspects.

Water Demand in most countries has shown a steady, relentless increase in the past and this has put increasing pressure on water resources globally. Chapter I, Water Supply Systems on the Threshold of the 21st Century, looks at water demand and some new concepts in meeting these demands: separate systems for potable and agricultural use, differing qualities of water for differing uses, protecting water sources and transmission lines, and waste water reuse, a subject which is coming under increasing scrutiny.

Increasingly, water is being moved over very long distances to balance areas of surplus and deficit, both in space and time, and Chapter II, Water Supply Conveyance and Control, looks at some examples of these systems and the associated problems. Firstly, traditional and new technologies in pipeline construction and structural requirements are considered and secondly two such systems, Brussels in Belgium, and Wessex in the UK, are looked at in detail to illustrate the kinds of technology now in use to monitor and control these large networks.

Over recent years new technology has affected water treatment quite radically. Firstly, new analytical techniques have revealed many complex organic pollutants to be present in raw and treated water (some formed as by-products of treatment), and some are associated with diseases such as cancer. The response to this has been a tightening of standards for a wide range of such parameters and, in consequence, major investment in new and innovative water treatment technologies.

Chapter III, New Technologies for Water Treatment, looks at these problems, providing an European and American perspective, and goes on to look in detail at the innovative approach adopted by Amsterdam Water Supply, in treating raw water derived from the river Rhine. An American view is provided which looks at technologies emerging in the United States, with particular reference to health effects, costs of different technologies, and research being conducted at Colorado State University.

Disinfection using chlorine has been used almost universally for decades but it is only in recent years that the presence of disinfection by-products, and their potential effects as carcinogens, has been recognised. This has led to new technologies for controlling and removing such complex chemicals, and research into alternatives. Chapter IV, Emerging Problems and Technologies in Water Treatment, looks at these technologies and also looks at new approaches to optimising older technologies for water clarification, coagulation and flocculation.

There have been numerous advances in the management and operation of existing systems and chapter V, Diagnosis and Rehabilitation of Existing Systems, looks at these technologies. Flow measurement, water balancing, network modelling, data logging, leak detection and pressure control have all seen major advances in the last decade along with new techniques for pipe rehabilitation. The current situation in the UK is examined in detail with some examples from around the world to provide a global perspective.

Network Analysis, or computer modelling of supply systems, is an old technique which is now finding new life because of advances in computer technology and advances in data management. Chapter VI, The Application of Modelling in the Management of Water Supply Systems, looks at the numerical methods underlying this technique and a new approach to the computing aspects. It also looks at the way that network modelling is now being integrated with telemetry and GIS to provide a live, up-to-date, dynamic model of the network which can be used as a real operational management tool, as well as in its more traditional role in planning and design.

In some special conditions, transient flows giving rise to water hammer can be very destructive if not controlled properly and Chapter VII, Transient Flows, looks at some of the technologies now being employed to examine and analyse these phenomena, including new numerical methods and computing techniques.

Geographical Information Systems (GIS) are relatively new computer systems now extensively used in the management, operation and analysis of water supply networks. Chapter VIII, The Application of GIS and Remote Sensing, looks at the theoretical basis of such systems, and how they can be used for spatial analysis and as digital maps for the management of underground assets in a water company environment. The potential of remote sensing for data capture and leak detection is also covered.

The social, economic and political factors affecting water supply systems vary enormously between countries and often have a profound effect on the management, operation and technological status of the network. Chapter IX, Institutional, Social and Economic Aspects, concludes the book by looking at some of these frameworks and the way in which such issues as charging, privatisation, levels of service, management structures and other issues are dealt with using some European examples.

The scope of this book is very wide, and indicates that "new technology" now affects almost every aspect of water supply. From resource management, conveyance, treatment, hydroinformatics, pumping, to telemetry, data logging and GIS, the modern water supply network now encompasses technologies from every part of the spectrum. The challenge is to learn how to apply and manage these technologies and to spread this knowledge throughout the global world of water.

This book and the ASI (Advanced Study Institute of the NATO Science series of Advanced Science Institutes) from which it is derived, has made a contribution to this process and has helped in spreading this understanding of new technology between the participating nations.

This is a reflection of NATO's objectives with the ASI programme - "to promote the dissemination of scientific knowledge, to impart experiences and foster professional contacts among scientists" - and it is to be hoped that this book will prove to be a worthy record of the ASI in Bulgaria 1994, and the hard work and commitment of all those who took part.

Čedo Maksimović
Francesco Calomino
John Snoxell
April, 1996

Acknowledgements

The Organizers wish to thank the following companies for their co-sponsorship of the NATO ASI in Bulgaria, 1994:

- *BERMAD Control Valves, ISRAEL*
- *BIWATER Industries, UK*
- *DEGREMONT ITALIA, ITALY*
- *I.T.T. Flygt, UK*

The Editors would also like to express their thanks and appreciation to the individual authors, chapter coordinators and all those who assisted in the preparation of this book.

Special thanks to prof. Vujica Yevjevich for his invaluable suggestions in preparation and ASI organization and running. We are also indebted to prof. Paolo Veltri who helped us a lot in the final phase of preparation of the text for publishing.

Finally, special thanks are due to Ms Zorica Todorović, B.Sc. for her outstanding work and assistance in transforming the individual contributions into a consistent and high-quality volume.

The Editors
April, 1996

List of Contributors

Daniel L. Brumagne
Compagnie Intercommunale
Bruxelloise des Eaux
Rue aux Laines 70, 1000 Bruxelles
BELGIUM

Francesco Calomino
Dipartimento di Difesa del Suolo
Università della Calabria
87040 Montalto Uffugo (Cs)
ITALY

Tory L. Champlin
Department of Civil Engineering
Colorado State University
Fort Collins, Colorado
USA

Thomas William Charnock
Department of Civil Engineering
Aston University, Aston Triangle
Birmingham B4 7ET
UK

Lisa M. Dunn
Department of Civil Engineering
Colorado State University
Fort Collins, Colorado

John Elgy
Department of Civil Engineering
Aston University, Aston Triangle
Birmingham B4 7ET
UK

A. Graveland
Amsterdam Water Supply
Vogelenzangseweg 21
2114 BA Vogelenzang
THE NETHERLANDS

Vladimír Havlík
Czech Technical University of Prague
Thakurova 7, 16629 Prague 6
CZECH REPUBLIC

Jan Peter van der Hoek
Amsterdam Water Supply
Vogelenzangseweg 21
2114 BA Vogelenzang
THE NETHERLANDS

Peter David Hedges
Department of Civil Engineering
Aston University, Aston Triangle
Birmingham B4 7ET
UK

Petr Ingeduld
Czech Technical University of Prague
Thakurova 7, 16629 Prague 6
CZECH REPUBLIC

Marko Ivetić
Institute of Hydraulic Engineering
Faculty of Civil Engineering
University of Belgrade
P.O. Box 895, 11 000 Belgrade
YUGOSLAVIA

Pavel Jiroušek
Charles University
Prague
CZECH REPUBLIC

Peter Kalinkov
Higher Institute of Architecture
and Civil Engineering,
1 Christo Smirnenski , 1421 Sofia
BULGARIA

Jean Louis Monzo
Higher Institute of Architecture
and Civil Engineering,
1 Christo Smirnenski , 1421 Sofia
BULGARIA

Čedo Maksimović
IRTCUW, Insitute of Hydraulic Engineering
Faculty of Civil Engineering
University of Belgrade
P.O. Box 895, 11 000 Belgrade
YUGOSLAVIA

Mario J. N. Neves
Universidade do Porto
Rua Jerónimo Mendonça 65
4200 Porto Codex
PORTUGAL

Dušan Obradović
WINS Ltd.
Tower House, Parkstone Road
Poole, Dorset BH15 2DH
UK

Dušan Prodanović
IRTCUW, Insitute of Hydraulic Engineering
Faculty of Civil Engineering
University of Belgrade
YUGOSLAVIA

Thomas G. Sanders
Department of Civil Engineering
Colorado State University
Fort Collins, Colorado 80523
USA

John Snoxell
WESSEX WATER
WESSEX House, Passage str.
BS2 0JQ Bristol
UK

Stanislav Vanecek
Czech Technical University of Prague
Thakurova 7, 16629 Prague 6
CZECH REPUBLIC

Adrian Verwey
IHE, Delft
P.O. Box 3015, 2601 DA Delft
THE NETHERLANDS

Paolo Veltri
Dipartimento di Difesa del Suolo
Università della Calabria
87040 Montalto Uffugo (Cs)
ITALY

Radisav D. Vidić
Department of Civil and Environmental
Engineering
University of Pittsburgh
949 Benedum Hall, Pittsburgh PA 15261
USA

Vujica Yevjevich
Department of Civil Engineering
Colorado State University
Fort Collins, Colorado 80523
USA

Evžen Zeman
Czech Technical University of Prague
Thakurova 7, 16629 Prague 6
CZECH REPUBLIC

Table of Contents

CHAPTER I:
WATER SUPPLY SYSTEMS ON THE THRESHOLD OF THE 21ST CENTURY

CHAPTER II:
WATER SUPPLY CONVEYANCE AND CONTROL

CHAPTER III: NEW TECHNOLOGIES FOR WATER TREATMENT

CHAPTER IV: EMERGING PROBLEMS AND TECHNOLOGIES IN WATER TREATMENT

CHAPTER V: DIAGNOSIS AND REHABILITATION OF EXISTING SYSTEMS

CHAPTER VI: THE APPLICATION OF MODELLING IN THE MANAGEMENT OF WATER SUPPLY SYSTEMS

CHAPTER VII: TRANSIENT FLOWS

CHAPTER VIII: USE OF GIS AND REMOTE SENSING IN WATER SUPPLY

CHAPTER IX:
INSTITUTIONAL SOCIAL AND ECONOMIC ASPECTS

Chapter I

Water Supply Systems on the Threshold of the 21st Century

Chapter I

Water Supply Systems on the Threshold of the 21st Century

INTRODUCTORY STATEMENT

The development and expansion of water supply systems are continuing problems for the highly industrialized countries as well as the transition countries on the threshold of the 21st century. Although the magnitude and the complexities of the problems will vary from country to country, the fundamental issues of demand, planning concepts, water sources, pollution control and wastewater reuse are universal. In this Chapter each one of these issues will be discussed.

The past and present demand of water systems in Europe and the United States are presented and reviewed. Water demands from as early as the 1900's in the US are presented as well as more recent demands illustrating the increasing variability of demand making the prediction of future demand, which is the basis for design, extremely difficult. In addition, many of the European countries' water demand are reviewed as well. The annual, seasonal, monthly, weekly, daily and hourly demands experienced in the US are summarized. As discussed, each has a value and is used in the design of water systems in the US. It is interesting to note that there appears to be an increasing trend in domestic or urban water use in the US, using data from New York City, Dallas, Texas and Los Angeles, California and the consumption per capita has increased. In Italy, the increase in consumption per capita has been small through the period 1975 to 1984. In Europe, the average domestic use of water ranged from a little over 110 l/p/day (liters/person/day) in Belgium to 250 l/p/day for Switzerland while in the US, the average urban use varied from 130 l/p/day to much greater than 800 l/p/day with a national average of 628 l/p/day.

As a result of the worldwide problem of diminishing sources of fresh water, the availability and selection of sources of water for future water supply systems is increasingly important and are summarized and discussed. Surface water sources include: springs, lakes and reservoirs, ponds, streams, estuaries wetlands and swamps. Groundwater sources include: alluvial aquifers, rock fissure water systems, karstic water systems, volcanic rock groundwater, artificially recharged groundwater and juvenile groundwater aquifers. Brackish and sea water sources are also mentioned. Criteria needed for selection of sources for water supply to be covered are, source characteristics, regional characteristics, water rights and markets, stage development and economics.

The Section of the Chapter covering new concepts for water supply systems reviews the existing and new concepts and investigates the issues associated with the safety, technology and economics of interconnecting adjacent water supply systems. New water quality standards and the concept of producing various qualities of waters for various uses is also discussed. The uses include: boiler feed water, drinking water, human hygiene and recreational water, irrigation water, industrial including mining, smelting, and cooling water and hydropower and navigation water. Each requires a specific water quality, from distilled water for boilers to water free from floating material for navigation. Other water uses such as wildlife protection to wastewater dilution will have specific water quality requirements as well. Although not a new concept, the idea of a dual water system is discussed also.

The protection of water supplies from pollution is a problem of continuous concern particularly as countries become more industrialized. The concept of the multiple barrier system which is the principle most water supply systems in the US is based upon is discussed. The protection in the multiple barrier system varies from restricting access to the source watershed and collection systems to the technology based water pollution control program promulgated by the federal government and enforced by the individual states. In many areas of the United States, physical access to source water supplies is restricted. This includes not only the watershed but also transmission works and storage structures. Land use restrictions and prohibition of specific pesticides use have been imposed and improved water quality has resulted. The water quality legislation since 1965 when a water quality based pollution abatement program was first introduced and in 1972 when the technology based program was instituted with a goal of "zero discharge" and all rivers feasible and swimmable has endeavored to improve water supply quality by diminishing pollution sources. The foundation of the 1972 legislation, the National Pollution Discharge Elimination System (NPDES), Best Available Technology (BAT), and the Best Management Practices (BMP) are discussed. This focus on pollution sources will continue with the emphasis in the future on nonpoint sources. In addition, there has been a renewed interest in new technologies in water treatment as a result of the presence of trace compounds in drinking water supplies and the continuous vulnerability of the water supplies to infectious agents. The advanced and new technologies are discussed in greater detail in Chapter III.

And finally, the use of wastewater as a viable source for urban water supplies is discussed in this Chapter. Since history was first written, man has been reluctant to use his wastewater, usually disposing of it in the nearest river, lake or ocean. And although it was not until the mid 1800's that the relationship between disease and contaminated drinking water was established, the reluctance in using it today continues even though the technology exists to treat wastewater to a quality exceeding existing drinking water standards. This is due in part to the concept of a multiple barrier system discussed in the Chapter where the source water for drinking should be unpolluted and, in particular, unpolluted by man's waste, and thus has prevented widespread consideration of wastewater as future water supply source. However, things are changing as the sources of new water supplies are decreasing and the cost of treating brackish and sea waters continues to increase. Examples of existing uses of wastewater and the costs are summarized.

1. Urban Water Demand

Thomas G. Sanders
Vujica Yevjevich
Department of Civil Engineering
Colorado State University
Fort Collins, Colorado 80523
USA

1.1. INTRODUCTION

Water demand generally referred to as future water requirements will be a function of the demanding entity. The demand characteristics such as trending, jumps, periodicity and stochasticity are different for an urban demand, irrigation and agriculture demand, industrial and hydropower demands. The variation in demand can be large as well. For example, urban water demand will vary hourly, daily, weekly, seasonally and annually. Whereas, irrigation demand in the Western United States, where appropriative water rights exists (first in time, first in right), although quite seasonal is much more predictable as most water rights specify a specific amount of water delivered during a specified period of time. If the demand exceeds the supply, the water rights having the highest priority will be supplied first and those rights having a lower priority may not receive any water at all.

Urban water demand which is probably the most complex of all the demands as it is a combination of domestic, public, commercial, industrial, thermal power and even irrigation demand, especially during the summer months, will be the focus of this Section.

Annual water demand is an indicator of the size and characteristics of the population served. The location of cities to a large extent have been a function of the annual characteristics of the nearby water resource; this availability of supply leads to the demand. As the population increases and water uses diversify, estimating demand becomes more complex. The factors that affect demand include not only the population demographics but also availability, climate, economic development, technological development, water pricing, possibly water metering and the level of embracing the environmental ethic of conserving resources to name a few.

NATO ASI Series, Partnership Sub-Series, 2. Environment – Vol. 15
Water Supply Systems – New Technologies
Edited by Č. Maksimović, F. Calomino, and J. Snoxell

Engineers use the annual demand statistics to determine the size of the water infrastructure to meet these demands. Components of the infrastructure would include transmission works, reservoirs and treatment facilities. Seasonal demand affects the operational schedule of the infrastructure: for instance, empty reservoirs in the Fall and full reservoirs in the Spring, use of less than full capacity of the treatment system during the winter but full capacity utilized in the summer. Monthly and weekly demands are associated more with the seasonal fluctuations. Variations in daily demand will affect production as water demand decreases on the weekends. And finally, hourly demand fluctuates sufficiently that pumping stations and the capacity and configuration of the pumps are sized and designed on the maximum hourly flow.

1.2. DEMAND VARIATIONS

1.2.1. Annual

The annual water demand in a survey of 40 cities in the United States varied a great deal in 1890. A high of 186 gallons per capita per day, gpcd, (704 l/p/day) was recorded in Buffalo, New York and a low of 13.8 gpcd (52 l/p/day) was recorded in Chippewa Falls, Wisconsin (Table 1.).

Fifteen years later the daily per capita use in Buffalo was 324 gallons (1226 l/p/day) and in Chippewa Falls 100 gpcd (378 l/p/day) indicating a 74% and 624% increase in demand respectively in 15 years. During this same period of time, the per capita use of water in Rochester, New York increased from 66 (250 l/p/day) to 88 gpcd (333 l/p/day), a 33% increase (Table 2.). In 1980, the consumption of water in Rochester increased to 451 l/p/day, a 583% increase from 1890. Of the 22 cities surveyed in 1980, Rochester had the seventh lowest consumption while Passiac, New Jersey had the highest at 807 l/p/day and St. Paul, Minnesota had the lowest at 277 l/p/day. Recently, in Fort Collins, Colorado where much of the municipal water is used for lawn irrigation in the summer time, the daily per capita rate was 187 gallons (708 l/p/day).

While the aforementioned data indicates trending, there is variation in the annual demands as well. With reference to Figure 1., the variation in annual demand can be substantial. Even as early as the late 1800's, the annual demand could shift over 100% in several years. For example, the demand in Detroit, Michigan went from 120 gpcd (454 l/p/day) to over 200 gpcd (l/p/day) and back to 125 (l/p/day) in a period of 20 years.

Table 1. Consumption of water in american cities and towns in 1890 and 1905 (Turneaure and Russell, 1908)

City	Population	Population per Tap	Per cent of Taps Metered	Consump. per Inhabitant Daily	Per cent of Taps Metered	Consump. per Inhabitant Daily
	1900	1890	1890	1890	1905	1905
Chicago	1,698.600	7.1	2.5	140	...	200
Philadelphia	1,293.700	6.1	0.3	132	1.0	230
St Luis	575.200	11.8	8.2	72	7.8	92
Boston	560.900	6.6	5.0	80	5.0	151
Cleveland	381.800	8.7	5.8	103	68	137
Buffalo	352.400	6.3	0.2	186	3	324
San Francisco	342.800	9.9	41.4	61	21	96
Cincinnati	325.900	8.5	4.1	112	12	130
Detroit	285.700	5.1	2.1	161	9	188
Milwaukee	285.300	11.1	31.9	110	80	91
Louisville	204.730	11.9	5.9	74	8	81
Minneapolis	202.720	16.5	6.3	75	47	76
Providence	175.600	9.4	62.4	48	86	68
Indianopolis	169.160	35.6	7.6	71	10	82
Kansas City	163.750	...	...	...	38	73
St. Paul	163.065	12.7	4.2	60	38	56
Rochester	162.600	5.4	11.4	66	41	88
Toledo	131.820	18.6	9.4	72	70	75
Columbus, O.	125.560	11.5	6.4	78	76	110
Worchester, Mass.	118.420	8.9	89.4	59	95	75
Fall River, Mass.	104.860	14.9	74.6	29	97	42
Memphis, Tenn.	102.320	11.9	3.7	124	20	100
Lowell, Mass.	94.970	9.2	22.9	66	69	58
Atlanta, Ga.	89.870	20.0	89.6	36	100	65
Dayton, Ohio	85.333	20.0	3.8	47	70	70
Nashville, Tenn.	80.870	14.9	0.8	146	52	148
Camden, N. J.	75.940	...	...	131	3	155
Yonkers, N. Y.	47.930	12.0	82.	68	99	115
Newton, Mass.	33.587	5.5	67.	40	86	58
Aurora, Ill.	24.147	8.2	19.3	40.7	36	56
Madison, Wis.	19.164	11.0	31.0	40	97	46
Ashland, Wis.	13.074	9.9	2.8	90	...	81
Champaign&Urbana, Ill.	14.826	7.3	2.5	43	...	45
Chippewa Falls, Wis.	8.094	7.4	6.6	13.8	...	100
Middleborough, Mass.	6.885	11.7	24.0	21	47	38
Beloit, Wis.	10.436	10.2	10.0	64	...	130
Foxborough, Mass.	3.266	8.7	34.0	44.0	46	63
Clinton, Ill.	4.452	4.1	3.0	27.0	1.1	99
Shenandoah, Ia.	3.573	15.5	15.5	39	...	35
Melrose, Mass.	12.962	4.2	1.7	71	3	112

With reference to Figure 2., the annual demand of New York City, increased from approximately 125 billion gallons per year (473 billion liters per year) in 1898 to 450 billion gallons per year (2044 billion liters per year) in 1968, a 260% increase.

Viessman and Hammer (1985) report an average water consumption in the United States of 628 l/p/day. As recently as 1991, the average per capita demand reported in the United States varied from 130 to 2000 l/p/day (McGhee, 1991).

Table 2. Recorded rates of consumption in some american cities (McGhee,1991)

City	Average daily per capita consumption, l *	Maximum 1-day consumption in a 3-year period, l *	Maximum in proportion to average, %
Rochester, N.Y.	451	637	141
Syracuse, N.Y.	728	917	126
Hartford, Conn.	671	889	132
Albany, N.Y.	671	860	128
El Paso, Tex.	447	739	165
Portland, Me.	572	773	135
Camden, N.J.	641	963	150
Albuquerque, N.M.	402	766	190
Winston-Salem, N.C.	447	580	130
Waterloo, Iowa	383	625	163
Passaic, N.J.	807	1016	126
South Gate, Calif.	550	891	162
Fort Smith, Ark.	474	652	138
Poughkeepsie, N.Y.	569	728	128
Tyler, Tex.	371	743	200
Monroe, La.	584	875	150
Spartanburg, S.C.	754	955	126
Pomona, Calif.	629	1092	173
St. Cloud, Minn.	277	701	254
Salina, Kan.	603	1357	225
Alliance, Ohio	796	1114	140
Ashtabula, Ohio	766	985	129

* Gallons = liters x 0.264

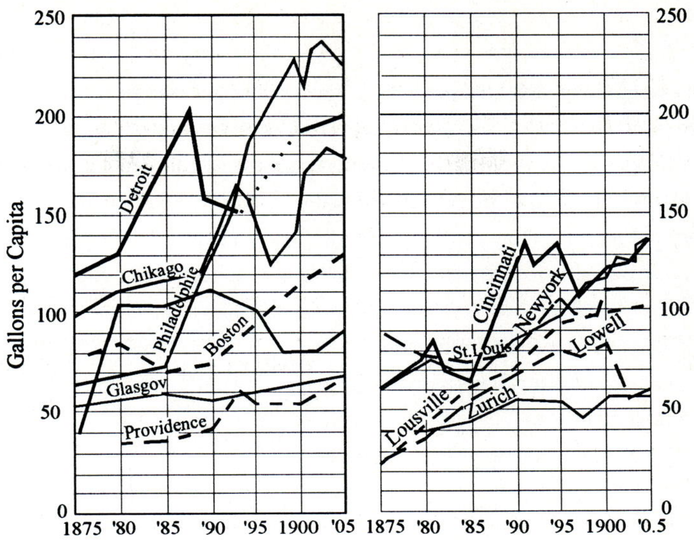

Figure 1. Variation of yearly rates of consumption (Turneaure and Russell,1908)

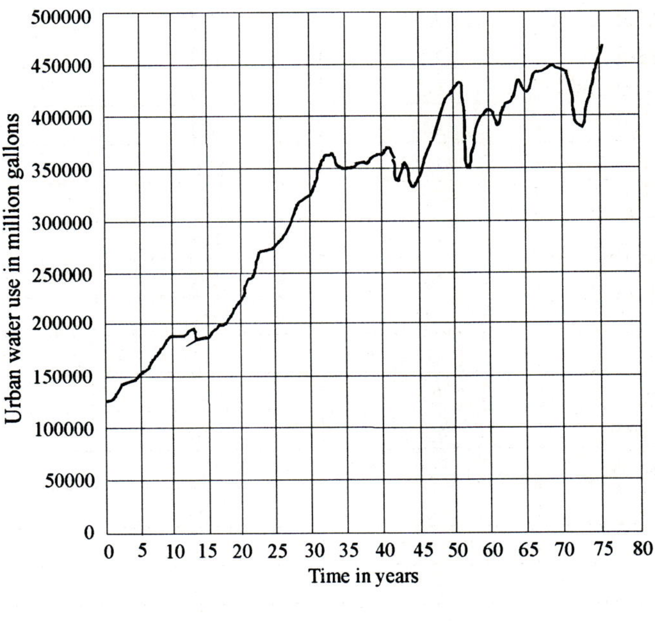

Figure 2. Annual water use for New York City, New York, for 1898-1968 (Salas and Yejevich,1972)

1.2.2. Seasonal

Seasonal water demand is characteristic of most municipal water supplies in the United States. With reference to Figure 3., during the period of 1950-1969 in Dallas, Texas in addition to the obvious trending of the mean, there is an appreciable seasonality of the monthly demand indicated by the annual peaks and troughs of the water demand. The fluctuations are not only occurring every year but they also appear to be increasing indicating a trending (nonstationarity) of the variance as well. A similar trending of the mean and variance can be seen in the water demand for Los Angeles, California for the time period of 1940 to 1965 (Figure 4.). The demand increased from about 4 billion gallons per year (15 billion liters per year) in 1940 to over 10 billion gallons per year (38 billion liters per year) in 1965 and seasonal fluctuations of a little more than 2 billion gallons per year (8 billion liters per year) and 4 billion gallons per year (15 billion liters per year) respectively.

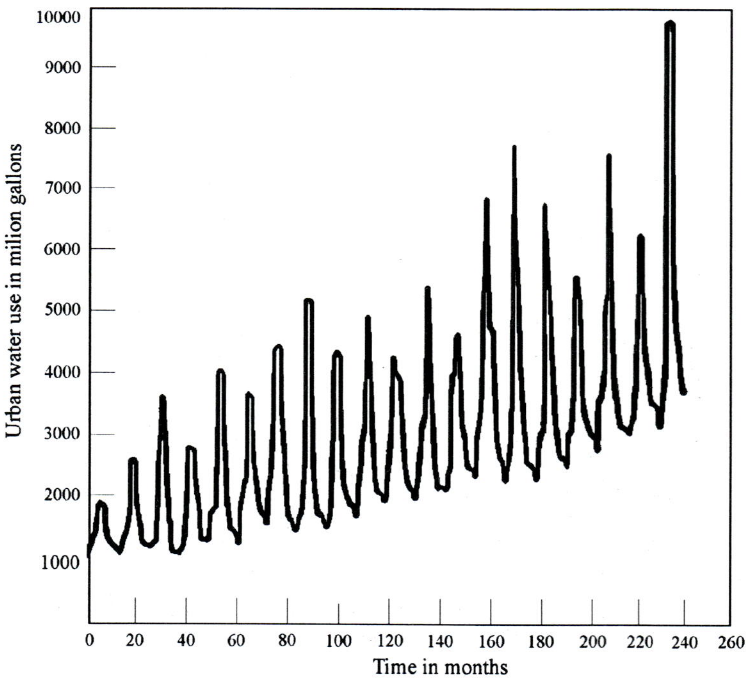

Figure 3. Monthly water use for Dallas, Texas, for 1950-1969 (Salas and Yevjevich,1972)

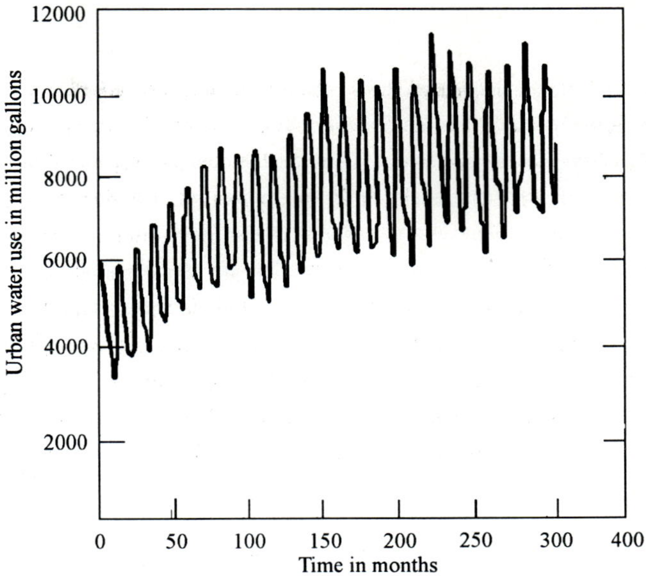

Figure 4. Monthly water use for Los Angeles, California, for 1940-1965 (Salas and Yevjevich,1972)

1.2.3. Monthly and Weekly

Monthly and weekly demands will be related to seasonal variations and not as important as the seasonal or daily fluctuations. However, weekly maximum have varied between 119 % and 167% of the average daily consumption (Turneaure and Russell, 1908). Tchobanoglous and Schroeder (1985) report daily averages in the maximum month to be typically 120% of the annual average and the daily average of the maximum week to be 140% of the annual mean.

1.2.4. Daily

Daily demands vary depending on the day of the week. Usually, the demands on Saturday and Sunday are less than during the week. An exception would be in areas similar to Fort Collins, Colorado in a summer without appreciable rainfall, the demand for irrigating the lawns may be greater than the demand during the week. Tchobanoglous and Schroeder (1985) report the maximum day to be typically 180% of the annual average.

1.2.5. Hourly

The hourly water demands show the greatest fluctuations because of the anthropogenetic diurnal cycle. Other factors which affect the hourly demand are seasonal and antecedent precipitation. As can be seen in Figure 5., the demand in gallons per dwelling unit approaches zero around 4 am and peaks at 10 am and 8 pm at approximately 400 gallons (1514 liters) per day per dwelling during a typical winter day. A maximum winter day was found to be over 2000 gallons (7571 liters) per dwelling per day. A similar fluctuation occurs during a summer day (90°F), with and without precipitation (Figure 6.). A low demand of approximately 50 gallons (189 liters) per day per dwelling and two peaks of approximately 350 gallons (1325 liters) per dwelling on days with rain were realized while demands as high as 1560 gallons (5905 liters) per dwelling on days without rain were realized. Tchobanoglous and Schroeder, 1985, report the maximum hour to be typically 270% of the annual average.

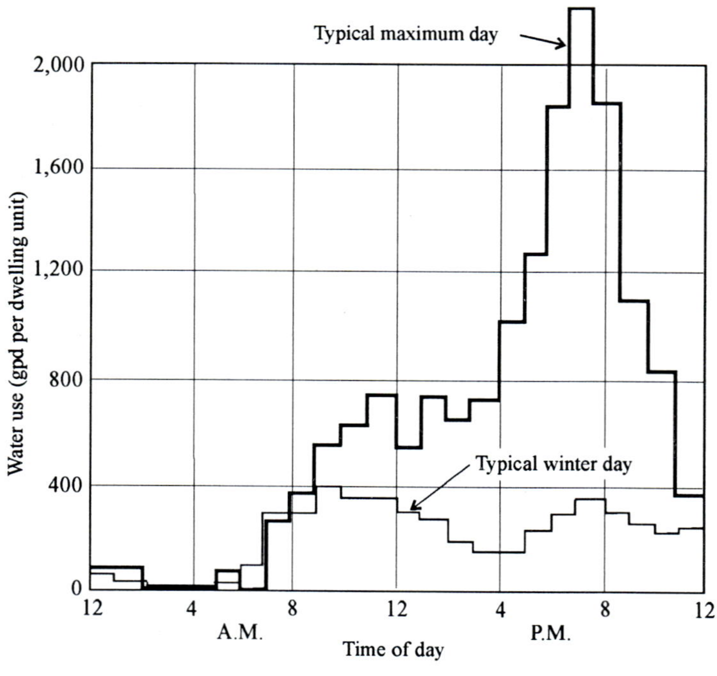

Figure 5. Daily water-use patterns, maximum day and winter day (from Residential water-use research project, Johns Hopkins University and Federal Housing Administration) (Viessman and Hammer, 1985)

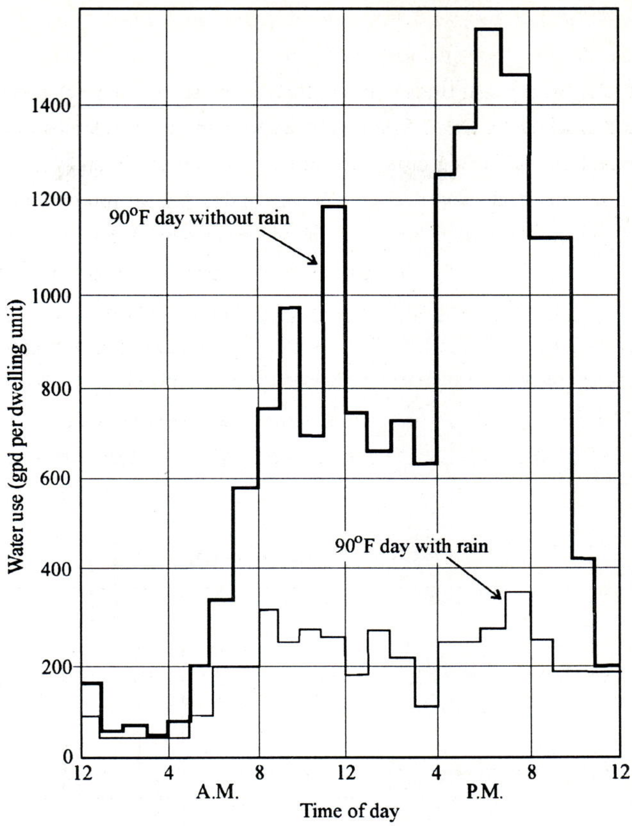

Figure 6. Daily water-use patterns in R-6 area, maximum day and minimum day (from Residential water-use research project, Johns Hopkins University and Federal Housing Administration) (Viessman and Hammer, 1985)

1.3. ANALYSES

In order to estimate urban water demand, the trends, periodicities and stochasticity of the data should be understood and then modeled. In a study of four cities in the United States and one city in Canada, Salas and Yevjevich, 1972, determined the polynomial trend model, the dependence model and the stochastic model for urban water use for each city. With reference to Tables 3. and 4., the annual trends were modeled with a quadratic with the exception of Colorado Springs, Colorado which had a cubic trend model. Fort Collins,

Colorado and Colorado Springs did not require dependence models as the residuals after removing the trends were independent. However, the remaining cities' residuals showed dependence requiring 1st and 3rd order autoregressive models. The independent residuals of each city was further analyzed and it was found that all could be modeled using a normal or in the case of Montreal, Canada, a lognormal model.

Table 3. Polynomial regression coefficients for the trend in annual urban water use (Salas and Yevjevich,1972)

Name	Unit	Order of trend	Regression coefficients A	B	C	D
Fort Collins, Colo.	m.g.	Quadratic	1,172.420	- 7.245	1.573	
Colo. Springs, Colo.	m.g.	Cubic	3,532.943	- 282.570	25.118	-0.276
Baltimore, Maryland	m.g.	Quadratic	13,088.346	302.060	7.244	
Baltimore, Maryland	g.p.c.d.	Quadratic	91.120	1.368	- 0.0071	
New York, New York	m.g.	Quadratic	101,737.57	8187.10	-49.577	
New York, New York	g.p.c.d.	Quadratic	104.389	0.844	0.0031	
Montreal, Canada	m.g.	Quadratic	44,792.895	1135.30	17.344	
Montreal, Canada	g.p.c.d.	Quadratic	114.120	- 0.2569	0.0267	

Table 4. Dependence model and distribution of the independent stochastic component for annual urban water use (Salas and Yevjevich,1972)

Name	Unit	Type	Dependence model Parameters ρ1	ρ2	ρ3	Distribution Type	Parameters * A	B	C
Fort Collins, Colo.	m.g.					Normal	0.0	204.29	
Colorado Springs, Colo.	m.g.					Normal	0.0	419.99	
Baltimore, Maryland	m.g.	First	0.825			Normal	0.0	0.563	
Baltimore, Maryland	g.p.c.d.	First	0.766			Normal	0.0	0.633	
New York, New York	m.g.	First	0.681			Normal	0.0	0.729	
New York, New York	g.p.c.d.	First	0.493			Normal	0.0	0.870	-7.54
Montreal, Canada	m.g.	Third	0.629	0.318	-0.23	Lognormal-3	2.019	0.084	-4.90
Montreal, Canada	g.p.c.d.	Third	0.725	0.413	-0.067	Lognormal-3	1.593	0.116	

* for Normal: A = Mean
B = Standard deviation

for Lognormal-3: A - Mean of ln (ξ-C)
B - Standard deviation
C - Lower boundary

Further analyses of the Fort Collins' weekly data yielded that the trend of the data accounted for 41.9 % of the variance while the periodicities accounted for 40.9% of the variance as well and the stochastic component accounted for 17.20 % of the variance (Salas and Yevjevich, 1972). Analyses of urban demand data for other cities will probably yield similar results pointing out the necessity of taking into account the trends, periodicity and stochasticity of the data when determining urban water demand.

1.4. CONCLUSIONS

Urban water demand has had and continues to have large variations as a result of many factors. These factors include population growth, water availability, climate, economic development, technological development, water pricing, water law (appropriative vs. riparian), and possibly water metering. As a result, urban water demand exhibits trends, periodicities and stochasticities each responsible for a proportion of the variance. Therefore, in order to be able to develop realistic and useful estimates of urban water demand, procedures and methodologies must be able to quantify and take into account these factors of urban water demand which are responsible for the variances.

1.5. REFERENCES

McGhee T.J. (1991) Water Supply and Sewerage. 6th edition, McGraw-Hill, New York.

Salas J.D., Yevjevich V. (1972) Stochastic Structure of Water Use Time Series. Hydrology Papers, Number 72, Colorado State University.

Tchobanoglous G., Schroeder E.D. (1985) Water Quality, Characteristics, Modeling, Modification. Addison-Wesley Publishing Company, Reading, Massachusetts.

Turneaure F.E., Russell H.R. (1908) Public Water Supplies. 2nd edition, John Wiley & Sons Inc., New York.

Viessman Jr. W., Hammer M.J. (1985) Water Supply and Pollution Control. 4th edition, Harper & Row, Publishers, New York.

2. Water Demand Has a Threshold

Paolo Veltri
Dipartimento di Difesa del Suolo
Università della Calabria
87040 Montalto Uffugo (Cs)
ITALY

2.1. INTRODUCTION

The limits to sustainable development in the supply and use of freshwater resources have been reached during the last decades in many parts of Europe: in spite of limited growth in specific domestic consumption in the most industrialised countries, many factors show that in the same field a lot of countries will demand in the next years more and more freshwater resources. As a matter of fact it is believed that all the countries with improving life conditions will use more water consumption facilities and, in addition, an increase in water demand for irrigation and industries has to be expected. These tendencies are however in conflict with the inadequacy of water resources in terms of both quality and quantity: this is true both in the Southern regions of Europe, and in the most industrialized parts of the Central and Northern Europe in terms of water resource degradation. The misuse and overuse of the resources have resulted in long term damage such as groundwater contamination, both by polluting concentrated sources, widespread agricultural pollution, heavy pollution of rivers and estuaries and salt water intrusion into groundwater.

2.2. WATER DEMAND - EUROPEAN EXPERIENCE

All freshwater demands interfere with the reasonable maintainance of the right ecological equilibrium: the terms of a sustainable development demand more cautious use of the water.

NATO ASI Series, Partnership Sub-Series, 2. Environment – Vol. 15
Water Supply Systems – New Technologies
Edited by Č. Maksimović, F. Calomino, and J. Snoxell

In the recent SAST Project n.6 of the UE (Schmitz, 1994) the following key figures are given for UE countries:

• domestic water consumption	10 %
• industry (processing and cooling) water consumption	56 %
• agricultural water consumption	33 %
• others	1 %

Groundwater resources represent 60% of sources in public water supply system; several member states depend on other countries for more than 50% of delivered water.

This global scenario must be verified case by case, in fact in Italy the 1990 updating of the National Water Conference (1990) provided a different estimate:

• civil water consumption	15 %
• industry water consumption	24 %
• agricultural water consumption	60 %
• others	1 %

Consumption value, disaggregated according to consumer category, have been also quoted in Great Britain (Archibald and others, 1984) as follows:

• category of consumer (consumption)	Ml/day	% of total
• households	6.000	54
• industry	3.050	27
• commerce and service industries	1.600	14
• agriculture	500	5
• miscellaneous	50	< 1

As regards individual consumption, Archibald and others found an evident relationship between this consumption and the number of householders. The range is from 140 l/p/day for single person apartments, to 64 l/p/day in the case of more than seven inhabitants.

In order to justify the different terminology in UE and Italian scenarios it is useful to refer (Veltri and Maiolo, 1994) to a civil consumption scheme (Figure 1.), in which losses are also quoted in evaluating the volumes to be delivered at the starting point of the drinking water supply system. The term domestic is intended as that necessary for drinking, washing, cooking and sanitary purposes.

The most relevant problem in assessing drinking water demand is that of finding the necessary requirement, avoiding misuse and limiting losses both by end-users and water distribution companies. In this respect it is possible to make analytical evaluations of water demand in a civil context, even if this one is dependent on many factors, among which climate, water consumption customs, continuous and unlimited resource availability, and last, but not least, the water price and the adopted tariff system. Severe measures and controls of the delivered water volumes and high water prices become the most significant factors discouraging water misuse and loss. Figure 2. (Veltri, 1994) shows the average domestic consumption in several countries, in which the average tariff (Figure 3.) plays an evident role. In any case comparisons are difficult because some tariffs include both drinking and dirty water charges.

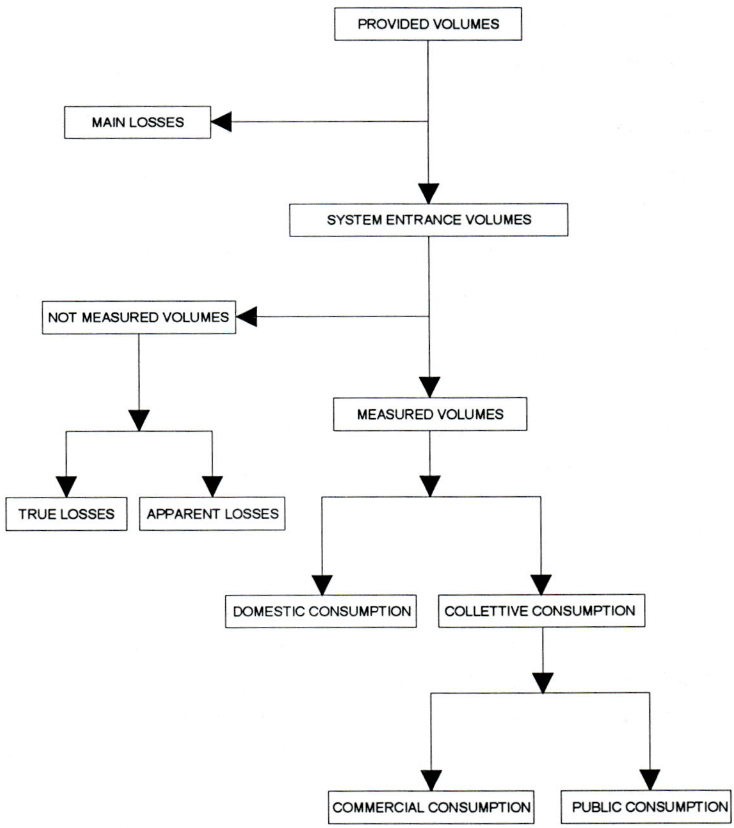

Figure 1. Civil consumption scheme

The trend of domestic and total civil consumptions in the case of Italy is also shown in Figure 4. (Maione and Orsi, 1990). It seems reasonable to assume that in the countries in which new economic management entities are going to operate, and consequently higher water charges will be applied, domestic consumptions will decrease.

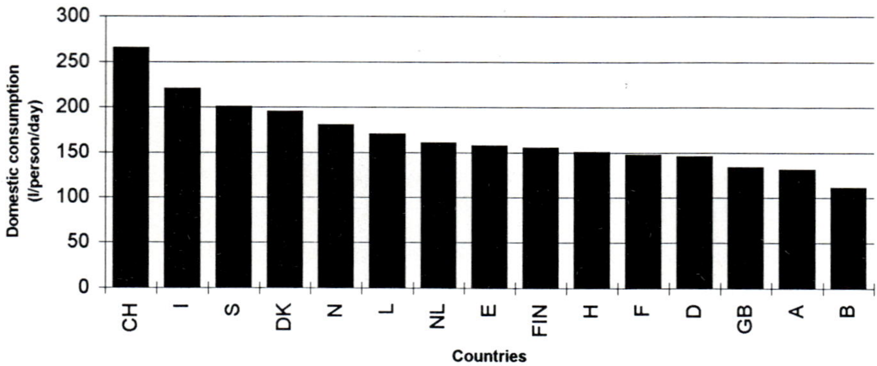

Figure 2. Average domestic consumption in several countries, year 1986 (Veltri, 1994)

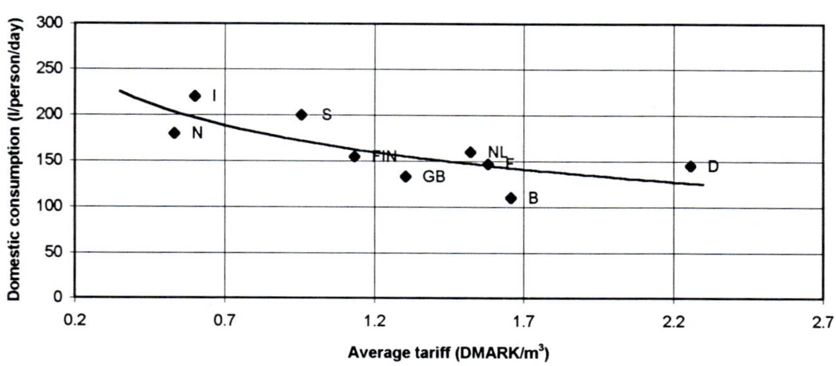

Figure 3. Domestic consumption versus average tariff in some european countries

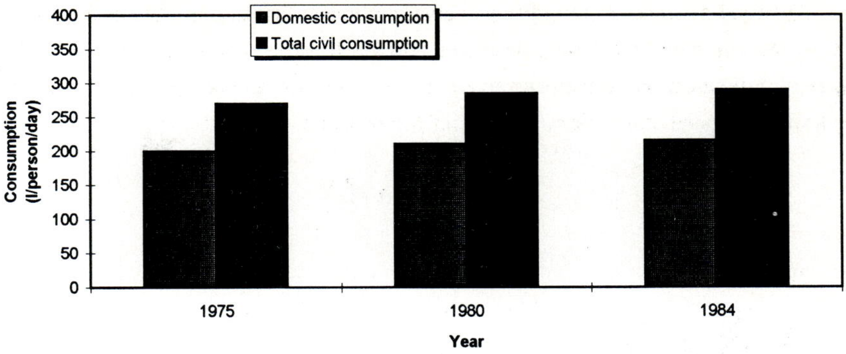

Figure 4. Trend of domestic and total civil consumption in Italy (Maione and Orsi, 1990)

Based on a study of civil consumptions of many Italian centres a very simple formula (Veltri and Maiolo, 1994) can be roughly used among the many existing criteria to forecast the highest daily consumption:

$$C_l = C_d + C_c + P$$

in which:

in which:	C_l	(l/p/day) is the specific civil consumption,
	C_d	(l/p/day) is the specific domestic consumption,
	C_c	(l/p/day) is the specific collective consumption,
	P	(l/p/day) are specific losses.

In the formula, C_d is weakly dependent on the city scale (it seems reasonable to assume that it has almost the same value both in large and small towns); even if its value is affected by the factors already highlighted, first of all water prices, it does not go below a reasonable physiological treshold, which could represent the individual requirement (for water). C_c is instead wrongly dependent on town dimension, in terms of public and collective water demanding services; it is possible both to determine it from historical consumption data and to assign an adequate value in terms of percentage of C_l. Finally P is the losses term and it must be assigned as accepted and not completely eliminable (it must be assumed to be not greater than 10% of C_l).

If the water demand for collective usages is not available and a reasonable value of specific losses is tolerated, it can be assumed that these are a percentage, α and β respectively, of C_l; by re-writing the formula, C_l becomes:

$$C_l = C_d / (1 - \alpha - \beta)$$

in which C_d can be assumed according to either existing data or on the basis of analytical summation of required usages, whereas α and β can be evalueted as follows: α between 0.10, for small towns, and 0.30, for large cities (and even higher for the largest ones), β not higher than 0.10.

2.3. REFERENCES

Archibald G.G., Kitson T., Waters B.D., Watson T.J. (1984) Consumer Requirements. In Water Practise Manual, vol. 4, Water Distribution Systems. T.W. Brandon editor, IWES, London, England.

Conferenza Nazionale delle Acque (1990) I problemi delle acque in Italia. Aggiornamento al 1989 dei risultati della Conferenza Nazionale delle Acque, Edizioni Agricole, Bologna.

Maione U., Orsi E. (1990) Consumi idrici delle realtà urbane. Convegno: La conoscenza dei consumi per una migliore gestione delle infrastrutture acquedottistiche, Sorrento, aprile, CUEN, Napoli.

Schmitz B. (1994) Research and Technological Development for the Supply and Use of Freshwater Resources. SAST Project n. 6, EUR 15645 EN, Brussels.

Veltri P. (1994) Analisi dei dati storici disponibili. In: Rapporto sui Consumi Idropotabili in Italia sul finire degli anni 80. A cura di P. Bertola e L. Taglialatela, Quaderno n. 1, CUEN, Napoli.

Veltri P., Maiolo M. (1994) Consumi annui dei centri urbani e turistici. In: Rapporto sui Consumi Idropotabili in Italia sul finire degli anni 80, a cura di P. Bertola e L. Taglialatela, Quaderno n. 1, CUEN, Napoli.

3. Availability and Selection of Sources of Water for Water Supply Systems

Vujica Yevjevich
Thomas G. Sanders
Department of Civil Engineering
Colorado State University
Fort Collins, Colorado 80523
USA

3.1. INTRODUCTION

Sources of water are systematically enumerated by the five major categories: (1) surface, (2) underground, (3) atmospheric, (4) recycled, and (5) brackish or saline waters. Each category of sources of water contains two or more specific kinds of water, with brief description of each. A total of 21 kinds of sources of water are identified. Similarly, criteria needed for selection of source(s) of water for the water supply systems are divided into the five major categories. They are: (1) criteria which characterize the source(s) of water and water itself, (2) regional characteristics of sources as related to the water supply systems, (3) water rights and market for water, (4) development of the system in stages, and (5) economics of water supply systems criteria of selection of water source(s). Each of these categories of criteria contains two or more kinds of criteria, with their brief description. A total of 15 criteria are identified. The principles of the application of criteria to selection of the source(s) of water for water supply systems are reviewed in the third part of the text. The last part of this text treats briefly the comprehensive and integrated development, conservation, control and protection of all regional sources of water.

This text on the availability and selection of sources of water for the future water supply systems covers first a section on the systematic listing of categories and kinds of water sources, with a brief description of major properties of each category, and of each kind of sources of water in a category. Examples of the categories of sources of water are: surface, underground, atmospheric, recycled and saline waters. Examples of the kinds of sources of water, in the category of the surface sources of water, are: springs, lakes, streams, reservoirs,

NATO ASI Series, Partnership Sub-Series, 2. Environment – Vol. 15
Water Supply Systems – New Technologies
Edited by Č. Maksimović, F. Calomino, and J. Snoxell

ponds, wetlands, etc. The third section treats criteria for the selection of sources of water in planning new or expanding the existing water supply systems. The fourth section covers the principles of selection of sources, including the future factors that may influence that selection. The fifth and final section is devoted to regional integration of all sources of water into a comprehensive solution of water supply problems.

3.2. CATEGORIES AND KINDS OF SOURCES OF WATER, AND THEIR PROPERTIES

3.2.1. Surface Sources of Water

The surface sources of water are characterized by an easy access for tapping them for the water supply systems. Their pollution is easy also either by slow processes and accidents in nature or by various human activities. The following kinds of surface sources of water are (see the book by O. Starosolszky, Surface Water Hydrology, 1987):

3.2.1.1.Springs. Springs are natural outlets of different types of underground water systems. They occur when the piezometric subsurface water table intersects the ground surface. The average quantity of water and the time variation in discharge and quality of this source of water depend highly on the type of underground system, especially for the springs of karstified water-bearing formations. In most cases, springs are easily tapped for water supply systems. The large majority of cities in ancient times were supplied by spring waters. The best examples are the cities of the ancient civilizations around the Mediterranean Sea. Quantity and quality of spring waters may change with time due to various natural or anthropogenic change-inducing factors. The quality of spring water is usually suitable for drinking purposes, provided it comes from the unpolluted underground water systems. Springs in most cases should be of a high water quality. Unfortunately, pollution of underground water systems and water itself leads also to pollution of springs, as the causal consequence.

3.2.1.2. Lakes. Lakes are likely to be the source of water with the second highest quality of water for water supply systems, behind the springs. Under the natural conditions, without the anthropogenic pollution, many lake waters are used by the native populations for drinking without treatment. Their water budget, and level and volume of discharge fluctuations determine the quantitative capacity for their use in water supply systems. Lake water is easy to tap, and also in most cases easy to treat to drinking water standards. The trends in population increase often lead to dense settlements around lakes with subsequent deterioration of the quality of water, making more expensive water treatment necessary. To get acquainted with the properties of lakes as the source of surface water for water supply systems, any good book on limnology, the science of lakes, would be sufficient.

3.2.1.3. Ponds. Ponds are small lakes or small reservoirs, either natural or man-made bodies of water, which usually serve a particular purpose (such as fish production, water supply, irrigation, recreation, etc.). Water quantity and quality of ponds are limited, and in many cases their waters need treatment to be a source of drinking water.

3.2.1.4. Streams (rivers). Streams are currently the most available source of surface waters. Because of the large fluctuation of discharge with time, the tapping of water from small rivers for water supply systems most often requires the building of reservoir(s) for storage of water. This is most certainly true as the water diversion points for water supply systems are very often located at the upper regions of the catchment where the fluctuation of flow is the greatest. Because of pollution by the natural processes as well as by the anthropogenic sources, river waters, nearly without exception, require treatment for drinking water purposes. Any classical book on rivers and river waters (potamology) will enable the reader to obtain enough information on them for the purpose of tapping this most frequent surface water source.

3.2.1.5. Estuaries. The most downstream sections of rivers, at their entrances into the seas or large level fluctuating salt lakes, may be mixed with waters of those receiving bodies of water. In the form of estuaries they also represent a potential source of surface water. The central problem becomes, how to avoid the mixing between the fresh and saline waters, which produces a brackish water unsuitable for drinking water purposes without desalination. By proper hydraulic barriers, it is possible to avoid that mixing. The tidal wave propagation upstream along the estuaries may also be eliminated by a hydraulic barrier in the form of a weir. An example of such a structure is in the estuary of the Thames River downstream from London. Estuaries may be transformed into reservoirs for water supply by similar weir barriers or by dams.

3.2.1.6. Reservoirs. More than 250 thousand of all types and sizes of reservoirs exist already in the world, of which about 55.000 are large and medium size reservoirs (according to the International Commission on Large Dams), and the rest are small and very small reservoirs and ponds. Many of them serve as the source of surface water for water supply systems.

Reservoirs have been and will continue to be one of the most important sources of water in the foreseeable future. Settlement of suspended river sediment and temperature stratification of water in the deeper reservoirs as well as the mineralization differences, make reservoir water usually easier to treat for drinking purposes than the river water, particularly if the water is drawn from the epilimnion (the surface layer of a stratified lake or reservoir). However, if the water is drawn from the hypolimnion (the bottom layer), as is the case for many reservoirs in the United States of America, the reverse may be true. The water may be highly mineralized, turbid and without oxygen, necessitating treatment prior to most uses.

3.2.1.7. Wetlands and Swamps. If wetlands and swamps are permitted to be partially or fully drained, they may be utilized as surface sources for water supply systems. Usually, the preservation of wetlands stemming from ecological concerns, or their use for tertiary treatment of wastes, may forbid the use for any other water resources purpose. They may be used for smaller water supplies, especially if the maintenance of wetlands does not require all the available water. The problem of fluctuating water quality will require treatment if water is used for drinking purposes.

3.2.2. Subterranean or Underground Waters

Usually, these waters are called groundwater. The products of atmospheric precipitation, which infiltrate into the different types of geological formations, create underground water systems. The quality of these waters are characteristic of the geologic setting and may require treatment before use in water supply systems. The exception to this rule is the juvenile groundwater, retained in the soil and rock formations during their formation (this water when used is classified as "mining water.") The example of mining of this type of juvenile water is the large water supply system in Libya, where such an aquifer in the desert is tapped and water is brought a long distance to the coastal region, however not as a renewable water resource. In general, the underground water resources can be of very good quality as long as there are few sources of pollution. The costs of pumping water in many cases prohibit the use for some price-sensitive water supply purposes. The variations in these underground water systems are many (see the book by G. Kovacs and Associates, Subterranean Hydrology, 1981), and their kinds are:

3.2.2.1. Alluvial Aquifers. The permeable deposits of silt, sand and gravel, with an adequate water supply, form most of the groundwater systems known as alluvial aquifers. They are found everywhere but particularly along, on the sides and beneath the rivers as well as along the lakes and seas. Their waters are usually clean and easy to tap for water supply. They are available directly for drinking purposes, provided that anthropogenic pollution does nor occur requiring water treatment before use.

3.2.2.2. Rock Fissures Water Systems. Many rock formations, either granular or solid, have connected faults and fissures, generated by the orographic movements and processes, so that they permit infiltration, accumulation and movement of water. These sources of underground water are not simple to tap, except under particularly convenient circumstances. Some of the best such systems are sandstones, with primary or granular porosity, and faults/fissures system of porosity. This latter system permits a fast movement of water in or out of the system, by slowly recharging or draining the significant primary, granular porosity.

3.2.2.3. Karstic Water Systems. The dissolution of carbonate rocks (limestones and dolomites) and gypsum creates not only the specific surface forms of karstic type formations, but also the systems of underground channels and caverns, thus creating the subterranean system of channels with free turbulent flow or turbulent flow under pressure, or both. (see the book by P. Milanovic, Karst Hydrogeology, 1981). These groundwater systems usually terminate either with a spring or with many outlets of several springs. The karstic waters may be of drinking quality during low flows, but often polluted during highest flows. This property requires treatment of water during the high flows at the springs. In other cases, channels are the part of the karstic aquifer, needing water pumping to tap it for water supply systems.

3.2.2.4. Volcanic Rocks Groundwater. The geologically recent volcanic rocks outflows have created the underground porosity and channels (often called tubes), which are in many aspects hydraulically and hydrologically similar to the karstic systems. Water from these rocks are of much lower pH than water from the karstic systems which tend to be alkaline (high pH). Water supply systems of the Hawaiian islands are the best examples of these systems, where the horizontal drainage galleries produce substantial flow discharges. This drainage system by constructing the long horizontal galleries are often referred to as the Maui type (after the Hawaiian Maui island where it serves as the water supply system).

3.2.2.5. Groundwater Drawn in Directly From Surface Water. In many cases, the groundwater tapped from aquifers are nothing more than the surface water (of rivers, lakes, ponds, reservoirs), drawn in by operating production wells at relatively small distances from the surface water sources. This source of water may then be considered as a combined surface/subsurface source of water. The conventional surface water treatment costs can be thus decreased substantially since the water is filtered through the bank alluvial or other porous media.

3.2.2.6. Artificially Recharged Groundwater. The underground water which is artificially recharged with surface waters using recharge wells, galleries, reservoirs, ponds or irrigation system of ditches, sprinklers and land inundations, is also to be considered a combined surface/subsurface source of water for water supply systems. The storage of water, purification by filtration through porous media and sometimes the prevention of saltwater intrusion are the reasons for recharge.

3.2.2.7. Juvenile Groundwater Aquifers. These aquifers are very old waters, locked into the subterranean porous media during its formation. This water most often is highly mineralized, which requires either treatment (usually desalination) or mixing with low mineralized vadose waters in order to be tapped and used for drinking water purposes. This water is most often used only for the limited time supplies, matching the life of water supply user's needs with the durability of the available quantity of juvenile water source. As already mentioned, an example of the use of this water is the tapping of the juvenile water in Sahara by Libya.

3.2.3. Atmospheric Waters

The atmospheric waters are tapped directly, either as precipitation (rainfall and/or snowfall) or as condensation of moisture from the humid air at the water collecting natural or artificial barriers. These waters are low in dissolved solids but carry air-borne pollutants. Atmospheric water may have low pHs, and is a source of usually small quantities of water. It is stored in the specially built tanks, most often in the arid and semi-arid regions of the world and on waterless islands.

3.2.3.1. Precipitation Collected on Impervious Surfaces. Precipitation on roofs, paved collecting surfaces or other impervious surfaces are collected and stored. This water is relatively aggressive, because of the lack of dissolved minerals. Many islands, especially those composed of karstified limestones and dolomites, or the highly pervious volcanic rocks, in arid and semi-arid regions, like at some Greek and Caribbean islands, are forced to use this type of collection. This water is often a cheaper solution than bringing water by boat-cisterns. A good example is at Gibraltar, where water was collected on a paved inclined surface in the times of conflict between England and Spain for the ownership and control of the "Rock." The way how this water is kept in storage, sometimes with crushed rocks (mainly carbonates), enables minerals to be picked up as dissolved solids. Before water is used for drinking and cooking purposes, it is filtered or boiled.

3.2.3.2. Directly Condensed Atmospheric Water. It is known that in the arid or semi-arid regions the passage of the moist air over the cold elevated land forms often leaves condensed water. This can be artificially enhanced, and small quantities of drinking water obtained.

3.2.4. Recycled Wastewaters

Wastewaters from urban and rural communities, industries, mining and other uses may be recycled as new sources of water. Different kinds of these waters are:

3.2.4.1. Recycled Urban and Rural Community Wastewaters. The modern technology of treatment of wastewaters can produce clean waters that even drinking water quality could be obtained. Psychologically, the solutions are such that the treated wastewater is traded with the other drinkable waters. This may be the primary water source when no other source is available. However, much effort must be taken as this source of drinking water has a high potential to propagate, and transmit, the communicative diseases such as cholera, dysentery and hepatitis. Therefore, as long as there are other, safer, sources of water, recycled wastewaters should not be a source of drinking water. It is no longer a question of treatment technology, it is a question now of reliability and psychology.

3.2.4.2. Industrial Wastewaters. The wastewater from industry may be recycled either to be reused for the same purpose, or it can be treated to such a level that it can be used even as drinking water. It will depend on which industrial purpose was originally served and how expensive is treatment for recycling the water to serve the same or a new purpose. Cooling waters may be so little changed from the original quality, that minimal recycling treatment would be necessary for drinking water purposes, if the original water used for cooling was of a good quality.

3.2.4.3. Mining Related Wastewaters. Basically three sources of contaminated water come from the mines and mining processes. They are: (1) drainage water from the mine galleries and shafts, either from the old, abandoned mines or from those in operation; (2) ore processing waters; and (3) waters leaching pollutants from the mine tailings. All of these sources can pollute the natural water courses and treatment or other expensive solutions become necessary. Unfortunately, the costs to clean up these waters may exceed the value of the mined element. This cost certainly appears to be the case for the costs to clean up the impacts of the summit gold mines in Colorado. The value of the gold was estimated approximately at 90 million dollars and the cost of cleanups in excess of a 100 million dollars (McWhorter, 1992).

3.2.5. Brackish and Saline Waters

Close to seas, in estuaries, salt lakes, or even in some groundwaters, one finds waters which contain too much salt or even some other undesirable minerals, that they can not be used for drinking purposes without treatment. They are:

3.2.5.1. Brackish Waters. These waters are often encountered in springs close to the sea in karstic and recent volcanic formations, in estuaries, close to salt lakes, and at the lower layers of some aquifers. The cost of desalination is a function of concentration of dissolved solids to be removed. Whenever one has to select between the desalination of brackish and saline water, the former usually is a cheaper solution.

3.2.5.2. Sea Water. This source of water is inexhaustible for the coastal regions. The major problem is the high cost of converting these waters to become drinking water. All depends on the cost of energy. Many examples of saline water conversion plants, on continents, islands and ships, testify that in the absence of other cheaper drinking water alternatives, the conversion of saline water to drinking water becomes the right and sometimes the only solution. Some conversion plants in the Persian Gulf Region are of the capacity of several cubic meters per second. The major value of saline water conversion may be on the ships that can move to and be used temporarily when conventional water sources are not available, during and after floods, hurricanes, fires or earthquakes.

3.3. CRITERIA FOR SELECTION OF SOURCES OF WATER FOR WATER SUPPLY SYSTEMS

The selection of water source(s) for a water supply system depends on the criteria chosen for that selection. They are grouped in five categories which are:

(1) criteria which characterize the source(s) of water and water itself,

(2) regional geographic properties of the relationship between the water supply system and the source of water (such as the distance, terrain characteristics, geomorphological and geological feasibilities for creating storage reservoirs, etc.),

(3) water rights and the availability of market for these rights,

(4) feasibility for development of the system in stages as the function of the quantity and quality of water in a changing water demand,

(5) economics of water supply criteria.

These five categories may be then classified in kinds of criteria, and discussed appropriately.

3.3.1. Criteria which Characterize the Source(s) of Water and Water Itself

The major kinds of this category of criteria for the selection of the source(s) of water, and related to the source itself, are:

3.3.1.1. Availability of Water. This criterion means how much and of what quality of water each considered source has. The parameters of variables describing water of the source, such as the mean discharge, the standard deviation, the periodicity and randomness describing functions and parameters, the time variation functions (models) and their parameters, etc., all are related to the availability of the water sources and may be used to describe these kinds of criteria. For example, the mean annual discharge may determine the size of the population to be served, while the standard deviation of annual flows is related to the storage capacity necessary to meet the maximal needs of the served population. The same measures apply to quality determining the amount of necessary water treatment. Simply, the state of these hydrologic or hydrogeologic characteristics make up the science of water quality hydrology.

3.3.1.2. General Longevity of Water Supply Sources. This criterion requires that very small chances exist for water of a source to be diverted or completely lost either by a natural process or man-made actions. It is known from the historical events that earthquakes have completely diverted water to go other directions, and sometimes created the new spring(s) in large karstic formations. Even the upstream uses of water by humans often imperil the quality of water supply downstream. In discussion of water conservation needs and measures, often one could hear that "upstream water conservation may mean downstream drought." Here one must distinguish and investigate both the natural processes of sudden or slow change, which are not under human control, and those changes that are man-induced and man-controlled. In fact, that is the major research challenge of the next century to distinguish those processes and results that are man-induced and thus controllable and those processes and results that are not.

3.3.1.3. Potential for an Easy Water Pollution. It is known that various sources of water for water supply systems have different potentials for pollution of water and the system's environment. Examples of an easy water quality deterioration are many, since all sources of water can be polluted. All surface and underground sources of water may be polluted. Karstic water sources are among those which may be most easily polluted with difficulties to clean them evident. The long history of attempts and costs of cleaning the polluted groundwaters and their systems in the United States testifies that polluting water is easy, fast and long-lasting, while cleaning the water environments is a long-range, difficult and costly endeavor. Estimates as high as exceeding 300 billion dollars may be required to clean up known contaminated groundwater sources in the USA (McWhorter, personal communication, 1994).

3.3.1.4. Access to the Source for Water Tapping Purposes. This kind of criteria is basically reflected in the cost and safety of access to the source and the easy way of which water can be tapped without the risk of either losing it or polluting it.

3.3.2. Regional Characteristics of Water Sources as Related to Water Supply Systems

Both the quantity and quality of water can be understood in many cases as a resource of the region. Obviously, waters near the ocean and industrial centers will have different quality characteristics and will have different economical values. The regional category of criteria relates basically to the distance of the source of water from the center of water demand, to the potential construction of the seasonal flow regulation along the connecting line between the source and the center of water demand, as well as the local potential disasters threatening the interruption of water supply. The kinds of these criteria are:

3.3.2.1. Distance of Source of Water From the Center of Water Consumption. The distance of the source of water from the center of water use of a water supply system is one of the major regional characteristics, which determines the cost and the reliability of connecting the source to the system. Since the long distance requires the main conduit(s) to traverse valleys and other geographic features, which may be disrupted or even destroyed during floods, earthquakes and other natural or man-made disasters, the safety of connecting lines becomes an important kind of criteria in selecting the source(s) of water.

3.3.2.2. Availability of Natural or Man-Made Water Storage Capacities. A geologically sound and spatially attractive conditions for seasonal water flow regulation along the connecting line from the water source to the water supply system, either existing or to be constructed, represents an asset for sources of water. The natural assets may be existing lakes and aquifers. The man-created assets may be reservoirs and ponds.

3.3.2.3. Availability of Elevated Terrain in the Vicinity or Within the Water Distribution Network of Water Supply Systems. Each water supply system, which is fed with nearly constant or low varying daily inflow from the source(s), requires daily water regulation because of highly variable water consumption within 24 hours of operation. Any convenient elevation of terrain enabling a relatively economical construction of daily regulation reservoir(s), instead of building towers with tanks all over the plain terrains, is an asset to the system. This is especially important if reservoirs also have the role of weekly flow regulation, namely to store the surplus inflow water during lower water consumption over the weekends, and to increase the corresponding average water supply during the working days.

3.3.2.4. Non-Aggressive Soils in Which the Main Conduits are Buried. The connecting conduit lines from the source of water to the water supply system's distribution network should not be laid down in a highly aggressive soil, because of the corrosion of conduits or the increased cost in special protection of conduits by the exterior coatings.

3.3.3. Water Rights and Availability of Market for Water

Water rights, their distribution and length of ownership will always play an important role in the development of water supply systems. Obviously, to commit large sums of capital, development system managers must have access and ownership for long periods of time. In Colorado and much of the West in the USA, once obtained, the right lasts forever, while in other regions of the world they are for a given time period. In either case the guarantee of access is vital and must be institutionalized. Legal and market aspects of water supply systems also impose some criteria in selecting a water source for that system. These kinds of criteria are:

3.3.3.1. Availability of Water Rights and Their Protection. For a secure future water supply, there should exist a governmental set up under which the owner or operator of any water supply system must have assurance of their water rights of at least 50-100 years in advance, with the guarantees for the future renewal or extension of those rights. If those rights are marketable, they may be purchased or sold as economics and reliability of a water supply system may require.

3.3.3.2. Availability of Market for Water. If there is a market for water, namely that holders of water rights (public, semi-public and private institutions, or individuals) have the right to sell partially or fully waters under those rights, it would insure that the highest need users of water can apply the market mechanism and economics to procure the water. When Thornton (a suburb of Denver, Colorado, USA) buys a ranch, not to raise cattle or cultivate crops, but to procure ranch water rights for this city's water supply system, this becomes a market vehicle to assure waters to the most important users. Unfortunately, having a free market system for water using means that agricultural use cannot compete with the many other uses. For example, an industry would have no problem paying $1000's for a unit of water (in this case an acre-feet, or about 1200 cubic meters) while a farmer would have great difficulty paying several hundred dollars for the same unit.

3.3.3.3. Availability of Water Market Among Countries. It is very delicate problem when water from a water rich country has to be administratively decided to be diverted to a water deficient country. However, if water is considered the market commodity, like electricity, to be sold to neighbors for a price, often it would be easier to find a solution. If the USA asks Canada to permit the diversion of water flowing into the Hudson Bay to the Great Plains of the US West, it would be difficult to find a solution. But, if somebody offers to buy that water, at a price per cubic meter or otherwise, definitely the Canadians would be at least interested to listen. This will be analogous to the selling of electric power, like the Canadian province of Quebec selling electric power to New England of the USA.

3.3.4. Development of Water Supply Systems in Stages

The development of water supply systems could usually be traced to minimal cost of new water, either as the system extension by using the surplus water of the already used water source, or by tapping the new source(s) of water. This extension most often starts with the nearby surface waters, then with subsurface waters, interbasin transfer waters, and finally recycling wastewaters. Most sources of water for the major urban areas are nearby surface and interbasin surface waters, while many smaller urban areas as well as rural areas use the nearby surface and subsurface sources of water. Recycling wastewaters is becoming a very viable option, especially in water-short California, where recycled waters are used for recharging aquifers and for irrigating crops that are processed before being consumed. With the advanced technology of recycling and recharging wastewaters, and their reliability, the re-use of waters will become more widespread as it is the last economically feasible option as a source of new water in many regions of the world. A feasibility to expand the water supply systems in stages, as well as the procuring of sufficient reserve water sources for the future, require some criteria to be considered in selecting a water supply system alternative. These kinds of criteria are:

3.3.4.1. Feasibility of Development of Water Supply Systems in Stages. Two factors affect the future growth in water demand, the increase of population to be supplied with water by that system, and the increase in standard of living of water users. This is usually translated into a larger daily average consumption of water per inhabitant of an increased population. Therefore, each system must be planned to be developed in stages, with each new stage also improving the performance of the previously developed stages whenever possible or feasible.

3.3.4.2. Procuring Water for Future Demand by New Water Rights. The wise policy in planning future developments of water supply systems in stages is to insure that sufficient sources of water for future demand will be available either through additional water rights, or by the purchase of water already allocated to other water users or purposes. The projected demand for water of 25-75 years, regardless of the low accuracy of that projection, will well serve the policy of insuring the future system's expansion. The city of Fort Collins, Colorado, USA, has accumulated enough water rights to meet the projected water demands in 2020. Other cities like Thornton, Colorado, mentioned previously, are securing water rights and designing water systems to meet their water demands at the turn of the century.

3.3.5. Economics of Water Supply Systems

Economics of a water supply system is important source of criteria in selecting the best alternative of that system. However, as households and other users of water may tolerate a relatively high price of drinking water, in comparison with the economics of the other purposes in developing water resources, the water supply systems are less affected by its economics than other water resources purposes, so the other criteria may have the same or a much higher weight that in the case of other water resources development purposes. The economic kinds of criteria are:

3.3.5.1. Measures of Economic Efficiency. The price of delivery of a gallon or cubic meter of water to users is the major measure of economic efficiency, especially when compared with the same prices of other water supply systems. The major measure of economic efficiency of the other water resources purposes (irrigation, hydropower production, navigation, flood control, recreation, etc), are most often measured by the benefit/cost ratio. This index of the purpose of water supply systems is not applicable, since drinking water unit price can be high and still people would accept it, if there were no other more attractive alternative. This unit price of water is usually adjusted for inflation with the time. The price serves as one of the criteria in selecting the best water supply system in the case of several available alternatives.

3.3.5.2. Upper Limit Cost of Unit Water Volume as a Criterion. Nearly any water price will be paid by people for drinking purpose if there is no other source of water. So, in planning water supply systems there is usually an upper price limit which in each case should not be exceeded. One type of these limits is the cost of good bottled drinking water which will be surely available on the market if that limit is exceeded. The other limit may be economically imposed, since the high cost of utilities in general decreases the standard of living in communities, and the future growth of communities may be affected. The job creating businesses may be shying away from communities with the exorbitant prices of utilities in general, and of water in particular. The third factor which imposes a kind of upper price limit on the unit volume of water is the tendency by the lowest economic strata of people to tap any available but cheaper local source of water, usually water found in communities themselves. Such cases are the excavating or drilling a well close to rivers or underlying aquifer, with that water being in principle of lesser quality, and less controlled by professionals, than water from the well designed, constructed, operated and maintained water supply system.

3.4. SELECTION OF SOURCE(S) OF WATER FOR A WATER SUPPLY SYSTEM

3.4.1. Use of Single or Multiple Selection Criteria

It is simpler to use only one criterion in selection of an alternative for a water supply system, than to embark on all intricacies of application of the multiple criteria of selection. To do that, all other criteria beyond the selected single criterion should be reduced to their effects on this selected criterion. In many water resources plannings and designs, the single criterion is usually monetary, say in the form of benefit/cost ratio or the difference between the expected annual benefit and the expected annual cost, in the case the average annual values of benefit and cost are being compared and used in decision making.

Many criteria described in Section 3. of this text may be partially or fully expressed in the form of their influence on the cost of water supply system, or as the contributor to the cost of unit volume of water delivered by the system. For instance, criteria of the impact of distance of water source from the center of water consumption, as well as the cost of seasonal, weekly or daily flow regulation either at the source itself, or on the line connecting the source and the system, or in the area of water consumption, can be expressed as the contribution to the cost. Monetary effects of some other criteria of Section 3. may be only approximately estimated. The monetary effects of the criteria related to the quality of water may be also determined by using the cost of treating water to the given water quality demand standards.

However, some criteria, especially related to availability and risk, may not be accurately applied through their monetary effects. Their influences on the selection of a water supply system alternative may be only feasible through professional judgements. An example would be the selection of a water supply source which can be controlled both in water quantity and quality by another country, but is much cheaper than an alternative.

3.4.2. Selection of Water Source or Several Water Sources

The single economic criterion, used for the selection of the alternative in cases where the planning of water supply system comes out with two or more alternatives, may decide in the selection of its source(s) of water. The minimum cost of delivered water of the given quality, per unit volume, rather than the application of all other criteria, may help decide on the final alternative. However, if the other criteria, especially the availability of water, reliability of supply, risk of major accidents and damages by disasters, are considered important, the professional judgement and the best outlook for the system's maintenance, operation, extension and future renovation may prevail in the decision, especially if the cost difference is not substantial, say not more than 5-20 percent of the cost of the cheapest alternative.

If the multiple criteria are used in selecting the best alternative, criteria may be compared by giving each a weight in the analysis of each alternative. The maximum total weight may give the preference to that alternative. Or, the other, already classical multiple criteria approaches of decision making may be used. The differences in the quality of supplied water between alternatives may be weighed heavily in comparison toward the minimum cost criterion.

3.5. COMPREHENSIVE AND INTEGRATED PLANNING FOR DEVELOPMENT OF ALL SOURCES OF WATER

The comprehensive water resources planning for utilization of all water resources in a region includes all water sources above described by using all the above outlined criteria. In that case the present and future water supply systems must compete for the available water with all other sources of water, however giving always some kind of preferences to the drinking water purpose. In most cases, but particularly in the developed countries the volume of drinking water is only a small percentage of all available sources of water (less than 1% of per capita water use in the United States of America). If it doesn't exceed a couple of percent of all regionally available waters, except in arid and some semi-arid region, drinking water has always preference.

The integrated water resources planning means the inclusion of all the interests and purposes in the development of water resources in the region. In that regard, the answer to the question "what kind of water to what purpose" becomes crucial in this comprehensive and integrated planning of water resources development, conservation, control and protection. This approach to regional planning makes water supply systems only a subsystem of a system of regional water resources. This fact often does not sit well with the leaders of the urban water supply systems, which by the importance of drinking water always tend to be autonomous and independent of the solutions proposed by the plan for all other water resources purposes. Though there is something vital in this attitude, economics, safety and quality requirements of water supply may be resolved adequately by the regional comprehensive and integrated water resources planning and resulting schemes.

The important aspects of this approach are:

(1) The planning of water resources development has passed many phases in the last 100-150 years. From a concept of single structure for a single water resources purpose and a single source of water, the planning has evolved to the multi-structure, multi-purpose and multi-source of water,

(2) The planning is more and more regional or basin-wide, so that the selection of the areal horizons become unavoidable, such as dividing the areas of the physical limits in planning, then the adjacent areas to them with which there will be an exchange or only one way transfers of water, and finally still a wider area which affects the plans by the influence of joint demand and the fulfillment of that demand,

(3) Planning has several time horizons to look at, such as the time horizon of the economic life of a system, the time horizon of its physical life, horizon of its life of obsolescence, and eventually the horizon imposed by the eventual climatic changes.

3.6. CONCLUSIONS

The following three major conclusions may be drawn from the above discussions:

(1) The selection of the source, or the sources of water for the urban or rural water supply systems is very complex process, because the quality of water for drinking purposes imposes several strict criteria to fulfill, that many water sources can not meet economically,

(2) With the above classification of five categories, and within them of 21 kinds of sources of water, a systematic approach to the investigation of all sources of water, and the eventual combination of those sources for water supplies, are indicated, in a comprehensive and integrated water resources planning for their development, conservation, control and protection,

(3) With the above classification of five categories, and within them of 15 kinds of criteria for judging the suitability of the sources of water and their selection for the water supply systems is facilitated, especially in the complex regional planning of water resources schemes and in the allocation of water rights for the required quantities and qualities of the available sources of water.

3.7. REFERENCES

Kovacs G. & Assoc. (1981) Subterranean Hydrology. WRP, 970 pages.

McWhorter D. (1992) Profession of Agricultural and Chemical Engineering. Report, Colorado State University, Fort Collins, Colorado, USA.

McWhorther D. (1994) personal communication.

Milanovic T.P. (1981) Karst Hydrogeology. WRP, 434 pages.

Starosolszky O. (1987) Applied Surface Hydrology. WRP, 821 pages.

4. New Concepts for Water Supply Systems

* What Kind of Water for What Purpose in the Future *

Vujica Yevjevich
Department of Civil Engineering
Colorado State University
Fort Collins, Colorado 80523
USA

4.1. INTRODUCTION

The most basic concepts in planning and design of water supply systems are already well shaped and currently widely used. The unique system provides the drinking quality water, regardless that many users of high quantity water would tolerate water of lesser quality. Sometimes users of large quantities of non-drinking water quality may construct a separate system, as an economical solution. Regardless of the availability of advanced technology, research and development in many areas promise the planning and design of better water supply systems in the future. These research and development areas are: safety, technology and economics of interconnecting the adjacent water supply systems; planning systems renovation in advance; design of new water quality standards based on advanced research; role of water quality and composition on human health and nutrition; use of new materials in construction and operation of systems; incrustation, erosion and corrosion in water supply systems; new materials for pipes and conduits; etc. In deciding what kind of water for what purpose may be used, distinction should be made between using the natural untreated good quality waters and the treated waters to various degrees of quality.

4.2. CONCEPTS FOR DESIGN AND PLANNING OF WATER SUPPLY SYSTEMS

4.2.1. Main Existing Concepts

The main concepts which are currently applied in design and planning of water supply systems are:

NATO ASI Series, Partnership Sub-Series, 2. Environment – Vol. 15
Water Supply Systems – New Technologies
Edited by Č. Maksimović, F. Calomino, and J. Snoxell

(1) Water demand, both by its quantity and quality requirements, is determined as accurately as the available information permits. This is done either for a new or an expanding water supply system. It is usually stated in the form of parameters of the required water demand variables, as these parameters change with time,

(2) All sources of water within, or in the vicinity of, the area covered by the water supply system are considered and investigated as the potential sources of water in planning a system or in a system's expansion,

(3) Criteria are selected and applied for the evaluation of the alternatives of water supply in planning the new or expanding an existing water supply system,

(4) Major criteria applied to water supply systems are: the availability of water, its reliability of supply (time of access, assured water rights, etc.), water quality, cost of the unit of water supplied by the system, and the future potential for system expansion,

(5 Safety of the system as a whole against various types of water pollution as well as natural or man-induced disasters, and the related overall water quality, must be acceptable to both the decision makers and the water users,

(6) Treatment of water to meet the quality requirements, the pollution-free transport and storage of treated water, and the distribution network are designed and planned according to the prevailing technologies and the best engineering practice at the time of decision making.

4.2.2. Composition and Characteristics by Quality and Quantity of Water Users in Water Supply Systems

Most water supply systems, especially the urban ones, have a variety of water users, often with requirements of quite different water quantity and quality characteristics. The major concept used for the contemporaneous water supply systems is that water treatment must meet the criteria of the user of the system with the highest water quality requirement. Because drinking water is usually the use with the most stringent water quality requirements, all other users which can tolerate the lesser water quality are generally supplied with, and use, this drinking quality water.

As an example, watering of lawns, shrubbery and trees around houses, in parks and around the public buildings and facilities in most urban water supply systems uses a much better, treated water than is necessary for such irrigation. And, this is often the user of the largest quantity of water of the system. This concept, that all water users of a system, receiving water of the user with the highest quality requirements, may be designated as the unique network water supply system. This means there is only one water distribution network of conduits and pipes, conveying the same quality of water for all users, regardless of their differences in required quantity and quality of water. This unique system imposes a possible waste of resources in overtreating water for many large urban or rural water users. Also, due to the uncertain quality of an untreated or insufficiently treated water, which may come into contact with the general population, the water supply system decision maker, trying to avoid public controversies, tends to prefer a single, high quality water supply system.

A case of departure from the rule in planning the general unique water supply system is when a large water user finds it economically and safety-wise acceptable to have a separate water supply system in a part of area covered by the unique water supply system. Such an example is in Fort Collins, Colorado, USA and the Colorado State University. The city is supplied with treated water from the Horsetooth Reservoir. However, irrigation of the large area of the Campus of the University is supplied directly from an untreated reservoir water, because it was a cheaper solution to have a separate system for that use. It was, therefore, considered that little risk for water-generated health problems existed using the reservoir water for watering of Campus lawns. This situation is not more likely to be a risk than is recreation (boating, swimming, picnicing, fishing) at the same reservoir. The city, however, often uses treated water for irrigation.

4.2.3. Ongoing Technology and its Potential for Improvements

In principle, one rarely encounters ongoing technologies of the complex infrastructural systems, like the large urban water supply systems, that research and development could not improve on the existing or create new technologies, which can successfully replace the existing ones. So, both the new concepts and the new technologies are worthy searching for in order to improve the water supply systems of the 21st century.

The goals of this research and development are versatile. They are:

(1) To increase the availability of water sources for a system,

(2) To make reliability of supply as high as the needs and economic criteria would require or permit,

(3) To continually increase work on water quality research and development, especially on the effects of water quality on humans, with new findings expected,

(4) To decrease the costs of water of the prescribed drinking water standards,

(5) To increase the potential for relatively economical expansion of existing water supply systems,

(6) To better plan and design how to overhaul and innovate the aged water supply systems in the future,

(7) To improve the monitoring of water supply processes and better maintain water supply subsystems; and similar goals.

4.3. FUTURE ALTERNATIVES FOR WATER SUPPLY SYSTEMS

4.3.1. Basic Dilemma in Selecting the Type of Water Supply System

The application of the principle that each water user in a water supply system should obtain the quality of water which meets only the necessary quality requirement of that user, means that the unique system of water distribution network must be changed, modified or abandoned. The alternatives are either two or more combined water supply systems, each serving the particular user, or a group of users, with the required water quality standard. The highest quality water user will be in the system supplying the drinking and cooking water for the population. The lowest water quality user happens to be usually the bulk large quantity water uses, such as cooling water, water for irrigation, water for washing of streets, and other public or private large quantity water using facilities. However, it must be noted that many industrial processes (food processing industry included) may require higher quality water than drinking water.

In the middle of the quality scale will be water quality needed for the basic, human hygienic needs. In a two-pronged water supply system, water needed for hygienic purposes (bathing, washing, toilet, flushing, etc.) will be served by the bulk water of lower quality, provided this quality is sufficient for those purposes. However, drinking water quality will be used if the quality of the bulk water supply is not acceptable for human hygienic purposes.

The immediate reaction of planners, designers and decision makers to such a multiple network water supply system alternatives would be that the increased cost in the double or multiple pipe networks will have to be compared with, and be smaller than, the additional cost of treating all water to the drinking water standards in the unique network alternative. This comparison by the economic analysis can be easily done since the cosst of treatment of water and the cost of construction of networks are readily available.

4.3.2. Alternatives for Future Water Supply Systems

The investigation of new alternatives of water supply systems with two or more separate pipe networks, one for each major, particular user, or the groups of users, faces several dilemmas. For example, should the best drinking water be of the same quality for all people? It is known that people have different acidities of the stomach juices (hypertonic, neutral, hypotonic, or of several more classes instead of only these three classes). Some may prefer pH of the water that best corresponds to their health and stomach tolerance, and all of them can not satisfy their preferences with the same drinking water quality at the same time. Only the industrially processed waters may produce several desired types of water, each of a different pH and different mineral composition, to be sold in food and/or beverage markets like the varieties of food selection that are offered currently. The alternatives of the future water supply systems, especially of the large and medium size systems, are:

4.3.2.1. Unique System. That means to continue planning the unique water supply systems, for which all water is treated to the prevailing drinking water standards.

4.3.2.2. Variety of Drinking and Cooking Water Provided at the Market. These types of water are provided by the market oriented to production of drinking water of various compositions and pH values, which can be regularly purchased in the food and beverages stores, and then to use the unique water supply system, with the quality of treated water to satisfy the people's needs for human hygienic purposes.

4.3.2.3. Dual Water Supply System. The dual system of water distribution networks will mean that two pipe networks are laid into the same ditch or gallery, with the smaller size pipes of the network serving the drinking, cooking and hygienic human purposes, while the larger size pipe network would serve all the other purposes, especially the users of the large quantities of water, with water treated only to be disinfected (safe from bacteria) and sterile standards (safe from viruses and protoviruses).

4.3.2.4. Multiple Network Water Supply Systems. The multiple network water supply systems will have three or more networks of pipes. Apart from the network for the users of water of drinking, cooking and hygienic quality standards, the urban irrigation, various industrial water users, plant cooling and other specific purpose uses, would have their particular networks or the individual network of pipes, while all other remaining uses as a group will form the separate pipe network. Then all these waters of different quality would each meet the water quality requirement of the use with the most stringent standards in each of the separate groups of uses.

4.3.3. Specific Topics for Investigation, which may Improve Future Water Supply Systems

There are several aspects of water supply systems which deserve continual active research and development, as well as a better attention and efforts in operation and maintenance than was done in the past and at present. They are:

4.3.3.1. Interconnection of Adjacent Water Supply Systems. Many large water supply systems are often close by distance, or the medium or small systems are located in between the two or more large systems. With the relatively short conduit interconnections, systems may be connected in such a way that water can be exchanged or shifted between the presently separate systems. This is not only important for the potential matching of surplus and deficit of water which occur in the interconnected systems, but also for the solution of the problems of drought deficit. This is analogous to the interconnected electric power grid systems. The more distant the geographic centers of water of the corresponding sources of water (rainfall, surface and subsurface sources of water), the smaller the probability that drought will cover both system sources with the same proportional deficit. This concept should be compared with the alternative concept of solving the drought problems by large reservoir capacities in both systems for the seasonal and/or interannual flow regulation. An example of such interconnection of the large systems may be the relatively densely populated urban areas between Boston and Washington D.C. in the United States. Many separate large and medium water supply systems between these two cities (including New York City, Philadelphia and Baltimore) can be easily interconnected for the benefit of all of them. A major problem still remains when one system produces a better quality water than another, even though both may meet the existing EPA (US Environmental Protection Agency) drinking water standards. An example of this is the quality of water of the city of Fort Collins, Colorado, and a water company nearby. The latter wants to be interconnected but the city of Fort Collins does not want the connection since the other water company does not produce as good a quality of water as the city does. However, if safety of water supply in emergencies and in droughts are the major reason for interconnection, the mixing of different quality of drinking water may be temporarily accepted.

4.3.3.2. Planning in Advance of System Renovation when Aged, or Deteriorated, or Obsolete. The ongoing renovation or overhauling of many large water supply systems in the world, especially those with deteriorated pipe networks, pose a simple question, namely how many of the existing water supply systems in their original planning and design have taken into account the fact that these systems will definitely age and will need renovation, or also when obsolete, be replaced? Therefore, one can justify future studies, with research and development, of these problems in order to incorporate into planning and design of the new or renovated systems also how to fight most economically, the aging process and when unavoidable to renovate or replace the systems.

4.3.3.3. Definition and New Standards for Water Quality. Regardless of the tremendous efforts to define water quality in general, and for the specific users in particular, the fact is that several hundred chemical compounds and dozens of biological species may be found in water. Apart from individual effects of these compounds and species, there are also many synergistic effects by which two or more compound or species have a larger effects on the water consumers than the sum of the effects of the individual compounds or species. As an example, recent research has shown a link of cancer in the stomach and intestines to the consumption of chlorinated water making a need for a better monitoring or disinfection of water, which do not contribute to increases of cancer (Canter, 1992). A huge potential is available for the systematic and long-lasting research and development, uncovering many synergistic aspects of water quality. This may change the ongoing definitions of water quality for the water supply systems and lead to the new, significantly better water quality standards.

4.3.3.4. Water in Human Health and Nutrition. It is well known that quality of water affects human health. Otherwise, there will be much less international and national efforts in the world, to supply water for drinking purposes, especially in developing countries of large populations. Less is known on how quality of water affects nutrition of people. Definitely, minerals and eventually benign biological matter, brought by water into the human bodies, must play some role in nutritional aspects, if not also in human health. There is a speculative theory, that water from rocks of the subsurface water systems brings with it some very rare elements, which serve as catalyzers (enzymes) in biological processes, especially in reproduction of cells. Waters from alluvial aquifers have less of those rare elements because the river water which deposited sediments of alluvial aquifers have washed out those elements into the sea. Research and development on the role of quality of water on human health and nutrition, likely will be a subject of many projects and new research approaches in the future. Maybe, that human longevity partially depends on these quality variables.

4.3.3.5. Use of New Materials in Construction of Water Supply Systems and in Their Operation. Materials used for water supply networks, pumps, water treatment structures and materials used in treating water, may affect the quality of water, sometimes decisively. The best example is the use of lead and copper pipes in water systems, even for the short connections and the ends of pipes bringing drinking water into households. Even a speculative theory goes with it, namely that the Roman Empire may have partially fallen due to the use of lead pipes. During that empire, water in water supply systems was mainly brought from springs to cities by the free-flowing water aqueducts. Only rich and politically leading and influential people could have afforded in houses and restricted public bath houses the internal water supply system under pressure with lead pipes As a poison, lead is a slow destructor of human tissues. Regardless of the speculative character of this theory, the fact that the lead and copper pipes are being removed from water supply systems in many countries, as dangerous material, testifies that the theory may have some validity. Also, the use of the chlorine in treating the drinking waters is not without controversies, because of the creation of its compounds which are related to cancer increases.

4.3.3.6. Incrustation, Erosion and Corrosion. Many water supply subsystems, especially their pipe networks, may exhibit three phenomena: (1) internal incrustation, particularly those which deposit calcium carbonates at the internal walls of pipes, with a decrease of effective flow area and an increase of roughness and head losses within the main conduits and pipe networks; (2) erosion of the interior pipe surface when the untreated water carries fine sediments, especially the fine quartz grains; and (3) corrosion of pipes either from outside or inside if water or soil are chemically corrosive, with this phenomenon most often occurring on outside rather than inside of the pipes. Research and development need to help in a better understanding of these three phenomena which decrease longevity of pipes and networks in water supply systems and decrease capacity for conveying waters. New pipe materials or new protection coatings, both inside and outside, may prolong the life of the systems and make their maintenance and operation more economical.

4.3.3.7. New Pipes and Main Conduits. The science of behavior of materials under various conditions is progressing rapidly by the end of the 20th century. It is likely that this progress will be maintained also in the first part of the 21st century. Therefore, special research and development on the new pipes and new conduits may help both the new and the renovating water supply systems. The stress in these special investigations should be on both the advantages but also the eventual disadvantages of any new material, being it in plastics, laminated materials or others, or insuring the anti-erosion, anti-corrosion and even eventually anti-incrustation properties of materials or coatings.

4.4. WHAT KIND OF WATER FOR WHAT PURPOSE IN THE FUTURE

In the treatment of the topic what kind of water to be allocated or used for what purpose in the future, a distinction may be beneficially made between the natural untreated waters of satisfactory qualities, and the treated waters to various degrees of quality, even pushing the treatment to the point of "manufacturing of special types of quality water for drinking purposes" by the industrial processes.

4.4.1. Natural Untreated Waters of Good Quality

The basic premise which must be made in using natural untreated waters for any purpose is that natural waters may be and will be protected from a substantial deterioration by pollution. Or, if these waters become polluted, that there will be the means and administrative or legal vehicles to force polluters to return the quality of these waters to the pristine natural original conditions.

In matching the users' demand for water quality and the quality of natural waters, historical developments and the recent experience may be the most useful guides in this task. The best quality water in nature, originally being with the springs and groundwater aquifers (used by excavating or drilling wells and galleries such as the ancient and even recent Persion-invented kanats), should be reserved primarily for drinking purposes. They need safe tapping, protected from any kind of local pollution. Because originally the quantity of these waters often exceeded the needs for drinking purposes, many spring and aquifer waters have been used for irrigation in agriculture, or for industrial, cooling and mining processing and washing purposes. With the increase of population and their standard of living, with both factors requiring more water of the best quality, pressures are usually exercised to use more spring and aquifer waters for drinking purposes, taking them away from agriculture, industry, cooling and mining purposes. Wastewaters from urban and rural communities, return flows from irrigation, wastewaters from industry, mining, cooling and other uses, all may be changed or deteriorated by use, either physically, chemically or biologically, or by any combination of them. This fact usually and in general threatens the quality of all spring and aquifer waters.

The recent experience has shown also that protection of the quality of naturally very clean drinking water is not an easy assignment. The Rocky Mountains "clear waters", advertised by a beer producer which uses those waters in beer production, may be now questioned. The fumes from copper smelters of Arizona, fumes from the coal-fired power plants of the Southwest and West of the USA, fumes from automobiles of Southern California, are all carried by the southwesterly and westerly winds over the Rocky Mountains. As these westerlies produce 6-15 snow or rainfall storms over the Rockies during the winter season, a lot of those air-borne fume materials are deposited over the Rockies. In some way, one can tell "good-by to the clear Rocky Mountains waters." Combined with the pollution by the mines of small rivers in the Rocky Mountains region (old mines drainage, leaching from tailings, and processing of ores of the ongoing mines), it is easy to demonstrate that the preservation of the quality of natural pristine waters for drinking purposes is a Sisyphusian task in the modern intensive agricultural, versatile industrial and broad mining activities.

The altered waters of springs and aquifers may become unusable for drinking purposes by the broad and versatile sources of pollution, especially by the chemical pollution. That pollution may not impede the use of those changed waters for agricultural or urban irrigation, for industries and mining operations, which tolerate the less quality waters than required by the drinking water purpose. If those spring and aquifer waters are first used in mining or in industry, and if their wastewaters become further polluted by these uses, they may not be suitable even for agriculture irrigation. Therefore, the priorities in using the present or previously pristine natural waters should be:

(1) for drinking water purposes,

(2) for special industries requiring water of very good quality,

(3) for agriculture irrigation or irrigation in urban areas, and finally,

(4) for many types of general industrial and mining water uses, which tolerate water of lesser quality.

The difficulties in carrying out this priority sequence arise from the facts that major mining water users are in mountains where mines and pristine waters are located. Industries and urban water users are often located in the wider downstream valleys, and irrigation users in the plains, usually in the lowest sections of rivers. This sequence of the location of water users does not follow, or coincide with, the sequence of the quality of water along rivers. Ideally, the pristine waters should be used first for urban and rural water supply systems as well as for the special industrial purposes requiring waters of a particular good quality, followed by agricultural and urban irrigation, and finally by the industry and mining tolerating in their use the waters of lesser quality (washing, cooling, ore separating, etc).

4.4.2. Altered Waters after Various Uses

Once water is used, it is usually either discharged into the rivers and underground, to be taken again later downstream by the next user, or it will be passed to the next user directly from the outlet of the previous user. If water is discharged into the rivers to flow freely along a river section and for a time, the natural purification capacity of rivers or underground water systems can be utilized to improve on water quality. Unfortunately, due to the limited power of this natural cleansing, either biological through the natural oxidation process of biological matter, or chemical by dilution of their concentration in river waters, the ability of many rivers has been exceeded many times over. This has caused massive pollution control problems especially during the low flows and the highest yearly temperatures. Also, it created the awareness and policies that rivers should not be considered in the future as organic and chemical waste transport and disposal sinks. If water goes directly from a user to the next one, planners and designers must cope with the altered water quality and changes in its quantity made by the previous user, when that water does not meet the water quality/quantity criteria of the next user. All this creates the interdependence among water users in a river system. The users can be a group or individual uses, cities, or even countries. Thus a chain of sequential uses results, which require most often water treatment between these uses.

A classical example is the Ruhr Region of Germany, with the chain of uses and treatments of water in between them, so that symbolically people used to say "that the Ruhr water passes through seven stomachs." However, because of the inflows of both natural and the used water in between the intakes of water for these users, water for any user is more a mix of the used and new fresh waters than only the upstream wastewater. Thus water is not only easier to treat, but also psychologically it does not imply that only the upstream wastewater is being used downstream after treatment.

4.4.3. Matching Kinds of Water and Water Use Purposes

When one considers what kind of water should be used for what purpose, first it is necessary to take into account the original characteristics of the source of water, before any anthropogenic changes in water quantity and quality have been made. Second, one must investigate whether the natural water source may be economically returned to its original pristine state. Third, if water quality may be returned to its original pristine state, and eventually also improved, the question to answer is whether the new characteristics of that source of water may be effectively protected from various sources of pollution in the future. The major sources of water meeting these properties are springs, aquifers and lake and reservoir surface waters. All other sources, such as direct use of the atmospheric water, recycled water, conversion of brackish and saline waters, may be considered as special cases to be considered if the major sources of surface and subsurface waters are no longer available.

4.4.3.1. Drinking Water. The original supply of ancient human communities has been first oriented to the use of water of perennial springs. They required aqueducts, usually with the free surface water flow, from the spring(s) to the water fountains inside the community. When low flows of these springs no longer satisfied the need for water, the next source and the priority in tapping water sources were the aquifers. They required usually the digging of a well or several wells, and then the power to lift water from the well (human, animal or wind lifting powers). In modern times, when the spring and aquifer sources of water have been already tapped for water supply systems, the attention was passed to the lake and reservoir surface waters. This latter alternative in most cases required treatment of water to be bacteriologically disinfected and free from viruses. In recent times, and due to an easy pollution of many types of springs and aquifers, the attention was paid to lake and reservoir waters. The special value was for the upper most reservoirs of smaller rivers, where no intensive agriculture, water polluting industries and mines existed. Grazing and silvaculture, with timber production, and the eventual recreation (wilderness areas, national forests, national and local parks and protected areas), enable an effective protection of water from high chemical pollution (with the exception of air borne pollutants), so that the necessary, classical water treatment is relatively inexpensive. When none of the above sources of water could satisfy the new water demands, especially in semiarid and arid regions, the drinking water may be directly collected on paved surfaces from the rain, of by converting brackish and/or saline water into the fresh water. The recycled wastewater usually is not directly used for drinking purposes. If that water is recharged into an aquifer, it may be pumped later and used as a source of water (with the new eventual treatment) for drinking purposes.

4.4.3.2. Water for Human Hygienic Needs and for Recreation. Since these uses of water have a less stringent water quality requirements than the drinking water, many surface and underground sources of water may satisfy these quality requirements, usually with the minimal treatment (say, by chlorinating such waters). Everybody knows that the swimming pool water is most often chlorinated to meet the minimal quality standard for such water use. Surplus spring and aquifer waters are directly used without treatment, even in the case of karstic springs which may carry turbid waters in floods. Lakes and reservoirs are the main sources of such quality waters.

4.4.3.3. Water for Irrigation. Nearly all sources of water in nature, which by its quality is acceptable for irrigation purposes, have been used one way or another. When springs are used for drinking purposes, if any water is left unused in the growing season, usually that surplus of water goes to irrigation. The same is valid for aquifers. However, water of rivers, lakes and large aquifers is predominantly used for irrigation, where irrigation is needed. From all the water resources development purposes irrigation requires the largest storage capacities in relation to the available mean annual water volume, because the best use of water in irrigation is by storing surplus water in wet seasons and wet years, and then using that stored water in dry seasons and dry years. The largest irrigation schemes in the world are developed by constructing large reservoirs. Among others, the US West is one of the best examples of those large irrigation schemes, not to mention also such large schemes in China, Russia, India and all over the world. Similarly, huge aquifers like the Ogallala formation in the US West feed irrigation projects in many states through millions of wells. As an example, only the state of Nebraska has nearly half a million of irrigation wells pumping water from this huge aquifer, so that the difficult problem of aquifer recharge looms at the horizon (Bittinger and Green, 1980). Contemporaneous agriculture uses lots of chemicals. Among them the nitrates are at the basis of artificial fertilizers. If irrigation is carried out over the aquifers, two modern phenomena occur. The lower layers of aquifer water are loaded with nitrates. By pumping water from these layers for irrigation, nitrates are recovered. The second positive aspect of this type of conditions, is that the use of irrigation water distribution network in non-growing season, aquifers may be easily and most economically recharged. Thus aquifers are not only used as sources of irrigation water, but also as storage capacity and the recovery of fertilizers.

Recycled water, or treated urban wastewaters have been used for irrigation of fast growing trees for paper industry, since one can control easily in such cases the contact of humans with these irrigation waters. In the watering of lawns, shrubbery and trees in urban areas, usually that water is of the quality of drinking water, so that health of the population is protected. If such irrigation uses a special independent network, the water usually must be disinfected.

4.4.3.4. Industrial Water Supplies. There is no uniformity in the industrial and mining water quality requirements. From food and pharmaceutical industries, which require drinking water quality, and sometimes even still better water quality which fits the need of a specific industrial process, all the way up to the washing water which doesn't even need to be desinfected and/or made sterile. The quality of water then should meet the needs of the particular industries, mines and smelters. Some industries may use waters of different qualities, the best for the specific process and the least quality water for simple washing. Close to the seas and salt lakes, the limited quantity of good quality fresh water may be used only for the industrial process, which can not use the lesser quality water, while the brackish and/or saline waters are used for the purposes which do not require the more expensive, quality water for such uses as washing floors, trucks, cars and machines. Water used for cooling purposes may be loaded with minerals, if its temperature is sufficiently low to meet the requirement of the cooling purpose. In classical thermal and in nuclear power plants the condensation process water should be as little mineralized as the aggressiveness of that water permits, while the cooling tower or flow-through cooling waters may be of a much lower quality as long as they accomplish the cooling purpose.

4.4.3.5. Water for Hydropower and Navigation. Because hydroelectric power plants use the energy head in passing water through the plant, the consumption of water (evaporation, leakages, losses, etc.) is relatively minimal. The only requirement on water quality is that it is not too aggressive for the structures through which water passes, so that they do not highly and rapidly corrode, erode or incrustate. A similar situation may be with navigation, where the quality requirements are such that water has no floating debris and should have sufficient navigable depths, or and sufficient volume of water to fill locks in the passage of ships and barges from lower to higher levels by using locks.

4.4.4. Six Basic Types of Water Uses

In summary, there are six basic types of water uses:

(1) Destilled water for use in boilers and uses in special industrial processes,

(2) Uses needing drinking quality water,

(3) Human hygienic and recreation water uses,

(4) Irrigation in agriculture and in urban areas,

(5) Industrial, mining, smeltering, cooling purposes uses, and

(6) Hydropower and navigation water uses tolerating the relatively low quality waters.

These basic uses may satisfy the water demands by quite different qualities of water, which depends on the character of the user's water quality requirements. All other water uses (wildlife protection, fish production, low flow augmentation, wastewater dilution, etc.) may be satisfied by one or more types of the above six water quality potential uses.

4.5. CONCLUSIONS

The following major conclusions of the above discussion are:

(1) The unique water supply system forces the treatment of water quality to the standards of drinking water, though the majority by the volume of the users may tolerate the lesser quality waters,

(2) The future investigations should resolve the dilemma whether the multiple network water supply systems have any safety and economic advantages, which may lead to different treatment levels of the multiple network water supply systems,

(3) There are many areas, with some discussed above in this text, which deserve the repeated research and development attempts, especially as it concerns the human health and nutrition related to the drinking water quality.

4.6. REFERENCES

Bittinger W. M., Green E.B. (1980) You Never Miss the Water Til (The Ogallala Story). Water Resources Publications, 116 pages.

Canter K.P. (1992) Cancer Risks Associated with Chlorination By-products. Water Review Technical Report, a publication of the Water Quality Research Council, USA, Vol. 10, No. 2.

5. Pollution Control for Urban Water Supply Systems

Thomas G. Sanders
Vujica Yevjevich
Department of Civil Engineering
Colorado State University
Fort Collins, Colorado 80523
USA

5.1. INTRODUCTION

The protection and prevention of contamination of an urban water supply system is a complex process which can be as simple as fencing off and preventing access to the contributing water catchment and as complicated as adding slow acting disinfectants in the treated water to protect it from contamination while it is being delivered through the distribution system. There are many other control strategies in fighting the contamination of water in an urban water supply, forming **the multiple barrier system** within that supply system.

The concept of **the multiple barrier system** is the cornerstone of the environmental engineering approach to producing safe drinking water. In order to protect the public water supply many measures are undertaken to eliminate, prevent and/or destroy pathogenic organisms. Each measure in itself could eliminate the pathogenic organisms but maintaining the multiple barrier system approach assures a high reliability of the entire system that the pathogenic organisms will be eliminated.

This strategy is the reason why the preferred source of the water is one where man's access is restricted, through the unit processes treating the water to remove the pathogenic organisms as well. In fact, the multiple barrier system reduces significantly the probability that something may go wrong in the prevention and protection of the drinking water against contamination, with all the consequences that the unremoved pathogenic organisms may produce on human health. Urban water supply systems can be divided into the following four componential parts (called here either the components or the subsystems) as related to the

NATO ASI Series, Partnership Sub-Series, 2. Environment – Vol. 15
Water Supply Systems – New Technologies
Edited by Č. Maksimović, F. Calomino, and J. Snoxell

protection and prevention of the contamination of the systems' waters: (1) collection (and/or tapping); (2) transmission (or transportation); (3) treatment; and (4) distribution.

The collection component or subsystem of a water supply system can be visualized as the catchment upstream from the point of diversion and/or the point of intake structures, used to collect, divert and/or control the proper inflow of the water into the man-made water supply system.

The other type of visualization are the reservoirs, as man-made structures, used to collect and store the water, as well as places where the water intake structures may be or often are located. The transmission or transportation component of a water supply system includes the main conduits, canals, aqueducts, tunnels, siphons and other water conveyance structures, which connect the diversion or intake structure(s) of the main source(s) of the water to the treatment plants or further to the local storage tanks or reservoirs for the treated water within the major water consumption areas. The basic concern is to prevent that no uncontrolled water inflow would penetrate the transmission conveyance line. The treatment works component or subsystem may only be the disinfection, or the conventional treatment of chemical coagulation, flocculation, and sedimentation followed by filtration and disinfection, or the conventional treatment followed by carbon absorption, reverse osmosis, etc. The final product of this subsystem will be the drinking water meeting the water quality standards selected by all the parties responsible for the public safety of the delivered water to consumers.

The distribution component of urban water supply systems is composed of a series of interconnected pipes which transmit the water from the treatment plant(s) or from the storage of treated water to points of use at a prescribed lowest water pressure. Most often, the distribution system of pipes uses the closed loops of pipes, so that each user of water may receive water from two directions, and that the local service interruption of a pipe does not interrupt also the water supply. Depending on the availability of head, standpipes or reservoirs located at elevated positions in the urban area or water pumping stations may be required to maintain pressure in the system and therefore would be also part of the distribution component.

5.2. PROTECTING SYSTEM'S WATER COLLECTION COMPONENT

5.2.1. Access to Catchment and Reservoirs

The protection of the collection component of the source of water in small water supply systems from contamination is carried out either by limiting public access to the catchment and reservoirs, or by permitting that access but in a regulated way. In the first case, only those people may be permitted access who are connected with the operation of reservoirs and the water supply system, and those who have the property or land utilization rights within the catchment. Those permits, however, must be clearly defined and regulated, so that the offenders may be prosecuted for violations. In the second case, which do not restrict public access, restrictions may be used for both motors on the reservoirs (sometimes permitting only the electrical, water non-polluting motors, operated on batteries), and allowing the use of sailboats, rowboats and canoes.

Whether the river, lake or reservoir is a source for water supply or not, disposal of all human wastes is usually prohibited. If wastewaters occur in the collection area, a strict regulation on the degree of their treatment before they are released into the catchment or reservoir must be enforced. Similarly, any other point or non-point source of contaminants must be identified and strictly controlled.

As the catchments get bigger and the restrictions for public access become unfeasible, even to reservoirs, the alternative is the restriction of land use. This could include the prohibition of the use of septic tank systems and require vaults for human wastes. Restrictions to allow only certain land uses is a viable alternative. An example would be to restrict or prohibit the use of pesticides, herbicides, insecticides and most of the artificial fertilizers thus discouraging intensive farming, or farming at all, in the catchment. If restricting land use is not feasible, aggressive point source and non-point source control of pollutants is the only alternative.

5.2.2. Control of Point Sources of Contaminants

Since 1972 a US national policy of zero wastewater discharge has been pursued to eliminate point source discharges from municipal wastewater treatment plants and from industries, without permits. This is a **technology based** program using the **best available technology (BAT**'s) for each discharger. Enforced by using the **NPDES** permits **(National Pollution Discharge Elimination System** permit), each point discharger is required to meet the permit's discharge requirements which will change as **best technology** changes. The individual states of the USA are charged with the enforcement of this policy and to periodically review and reissue new permits. The previous US legislation of 1965 was a pollution control program based upon a **water quality based** program. Each navigable stream in the USA was classified according to existing use and water quality. The idea was to not allow any river to get worse (antidegradation policy) with the eventual improvement and protect the water quality in courts. Unfortunately, it did not work as it was based upon legal enforcement of stream standards where the source of the violations could not, in most cases, be proven in courts.

5.2.3. Control of Non-Point Sources of Contaminants

Presently the focus of the US national policy on pollution control is directed toward non-point sources of pollutants. Primarily, looking at urban non-point sources at present, the program will eventually expand to non-point sources of contaminants from agriculture, silvaculture, pastures, roads and highways, airports, etc. Developing the concept of the **best management practices**, **BMP**'s, which are analogous to the **BAT**'s for point sources of pollution, will be the thrust of research and development of technologies for the next decades. Unfortunately, the BAT's are being recommended and may eventually be required without knowing whether they work or will have any kind of impact in the reduction of pollution. An example is the recommendation to treat surface runoff by using detention ponds, followed by filtering it through sand to remove phosphorus. The detention ponds will trap the phosphorus absorbed to sediment particles, which could be as high as 70% of the phosphorus load but there is no indication that sand and smaller particles sediment will remove significant amounts of dissolved phosphorus (Mestas, 1994).

Some control of quality of the water may be accomplished by using the proper types of intake structures at reservoirs. The multiple outlet water intake structures at reservoirs permit the drawing of water from the particular layers of stratified reservoirs and lakes, as the best management of water quality may require.

Thus, some reservoir inflow waters can stay longer in the storage giving them opportunity to eventually deposit with the suspended sediment as much phosphorus as particles can accommodate and store at the bottom. This is the case especially with the mud lake at the lowest part of reservoirs, close to the dam sites. These mud deposits may be washed out during floods, provided downstream conditions permit and there is the corresponding bottom outlet(s) at dams.

5.2.4. Protecting Water from Pollution at Intake Structures

Water intake structures are located at the transition between the collection and transmission components of water supply systems. The tapping of the spring and well waters requires special attention because of the potential of penetration of polluted waters. The most sensitive structure at the spring intakes is the spill-over part, due to the fact of the most frequent condition that only part of spring discharge is taken for the water supply system, producing an overflow of water at the spring and flow its natural stream or go underground. Spillways at such intake structures are a sensitive part of the intake, since animals and impurities may penetrate the intakes through the spillway or other surplus water outlets. In case of water wells, the penetration of polluted surface water along the well casings must be often specially treated.

5.2.5 Protecting Underground Water Collection Component from Pollution

It is much more complex and expensive to protect the underground water systems from pollution than it is the case with the surface waters. Special measures and technologies may be required to protect the karst or highly permeable lava deposit water systems from contamination, than is the case with the alluvial aquifers, though these aquifers may also be easily polluted, especially if they underlay the permeable and agriculture-used soils. Special techniques are needed and are available for building on, and using, the karst surfaces for various purposes. Urban and industrial areas require protection measures from an easy infiltration and/or sinking of the polluted water into karst formation.

A similar case may be with the water systems in the recent lava deposits (recent in the geological sense). These measures require that the permeable surfaces are paved over, or made impervious by other methods, urban and industrial drainage water carefully collected, and wastewaters of both treated to the point that they may be permitted either to be disposed in the sinks or permeable terrains, or if not sufficiently treated, to be disposed outside of these geological formations (Turk, 1975).

The protection of alluvial aquifers from contamination requires a careful control of the aquifer's levels in order to avoid drawing in the polluted surface waters, basically through the large side infiltrations. The agricultural pollution due to infiltration of various chemicals is part of the non-point sources of pollution.

For the aquifers under the cultivated land, often nitrates and pesticides collect in the lower layers of the aquifer. If the aquifer is used for irrigation, a method of solving the problem of this pollution is to pump water for irrigation purpose from the deep, the high nitrate concentration water layer, and not only to protect it from the expansion of nitrate loaded water layers but also to recover fertilizer.

5.3. PROTECTING SYSTEM'S WATER TRANSMISSION COMPONENT

One should distinguish between the protection of the water from contamination of an open and a closed conveyance water transmission. A closed conveyance may work either with the free surface flow or under pressure. The open conveyances are: natural channels, canals and open aqueducts. The free surface and/or under pressure flow conveyances are tunnels and closed aqueducts. Flow under pressure only is usually in the conduits. They may be on the surface and easily accessible, or they may be underground or buried into the soil, therefore less accessible.

5.3.1. Protecting Open Conveyances from Water Pollution

If the reservoir water is released into its natural channel, and the diversion or intake structure is located sufficiently downstream from the reservoir outlet, the protection of water from contamination will depend on the water users in the catchment between the reservoir and the diversion or intake point. This case is often encountered in mountains, where the optimal points of dam and diversion locations do not coincide. There are many such examples in the Rocky Mountains in Colorado. Simply, the problem of protection from pollution of the water becomes the same as it is for the collection component of the system. The only difference is that the water at the diversion point is a mixture of the reservoir stored and quality transformed water and the natural or man-affected waters inflowing into the stream in between.

Canals are the special case, very often when water is conveyed for two or more purposes, say for irrigation, hydropower, navigation, water supply, etc. Practically the diverted river water, liberated from floating and bedload materials, flows through the canal to the diversion or intake structure for the water supply system. Transmission canals are generally restricted to access to prevent contamination, if water is predominantly used for drinking water supply. In some areas of the high water power heads with relatively small discharge, water of the hydroelectric power plants can be also used for drinking water supply, after the last power plant. In this case, headrace and tailrace canals of hydroelectric plants are strictly restricted to access, or are covered.

Protection of open aqueducts is simple, because it is easy to completely restrict access to everybody except the maintenance crew. If the aqueduct carries clean, spring drinking water, usually it is covered to avoid the wind blown dust to reach the water.

5.3.2. Protecting Closed Conveyances from Pollution and Interruption

Two dangers threaten the closed transmission conveyances in operation, the penetration of contaminated water and the interruption of the line, meaning the out-of-operation state of the system. Rarely in the medium and large water supply systems is the safety secondary conveyance line economically feasible. The easiest way to protect a closed conveyance from direct pollution is to build a tunnel, to lay down a good conduit, or to cover an aqueduct. And it is done in such a way that protection from penetration of any outside, potential polluting water is well assured, except in case of disasters (such as earthquakes, avalanches, landslides, floods, wars, revolutions, etc.).

Protection from interruptions is done by the professional investigation of the terrain over and in which a line is positioned. Geotechnical and geological factors, as well as hydrological (traversing the stream and creek valleys) are most important, since they are determining the stability of any such conveyance structure. In some cases, interruptions may be also connected with the potential pollution.

5.3.3. Monitoring Conveyances for Safety against Pollution and Interruption

It is necessary to properly design a monitoring system for both the protection of conveyances from penetration of polluted waters as well as for location of potential sources of interruptions (a weak point in the line, a slow evolving erosion or corrosion of conduits). Either the pollution or interruption of water supply by the transmission line represents major accidents in the system's operation. Monitoring this component by a well designed plan and program may be paid off very many times over since either of the two types of accidents may be very costly, highly inconveniencing water users.

5.4. PROTECTING SYSTEM'S WATER TREATMENT COMPONENT

5.4.1. Types of Water Treatment and Protection of Quality

The treatment could be as simple as just disinfecting water using chlorination or more sophisticated approach using chemical coagulation, flocculation and sedimentation, followed by filtration and disinfection. Each unit process is part of the multiple barrier system by having the particular capability of removing pathogenic organisms. The addition of chemical flocculants deactivates bacteria due to the pH change and the actual entrapping and precipitation out of solution. The filtration, be it rapid sand filtration or slow sand filtration, physically removes these organisms. It should be noted, however, that this process is much more complicated than just physical removal. Research and practice have shown that without chemical pretreatment, rapid sand filtration is not that effective in removing some organisms, such as Giardia lamblia cysts and Cryptosporidium oocysts.

The final process of disinfection will deactivate any remaining bacteria and higher concentrations and longer detention times then are used in conventional treatment, apparently deactivates viruses as well. Chlorine, when added with ammonia, forms various species of chloramines (combined chlorine, NH_2Cl, $NHCl_2$, NCl_3) which are bactericidal but tend to be longer lasting than free chlorine (hypochlorous acid-chlorate ion, $HOCl$-OCl^-) and will remain active for several days in the distribution subsystem prior to consumption. If sufficient amounts of chlorine is added with ammonia to drive the reaction to completion creating NCl_3, a gas, the process is called breakpoint chlorination and can be effective for denitrification and destroying some of the other organic compounds, but unfortunately creates trihalomethanes (THM's), known precursors to cancer.

5.4.2. Potential for Increasing Future Protection

Future processes to be incorporated into the conventional water treatment train, which can remove chlorinated hydrocarbons, dissolved inorganics and possibly viruses include: granular activated carbon, reverse osmosis, ion exchange and ultra filtration. Other disinfectants besides chlorination are being considered as well since they have the capability of not forming THM's. They are ozone and chlorine dioxide (Sanders and Champlin, 1994). Unfortunately the use of chlorination and alum (aluminum sulfate), the most widely used chemicals in the water treatment industry for disinfection and flocculation, respectively, are being questioned due to known or speculated health consequences. There are indications that drinking a chlorinated water for many years increases an individual's chances of getting bladder cancer 9% and rectal cancer 15% versus individuals drinking unchlorinated water (American Journal of Public Health, 1992). And there has been some speculations that the consumption of drinking waters treated with alum is related to memory loss.

In order to provide more protection for future drinking water, the U. S. Environmental Protection Agency (EPA) has proposed the **Information Collection Rule (ICR)** in February 1994.

The rule requires utilities to collect information which will become a data base for the **Enhanced Surface Water Treatment Rule (ESWTR)** and the stage Two **Disinfectants/Disinfection By-Products Rule (D / DBPR)**. The study is nationwide, primarily to quantify the occurrence of contaminants and their removal, and will lead to regulatory controls for microbiological and disinfection by-products contamination (Black and Veatch, 1994).

Besides keeping records on the disinfection by-products, **ICR** mandates routine analyses for Giardia lamdia, Chryptosporidium, and culturable viruses. The expected cost to implement **ICR** for a typical water treatment plant is estimated to be $145,500 (Black and Veatch, 1994).

5.4.3. Monitoring Treatment Processes

Real time monitoring of the treatment processes using turbidity, pH, and residual chlorine, and effluent monitoring for coliform bacteria at the treatment plant and predetermined sites throughout the distribution system maintains the best quality control. The statistics-based monitoring programs for measuring the bacterial quality of the drinking water are related to the size of the population served and are the last barrier before water enters the storage or distribution system to be consumed. If viable organisms are found in the tests, immediate action is taken to find and eliminate the origin and cause of the problem.

5.5. PROTECTING SYSTEM's DISTRIBUTION COMPONENT

5.5.1. Storage of Treated Water and its Protection before it is Released into Pipe Network

The storage of treated water in tanks or reservoirs, for matching daily and/or weekly water supply and water demand, between the treatment plant and the beginning of distribution network, is not only needed for this regulation of flow but also for maintaining the pressure in the distribution network. In recent years there has been a tendency in water supply systems to cover these storage facilities. This is done primarily to eliminate problems with atmospheric deposition but also to prevent public access to the already treated water. These reservoirs also help for gases, especially those linked to chlorination, to leave water and removed from the tanks or reservoirs. Attention should be paid in design of reservoirs for all water to slowly move through reservoirs most of the time, so that no eventual solids precipitate and accumulate in dead angles of some types of water supply reservoirs.

5.5.2. Protecting Distribution Subsystem from Contamination

In many instances the storage of treated water in reservoirs at elevated locations provides sufficient head to maintain at least 40 psi (pounds per square inch) throughout the entire distribution subsystem. The minimum of 40 psi is primarily needed for fire fighting purposes but it also lessons the probability that very low or negative (sub-atmospheric) pressures occur in the distribution subsystem where untreated water may enter the subsystem. This can be a real problem with most distribution components of water supply systems during a fire. Once a large fire truck is connected to the distribution subsystem, very low or even negative pressures in the subsystem can occur allowing untreated water to enter into the distribution network.

An example of this occurring happened at Holy Cross College, USA. It was a hepatitis outbreak which infected the entire Holy Cross football team. In epidemiological analysis the outbreak was traced back to a sprinkling irrigation system cross connection (a physical connection of a potable water supply to a non-potable water supply) which during a large fire nearby lost pressure which became negative and sucked water in having urine from a boy carrying hepatitis. This was, by the way, one of the first proven cases that a hepatitis virus can survive and be transmitted in water.

The cross connections in the distribution subsystem are a continuing problem and is addressed by local authorities. Most water supply systems in the USA have laws or ordinances that ban cross connections. If they do occur, one way valves and pressure relief valves are required. All faucets for sinks and bathtubs are well above them so that if the sinks or tubs fill up and overflow, there will be no contact with the faucets. Although all faucets, which are threaded, permitting a hose to be connected, should have a one-way pressure relief valve, but in practice many do not (or do not function properly), constitute continuous potential problems with cross connections. Ironically, hospitals are among the major sources of this type of cross connection.

A recent phenomenon, a rapid proliferation of home underground sprinkler systems in the USA has created many new potential cross connections. Many of the cities where this is occurring have been slow to pass ordinances and/or enforce proper designs which required one-way valve protection.

Another design in the distribution subsystem to protect the quality of the water is the absence of dead ends of pipes where water can stand for many days before being used. Most distribution subsystems are interconnected in the sense that water can come from two directions (no pipe dead ends). This concept of design also allows a more even distribution of network pressures, and decreases the number of users losing access to water when a pipe has to be repaired.

Because there is an increasing fear of the general public that their drinking water from the tap is not safe, there is a new product on the market that connects directly to the faucet usually consisting of activated carbon, that is finding wide acceptance. It may help in the reduction of trace organics and particles. But if not properly used and replaced periodically, it constitutes an increased health menace if a break through occurs.

5.5.3. Aging and Corrosion of Buried Pipes as Related to Water Pollution

Buried pipes if not corrosive from inside is more likely to be so from outside, whenever pipes are buried into acid soils. Besides the problems may arise with leaking joints for whatever reasons. Leaking pipes buried into less permeable soils may maintain relatively large pressure outside the pipe in comparison with the inner water pressure. Because of the regular fluctuation of pressure at leaking points, water flow out and in the pipes frequently. These aspects are very important especially if polluted water infiltrates the soils in which pipes are buried. This is an occurrence in many old distribution networks, requiring precise pipe discharge measurments and monitoring, or even uncovering pipes at test sites for checking their conditions.

5.6. CONCLUSIONS

The above discussion of the safety of water supply systems against pollution of the water leads to the following conclusions:

(1) Security against contamination of the water within the water supply systems is highly enhanced by a strict application of the multiple barriers system concept,

(2) Systematic investigations of problems with the water contamination inside the water supply systems and the studies of measures for counteracting these dangers, will be much more successful if they are done by the four major system's components, collection, transmission, treatment and distribution,

(3) Major measures in the collection subsystem are those which limit human access to collecting areas, and when it can not be done fully, to regulate access and control both the point and non-point sources of pollution,

(4) Major measures in protecting transmission subsystem from water contamination are related to intake structures, non permitting access to uncovered conveyances, and by avoiding interruptions of water conveyance lines,

(5) Principal measures in protecting water treatment subsystem from contamination is to follow the principle of multiple barriers system and to apply the relatively rapidly evolving, new water treatment technologies,

(6) Major actions in protecting water distribution subsystem from contamination to insure the safe water storage of already treated water and to maintain the minimum prescribed water pressure in the network of pipes all the time.

5.7. REFERENCES

American Journal of Public Health (1992) V. 82, p. 955.

Black & Veatch (1994) SWDA Information Collection Rule. June.

Mestas P. (1994) The Efficiency of Dissolved Phosphorus Removal from Urban Runoff Sand Filtration. Master's thesis in partial fulfillment of the requirements for Master of Science of Civil Engineering at Colorado State University, Fort Collins, Colorado, USA.

Sanders G.T., Champlin T.L. (1994) Treatment of Drinking Water in the United States of America on the Treshold of the 21st Century. NATO Advanced Study Institute, Varna, Bulgaria, October 24 - November 4, 1994.

Turk L.J. (1975) Predicting the Environmental Impact of Urban Development in a Karst Area. Karst Hydrology and Water Resources, V.Yevjevich (ed.), WRP, Vol. 2, p. 681-702.

6. Wastewater Reuse - a Worldwide Issue

Thomas G. Sanders
Lisa M. Dunn
Vujica Yevjevich
Department of Civil Engineering
Colorado State University
Fort Collins, Colorado 80523
USA

6.1. INTRODUCTION

Water resource shortages are a problem that are plaguing the world. Wastewater recycling and reuse may be the solution to this worldwide problem. Recycled and properly treated wastewater is a safe alternative source for irrigation, industry, groundwater recharge, and miscellaneous subpotable uses. The biggest problem with wastewater recycling and reuse is the concern for potential health risks. Currently, there are wastewater recycling and reuse projects in Singapore, Israel, Colorado, California, Florida as well as others. While California leads the way in developing regulations for wastewater recycling and reuse and the US EPA has developed a manual on water reuse criteria, much debate exists regarding regulations, uses and appropriate levels.

Three fifths of the earth is covered by water. While this may appear to be an abundance, most of this water is not available for most uses. The oceans account for 97.2% of the earth's water (Warner, 1994). Because of the high salinity, this water is of no use for many water users. The remaining 2.8% of the water consists of surface, subsurface and atmospheric water. Planning and managing this 2.8% has become a great challenge. Since 1900, the United States' population has increased by 200%. Yet, the per capita water usage has increased by 800% (Arber, 1991). This type of increase in demand through out the world is forcing countries to develop new water sources, while protecting the existing ones, as well as implementing conservation measures.

The continental United States receive an average of 762 mm of rain annually. Approximately two thirds of this water is lost through evaporation (Asano et al., 1991).

NATO ASI Series, Partnership Sub-Series, 2. Environment – Vol. 15
Water Supply Systems – New Technologies
Edited by Č. Maksimović, F. Calomino, and J. Snoxell

Agricultural activities and steam electric plants use over 75% of all the fresh water withdrawn in the United States (Asano et al., 1991). Currently there are water shortages in the Rio Grande, the Texas-Gulf, Upper and Lower Colorado, Missouri, and California (Asano et al., 1991). Most of these regions are using water in excess of their presently sustained renewable water resources. Others are entirely dependent on mining groundwater.

Water resource shortages are not limited to the United States; they are a world-wide problem. In Jordan, water is the limiting factor for industrial and agricultural growth and development (Gur and Al Salem, 1992). Not only is water scarce, it also needs to be transported long distances. As a result, their groundwater resources have been utilized to the extent of overpumping, leading to a deterioration in the quality.

Israel is also faced with a severe water shortage and pollution threat to its water resources. In order to service its concentrated urban population and its large agricultural demands, the groundwater resources are currently being overused beyond their natural recharge (Kuttab, 1993).

Singapore is faced with not having enough water for its growing population. The main island of Singapore, with a total area of 620 km^2, is home to 2.69 million people. The annual precipitation is 2400 mm. However, 1600 mm/year is lost in the form of evaporation (Shelef, 1991).

In California, a five year drought has cut agricultural water supplies by 75% (Asano, 1992). Under these circumstances, using potable water for landscape irrigation appears unreasonable and wasteful.

6.2. NEW SOURCES OF WATER

With the population continuing to grow, the water demand is exceeding the supply capabilities in many places. As a result, it is necessary to find and develop new sources of water.

Wastewater recycling and reuse may be a solution to this world-wide problem. Arid and semi-arid areas can augment 15-20% of the water supply through wastewater reuse (Shahalam, 1989).

This wastewater is best used, but not limited to agricultural purposes. Besides providing moisture and nutrients to the soil, the reuse of this water would reduce wastewater pollution and disposal costs.

6.3. RECYCLED WASTEWATER USES

6.3.1. Irrigation

Agricultural and landscape irrigation are the largest users, current and future, of recycled wastewater. Some potential constraints to wastewater reuse are the effect of the water quality on soils and crops, public health concerns, and the marketability of the crops (Asano et al., 1991). Yet, many times the need for additional water far outweighs the potential negative effects. The recent drought in California has increased the incentives for using recycled water because many cities had passed drought control ordinances that prohibited the use of potable waters for construction, dust control and compaction. Also recycled water is to be used on landscapes if it were available (Newby and Hough, 1994). In Santa Clara, California, recycled water was utilized for golf course irrigation. Not only did this conserve water, it additionally saved money. The reused water rates were set at 90% of the retail rate (Smith and Walker, 1991). Reuse of wastewater in Florida for agriculture irrigation is promoted as a means of decreasing the nutrient loads into adjacent streams and reducing treatment costs. The treatment requirements are increased if the wastewater is discharged into streams when it cannot be used for agriculture (Wright,1994). The country of Jordan supplements its irrigation supply, approximately 4 MCM/yr, by using wastewater for producing fodder, cereals, fruit trees, and vegetables which are cooked before being eaten (Gur and Al Salem, 1992).

6.3.2. Industry

A second category for reused wastewater is industry, where it is used primarily for cooling and processing. In Singapore, industrial demand is increasing as a result of restriction of potable water to industrial users and a preferential water tariff system to encourage the industry to use recycled industrial wastewater. Reused water costs industry $0.24/m^3 in Singapore, while potable water costs $1.1/m^3. In addition, to cover sewer costs in Singapore, a minimum of $50/month is charged to industrial users (Chin and Ong, 1992). The program is working so well, current demand for industrial water has exceeded the capacity of the Jurong Industrial Water Works. Expansion is being done to increase yield by 20,000m^3/day. The ultimate capacity of the plant is 91,000m^3/day (Chin and Ong, 1992). Potential constraints to industrial use are public health concerns, and scaling and corrosion of the pipes (Asano et al., 1991).

6.3.3. Groundwater Recharge

Another way of using wastewater involves groundwater recharge. For example, in Jordan, groundwater recharge uses wastewater from cesspools and effluent infiltrating through the layers of ground, ultimately reaching and recharging groundwater aquifers (Gur and Al Salem, 1992). Potential constraints include toxicological effects of chemicals in the wastewater, dissolved solids and pathogens (Asano et al., 1991). For many years, recharge of treated wastewater has been used in Southern California to prevent saltwater intrusion.

6.3.4. Miscellaneous Subpotable Uses

The last category is miscellaneous subpotable uses. These include recreational lakes, aquaculture, and toilet flushing. "Recreational water used for body contact sports must be aesthetically acceptable, must not contain substances that are toxic upon ingestion or irritating to the skin, and must be reasonably free from pathogenic organisms" (Culp and Culp, 1971). This category accounts for less than 5% of total wastewater reuse. Potential constraints to this category are mostly regarding public health concerns (Asano et al., 1991).

6.4. HEALTH RISKS

Common to all the categories of wastewater reuse is the concern for potential health risks. Most of the potential health risks associated with the reuse of wastewater are due to pathogenic organisms in wastewater. These pathogens can be spread three different ways (Shuval, 1991).

(1) Through wastewater contamination of food, mainly eaten uncooked to the public at large,

(2) To the farm workers who have direct contact with the wastewater,

(3) To population groups living near wastewater irrigated fields possibly by aerosolized pathogens from sprinklers irrigation or by direct contact of children playing in or near wastewater irrigated fields.

There are several factors that increase the relative effect of various groups of pathogens to cause infection in humans (Shuval, 1991a).

(1) Long persistence in the environment,

(2) Low minimal ineffective dose,

(3) Short or no immunity,

(4) Minimal concurrent transmission through other routes such as food, water and poor personal and domestic hygiene,

(5) Long latent period and/or soil development stage required.

Based on these criteria, the worm (helminth) pathogen will be the most effectively transmitted through irrigation of wastewater. Enteric virus disease should be least effective.

What does all this mean to people? How does this effect people's health and activities? Depending on the method of transmission, different diseases can be linked to raw or poorly treated wastewater (Shuval, 1991a):

(1) for the general public, eating salads or vegetable crops irrigated with raw wastewater: ascariasis, trichuriases, cholera, tapeworm,

(2) for wastewater irrigation workers: ancylostomiasis (hookworm), ascariasis, cholera,

(3) for the general public nearby wastewater irrigation: there is little or no transmission even when sprinklers are used.

6.5. TREATMENT PROCESSES

In order to protect public health, while making use of this new water source, it is necessary to utilize various treatment processes. Many different agencies have studied the use of wastewater reuse and the safety of the treatment process, each with different results and limits. See Table 3. for regulations. In order to understand the different treatments, it is necessary to understand the processes.

6.5.1. Primary Treatment

The available treatment processes can be classified into three group, primary, secondary and tertiary. Primary treatment produces the lowest quality effluent. It is composed of physical processes, primarily sedimentation followed by disinfection with chlorine (Shahalam, 1989). Primary processes are septic tanks, Imhoff tanks, primary sedimentation, racks, screens. Primary treatment is typically capable of removing 35% Biochemical Oxygen Demand (BOD), 10% total coliform bacteria, 35% fecal coliform bacteria, and 55% suspended solids (Shahalam, 1989).

6.5.2. Secondary Treatment

Primary treatment is followed by secondary treatment which is a biologically based process or its equivalent. Its main purpose is to remove dissolved organic materials (Shahalam, 1989).

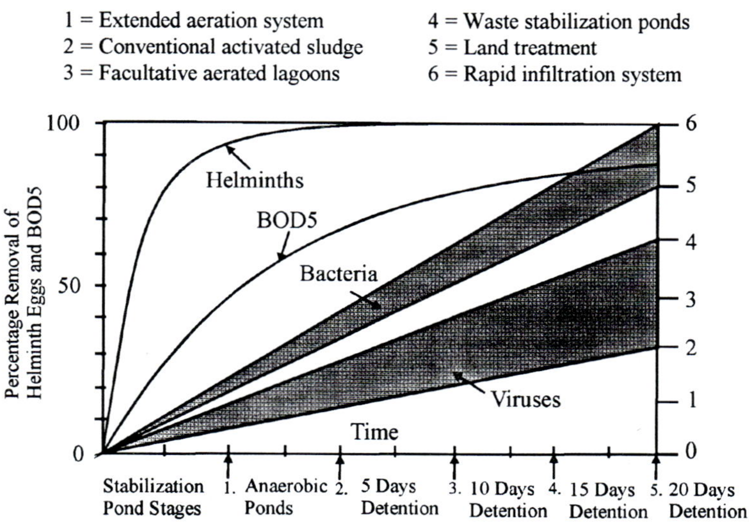

World Bank

Figure 1. Generalized removal curves for BOD, helminth eggs, excreted bacteria, and viruses in waste stabilization ponds at temperatures above 20^0 C (Shuval, 1991a)

Table 1. Performances of secondary processes treating domestic sewage

ITEM	1	2	3	4	5	6
PERFORMANCE						
BOD Removal %	90-95	85-93	70-90	70-90	80-90	99
Nutrient removal %N	15-30	30-40	-	40-50	80-90	80
Nutrient removal %P	10-20	30-45	-	20-60	80-99	80
Suspended solids %	88	88	50	88	92	-

There are many systems that perform secondary treatment. They include oxidation ponds, stabilization ponds, oxidation ditches, aerated lagoons, trickling filters, and activated sludge processes. The removal of BOD, nutrients and suspended solids as well as pathogenic organisms vary widely depending on the system used and its detention times (Table 1. and Figure 1.).

6.5.3. Tertiary Treatment

Tertiary treatment is made up of generally more chemically based processes and is usually preceded by primary and secondary treatment. Examples of tertiary treatment processes include nitrification denitrification, phosphorus precipitation, activated carbon adsorption, ionexchange, reverse osmosis, desalination, coagulation, sedimentation and filtration.

What treatment is the best and most cost effective will be the source of considerable research in the future. The Sanitation District of Los Angeles County's Pomoana Research Facility conducted a study in 1976-1977 to compare virus removal capacity of an alternative tertiary process, contact filtration, with the full treatment process. The full treatment process includes primary and secondary treatment followed by polymer and alum addition, flocculation, sedimentation, granular medium filtration and disinfection. This is the most stringent treatment process specified by criteria. The alternative process was similar to the full process except it did not include the flocculation and clarification steps. Full treatment is an expensive process due to the high costs associated with coagulant chemicals, sludge handling, and tertiary tanks (Asano, 1992). The study found that with high chlorine residuals (10 mg/l) following both processes, no difference was seen in the overall removal or inactivation of the seeded poliovirus. With low residuals (5 mg/l), a slight difference was observed; the full treatment was better (Asano,1992). The total annual wastewater reclamation cost ranges from \$200/acft(\$0.16/m^3) to(\$1300/acft(\$1.05/m^3). The ratio of full treatment to direct or contact filtration is 2.0 to 2.4 for capital costs, 3.9 to 5.6 for operations and management, and 2.4 to 2.9 for total life cycle cost. Although the treatment costs may be high, the distribution system, for reclaimed water is the principle costs for reuse projects (Asano, 1992).

6.6. EXISTING PROJECTS

To offset the rapid increase of potable water demand in Singapore, the Jurong Industrial Water Works (JIWW) uses treated effluent from the Ulu Pandan Sewage Treatment Works. The JIWW has a design capacity of 45.000 m^3/day (Chin and Ong, 1992). Singapore has six treatment works, which treat more than 700.000 m^3/day. The JIWW treatment process includes screening and pre chlorination, chemical clarification, rapid gravel filtration, aeration, and postchlorination. Effluent water quality for treated water is shown in Table 2. The estimated cost of treating water is \$1.76/m^3 prod. water (US) (Chin and Ong, 1992)

Table 2. Water quality (Chin and Ong, 1992)

Variables	
BOD5	1-3 mg/l
DO	5-6 mg/l
Suspended solids	1.5-3.5 mg/l

In rural areas the question of how to provide sewage drainage/treatment facilities is as important as how to make water available. Save the Children/USA has utilized a wastewater treatment/reuse system, Subsurface Drainage Technique (SDT) (Kuttab, 1993). The system is being used in Israeli occupied villages on the West Bank and in the Gaza Strip. The system can be used by any household with enough land in the immediate vicinity. The wastewater is drained from the house in sewer lines to a watertight septic tank, where the sewage undergoes treatment utilizing sedimentation and stabilization by anaerobic bacteria. The sewage undergoes secondary treatment in the drainage field. The total cost for the system and labor is about \$500 (US) (Kuttab, 1993).

In Colorado, Denver's Potable Water Reuse Demonstration project created and provided answers on product quality, public health, technological and economic feasibility, conservation and consumer and regulatory agency acceptance. In 1984, the Potable Water

Reuse Demonstration Plant began operation. The plant used unchlorinated secondary effluent from the metropolitan Denver Wastewater Treatment Facility as raw water supply (Rogers and Lauer, 1992). The plant processes included high pH lime treatment, single or second stage recarbonation, pressured filters, selective ion exchange for ammonia removal, two stages of activated carbon adsorption, ozonation, reverse osmosis, air stripping, and chlorine dioxide disinfection. The plant is designed with multiple barrier protection. For example, viruses and bacteria can be deactivated with high pH lime clarification, ultraviolet irradiation, reverse osmosis/ultrafiltration, ozonation, and chlorine/chloramination. The ultimate goal in design was to eliminate unreliable or unnecessary treatment steps without jeopardizing water quality or consumer safety (Rogers and Lauer, 1992). In all, the plant has shown that the advanced wastewater treatment technology can produce an effluent with water quality exceeding that of approved drinking water supplies.

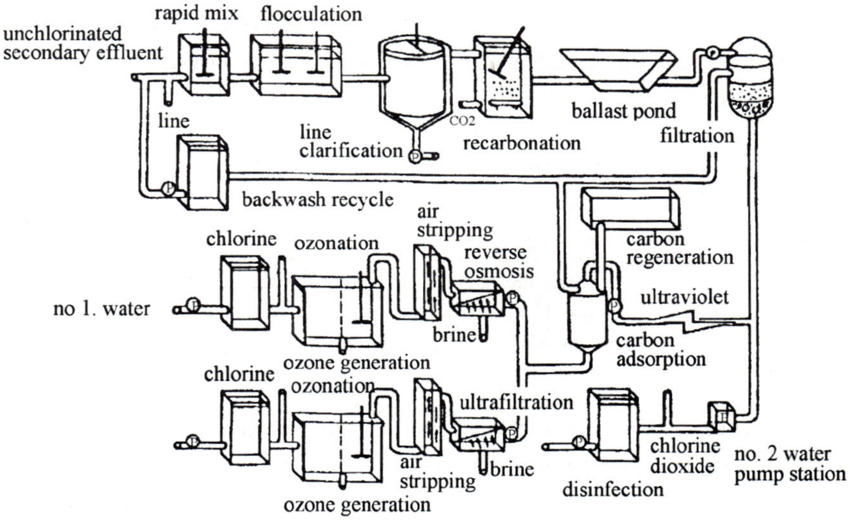

Figure 2. Process train for the Denver potable water reuse plant (Rogers and Lauer, 1992)

6.7. REGULATIONS

With the worldwide importance of wastewater reuse, how are regulations determined? Currently there are no worldwide regulations. In fact, there are no regulations specific to the

United States of America although USEPA has issued guidelines (EPA, 1992) for water reuse for states that have no standards or are revising their existing standards.

Regulations are set by the standard of living and the acceptance of the people in the area. For example, the World Health Organization report offers a minimum guideline for developing countries. In contrast, the State of California has one of the most stringent criteria, which requires extensive wastewater treatment, frequent waterquality monitoring, and strict use-area controls.

California lead the way by developing the first wastewater reuse regulations in 1918. The regulations were formed out of the growing belief that a new science of hygiene and sanitation could create an "antiseptic", bacteria free environment, that would protect the public from infectious diseases (Table 3.). These limits were set despite the lack of epidemiological evidence and were based on the safe drinking water level (2.2 coliforms/100ml), justifying it must be safe for irrigation and thereby conforming to the "zero risk" attitude of society (Shuval, 1991b).

California's regulations were later revised to include standards for landscapes and recreational impoundments. See Table 4. for California's current regulations.

While California set the pace for wastewater regulation, many have disagreed with their "zero risk" attitude. The strictness of the codes makes it economically unfeasible for developing and transition countries. In 1973, the World Health Organization (WHO) developed their own regulations for developing countries. Their regulations called for primary and secondary treatment and disinfection before use on raw edible crops (Table 3.).

In July of 1985, WHO, along with the World Bank, UNEP, UNDP and the International Reference Center of Waste Disposal reviewed and set new guidelines, the Engelberg Report. (Table 3.). The new WHO guidelines are easily achieved at a low cost through the use of robust stabilization pond systems, which are well suited for developing countries. These requirements can also be met with other processes for areas not suited for stabilization ponds (Culp and Culp, 1971).

Israel is faced with a severe water shortage and a pollution threat to its diminishing water resources. In addition, it has a concentrated urban population and large agricultural demands. Their groundwater resources are currently overused beyond the natural recharge. Wastewater reuse is a solution to their problems. In 1973 the Israeli Minister of Health appointed an advisory committee on quality requirements for wastewater reuse in Israeli (Shelef, 1991). See Table 3. for the regulations. In 1978 those regulations were still guidelines for wastewater reuse in Israel.

In Jordan, water is both scarce and far from the people. As the population grows, the demand for water increases, making- wastewater reuse vital. In 1982 the first wastewater reuse legislation was adopted, Martial Law No. 2. This brief legislation defined types of crops that could be irrigated with treated effluent. It did not allow any use of treated wastewater on cereals and vegetables (Gur and Al Salem, 1991). Martial Law No. 2 was revised in September of 1989 (Table 3.).

Table 3. CA 1918, WHO, Engelberg Report Regulations (Shuval, 1991b)

REGULATIONS					
	CA 1918	WHO	Engelberg Report (WHO)	Israel	Jordan
Raw Crops	2.2/100 ml	<100/100 ml in 80% of samples	geometric mean 1000/100 ml ⇐ 1 helminth egg/l	12/100 ml 80%	NA *
Processed Produce	23/100 ml			250/100 ml	
Cereal crops, industrial crops, fodders, trees			⇐ 1 helminth egg/l		<1000/100 ml
Landscape: public, parks			⇐ 1000/100 ml ⇐ 1 helminth egg/l	2.2/100 ml 50%	<200/100 ml
Sprots fields			⇐ 1000/100 ml ⇐ 1 helminth egg/l	250/100 ml	<200/100 ml
Lakes	23/100 ml				
Rec. Lakes	2.2/100 ml				
NA* - Not allowed					

Colorado has always enjoyed the luxury of having fresh, clean snowmelt from the mountains to irrigate their crops. With the increasing population, and growing need for potable water, Colorado is turning towards wastewater reuse for crop irrigation. The Northglenn Colorado Water Reuse Program utilizes this concept. The city Northglenn Colorado borrows the farmer's irrigation water for domestic purposes. The water is then returned along with 30% makeup water plus a 10% bonus (Sanders, 1981). See Table 5. for Colorado's regulations. Unfortunately, Colorado has very little control over wastewater reuse. As long as the sewage effluent is in a natural watercourse, even only for a few meters, it can be used to irrigate any crop, even raw edibles.

Table 4. California Regulations (Shuval, 1991b)

CALIFORNIA REGULATIONS				
Application	Minimum Characteristics			
	Primary	Secondary and disinfected	Secondary coagulated filtered and disinfected	Coliform MPN/100 ml Median (daily sampling)
Irrigation	X			NR
Fodder Crops	X			NR
Fiber	X			NR
Seed crops	X			
Produce eaten raw, surface irrigated		X		2.2
Produce eaten raw, spray irrigated			X	2.2
Processed Produce, spray irrigated		X		23
Landscapes: golf course, cementeries, freeways		X		23
Landscapes: parks, playgrounds, schoolyards			X	2.2
Recreational Impoundments No public contact		X		23
Boating and fishing only		X		2.2
Body contact				

Table 5. The Code of Colorado Regulations (1994)

COLORADO REGULATIONS			
Parameter	Parameter Limitation		Instantaneous
	7-Day Average	30-Day Average	Maximum
BOD5	45 mg/l	30 mg/l	NA
Suspended solids	45 mg/l	30 mg/l	NA
CBOD5	40 mg/l	25 mg/l	NA
Residual chlorine			0.5 mg/l single grab
pH			6.0-9.0 mg/l single grab
Oil and grease			10 mg/l single grab

6.8. CONCLUSIONS

The need for new sources of water supplies in many regions of the world is increasing. Aggravated by recent droughts and the increasing demands, reuse of wastewaters for nonpotable uses appears to be a viable source of new water in developed, transitional and developing countries. However, reusing or reclaiming wastewater should only be used for nonpotable uses. Much more needs to be done in research on the health based risks and criteria and technology reliability before potable uses can be safely met. Using recycled wastewater for only nonpotable uses will increase available water supplies and may, in fact, decrease costs of wastewater treatment since less treatment is required when compared to treatment costs for potable waters.

6.9. REFERENCES

Arber H., Richard P. (1991) Wastewater Reuse: Thoroughly Examined. Water Engineering & Management, 138: 5, pp. 36-38.

Asano T. (1992) Evolution of Tertiar Treatment Requirements in California. Water Environment and Technology, Feb., pp. 36-41.

Asano T., Tchobanoglous G. (1991) The Role of Wastewater Reclamation and Reuse in the USA. Water Science Technology, 23: pp. 2049-2059.

Chin K.K., Ong S.L. (1992) Experience of Non-Potable Reuse of Wastewaters. Water Science Technology, 26: 78, pp. 15651571.

The Code of Colorado Regulations (1994) 5 CCR 10023, 17CR 6,6- 94, 10.1.4.

Culp R.L., Culp G.L. (1971) Advanced Wastewater Treatment. Van Nostrand Company, New York.

Gur A., Al Salem S.S. (1992) Potential and Present Wastewater Reuse in Jordan. Water Science Technology, Vol. 26: 78, pp. 1573-1581.

Kuttab A. (1993) Wastewater Treatment/Reuse in Rural Areas. Water Science Technology, 27: 9, pp. 125-130.

Newby J.P., Hough S.G. (1994) Urban Retrofit on a Grand Scale. Journal of Water Environment & Technology, Vol. 6, No. 11, Nov.

Rogers S.E., Lauer W.C. (1992) Denver's Demonstration of Potable Water Reuse: Water Quality and Health Effects Testing. Water Science Technology, 26: 7-8, pp. 1555-1564.

Sanders T.G. (1981) Recycling Wastewater in Colorado. The Issues, the Problems and the Solutions. Industrial Wastex, Proceedings on the Thirteenth mid-Atlantic Conference, June 29-30, 1981, pp. 588-595.

Shahalam A.B.M. (1989) Wastewater Effluent vs. Safety In Its Reuse: State of the Art. The Journal of Environmental Sciences, Sept./Oct., pp. 35-41.

Shelef G. (1991) The Role of Wastewater Reuse in Water Resources Management in Israel. In Water Resources Management on Israel, 23: pp. 2081-2089.

Shuval H.I. (1991) Health Guidelines and Standards for Wastewater Reuse in Agriculture Historical Perspectives. Water Science Technology, 23: pp. 2073-2080.

Shuval H.I. (1991) The Development of Health Guidelines for Wastewater Reclamation. Water Science Technology, 24: 7, pp. 149-155.

Smith R.G., Walker M.R. (1991) Water Reclamation and Reuse. Research Journal WPCF, 63: 4, pp. 428-430.

United States Environmental Protection Agency (1992) Guidelines for Water Reuse.

Warner J.W. (1994) Class notes. Colorado State University, Fort Collins, Colorado USA

Wright R.R. (1994) The Best of Both Worlds. Journal of Water Environment & Technology, Vol. 6, No. 11, Nov.

Chapter II

Water Supply Conveyance and Control

Chapter II

Water Supply Conveyance and Control

INTRODUCTORY STATEMENT

Worldwide demand for water has increased almost continuously in the recent past and this has increased the pressure on available resources and has led to increasingly innovative methods for moving and controlling water over very long distances.

Examples of such schemes include the Eastern Cross Harbour Main in Hong Kong which involved the construction of major transfer mains across a harbour and innovative methods of pipeline construction, installation and corrosion protection.

The Az Zour Water Distribution scheme in Kuwait involves the treatment and distribution of up to 720 Ml/d of desalinated water over a distance of 100 km, with particular problems of maintaining disinfectant, controlling corrosion and avoiding taste problems.

The London Ring main is an example of a major scheme carried out below one of the world's major capital cities with enormous potential problems of disruption to existing services.

This Chapter looks at some of these problems and how technology is employed to provide solutions. Firstly traditional and new technologies in pipeline construction are examined and then two examples of long distance water conveyance systems, in Brussels, Belgium, and Wessex Water, Great Britain, are used to illustrate some of the engineering techniques and information technology now being used to understand and control such large integrated networks.

1. Construction Methods and Pipe Structural Requirements

Francesco Calomino
Paolo Veltri
Dipartimento di Difesa del Suolo
Università della Calabria
87040 Montalto Uffugo (CS)
ITALY

1.1. INTRODUCTION

In pipeline construction, the use of new technologies is by far less impressive than in pipeline appraisal and design.

This is true if one considers that every innovation, before it becomes widespread, needs to be tested for years devoted to practice, documentation of results, discussions, publicity, and official acknowledgement.

Moreover, if on one hand it is difficult to convince engineers of the advantages of using new technologies in their projects, on the other hand it is even more difficult to convince contractors which have to make practical use of these technologies.

Among several aspects involved in pipeline construction, this Section will deal with current pipe production and main laying technologies. Computational aspects will be briefly considered for the assessment of buried pipe structural requirements, highlighting those that could bring real changes in perspective.

NATO ASI Series, Partnership Sub-Series, 2. Environment – Vol. 15
Water Supply Systems – New Technologies
Edited by Č. Maksimović, F. Calomino, and J. Snoxell

1.2. PIPE MATERIALS

1.2.1. General

Several materials are in use today for pipe manufacture; some of them are classical materials, used for decades in this field: ductile iron, steel, asbestos cement, prestressed concrete. Plastic materials like PVC and PE are more common today: for relatively small diameters and low pressures, they have completely substituted the former ones in most countries.

The use of plastics is more and more important for internal lining and external protection, in connection with other materials, and the union constitutes really new composite pipe materials.

1.2.2. Ductile Iron

These pipes should be mentioned before all others. Pipes made up of its ancestor, cast iron, were used since the XVI century in Europe. Today, ductile iron is the dominant material for pipelines with 250 to 800 mm diam (Osborne, 1994).

In the past, spun iron pipes were obtained by a centrifugation process of an alloy of Iron and Carbonium (Twort et al., 1974, The Pipeline Industries Guild, 1994).

With the addition of a small quantity of Magnesium, and using centrifugation, the Ductile Iron (DI) is today manufactured, whose internal structure is characterised by the presence of rounded crystals (spheroidal graphite iron). Fragility of the external surface is prevented by means of a thermic process, giving at the same time a ferritic structure to the material.

ISO 2531 gives the guidelines for ductile iron pipes, fittings and accessories for pressure pipelines (Brandon, 1984).

The pipes are internally lined by a spun cement mortar layer; externally, they can be painted with zinc and a second bitumen layer. An external polyethylene sleeving can be used before laying the pipes in particularly aggressive soils.

Currently, DI pipes on the market have diameters from 60 mm to 2000 mm (of which, of course, the largest ones are produced to order), and are 6 to 8.3 m long.

Wall thickness should be in agreement with ISO 2531.

Working pressures are relatively high, ranging from 20 to 40 bar.

The pipes are normally provided with socket-and-spigot joints, although a number of proprietary types of joints are known and can be used in different situations; a rubber ring usually makes the joints watertight. A small angular deviation (up to 5° for smaller diameters) may be allowed by some joints.

Internal lining and external coating, in addition to electrical discontinuity at the joints, are usually good measures against corrosion. The wall thickness is able to provide high pressure performance; the weight of the pipes is, on the other hand, high, and the transport and laying costs are usually high. The need for thrust blocks at angular deviation points is also to be taken into account, when not using specific joints.

1.2.3. Steel

Steel pipes have been used since last century, in preference them to cast iron pipes for their lightness and improved resistance, due to the low carbon content. The most widespread manufacturing procedure consisted in piercing a bar of hot steel and then obtaining the right dimensions by rolling it (Mannesmann).

Nowadays there are high-strength weldable steels suitable for pipe manufacture (The Pipeline Industries Guild, 1994), and a steel pipe can be obtained by bending a steel plate and welding the two edges, or by welding a spirally bent plate (Twort et al., 1974, The Pipeline Industries Guild, 1994).

These are usually referred to as "welded" or "seamless" pipes.

Steel is the dominant material for pipelines installed over 800 mm in diam in the UK (Osborne, 1994).

ISO R559 gives the standard for steel pipes. Fe 360, Fe 420 and Fe 510 are the types of steel in use.

Pipes in current production range from 40 to 500 mm in diam. for longitudinally welded pipes, to 900 mm for seamless pipes. Different wall thickness are produced for the same nominal diameter, according to the working pressures, which gives the designer a larger choice. Working pressures can be as high as 150 bar.

The first steel pipes had run-lead joints to be used with socket-and-spigot. Today joints are normally prepared for welding; spheric, cylindric and butt welded joints are the most common types.

The disadvantage of steel is corrosion; protection against corrosion is provided by means of external glass fiber and bitumen wrapping; internal bitumen lining is in many cases insufficient and active (cathodic) protection has to be chosen (Brandon, 1984).

Polypropylene for external pipe protection can be used; epoxy paint up to thickness of 0.3 mm can be used for internal protection.

In case of severe corrosion problems, clad steel pipes may be used; these consist of an outer body of lower grade steel with a liner of a corrosion resistant steel alloy (The Pipeline Industries Guild, 1994).

1.2.4. Prestressed Concrete

For diameters larger than 800 mm prestressed concrete pipes have been largely used, for the last 50 years, on the grounds of cost (McGhee, 1991).

Pipes are manufactured by winding a spiral of steel wire around a core, made up of a reinforced concrete or a steel cylinder. The concrete core is longitudinally prestressed by means of steel bars, vibration and centrifugation are employed to produce the cores. After stressing the wire, the external concrete layer is completed. Very large diameters can be produced this way, up to 3800 mm; working pressures usually range from 4 to 12 bar, up to 30 bar in some cases.

Standards for these pipes are given by BS 4625 and ANSI/AWWA C301-92 and C304-92 (Brandon, 1984).

The joints are of the socket-and-spigot type, with the use of a rubber ring. For steel-core pipes, the steel cylinders may be butt-welded.

The pipes must be filled and pressurized for a few days before testing to allow absorption of water by the pipe wall (The Pipeline Industries Guild, 1994).

1.2.5. Plastics

Many kinds of plastic pipes are today on the market: Polyvinyl chloride (PVC), Polyethylene (PE), Polypropylene (PP), Acrilonitrile-butadiene-styrene (ABS), etc. Thermoset plastics are cross-linked polymers; while thermoplastics consist of long chains of molecules. The first material shows greater ductility and capacity to withstand strain. Petroff (1993) reports the mechanical characteristics of plastic materials. The most common materials for pipeline construction are discussed below.

1.2.5.1. PVC. PVC, a blend of polyvinyl chloride polymers, was introduced as a pipe material in the 60's. For this purpose, no plasticizers are added to the material, so it is also called unplasticized PVC (uPVC) (Brown, 1994).

The pipes are manufactured by means of an extrusion process. Current diameters range from 60 mm to 600 mm; the pipes are usually 6 m long. According to working pressures, the pipes are usually produced in 3 classes: PN 9 bar, PN 12 bar, PN 15 bar.

Guidelines for pipe characteristics are given by ISO R1165 (Brandon, 1984).

Joints are of the socket-and-spigot type, with a rubber sealing ring, or welded by the use of a sealing cement.

One of the disadvantages of uPVC is the tendency to produce local breaks traveling through the pipeline (Brandon, 1984).

A new material derived from PVC is MOPVC (Molecular Oriented PVC), obtained by expanding a uPVC pipe by internal pressure. In this way a re-orientation of the molecular structure is achieved, and the pipe working pressure is increased significantly. Moreover, the material shows a good resistance to break traveling (Brown, 1994).

Pipe diameters range from 90 to 315 mm; working pressure classes are 8 bar, 12.5 bar, 16 bar.

The latest finding in the field seems to be MPVC (modified PVC); these pipes are made up of a plastic alloy of great toughness and durability (blend of chlorinated PE and traditional PVC).

The pipe length is 6 m, the diameters range between 110 mm and 630 mm, working pressure classes are 10 bar and 16 bar.

1.2.5.2. Polyethylene. Polyethylene can be manufactured as LDPE (Low Density Polyethylene), MDPE (Medium Density Polyethylene) and HDPE (High Density Polyethylene) (Beech, 1994).

LDPE pipes are used - in water supply systems - only for the very small diameters. MDPE is the most largely diffused, and it is reported to be the dominant pipe in the UK for diameters up to 250 mm (Osborne, 1994).

Due to the lower strength, a MDPE pipe has a wall thickness 2-3 times greater than an equivalent PVC pipe. The advantage of MDPE is the resistance to long term break propagation.

Current diameters range between 60 and 600 mm, working pressure classes are 6 bar and 9 bar.

The jointing system can be butt-fusion jointing and electro-fusion jointing, both performed by means of the appropriate fusion equipment.

HDPE is a stiffer material, able to withstand larger pressures, but with a tendency to rapid crack propagation. Nominal diameters usually range between 100 and 300 mm. The pipes can be rolled in coils (up to 180 mm) in lengths up to 150 m. Working pressures are 6 bar, 10 bar and 12 bar.

1.2.5.3. GRP. Glass reinforced plastic (GRP) pipes are made up of polyester laced with glass fibers brought about in successive layers by centrifugation, the remaining space being filled by resin and fillers and, in the reinforced-plastic-matrix pipe, sand. The sand is incorporated into the wall structure to increase the stiffness of the pipes, making it possible to handle, transport and instal them.

Joints are of the spigot-and-socket type with sealing ring.

BS 5480 gives guidelines for GRP pipes (Brandon, 1984).

The advantages of these pipes are the very high resistance to corrosion, smooth internal surface, easy transport and laying.

The pipes have lengths up to 6 m, with diameters up to 2000 mm.

1.2.6. Mixed-Structure

In this category there are pipes made up of a matrix including different fibers; these are asbestos cement and diffused reinforcement concrete.

Asbestos cement was for over 50 years one of the most popular pipe materials; pipe manufacture was firstly developed in Italy around the 20's (The Pipeline Industries Guild, 1994), the matrix is Portland cement, including asbestos fibers. Pipes are manufactured in diameters up to 900 mm; ISO R160 establishes the standard for asbestos cement pipes (Brandon, 1984).

Recent studies demonstrated that asbestos is carcinogenic when the fibers are inhaled, implying that such effect could affect also fibers possibly released from pipes (McGhee, 1991).

In fact, asbestos cement is now not in widespread use; in Italy, a 1992 law states that asbestos cement could no longer to be used as a construction material.

Diffused reinforced concrete is one of the most interesting findings in the reinforced concrete technology, when reduced thickness and weight are needed. This composite material benefits from a larger specific surface at the contact between concrete and reinforcement, to increase the pipe strength.

The pipes are manufactured by winding tiny steel wires around a rotating cylinder, where concrete layers are formed. A longitudinal reinforcement also made up of steel wires is placed between successive concrete layers.

Pipes are up to 6 m long, with diameters from 400 mm to 1200 mm. Working pressures are up to 15 - 20 bar.

As for prestressed concrete pipes, joints are of the socket-and-spigot type with sealing rings.

Advantages of such pipes are the relatively reduced wall thickness, the resistance to internal and external corrosion, the smooth internal surface.

1.3. CONSTRUCTION METHODS

1.3.1. General

Almost all of the length of a pipeline is made up of buried pipes, although external parts may exist, at crossings etc., and in some urban conditions the distribution network, as well as other services, can be put into underground tunnels. The following sections relate mainly to buried pipelines and their construction.

Current construction methods make use of the appropriate machinery for different operations. As this changes more rapidly than the pipe materials and characteristics, the pipeline installation practice is subjected to rapid improvements.

Moreover, increasing interest is nowadays devoted to new - trenchless - techniques.
In the following sections traditional and trenchless construction procedures will be separately treated.

1.3.2. Traditional Methods

Traditional construction operations consist of preliminary operations, trench excavation, bedding, pipe lying, backfilling (Brandon, 1984, ASCE, 1994).

1.3.2.1. Preliminary Operations. Preliminary operations are topographic operations and site preparation. As to topographic operations, the current practice for pressure pipelines consists in laying the pipes at a given depth under the soil surface, without following a given grade, although it is necessary to be aware of the elevation of each point in order to obtain a correct position for air valves and washouts. Site preparation may involve removal of vegetation and the opening of a track for the excavation machines. It is always important to locate exactly the services interfering with the pipeline under construction, for example gas pipelines, underground electrical and communication cables, sewer pipes.

1.3.2.2. Trench Excavation. Trench excavation is usually performed by excavators; these machines are suitable for widths over 50 cm and depths down to 5-6 m. Small trenches in urban centres require very small excavators with reduced bucket width, but in many cases hand excavation is the only solution. The usual trench depth is the one leaving 1-1.5 m cover on top of the pipe. The bottom width is equal to the pipe diameter plus 50-60 cm; in general it is the minimum which will allow laying and jointing operations. The trench upper width should be established taking into account the bottom width, the nature of the soil and its angle of repose.

Excavation of vertical trenches is often possible without any sheeting and bracing. In unstable soil it is necessary to support excavations, and for this purpose different methods are used, by means of timber or steel shields (McGhee, 1991).

Trench side slopes, as suggested by the Occupational Safety and Health Administration (O.S.H.A.) and other suggestions concerning trench escavation, are reported by Brewer and Hurd (1993).

1.3.2.3. Bedding. The trench bottom has to be accurately prepared to support the pipe. Foundation can be requested in order to maintain a stable trench bottom. Bedding will influence the distribution of reaction against the bottom of the pipe. Usually, a suitable bed granular material, sand or gravel,not self-compacting, 10-20 cm deep, is set on the trench bottom; in some cases a concrete bed is judged to be more suitable, depending upon the pipe and backfill characteristics. ANSI A 21-50 1981 (AWWA C 150-1981) shows standard pipe bedding conditions (ASCE, 1994).

Gravel or sand beddings are suggested for D.I, steel, GRP, uPVC, PE pipes; concrete bedding may be suggested for concrete pipes and for large diameter plastic pipes.

1.3.2.4. Backfilling. Terminology adopted by ASCE (1994) for backfilling of pipes will be useful (Figure 1.); foundation and bedding are described in the above paragraph. Haunching will influence the performance of the pipe; its compaction is essential for flexible pipes, whereas for rigid pipes it will ensure better distribution of forces on the pipe wall; this operation is often performed manually. Initial backfill has the function of protecting the pipe from damage by subsequent backfill. It should consist of a suitable granular material. Final backfill is important for trenches in traffic areas, where it is usually made of material that is easily densified; otherwise it can consist of excavated material, which should be lightly compacted and left mounded over the trench.

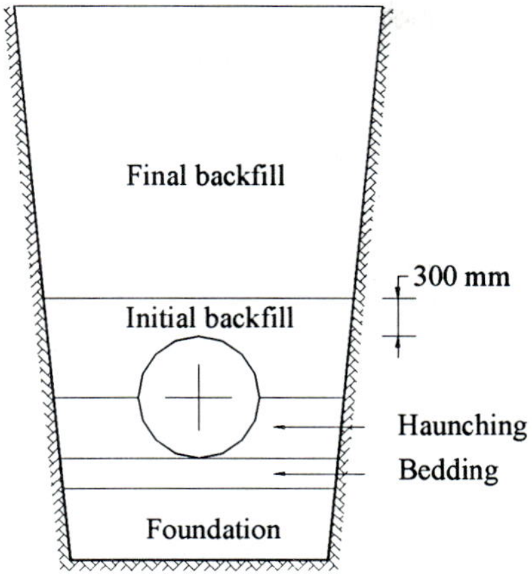

Figure 1. Trench cross section and backfilling

Brewer (1990) and Brewer and Hurd (1993) emphasize the utility of using Controlled Low Strength Material, Controlled Density Fill (CLSM-CDF) in filling trenches under road pavements. This material is nothing else than a sort of concrete, with a low content of cement.

The advantage of using this mixture is the possibility of easily filling the trench directly from the concrete truck and the fact that compaction is not required. Like concrete, the mixture dries achieving cohesion; up to 70 kg/cm^2 of compressive stress is the value at 28 days.

1.3.3. Trenchless Installations

Trenchless construction techniques have become more and more frequent in underground services like water, sewerage, gas, electricity, communications in urban environment.

The use of these techniques for pipelines and distribution networks is limited, when compared to sewers: the reason for this is economic, because the pipes are usually laid under a one-meter cover, and pipes for trenchless installation need to be specially manufactured. Trenchless techniques are on the other hand especially suitable in case of railway or motorway crossings (White and Tregoing, 1992).

Two main techniques exist: boring and jacking (ASCE, 1994).

Boring is performed by means of special machines, that are able to bore a hole into the ground, more or less horizontally, working from a pit; the pipe is then pushed into the hole (van Kesteren, 1992).

Excavation is performed by mechanical, hydraulic, or compressed-air techniques (for very small pipes of up to 60 mm). The excavated ground is taken away from the hole by means of mechanical devices, or mixed with water and evacuated through slurry pipes; when compressed air is used for boring, vacuum techniques may be used. Temporary pipes may be employed, equipped with internal cables and conduits, and then removed and substituted by the final pipes. This technique is referred to as microtunneling (Orchard, 1992).

Jacking is performed by means of jacks pushing a steel or reinforced concrete pipe through the ground. Behind the jack, strong backstops distribute the load to the soil. The pipe head has a cutter or a shield, and the excavated soil is evacuated through the pipe (Orchard, 1992, Stevens and Girka, 1992).

Two main possibilities exist today: the first one consists in the use of a steel or reinforced concrete pipe as a housing, into which the pipeline is inserted; the second one is the boring or jacking of the final pipe directly. In the latter case, pipes are especially manufactured to have good resistance to axial thrust; their joints must be able to transmit this force (Henry, 1992).

A new microtunneling technique has been recently presented by Najafi et al. (1993). The primary thrust force is transmitted through a steel linear casing running inside the pipe. An air-inflatable internal gripper enables PVC or other flexible pipes to be jacked to long distances. The frictional resistance force is transmitted through the gripper mechanism which binds the internal surface of the pipe to the linear casing, requiring less axial compressive forces than with conventional microtunneling.

1.4. STRUCTURAL REQUIREMENTS

1.4.1. General

Structural performance evaluation is today requested by many national regulations for pipeline construction, and this aspect has become nowadays more and more important nowadays with the increasing use of flexible pipes (Katona, 1993).

Conditions to be taken into account are internal pressure and external loads, together or separately according to the mechanical characteristics of the pipes.

Stress analysis or indirect methods, the latter based on the experimental strength of the pipe, can be carried out.

1.4.2. Internal Pressure

The model is that of the thin elastic ring embedded into an elastic environment; traditional stress analysis can be easily developed for internal pressure (Prevost, 1994).

Internal pressure p, according to the membrane theory, will result only in stresses parallel to the surface of the membrane (membrane stresses); hoop stress σ_1 and longitudinal stress σ_2 are given by, as is well known:

$$\sigma_1 = r\frac{p}{t} \; ; \quad \sigma_2 = r\frac{p}{2t}$$

where r is the pipe radius and t is the wall thickness (McGhee, 1991).

It is worth noting that classical results for maximum pressure values derived from classical water-hammer analysis are today under question in many cases. Indeed, in the classical water-hammer theory, the structure is assumed to be rigidly supported; wall thickness and elasticity affect only the celerity.

On the other hand, as pointed out by Kruisbrink and Heinsbroek (1992), pipe wall inertia effects related to pressure changes must be taken into account, as well as pipe motion caused by friction and elbows, tee and end effects; all these influence structural conditions.

A study carried out by the above Authors by means of the Fluid Structure Interaction (FSI) model, and supported by laboratory experiments on PVC pipes, shows that the classical values of maximum water-hammer pressures can be exceeded by more than 30% and the frequency of pressure waves is increased.

Moreover, the presence of air pockets can either damp or amplify the pressure peaks, depending on the character of the transient (Larsen and Burrows, 1992). In special cases severe overload and failure of the pipeline may occur.

1.4.3. External Loads

External loads are earth load (dead load) and superimposed load (live load). Pipe weight and water weight are usually taken into account, although in many cases they can be neglected. Thermic and seismic actions should be correctly considered.

When subject to external loads, the pipes may show different behaviour: if E' and E are the elasticity moduli of soil backfill and pipe material, r is the pipe radius and I the inertia of the longitudinal section of length 1 ($I=t^3/12$, where t is the pipe wall thickness), when the value of E' is less than 250 EI/r^3, the pipe should be considered rigid. In general, metallic (steel and D.I.) and plastic pipes result to be flexible.

For rigid pipes, evaluation of earth loads in trench conditions (Figure 2.) is possible by the classical Marston formula (ASCE, 1994):

$$W = C \gamma B^2$$

where W is the load per unit length of pipe,
B is the width of trench at the top of the pipe,
γ is the specific weight of backfill,
C is a dimensionless coefficient.

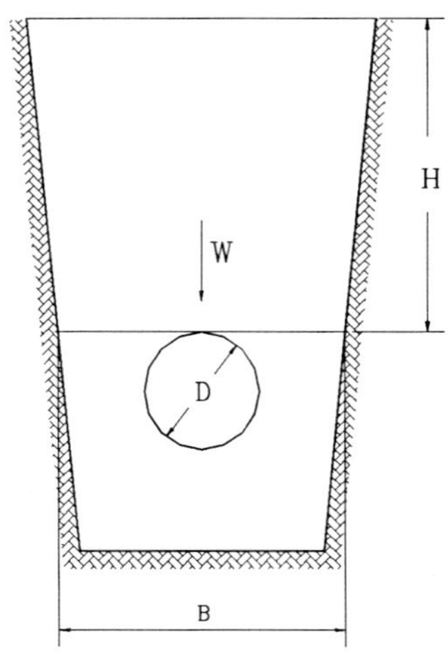

Figure 2. Variables for load computation

In practice, the coefficient C takes into account the actions between the backfill and the sides of the trench, and it is a function of the ratio H/B and of soil characteristics. The coefficient C is evaluated by:

$$C = \frac{1 - e^{\left(-2\,k\,m'\,\frac{H}{B}\right)}}{2\,k\,m'}$$

where $k=\tan^2(\pi/4 - \varphi/2)$, ($\varphi$ is the angle of internal friction of backfill and $m' = \tan\theta$ = coefficient of internal friction between backfill and sides of the trench).

Values of coefficient C are well-documented, such as that shown by ASCE 1994. For shallow trenches of depth H, it can be assumed that C = H/B; then

$$W = \gamma\,H\,B$$

that is, the pipe will carry the load of the solid backfill above its crown. Conservative design practice is based on this principle.

In case of flexible pipes, the soil at the sides will compact and the pipe will deform more than the soil; in this case the backfill at the sides of the pipe will carry a part of the load and the load carried by the pipe itself will be reduced.

The Marston formula in this case will become (ASCE, 1994):

$$W = C\,\gamma\,B\,D$$

where D is the pipe diameter and C is the trench load coefficient.

The expressions for this coefficient takes into account soil friction on the sides of the trench; but there are opinions that this cannot always be relied upon (Stephenson and King 1994). For shallow trenches, assuming C=H/B, is

$$W=\gamma HD$$

Appropriate coefficients are developed for jacked pipes, including cohesion of soil (ASCE, 1994) and for embankment conditions.

Superimposed loads are evaluated according to Boussinesq theory, and taking into account road loads for appropriate traffic conditions.

It is frequent practice to use concrete slabs as a pavement foundation, when a pipeline is located under a road. In this case, the pressure p on the pipe crown is reduced to a small fraction of the traffic pressure. For complete information, see ASCE, 1994.

1.4.4. Rigid Pipes Computation

For rigid pipe design, two methods are today in use: the Stress analysis method and the Load Factor Method.

For rigid pipes, the stress analysis method is based on the superposition of effects (Figure 3.); the moments and stresses in different points of the section are computed; eventually the maximum stress will be evaluated. Usually, this concerns the invert section of the pipe. Maximum stress should be not greater than the ratio of the modulus of rupture of the pipe material and a safety factor.

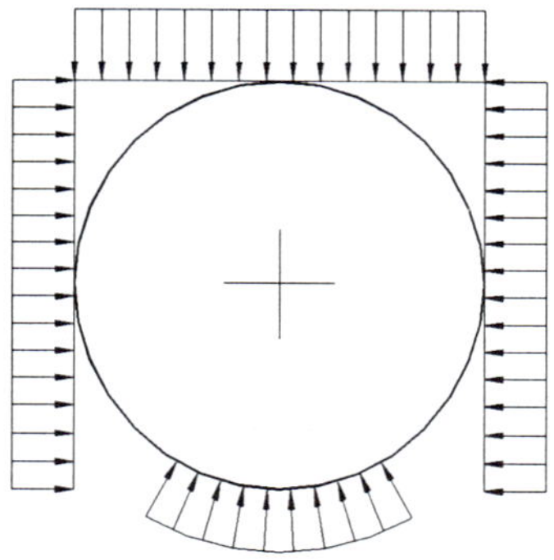

Figure 3. Direct method for rigid pipes (Pressure distribution)

The indirect method or Load Factor method is based on three edge bearing tests (Figure 4.), performed at the manufacturing plant. The field strength will be equal to the three edge bearing strength times the load factor or bedding factor B_f.

This factor is defined as the ratio of the strength of a pipe under any stated conditions of loading and bedding, W, to its strength measured by three-edge test, S (ASCE, 1994):

$$B_f = \frac{W}{S}$$

Values for this factor were determined by the classical experiences of Spangler at Iowa State College.

The required three-edge bearing strength for a given bedding factor, R, can be evaluated taking into account a safety factor S_f:

$$R = W \frac{S_f}{B_f}$$

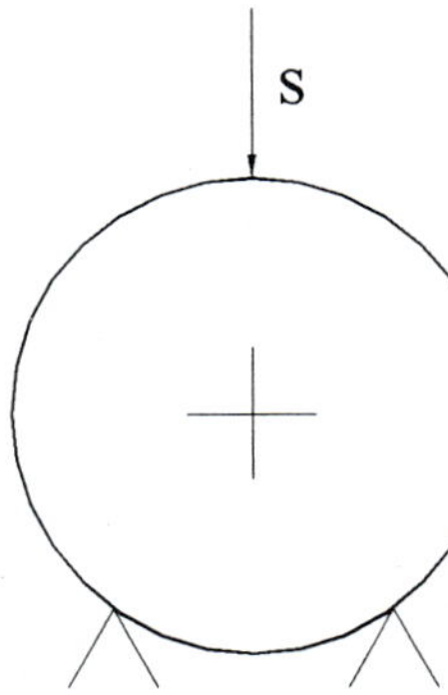

Figure 4. Indirect methjod for rigid pipes (Scheme of three-edge bearing test)

Although the usual procedure in the direct method consists in superposition of stresses computed by external loads and internal pressure, there is evidence that the pipe strength is less when the pipe is pressurized.

This effect is consequent to the fact that a pipe under a vertical load is subject to a deflection, the section passing from a circular to an elliptic shape; the added internal pressure tends to magnify this deformation, creating stresses larger than the ones computed by superposition of effects.

Schlick reported the following experimental equation

$$\frac{W_p}{W} = \sqrt{1 - \frac{p}{p_0}}$$

where p is the internal pressure

p_o is the bursting pressure with no external load

W is the pipe strength with no pressure

W_p is the pipe strength when internal pressure p is applied.

The equation was demonstrated to apply to cast iron, asbestos cement, plain and prestressed concrete pipe.

Since the indirect method can correctly take into account the internal pressure action, it is today preferred to the direct method (Prevost, 1994). Traditional and new methods for reinforced concrete design are reviewed by McGrath (1993). The Author shows the pressure distributions obtained by traditional methods as well as those obtained by means of Finite Elements.

1.4.5. Flexible Pipes Computation

Flexible pipes deform under external loads, and a part of the trench load is carried by the side fill; the rest of the load is carried by the pipe, that bends and arches under its action.

This means that the vertical load will partly deform the elastic ring, showing bending actions in its wall; and it will be partly transferred to the side fill, giving a compressive action to the ring longitudinal section (Gabriel, 1990, Hashash and Selig, 1990).

The combined effect in terms of deflection is usually accounted for by means of the classical Spangler formula.

This formula gives the horizontal pipe deflection x (Figure 5.) as a function of soil and pipe characteristics:

$$x = \frac{D_t K_b W}{\frac{EI}{r^3} + 0{,}061\, E'}$$

where D_t is a deflection lag factor, consequent to the time consolidation characteristics of the soil; suggested values 1.25-2.5.

K_b is a bedding factor

W is the load per unit length

r is the pipe radius

E is the pipe modulus of elasticity

I is the pipe wall moment of inertia per unit length

E' is the modulus of soil reaction.

The bedding factor K_b depends upon the width of the pipe bedding.

Values of soil reaction modulus E' were experimentally found by Howard (1977) depending on the backfill characteristics and compaction. In reality, these values come from back-calculation using Spangler formula, but they can not be directly measured.

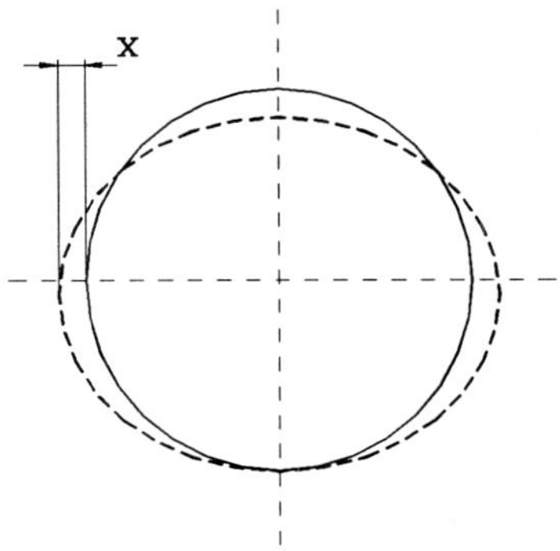

Figure 5. Flexible pipe deflection

More recently, Hartley and Duncan (1987) found a dependency of the modulus E' on the cover depth and published a table of values updating the original Howard table. The same Authors show an interesting relationship between E' and the Young modulus of the soil E_s.

It should be noted that the modulus of elasticity of plastic materials tends to reduce in orders of magnitude with time; it is then necessary to take into account its final value.

The theoretical background of the Spangler formula is today under question. Gabriel (1990, 1993) points out that at least two points should be criticized: first, the ring compression stiffness is important for plastic pipes and is to be considered in addition to ring bending stiffnes; and second, the composite stiffness of pipe-soil coupling should be obtained in a more effective way than by simple addition.

Currently, the pipe computation is performed by assessing the value of vertical deflection y equal to horizontal deflection x (for small values).

Moreover, when dealing with large diameter pipes, the term EI/r^3 may be much smaller than 0.061 E'. It is recommended that the first should never be less than 10-15% of the second one, so that the pipe can have sufficient local strength in bending and thrust to develop and utilize passive resistance of the soil (ASCE, 1994).

The flexible pipe computation is based on deflection. Limits to vertical deflection are suggested for different pipe materials. It is important to point out that superposition of all loads, including internal pressure, is the rule, except for material not sensitive to strain (steel, d.i., pvc); in these pipes, the computation can be carried out separately, for external loads and for internal pressure (Prevost, 1994). Anyway, as clearly pointed out by Petroff (1993), wall strains should be evaluated, especially for strain sensitive materials, and compared with expected design strains. To evaluate strains, the current formula is

$$\varepsilon = f \frac{y}{D} \left(\frac{t}{D} \right)$$

where f is a shape factor (Ciaponi and Papiri, 1995).

Buckling is an extreme phenomenon occurring when vertical deflection is critical; in this case, the sides of the pipe pull inward, the lateral soil support is largely reduced and this brings to the rapid collapse of the pipe (Prevost, 1994; Petroff, 1993). Kienow and Prevost (1989) have shown the possibility that, in the case of low pipe stiffness, installation and compaction can result in irregular pipe deformation (squaring), reducing the resistance of the supposed elliptically deformed pipe and leading to its collapse. Some manuals of practice suggest how to evaluate the maximum buckling pressure (Ciaponi and Papiri, 1995).

Today, there are models in which soil-pipe interactions are evaluated by means of finite elements. Pipe structural behaviour is also considered by more refined structural models, based on finite elements analysis. Zaman and Laguros (1990), among others, show a finite element procedure for studying both the deformations (strains) and stresses in a buried pipe and surrounding soil medium. Pipe material and soil medium are treated as elastic bodies and the interface elements are assumed to exhibit non-linearity.

The study includes the analysis of the interface between the pipe and the surrounding soil. It is recognized that the interface conditions can noticeably influence the values and distribution of normal and shear stresses in the pipe and in the surrounding soil.

It is to be said that, at the moment, current design computation of both rigid and flexible pipes is carried out by classical methods. F.E., although a promising tool, is considered in many cases too sophisticated for the design scope, because the in situ soils are frequently not adequately characterized to take advantage of the precision of Finite Element analysis (McGrath, 1993).

Although Spangler type consideration of soil-pipe interaction is questionable, as discussed above, simple and effective developments recently shown by Stephenson and King (1994) lead to a validation of this approach.

1.5. CONCLUSIONS

The Section highlights some classical as well some innovative aspects in pipeline construction.

New materials for lining and coating of traditionally manufactured pipes are examined and progresses in plastic pipe production are discussed.

As to pipe installation, the most common recommendations concerning buried pipes are taken into account; trenchless techniques are considered as an interesting development.

As to pipe structural requirements, buried pipe analysis is discussed on the basis of classical assumptions, pointing out the necessity of more studies on pipe-soil interaction, soil pressure distribution and real pipe deformation.

1.6. REFERENCES

Beech S.H. (1994) Polyethylene Materials for Pipeline Systems. 2nd International Conference on Water Pipeline Systems. Edited By D.S. Miller. BHR Group Publication n. 10, MEP, London.

Brandon T.W. (1984) The Institution of Water Engineers and Scientists (IWES). Water Practice Manuals. 4., Water Distribution Systems.

Brewer W.E. (1990) The Design and Construction of Culverts Using Controlled Low Strength Material-Controlled Density Fill (CLSM-CDF) Backfill. Structural Performance of Flexible Pipes. Edited By S.M. Sargand, G.F. Mitchell and J.O. Hurd, Balkema, Rotterdam.

Brewer W.H., Hurd J.O. (1993) Controlled Low Strengthh Materials - Controlled Density Fill (CLSM-CDF) as a Backfill Around Flexible Structures. Structural Performance of Pipes 1993. Edited by M. Sargand, G.F. Mitchell & J.O. Hurd, Balkema, Rotterdam.

Brown J.E. (1994) Developments in Thermoplastics Pressure Pipe Systems - Molecular Oriented PVC Pipe Systems. 2nd International Conference on Water Pipeline Systems. Edited By D.S. Miller. BHR Group Publication n. 10, MEP, London.

Ciaponi and Papiri (1995) Verifica statica delle condotte a pelo libero o in pressione. Sistemi di drenaggio urbano, manuale di Progettazione, costruzione e gestione. Centro Studi Deflussi Urbani, Politecnico di Milano, I ediz. (draft) (in Italian).

Gabriel L.G. (1990) Keynote Address: Pipe Deflections- A Redeemable Assett. Structural Performance of Flexible Pipes. Edited By S.M. Sargand, G.F. Mitchell and J.O. Hurd, Balkema, Rotterdam.

Gabriel L.H. (1993) Checking the Premise-Deflections of Flexible Pipe are Predictable. Structural Perfomance of Pipes 93. Edited by M. Sargand, G.F. Mitchell and J.O. Hurd, Balkema, Rotterdam.

Hashash N., Selig E.T. (1990) Analysis of the Performance of a Buried High Density Polyethylene Pipe. Structural Performance of Flexible Pipes. Edited By S.M. Sargand, G.F. Mitchell and J.O. Hurd, Balkema, Rotterdam.

Henry Ph. (1992) Pipes for Microtunnels: Suitability of Use. No Trenches in Town. No-Dig 92 Paris. Edited by J.P. Henry & M. Mermet. Balkema, Rotterdam.

Howard A. (1977) Modulus of Soil Reaction Values for Buried Flexible Pipe. Journal of the Geotechnical Engineering Division, ASCE, January.

Katona M.G. (1993) On the Analysis of Buried Conduits- Past, Present and Future. Structural Perfomance of Pipes 93. Edited by M. Sargand, G.F. Mitchell and J.O. Hurd, Balkema, Rotterdam.

Kienow K.W., Prevost R.C. (1989) Pipe/Soil Stiffness Ratio Effect on Flexible Pipe Buckling Threshold. Journal of Transportation Engineering, ASCE, vol. 115, March.

Kruibrinsk A.C.H., Heinsbroeck A.G.T.J. (1992) Fluid-Structure Interaction in Non-Rigid Pipeline Systems - Large Scale Validation Tests. Pipeline Systems. Edited by B. Coulbeck and E. Evans, Kluver Academic Publishers, Dordrecht.

Larsen T., Burrows R. (1992) Measurement and Computations of Transients in Pumped Sewer Plastic Mains. Pipeline Systems. Edited by B. Coulbeck and E. Evans, Kluver Academic Publishers, Dordrecht.

McGhee T.J. (1991) Water Supply and Sewerage. McGraw-Hill International Editions. Civil Engineering Series.

McGrath T.J. (1993) Design of Reinforced Concrete Pipe - A Review of Traditional and Current Methods. Structural Perfomance of Pipes 93. Edited by M. Sargand, G.F. Mitchell and J.O. Hurd, Balkema, Rotterdam.

Najafi M., Iseley D.T., Pumphrey N.D.Jr, Nishida H. (1993) 'LLB', an Innovative Microtunneling Propulsion System. Structural Perfomance of Pipes 93. Edited by M. Sargand, G.F. Mitchell and J.O. Hurd, Balkema, Rotterdam.

Orchard S. (1992) New Developments in Microtunnelling Techniques. No Trenches in Town. No-Dig 92 Paris. Edited by J.P. Henry & M. Mermet. Balkema, Rotterdam.

Osborne J.S. (1994). North West Water and the Use of Plastics Pipes for Larger Diameter Mains. 2nd International Conference on Water Pipeline Systems. Edited By D.S. Miller. BHR Group Publication n. 10, MEP, London.

Petroff L.J. (1993) Ring Bending Stiffness and the Design of Flexible Pipe. Structural Perfomance of Pipes 93. Edited by M. Sargand, G.F. Mitchell and J.O. Hurd, Balkema, Rotterdam.

(The) Pipeline Industries Guild (1994) Pipelines: All You Wanted to Know...

Prevost C. (1994) Buried Pipe Structural Design. 2nd International Conference on Water Pipeline Systems. Edited By D.S. Miller. BHR Group Publication n. 10, MEP, London.

Stephenson D., King. A. (1994) Pipe-Soil Interaction with Flexible Pipe. 2nd International Conference on Water Pipeline Systems. Edited By D.S. Miller. BHR Group Publication n. 10, MEP, London.

Stevens W.W., Girka A. (1992) Ductile Iron Pipe for Trenchless Techniques. No Trenches in Town. No-Dig 92 Paris. Edited by J.P. Henry & M. Mermet. Balkema, Rotterdam.

Twort A.C., Hoather R.C., Law F.M. (1974) Water Supply. Edward Arnold Ltd, London.

van Kesteren W.G.M., Bisshop F. (1992) The Excavation Behaviour of Soils in Trenchless Tunnelling Techniques. No Trenches in Town. No-Dig 92 Paris. Edited by J.P. Henry & M. Mermet. Balkema, Rotterdam.

Water Environment Federation - American Society of Civil Engineers (1994) Design and Construction of Urban Stormwater Management Systems. ASCE Manuals and Reports of Engineering Practice n.o.77. WEF Manual of Practice FD-20.

White H., Tregoing C. (1992) Microtunnelling Advances in UK. No Trenches in Town. No-Dig 92 Paris. Edited by J.P. Henry & M. Mermet. Balkema, Rotterdam.

Zaman M., Laguros J.G. (1990) Predicting the Behavior of Buried Pipes Using a Mixed Finite Element Approach. Structural Performance of Flexible Pipes. Edited By S.M. Sargand, G.F. Mitchell and J.O. Hurd, Balkema, Rotterdam.

2. Water Transfer Within Brussels Water Services

Daniel L. Brumagne
Compagnie Intercommunale
Bruxelloise des Eaux
Rue aux Laines 70, 1000 Bruxelles
BELGIUM

2.1. INTRODUCTION

In common with many large city water departments, the Brussels Water Service, Compagnie Intercommunale Bruxelloise D'Eaux (CIBE) now uses water sources a considerable distance from the point of demand; in this case some 90 km from Brussels, capital of Belgium (Figure 1.).

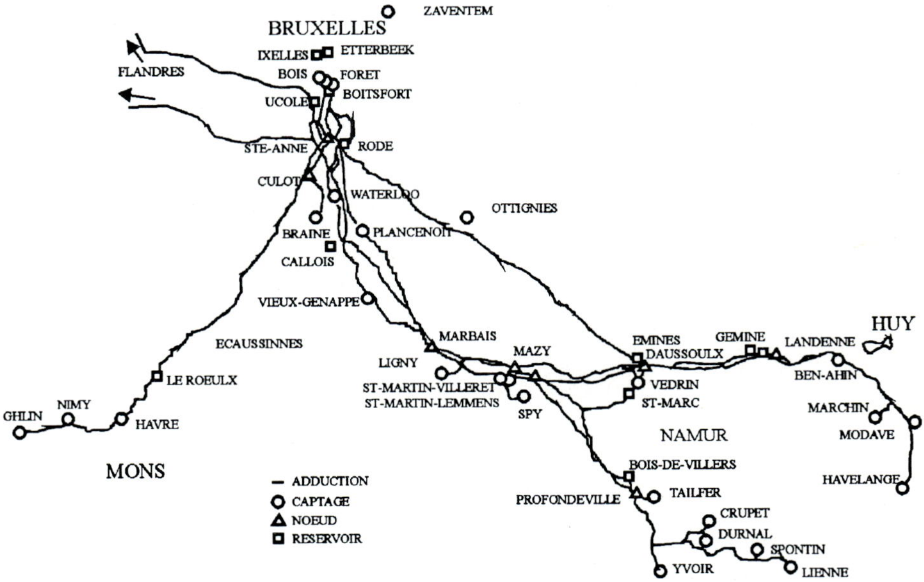

Figure 1. Schematic map of the adduction of CIBE, The Brussels Water Supply

NATO ASI Series, Partnership Sub-Series, 2. Environment – Vol. 15
Water Supply Systems – New Technologies
Edited by Č. Maksimović, F. Calomino, and J. Snoxell

The current system is very complex, with many network interconnections, reservoirs and sources which have varying yields throughtout the year, while demand on the system also varies, but in a different way.

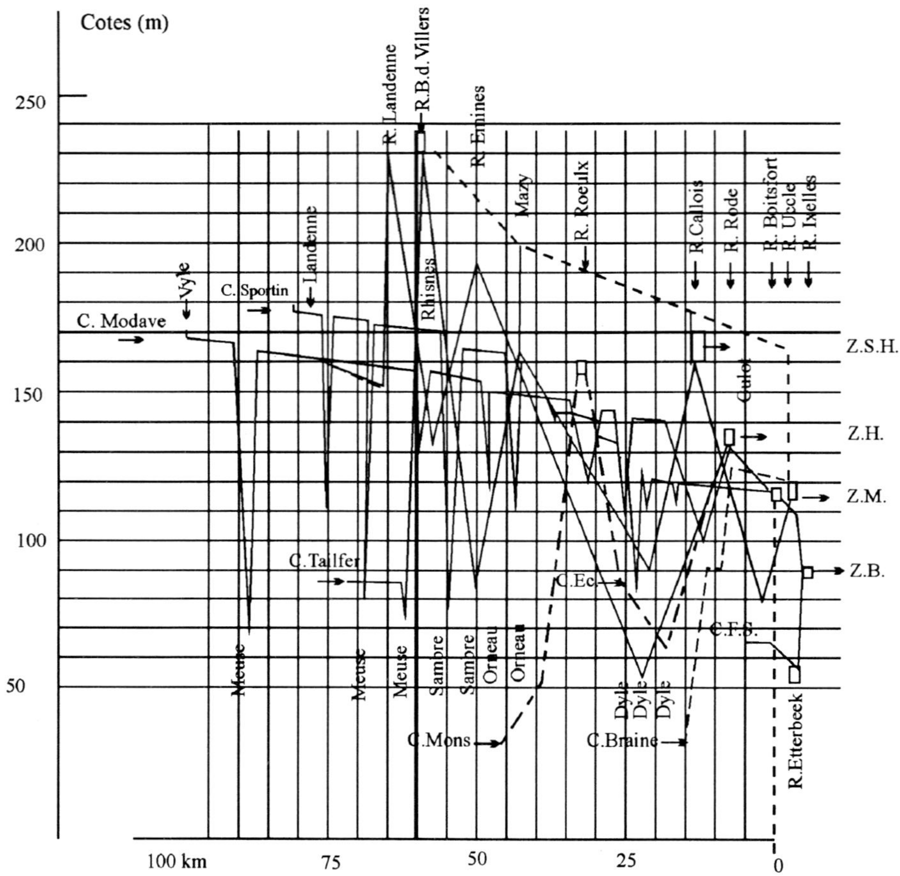

Figure 2. Schematic profile of the adduction of CIBE

The remote water sources need adequate mains with capacity for current and future needs and also for the capacity of the source itself. The paradox for the water services is that demand is greatest when resources are scarce, very often during hot dry summers.

There are obviously daily, weekly and yearly variation in water demand, and a system of transfers, regulation works and reservoirs can have great complexity. This is the case for Brussels (Figures 1. and 2.). The variation in the yield of the different sources of ground or surface water can also be important and it is necessary to search for increasingly remote resources.

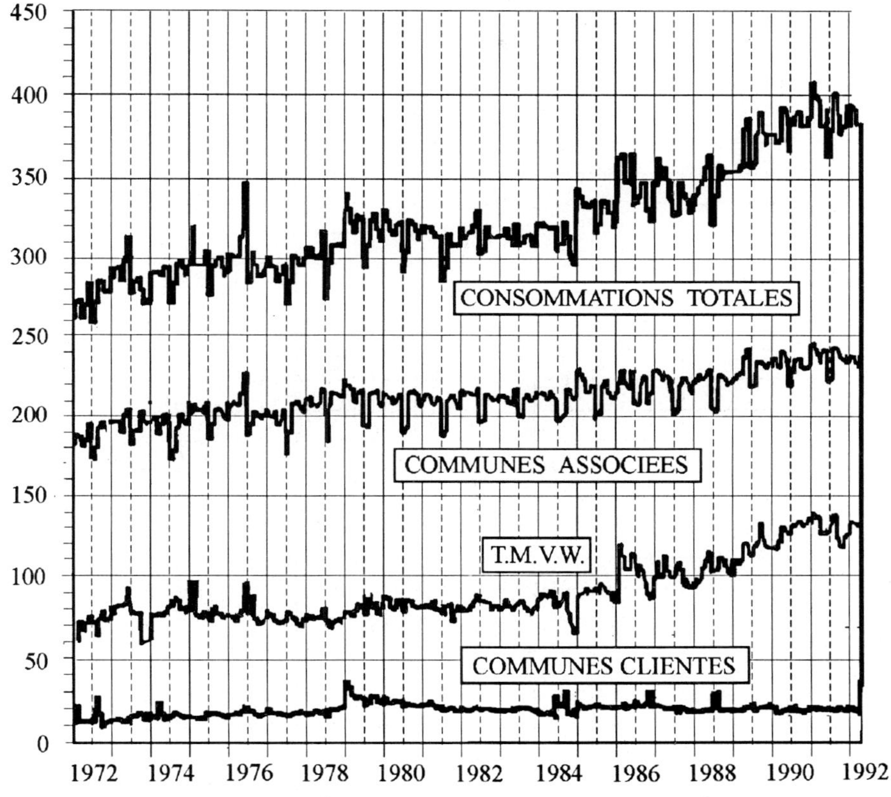

Figure 3. Diagram of the total monthly water consumption of the CIBE from 1972 to 1992

Groundwater sources often have less viable yields than surface resources: the catchment of Modave, into the Dinantial limestones, has a medium yield of 60.000 m^3 /d, with a minimum of 45.000 m^3 /d and a maximum of 85.000 m^3/d, while the river Meuse, at its entry in Belgium (near Givet) has a yield varying between 10 and 1.000 m^3/s.

For peak periods of consumption, it is often useful to have a big surface or underground reservoirs, such as old quarries or mines but the underground reservoirs are the cheapest and most important when their capacity is big enough. The transfer lines of the CIBE are of following types: aqueducts, open air channels, feeders and water pipelines.

2.2. THE AQUEDUCTS

The aqueduct is a very ancient system used for the transfer of water. The Romans built a lot of famous aqueducts all around the Mediterranean. There are also very old quanats in Iran or in Maghreb.

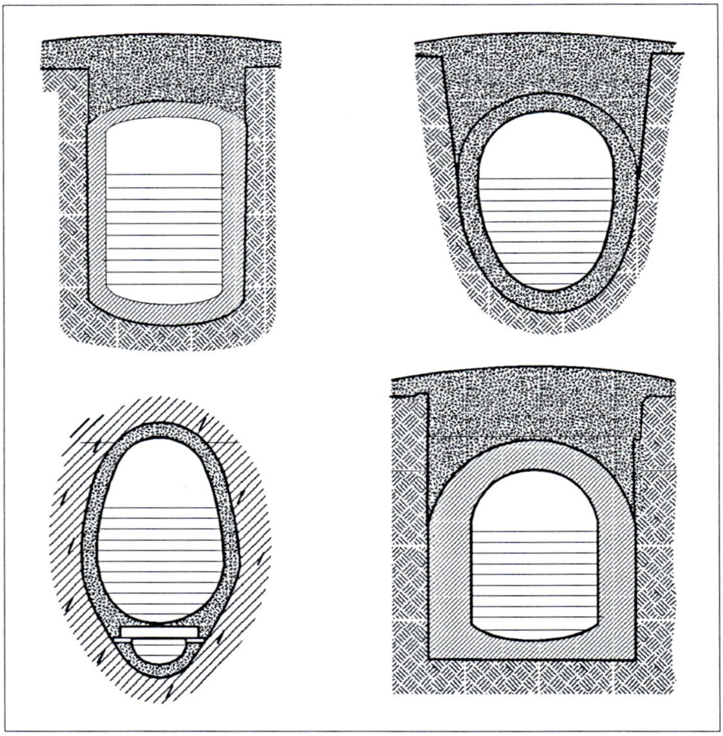

Figure 4. Cross sections of some aqueducts

Their gradient is very swallow to avoid unacceptable velocities. They are always built with local material: brick or concrete. The valleys were crossed with bridges, such as the 'Pont du Gard' in France (Figure 5.), they are now crossed with pressure pipelines.

Figure 5. The Mont-Saint-Pont aqueduct, Braine-l'Alleud **Figure 6.** Internal visit of an aqueduct

Sudden and big variations in yield are to be avoided. Therefore the size of an aqueduct depends on the importance of the source itself, often spring water. If the spring is big enough it is better to catch only the minimum available yield, stable all year round. As this is often not the case, the size of the channel depends on a yield that can be reached, say at least 10% of the time and, when the real yield falls below this value, level and velocity need to be regulated in the channel.

An aqueduct is a vulnerable system when it crosses some aquifers, but quality control can be easily achieved; at some valley crossings, for instance, where polluted water can be diverted.

The stability of aqueducts is usually good, especially underground and at depths of >30 m: the Trajan aqueduct, for instance, is still in service for Rome. The aqueducts of CIBE were built in 1855, 1895 and 1920. Their maintenance is easily carried out during visits made every 3 to 6 years (Figure 6.).

However, some problems exist when these works cross karstic areas.

Except for the largest yield (>10m^3/s) it is good practice to cover the aqueducts with more than 1 m of ground: evaporation is reduced, temperature is regulated, pollution by air born pollutants is also avoided.

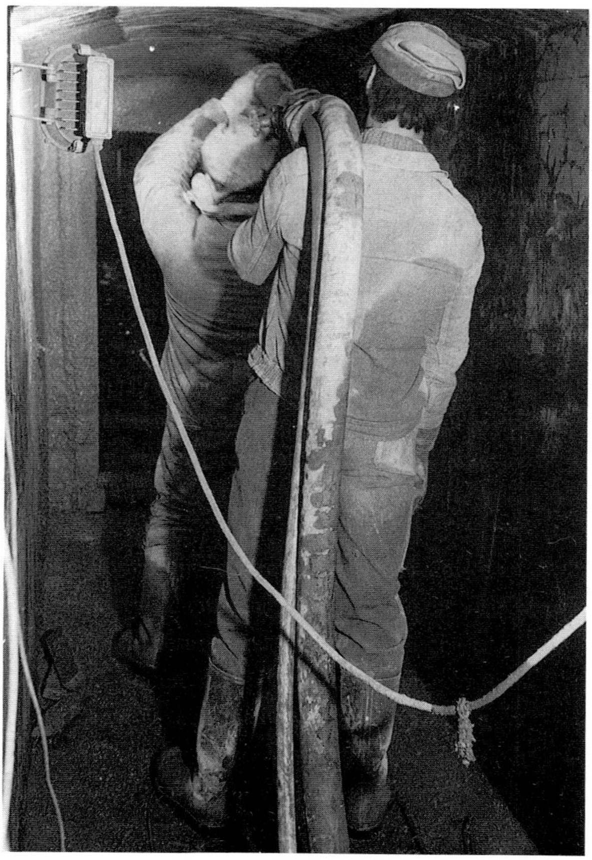

Figure 7. Concrete gunitiny in the aqueducts Modave-Mazy

Two new techniques are useful in the construction of aqueducts:

- microtunelling for new aqueducts at big depths,
- concrete gunite for new linings up to 700m of access shaft, for old aqueducts (Figure 7.).

2.3. THE FEEDERS

From the beginning of this century it has been possible to use pipes of bigger diameter (1 m) for the new transfers; these pipes are made in cast iron, steel, asbestos-cement or concrete, reinforced or prestressed. Prestressed concrete is very good for large pipes (>1.2 m) (Figure 8.).

Figure 8. Trenching and laying of the feeder

Obviously feeders are always to be placed below the piezometric line. In most cases, they join two big reservoirs, the first situated near the source, the second one near the consumers. That is the option chosen by the CIBE. The pipes generally follow the ground topography; air valves have to be placed at all high points, discharge valves at the low ones in the valleys. This pipe system has better flexibility than aqueducts above all when it serves some different reservoirs because it is easy to connect two feeders through a valve complex.

Feeders are less vulnerable to pollution when they are pressurised unless there is an accidental pressure fall.

During the layout of a feeder, the biggest problem for the builder is often the provision, transport and the placement of its big pipes.

The pipes of each feeder are to be designed to sustain the maximum internal pressure of water, taking into account eventual surge pressure (water hammer). They have to resist all external forces, vertical or other (ground pressures, charges of traffic load) and also longitudinal forces (caused by temperature variations): this is the case for pipes used for the transfer of surface water the temperature of which may vary between 2° C and 25°C.

Feeders allow rapid and important velocity variations; this is useful in adjusting the yield to demand. For instance, in less than one hour, the yield of the Hainaut feeder can vary from 1.000 to 4.000 m^3/h. This gives great flexibility to every system for water transfer, especially in the case of accident on an aqueduct or feeder; a good water service has to take into consideration the probability of such accidents.

Figure 9. Horizontal boring below an express way near Mazy

2.4. SPECIAL WORKS

Aqueducts and feeders often have to cross rivers, canals, roads, speedways and railways. The building of these crossings raise the same problem for both transfer systems. Techniques used for crossings are quite traditional now except for the horizontal boring ordinarily used in soil, but also in hard rock (Figure 9.).

For these crossings, especially when the soil is locally compressible, it is often necessary to place the water pipe into a deformable channel (an external pipe), almost always in reinforced concrete. This technique is used for the crossing of a canal in the surroundings of Mons (Figure 10.) and also for the crossing of a new expressway south of Brussels (Figure 11.). These protection structures are highly useful in the case of pipe rupture and in allowing the repair of the failing pipe (Figure 12.).

The connections between various transfer lines, aqueducts or feeders, requires the building of complex valve chambers. At the upper Mazy connection (Figure 13.), a regulation tower was built, the aim of which was to avoid excess pressure in the downstream part of the feeder (Figure 14.). To avoid surge pressure in the biggest feeders, with great yields and high static pressure it is useful, as in the Tailfer works (Figure 15.), to reinforce the pipe itself and to connect this pipe to a special purpose reservoir.

Figure 10. Deformable channel below the 'Canal du Centre' near Mons

Figure 11. Channel below an express way near Waterloo

Figure 12. Replacement of a steel pipe in a channel below an express way, Wauthier-Braine

Figure 13. The Mazy valve chambers

Figure 14. The Mazy regulation tower

2.5. REGULATION AND CONTROL

Spatial and temporal transfer of drinking water requires accurate information - quantity and quality. It is thus essential to place sensors and fittings at the upstream and downstream ends of a waterline, also often in the middle of them, to measure flow and pressure, turbidity and chlorine residual (when chlorine is added). The CIBE has consequently equipped its waterlines with flowmeters, pressure gauges, turbidimeters, chlorine monitoring in its catchments, along its aqueducts and feeders, and in its regulation reservoirs.

In order to manage all this information a dispatching system was built in Brussels in 1973 and recently (1993) renewed. All the information is updated at least every two minutes.

Figure 15. Pumping station at Tailfer, near Namur

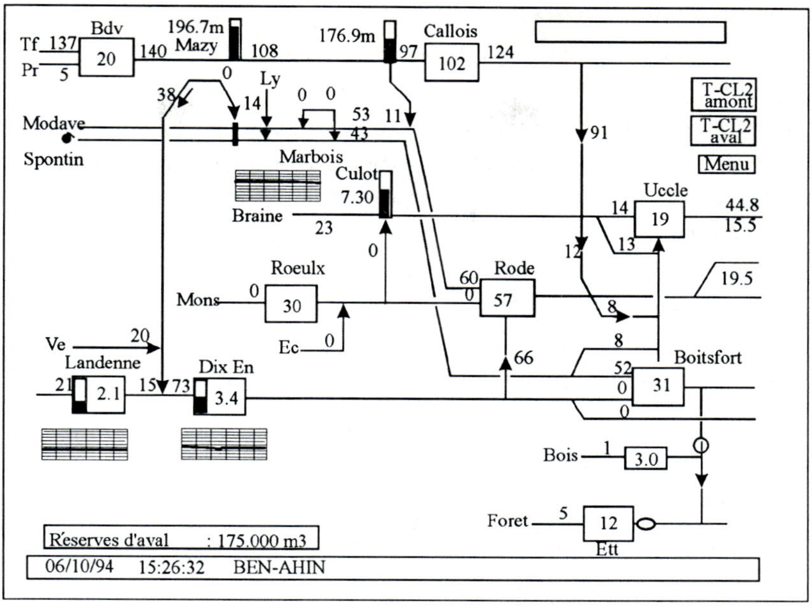

Figure 16. Schematic view of the adduction of the CIBE as to be seen on the screen of its central dispatching

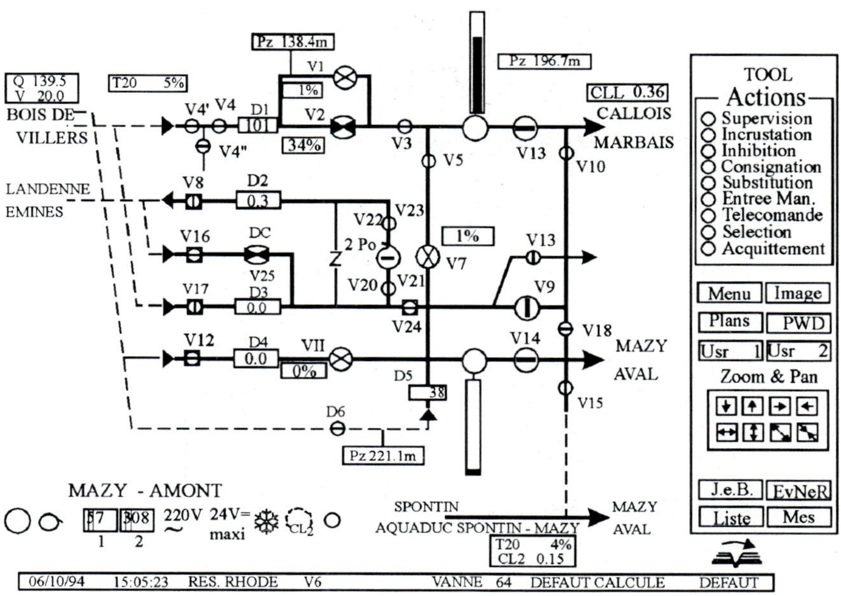

Figure 17. Schematic of the Mazy chamber as to be seen on screen

This new computer system is controlleed by six dispatchers who relieve one another permanently (Beal, 1987). The dispatcher in charge can always see the entire complex of transfer lines (Figure 16.) and choose to zoom on a specific installation as, for instance, the upper Mazy connection (Figure 17.) or the Callois Reservoir (Figure 18.). From Brussels the operator is thus able to check all the waterlines, command valves and pumps (Beal, 1987). In the case of pollution in a catchment, the operator has the option of opening the discharge valves and closing the line valves, before the entry of polluted water into aqueducts or feeder. But in all such emergencies, of course, chemists and engineers in charge of the production service are informed and can override if necessary.

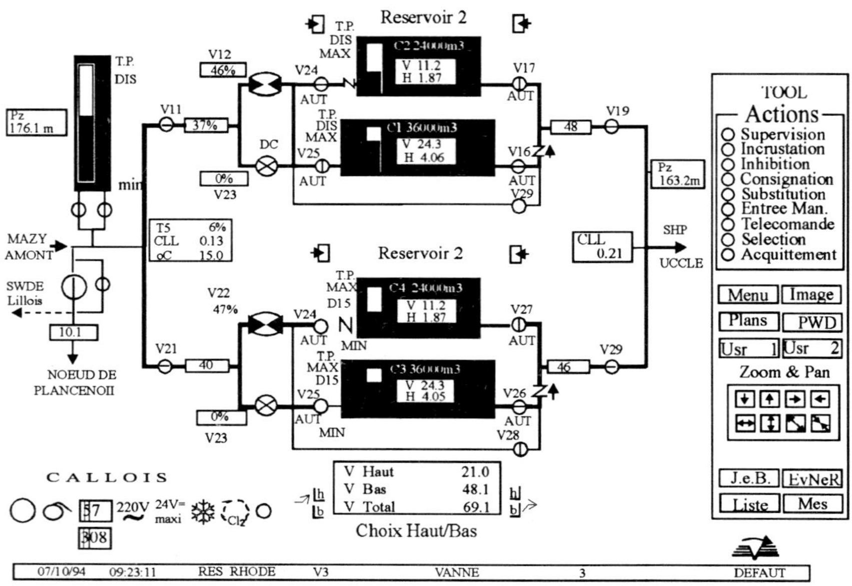

Figure 18. Schematic view of the Callois complex as to be seen on screen

2.6. DURATION OF THE TRANSFER

As the mean average velocity of water is about 1 m/s, water from each remote catchment takes less than one day to reach one of the regulation reservoirs built upstream of Brussels, in the southern part of the complex. One or two days later, water can reach the customers tap. That is the best situation, when all the feeders and aqueducts are nearly at their optimum operation. But, when a new waterline is put into service, waterflow and speed are low, the duration of the transfer is high and the quality of the water still has to be checked. To avoid low yields it is often necessary to balance water between resources.

The temporal transfer of groundwater does not create problems because such sources are diverted into artificial channels, below the ground level up to closed reservoirs, distribution network and taps. But, when one has to impound, treat and transfer surface water, and mix those with groundwater one may have some problems with the growth of bacteria or other micro-organisms: surface water, even completely sterilised, contains nutrients for micro-organisms always present in groundwater. One therefore always has to check the chlorine content of those mixed waters and, if necessary, adjust this by adding more chlorine, and this is usually carried out before the entry in to a regulation reservoir, and sometimes in the city network.

2.7. REFERENCES

Beal D. (1987) The Implementation of Information Technology to the Year 2000 - A Realistic Approach. Symposium Water/Eau 2000, Nice France, pp. 3-1-1 to 3-1-7.

Beal D. (1988) Integrated Distribution Management Systems. Aqua, Great Britain, No. 5, pp. 267-276.

3. Transfer of Water and Data Management in Wessex Water - UK

John Snoxell
WESSEX WATER
WESSEX House, Passage str.
BS2 0JQ Bristol
Dušan Obradović
WINS Ltd.
Tower House, Parkstone Road
Poole, Dorset BH15 2DH
UK

3.1. INTRODUCTION

Wessex Water supplies water to some 1.1m customers in the region between the English Channel and Bristol Channel, see Figure 1. Water is captured at 153 sites of which around two thirds are in constant use,with the rest at standby or used for stream support. The distribution system has some 350 pumping stations and over 400 reservoirs, arranged in several hydraulically independent systems - the linking of these systems is not yet fully completed. The total average demand reaches 450 Ml/d.

In the last ten years a number of major trunk mains have been introduced to link areas of surplus and deficit - these are shown in Figure 1. The major link was between the surface sources in the west and the groundwater (chalk and limestone) in the east. This was completed in 1990, when the south of the UK was suffering from drought and was one of the reasons why Wessex did not have to impose customer restrictions during the period.

With the gradual integration of resources, management has become increasingly centralised and an efficient data management system for the regional network is essential (Huntington, 1984, 1990). This Section describes the development of such a system and the way in which telemetry data, GIS, billing, and other data sources have been integrated to provide detailed information for operational management, leakage control, levels of service and so on.

NATO ASI Series, Partnership Sub-Series, 2. Environment – Vol. 15
Water Supply Systems – New Technologies
Edited by Č. Maksimović, F. Calomino, and J. Snoxell

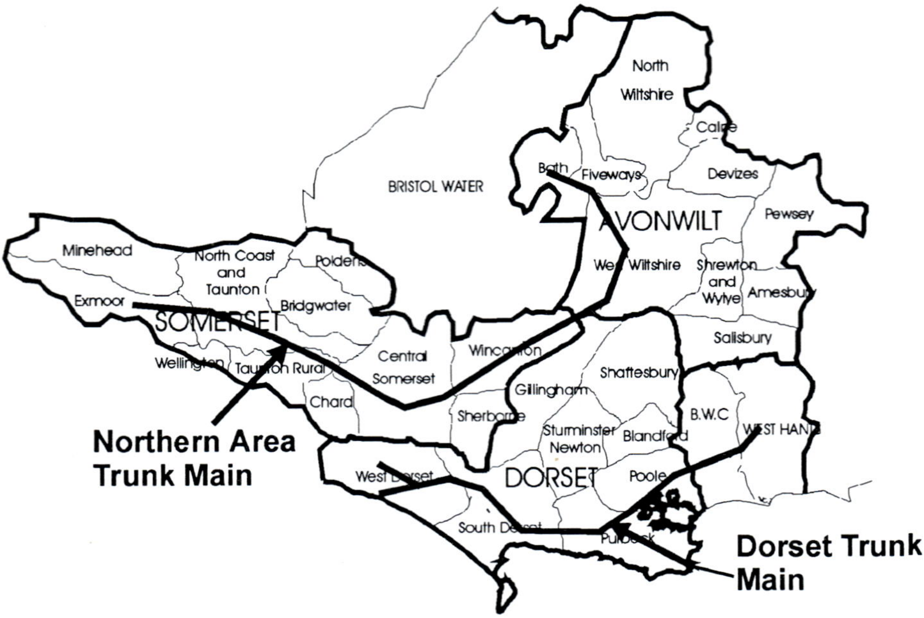

Figure1. Wessex regional area

First, the region was organised into three Areas (Dorset, Somerset and Avon/Wilts), which were in turn subdivided into some 29 Supply zones as shown in Figure 1. Two further layers were created below supply zone level:

- District Metered Areas (DMAs),
- Control zones.

DMAs are the basic unit of territory now used in the UK for monitoring leakage and are relatively stable- in Wessex there are now 335 DMAs which are monitored continuously by data logger or telemetry. Below DMA level there are numerous more Control Zones which are monitored for changes in minimum night flow etc as a direct aid to leak location.

Water source information was collated and checked and detailed schematics were prepared under a quality management system (BS5750); an example is shown in Figure 2.

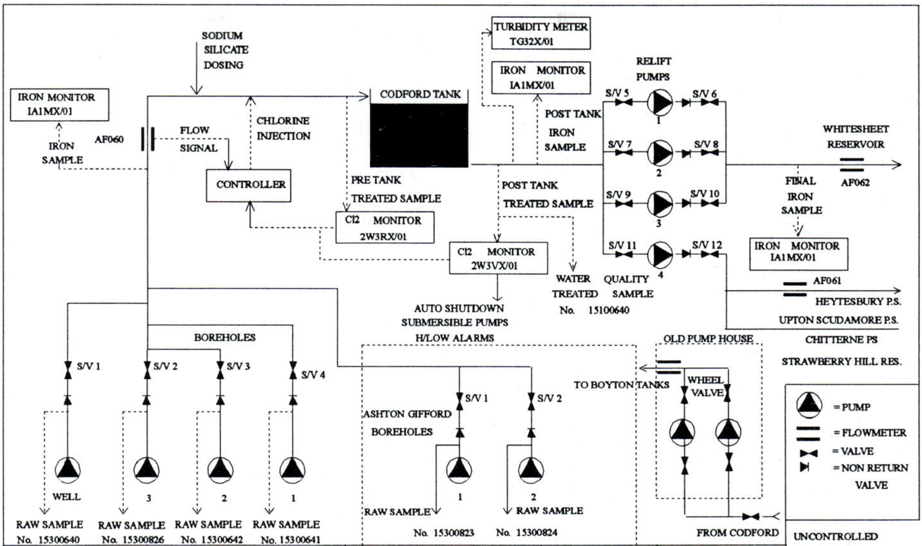

TITLE OF SECTION : SITE DATA	DOCUMENT No :	ISSUE No :	PAGE No :
DOCUMENT DESCRIPTION : CODFORD SITE SCHEMATIC	SECTION NO :	DATE :	VERSION No :

Figure 2. Site schematic produced for BS5750 certification

3.2. REGIONAL TELEMETRY

At the same time that the supply network was being integrated the existing telemetry system, which dated back to the mid-70's and was a distributed system, was also centralised and rationalised (Huntington, 1993); this presented the opportunity to collect all data in one central telemetry database and establish regional systems for the identification and naming of flow meters, level sensors, pressure transducers and so on, all essential steps in the creation of a centralised management information system.

The telemetry system is very large, collecting some 1500 analogue signals and expanding fast towards its design capacity (6000 signals).

All operating data are collected in VAX Central Telemetry computers (duty and standby) and refreshed continuously. Operators in the Regional Control Centre (R.O.C.) can monitor any part of the system from one central point and can intervene with specific control capabilities where appropriate.

Other users can have the access through dial-up computers, see Figure 8. (Huntington, 1993). Every 15 min the key data (mostly flows, pressures and water levels) is transferred to a PC File-server where it is available for all analysis and archiving purposes (Snoxell at al., 1989, Snoxell 1993a, 1993b, 1994).

The telemetry covers practically all sources, transfers between Supply zones and main distribution trunks, so these data permit computation of:

- Source outputs,
- Total demand (consumption and losses) in the Region, Areas and Supply zones.

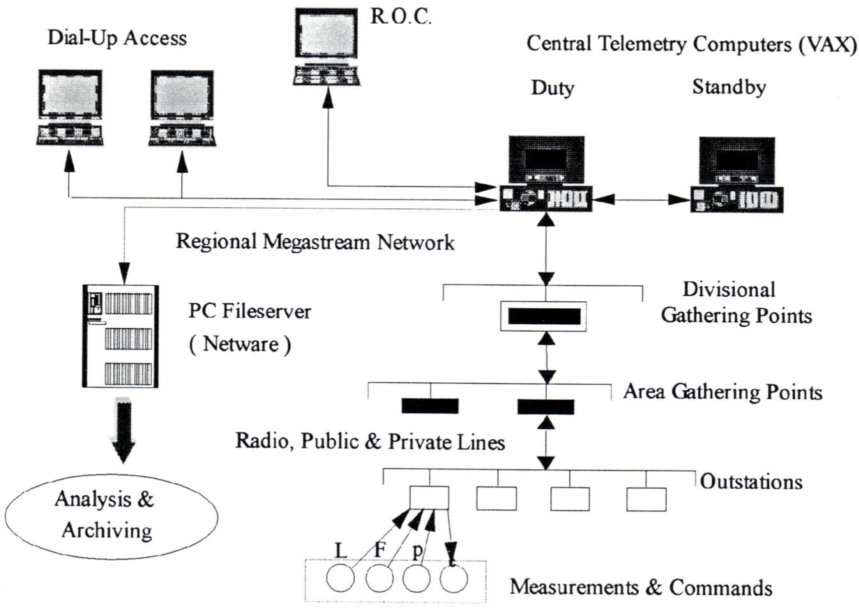

Figure 3. Regional telemetry

The overall picture is completed with other data, as shown in Figure 4.

Billing data is regularly transferred to the same PC File-server as Telemetry data from the billing centre. There are three type of files:

- monthly-read customers (roughly 2000), the largest and the most important ones whose water meters are read every month,

- commercial users (40 000 approximately), controlled twice per year,

- domestic users (over 23 000), also controlled two times every year.

These data are sorted using the UK **Post Code** System, established and maintained by Royal Mail. Briefly, a Post Code denotes a small area with up to 50 addresses clustered together, with known coordinates. These coordinates are linked to the Demand Management Districts through GIS, as described below, so the total metered consumption can be computed for each District and hence for Supply Zones, Areas and ultimately the Region by simple summation (note, however, that this in **not** the total consumption of water because the system is not fully metered).

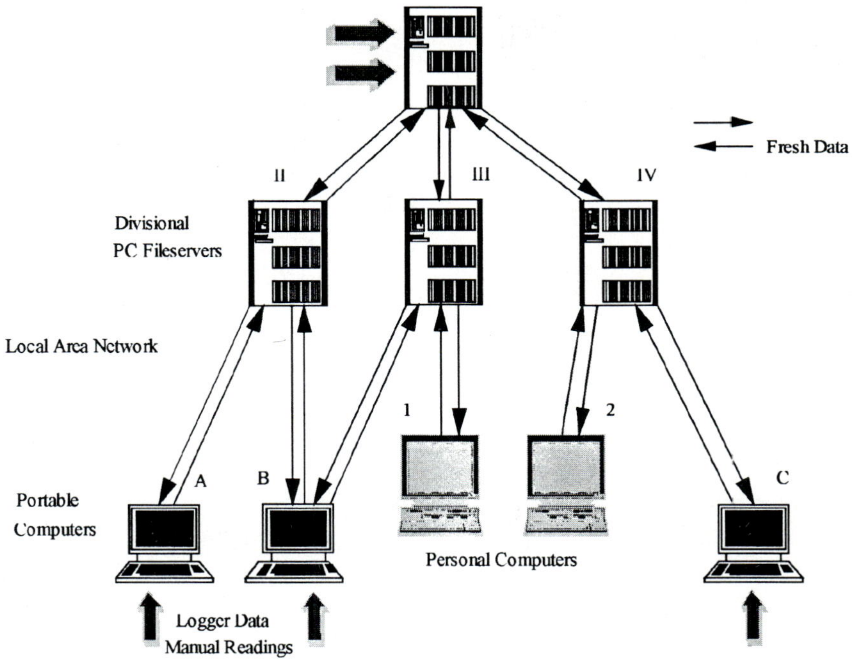

Figure 4. Merging all operational data together

The data from Telemetry (VAX computers) and Billing system (a mainframe) are easily processed and stored in the Master Database, on the PC Fileserver I (Figure 9.) because they come from a single source. The other data come from over 1000 flowmeters and watermeters located all over the Region. Some of these instruments are equipped with portable data loggers (350+) and others have to be read on the spot (manual readings).

Earlier attempts to organise centralised collection & processing of this valuable information ran into difficulties and it was decided to organise **distributed data processing** instead.

Operational staff equipped with notebook PCs can see the maps provided by GIS, as well as operational data for the last 100 days for their respective in great detail. The inspector then can enter manual readings and/or download data from the data logger, for all the sites which he has to cover.

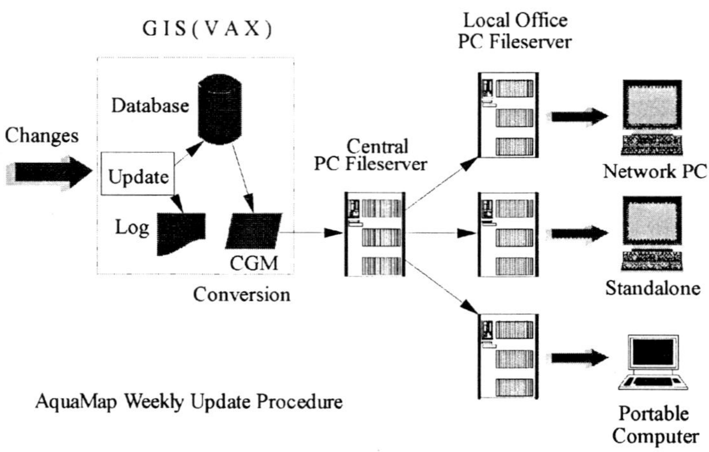

Figure 5. Regular update of GIS data

The inspector regularly (once per week) reports to a divisional headquarters where a file server keeps the data base updated every night by the Master database (Figure 9.). Once he connects his notebook computer to the divisional PC fileserver two tasks will be carried out automatically:

- fresh data from his notebook will be taken by Divisional PC fileserver,

- the database on his notebook will be updated with new telemetry and billing data, and also data from other sites.

The process takes a few minutes only. The field data are kept in the Divisional PC fileserver until night when they are transferred to the Master database for final processing.

Divisional PC fileservers (now six) provide additional capacity and flexibility for the system; even in the worst case - when all communication with the centre is lost- the local staff can continue to work unhindered, nothing will be lost.

The software on notebooks is very user-friendly and specifically tailored for field use. It is very robust and fool-proof. No special training is needed. Note also that all the hardware is standard PC configuration.

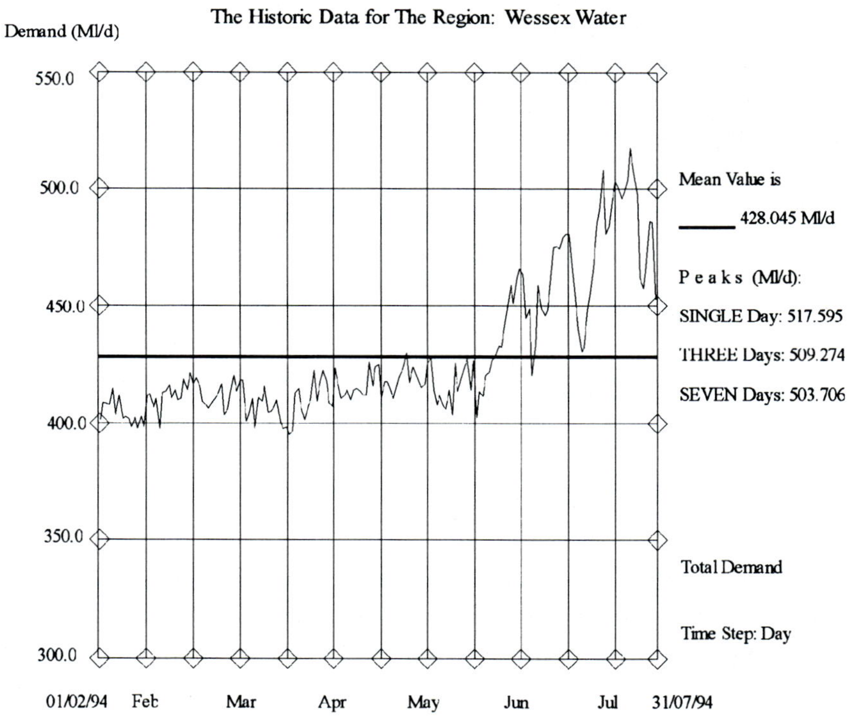

Figure 6. Wessex Water total production

3.3. MAPPING OR GIS (Geographical Information System)

Wessex started to computerise its asset information in the mid-80's and by the time that the supply network was integrated and the telemetry system centralised and rationalised, this process was similarly reaching completion.

One major step forward in the last 2 years has been the opening up of this data to personnel across the company using PC hardware, both on the desktop and with portable equipment for field use (Glasbrook, 1993, Snoxell, 1993).

The GIS department keeps records for all assets of the company, using highly specialised equipment for digital processing.

The changes in the water supply system are regularly updated and this updated information is kept in the GIS database. Once per week the updated files are transferred to the Central File server from where they are disseminated to several local file servers, see Figure 5.

The final users have access to the updated files through their networked personal computers, standalone computers or portable computers.

So far the paper has described the major management information systems that were put in place over the last 5 years to improve the control of a supply network which was becoming increasingly centralised and integrated and involved moving water over increasing distances via an increasingly complex trunk main system.

The following section describes some examples of the information which is now in common use for the operational management of that network.

3.4. ON-LINE ANALYSIS OF OPERATIONAL DATA

3.4.1. The Production of Water

The data captured by telemetry permit a rapid calculation of total production of water - see Figure 6., as well as demand (including losses) in all Supply zones, Areas and Region.

The current production of individual sources could be also seen and compared with historic data and abstraction license (Figure 7.).

This information is usually available for the previous day, but in a crisis it can be updated every 15 minutes giving vital data to the manager.

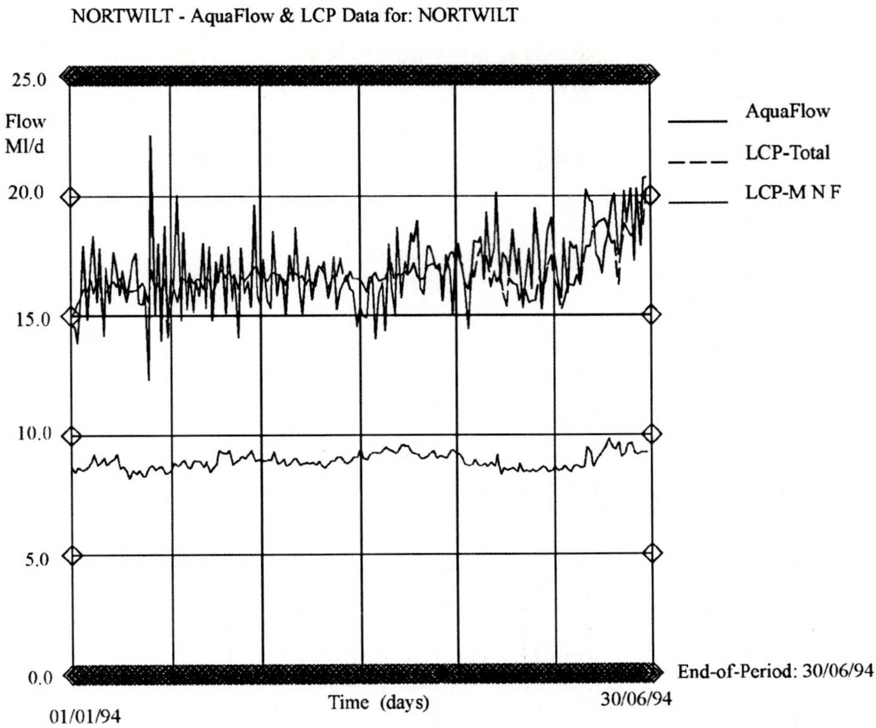

Figure 7. Production and consumption data for North Wiltshire

3.4.2. Monitoring of Leakage Losses

The demand in all Districts is monitored either by telemetry or by data loggers. The introduction of dial-up loggers is planned for the end of 1994. The data permit:

- computation of losses in the conveyance system (primary) as the difference between quantities produced at sources and those delivered to the Districts,
- evaluation of distribution losses within individual Districts taking into consideration population and billing data as well.

The analysis can be performed at each level from the Region down to individual Districts. Figure 7. shows results for a Supply zone (North Wiltshire).

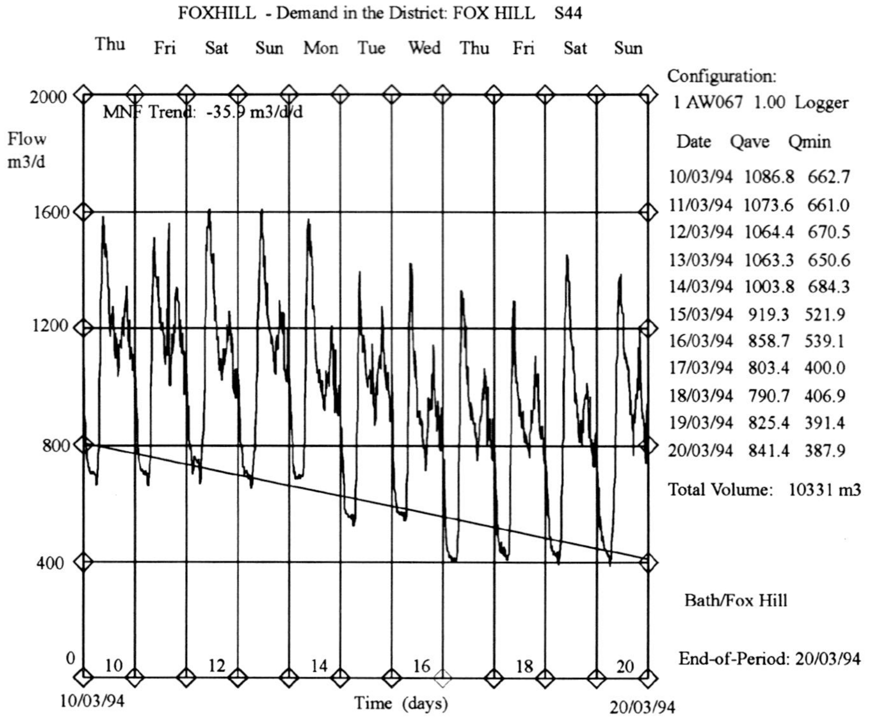

Figure 8. Variations of water demand in a district

Note that Production (full line) and Demand (broken line) data are very closely matched, showing that there are no significant losses in the primary system in that zone. The stronger oscillations of Production data are due to changes in storage (not taken into account).

Note also the Minimum night consumption line for the whole Region - it is quite steady, although a bit too high.

The distribution losses in a District could be estimated by analysing the variation of local demand - see an example in Figure 8. for Fox Hill District (Bath Supply zone, Avon & Wiltshire Area).

Note the sudden decrease in demand after 14 and 17 March 1994, after successful repairs, with obvious savings. However, the leakage losses are still relatively high, as could be proved by examining the data in a longer time period (Figure 9.).

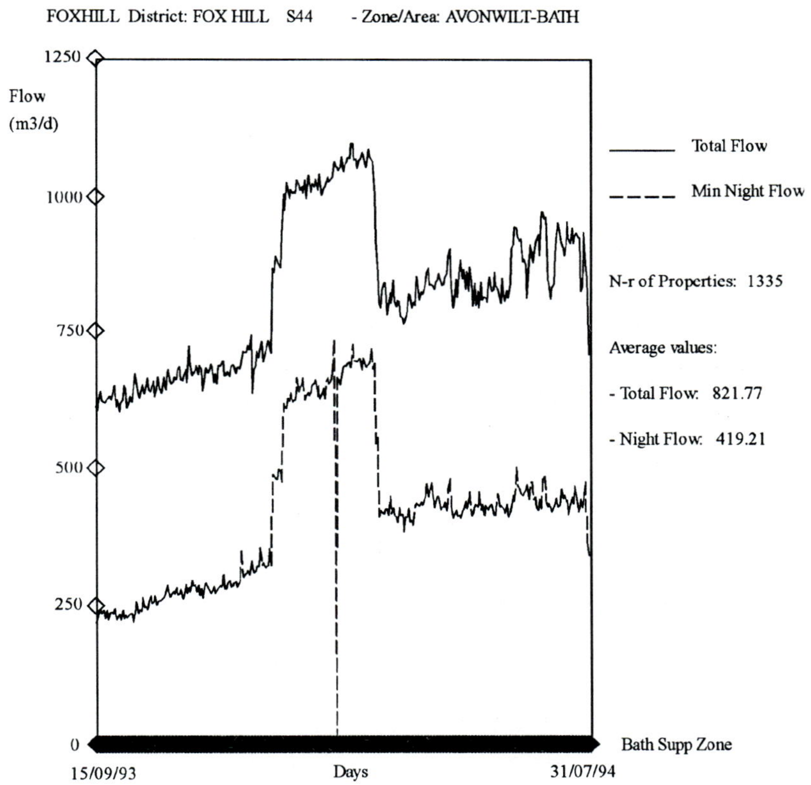

Figure 9. Historic perspective for the same district

The present level of night time demand (about 400 m^3/h) is still higher than in September 1993 (below 250 m^3h) - therefore the work is not finished! The significant difference between actual night line and legitimate night use (computed from water bills and average domestic water use) also points in the same direction.

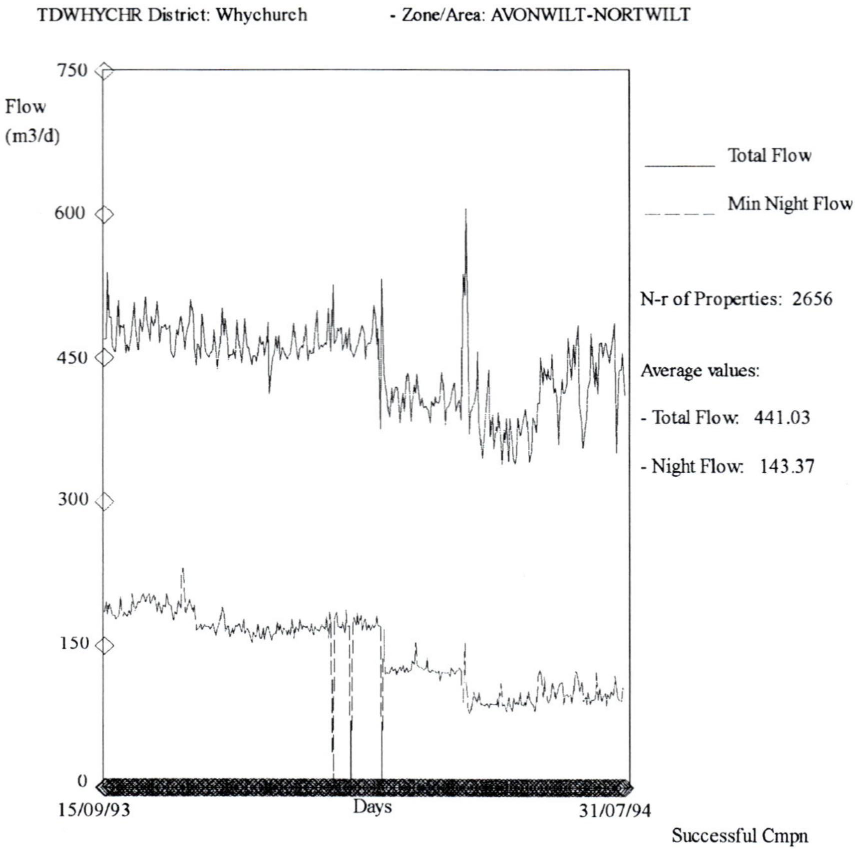

Figure 10. Reducing the leakage levels steadily ...

The campaign against leakage losses can never be completed, but continuous efforts can achieve substantial savings, as shown in Figure 15. for another District (Whychurch). Note the steady decline of night line during these ten months, even when the total demand was increased during a dry and warm period (June-July 1994). The sustained effort in pressure management will also help in keeping the losses within acceptable limits (Lonsdale, 1984, Tustin, 1994).

The analysis is made easier by providing tables for Supply zones, Areas or the region where Districts are listed in order of high leakage losses (by trend or by average).

3.5. DESIGN AND PLANNING

The archived data offer great advantages for all design and long-term planning jobs within the Company, providing a complete and accurate database. A few examples are included here showing:

- Figure 11.: Variations of water demand in Poole area in the last 14 years - note the rising trend, decline during the recession period 1990-93 and recovery in 1994; also the repeatability of seasonal patterns,

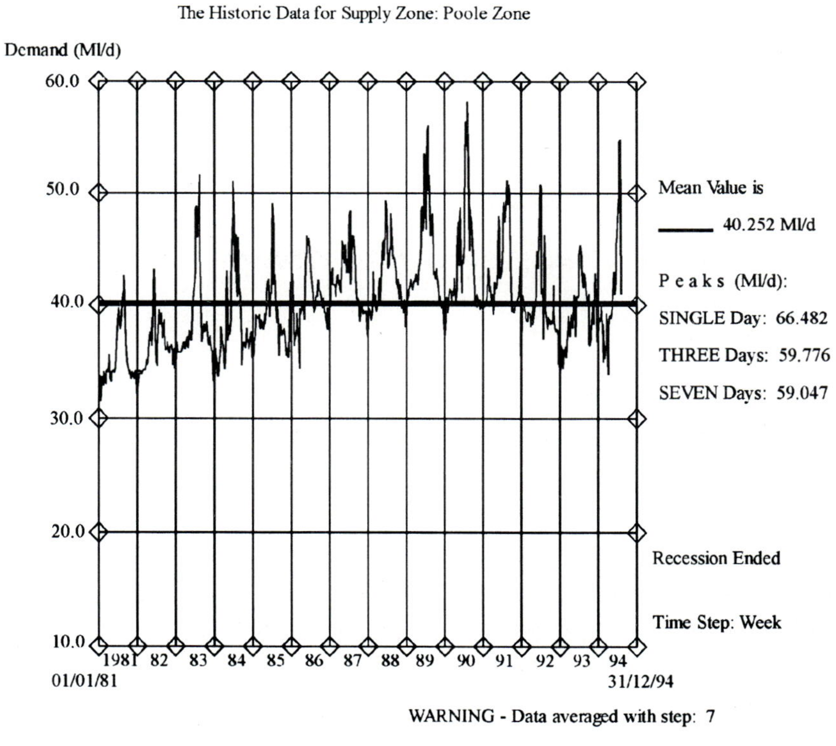

Figure 11. Variations of demand in Poole zone

- Figure 12.: Total demand at the regional level in August since 1981 set against rainfall data - note the relative increase in dry years (1983, 1989, 1990) and decrease in wet years (1985, 1986, 1992),

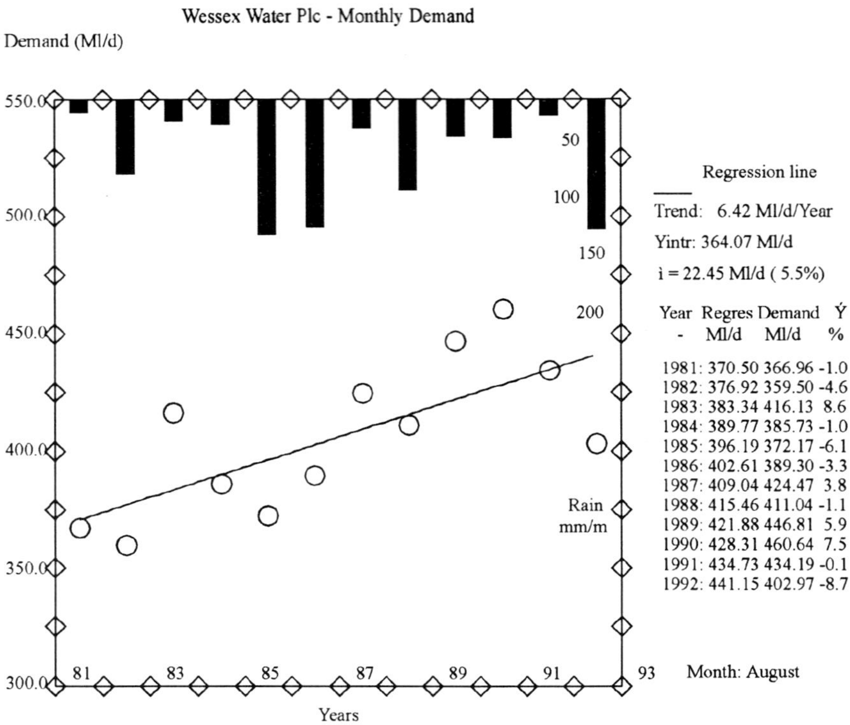

Figure 12. Effects of rainfall on the total demand

- Figure 13.: Source output for Ivyfields pumping station and limits set by Daily & Annual licence,

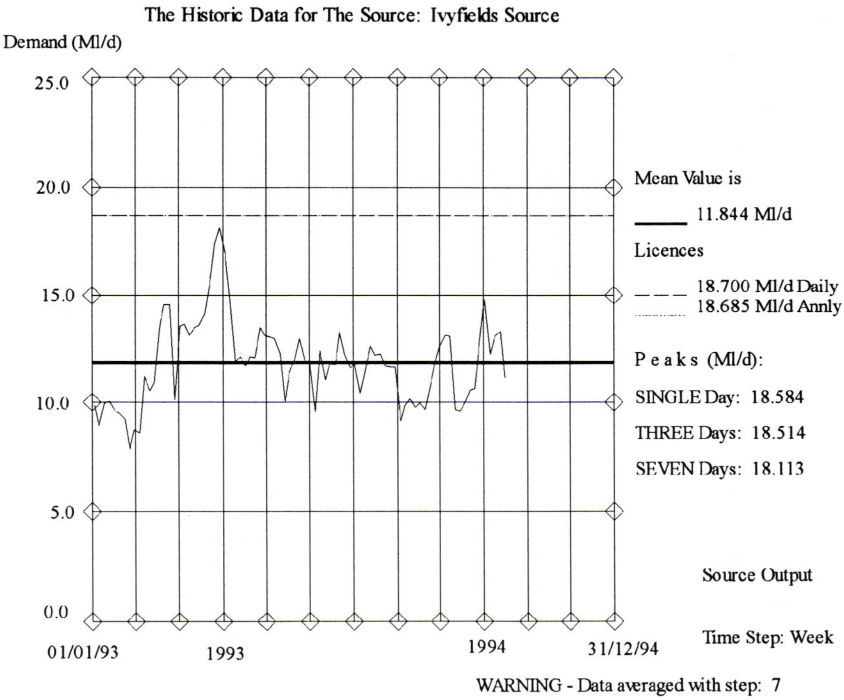

Figure 13. Current output of Ivyfields source

- Figure 14.: Consumption registered by a watermeter belonging to a large hospital in 1992-1994 period (billing data) -the artificial peak in July 1993 should be halved to cover the gap in June 1993 (human error)

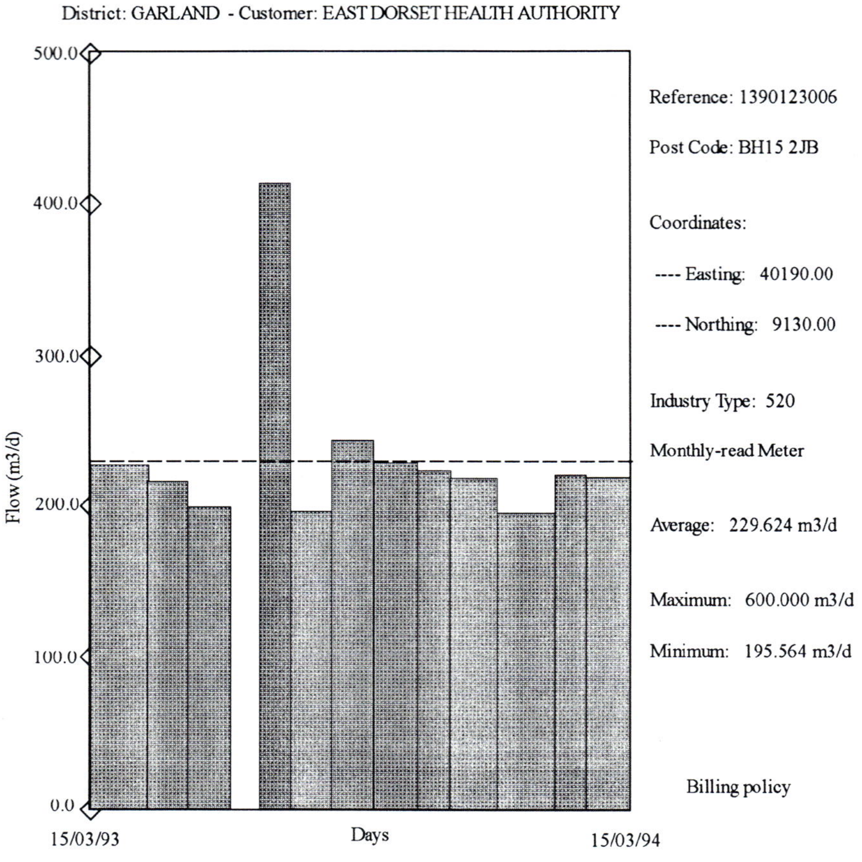

Figure 14. Billing data for a large customer

- Figure 15.: A week's data for the same meter captured by a data logger (temporarily installed) - the data show the real night consumption which is quite low in this case.

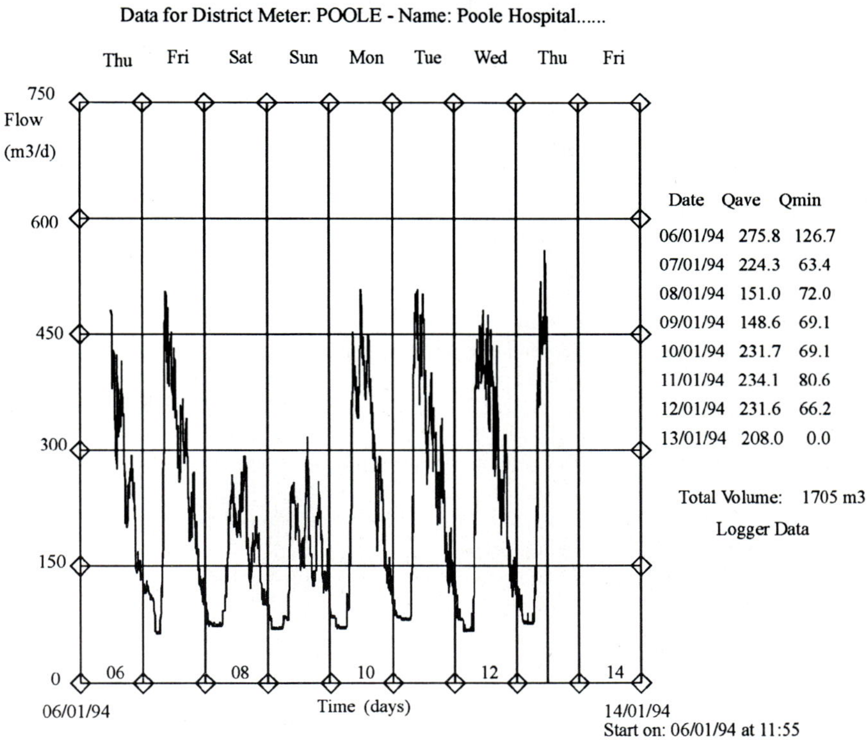

Figure 15. Monitoring the consumption of a large user

3.6. REFERENCES

Glasbrook D. (1993) The Practical Application of GIS, and its use for AMP2. IWO Conference 1993, Cardiff, Great Britain.

Huntington R. (1984) Improvement of Cost and Performance of Distribution Systems. Proceedings of XV Congress of International Water Supply Association, Spec. Subject No.9, Monastir-Tunisia.

Huntington R. (1990) WESSEX WATER's Integrated Water Distribution Management System. Proceedings of the 5th IAWPRC Workshop, Japan.

Huntington R. (1993) Updating and Development of A Regional Telemetry Scheme. 6th IAWQ Workshop on Instrumentation, Control and Automation of Water & Wastewater Treatment and Transport System, Canada.

Lonsdale P.B. (1984) Automatic Control of Pressure in Distribution Systems. Wessex Water Co., Bristol, Great Britain.

Snoxell J.D., Obradović D., Čavor R. (1989) The Use of Real-Time Telemetry and Modelling in Conjunction for the Operational Management Water Supply Systems. Computational Modelling and Experimental Methods in Hydraulics, HYDROCOMP '89, Dubrovnik, Elsevier Applied Science, London 1989, pp. 301-310.

Snoxell J.D. (1991) New Technology in the Water Industry - Making It Stick in the Real World. Proceedings of the Pipeline Industry Guild, England.

Snoxell J.D. (1993) Network Management in Water Supply. International Conference on Computer Applications in Water Supply and Distribution, De Monfort University, Leicester, England, pp. 331-345.

Snoxell J.D. (1993) The Use and Value of GIS in the UK Water Industry; Experience in WESSEX WATER Services Ltd., IWSA Specialised Conference on GIS, Lyons 1992.

Snoxell J.D. (1994) Management of a Widespread Multi-Plant Water Supply System. IWSA Regional Conference, Zurich 1994.

Tustin J.R. (1994) Economic of Leakage in a Water Supply System. Presented to Royal Society of Civil Engineers, Feb 1994, London, England.

4. Control Valves - Principles of Operation

Marko Ivetić
IRTCUW, Institute of Hydraulic Engineering
Faculty of Civil Engineering
University of Belgrade
P.O. Box 895, 11 000 Belgrade
YUGOSLAVIA

4.1. INTRODUCTION

The safety and reliability of operation of many water supply systems depend to a great extent on how well they can be controlled. In the first place, it is a hydraulic problem, and after that of control theory, system analysis, signal processing, etc.

Many hydraulic systems, which do not respond well to control actions, have mismatches in selected equipment (pumps, valves etc.) or operating regimes are in collision with system dynamics. The value of appropriate control equipment should not be underestimated in cost analysis. However, compared to very high capital costs of complex distribution systems, reservoirs, pumping stations, etc. and the required reliability and trouble-free operation of the system, the cost of control equipment is minimal. In selecting control valves for a system, the engineer has to consider valves different from those usually used in distribution systems (gate valves, butterfly valves, standard globe valves, etc.).

There are several tasks which are performed by control valves (Obradović, 1991). Usually, the same basic valves are used for several different purposes, with one or more control devices which are responsible for the desired control action. The basic valve, when combined with appropriate control loops, can cover a wide range of applications (Shinskey, 1988). Several types of control valves can be identified: pressure reducing valves, pressure sustaining valves, pressure relief or back pressure valves, pump control valves, check valves, flow control valves, electric remote control valves, etc.

There are basically two types of action of control valves, permanent and non-permanent. Control valves which are used for flow modulation, with the aim of providing desired flow conditions within the system, are of the permanent type. The other group,

NATO ASI Series, Partnership Sub-Series, 2. Environment – Vol. 15
Water Supply Systems – New Technologies
Edited by Č. Maksimović, F. Calomino, and J. Snoxell

for example check valves, float, pressure relief valves, and especially air valves, with non-permanent operation, are sometimes not considered as control valves.

4.2. HYDRAULICS OF CONTROL VALVES

4.2.1. Flow Control

The principle of flow regulation will be explained with an example of flow between two reservoirs through a relatively long pipeline (Figure 1.). Flow rate can be determined from

$$\Delta\Pi = \left(\Sigma\xi_L + \lambda\frac{L}{D} + \xi_Z \right)\frac{Q^2}{2gA^2} \tag{1}$$

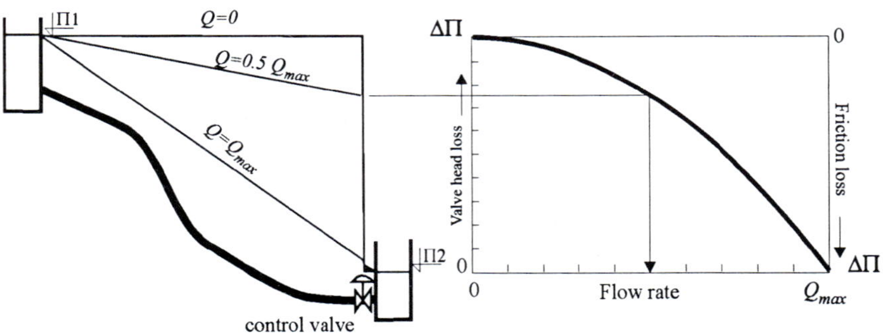

Figure 1. Flow control in a long gravity pipeline

where, ξ_Z is variable head loss coefficient of control valve, and, $\xi_\Phi = \Sigma\xi_L + \lambda\frac{L}{D}$ is used for energy loss coefficients which can not be changed, and which does not depend too much on flow rate. When the pipeline is longer than 1000 D, this part of the energy loss is mainly due to friction. This assumption is correct for fully developed turbulent flow in rough pipes. Equation 1. can be written as

$$1=\left(1+\frac{\xi_Z}{\xi_\Phi}\right)q^2 \tag{2}$$

where, $q = Q/Q_{\max}$, and $Q_{\max}$ is defined from

$$\Delta\Pi = \xi_\Phi \frac{Q_{\max}^2}{2gA^2} \tag{3}$$

Flow rate of the system depends on the relative head loss coefficient of the control valve scaled with ξ_Φ. This simple, but important relation, is given in diagram (Figure 2.). For $\xi_Z/\xi_\Phi = 1$ flow rate is equal to $0.7Q_{\max}$, for $\xi_Z/\xi_\Phi = 3$, flow rate is equal to $0.5Q_{\max}$ etc., while in order to get $0.1Q_{\max}$, head loss at the valve has to be 100 times greater than the friction loss. It should be noted that flow regulation turns out to be intentional dissipation of the energy by control valve. For maximum flow, energy loss is uniformly distributed along the pipeline, while for $q < 0.7$ a greater portion of the available energy, $\Delta\Pi = \Pi_1 - \Pi_2$ has to be lost at the control valve.

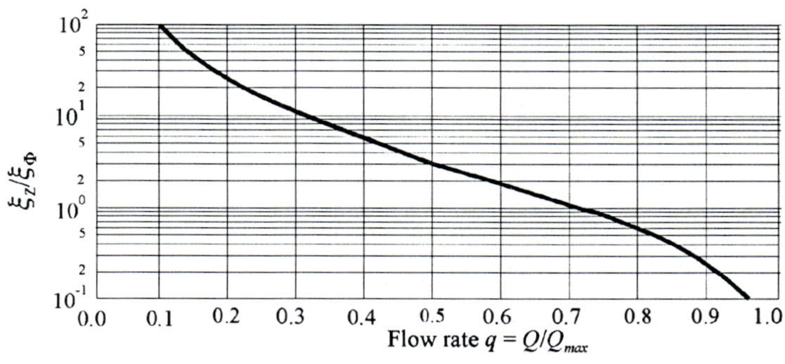

Figure 2. Flow control requirement in a long gravity pipeline

4.2.2. Head Loss Coefficient

The control valve is an obstacle to the flow. It reduces the area of flow, causing a significant increase in local velocities, and local pressure drop. Further downstream the flow area is expanded, water fills that space, and in highly turbulent conditions dissipates most of the locally increased kinetic energy. With the same diameter of pipe upstream and downstream of the valve, as a result of the valve action, there is a head loss , $\Delta\Pi = \Pi_1 - \Pi_2$. The valve is also responsible for a local pressure drop equal to, $\Delta\Pi_{max} = \Pi_1 - \Pi_{min}$ (Figure 3.).

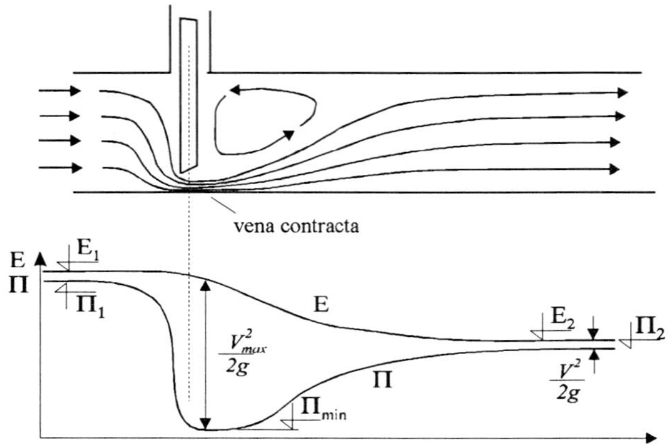

Figure 3. Variation of piezometric head and total energy in a valve

Head loss of the valve depends on the valve type, opening and also on the valve size, actually on the Reynolds number ($Re = VD/\nu$ where V *is* mean velocity, D pipe diameter, ν - kinematic viscosity of water). It is related to flow characteristics, velocity head, $V^2/2g$, via head loss coefficient, ξ_Z

$$\Delta\Pi = \xi_Z \frac{V^2}{2g} \tag{4}$$

4.2.3. Discharge Coefficient

The value of ξ_Z varies from a very small value (for open valve) to infinity (closed valve). That is why it is replaced by a more convenient coefficient of flow, or discharge coefficient, C_D,

$$C_D = \frac{1}{\sqrt{1+\xi_Z}} \tag{5}$$

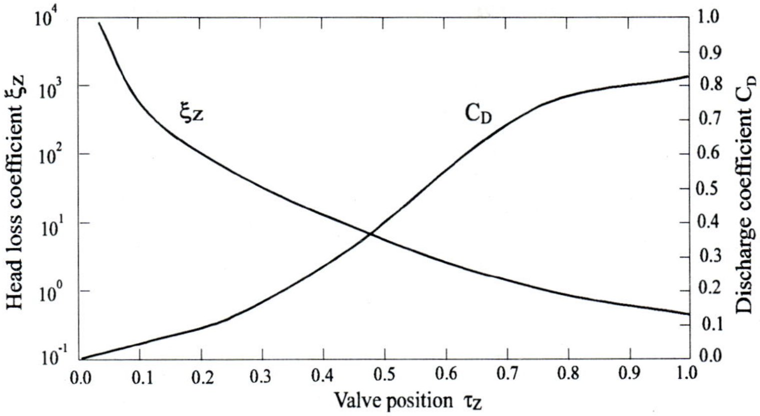

Figure 4. Characteristics of a valve: head loss coefficient and discharge coefficient

A diagram with relationships of these coefficients to the valve opening, called *valve characteristics*, is given as an example in Figure 4. The valve opening, τ_Z, is given in relative terms, 0.0, valve closed, 1.0, valve completely open. Some manufacturers prefer to use the dimensional presentation of valve characteristics (KUBOTA-NEYRTEC, 1986, etc), like *specific flow*

$$q_{11} = \frac{Q}{D^2\sqrt{\Delta\Pi}} \tag{6}$$

or, in terms of head loss coefficient, ξ_Z

$$q_{11} = \sqrt{\frac{\pi^2 g}{8\xi_Z}} \tag{7}$$

It is defined as the flow passing through a one meter diameter valve at a head loss equal to one meter of pressure head.

4.2.4. Cavitation

The ratio of $\Delta\Pi/\Delta\Pi_{min}$ called *the coefficient of recuperation* of pressure, is always smaller than 1. Those valves which have higher coefficients of recuperation are better in respect of cavitation danger. Cavitation might cause serious problems in the operation of control valves. The troubles start when local pressures drop to the value of vapour pressure at ambient temperature (p_v, around 2-3 kPa). Then, small bubbles of water vapour (cavities) are formed. Further downstream pressure is recovered, causing cavities to implode. If this is happening in the free stream, nothing but high level noise is generated, but if it is happening near the pipe wall, small jets of high velocities may damage the surface.

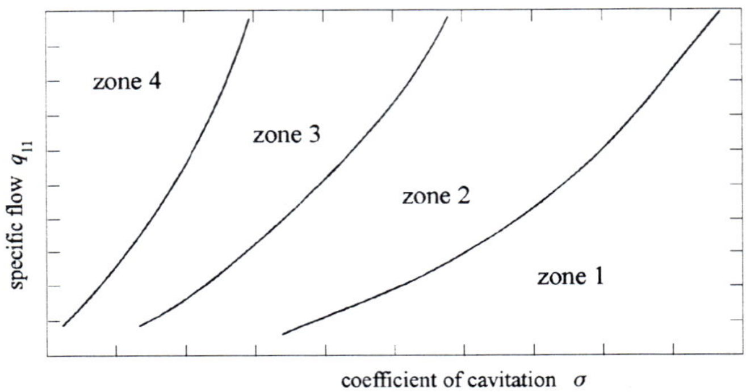

Figure 5. Cavitation diagram for a control valve
zone 1: no cavitation; zone 2: incipient cavitation, risk of damage small;
zone 3: noticeable cavitation, damage of valve after several hours/days of operation;
zone 4: operation is not possible.

Obviously, cavitation starts at the place where minimum pressure occurs (Π_{min} in Figure 3.). There are two ways to avoid cavitation: to choose valves with relatively low difference between Π_{min} and Π_2, and to keep minimum pressure higher than vapour pressure.

Cavitation characteristics of a valve have to be determined by experimental investigation. Results of these measurements are usually given via *coefficient of cavitation,* as a function of the specific flow, in diagrams like in Figure 5. The coefficient of cavitation is defined as

$$\sigma = \frac{P_1 - p_v}{P_1 - P_2} \tag{8}$$

Absolute pressure P_1 is measured one pipe diameter upstream of the valve, absolute pressure P_2 is measured 10 pipe diameters downstream of the valve and corrected for friction losses between sections 1 and 2. The available cavitation number has to be calculated for all operating regimes of the valve, by Expression (8). It is then compared with the required cavitation number of the selected control valve (Figure 5.). Depending on the intensity of cavitation, there are several zones where calculated cavitation number can fall.

4.3. DESIGN CONSIDERATION

There are several types of valves which are used for control purposes. They are characterized by considerable energy loss, accurate positioning, durability and high reliability, ease of manipulation, cavitation free operation etc. Examples of valves with good control characteristics are given in Figures (7.), (8.) and (9.). The valve in Figure 7. consists of valve body and two perforated plates, one of which is fixed and the other is moving. The valve has excellent cavitation characteristics, simple design, full range sensitivity etc.

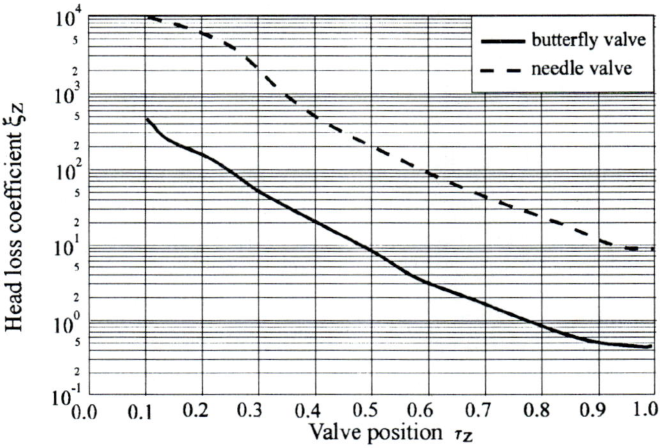

Figure 6. Head loss coefficients for butterfly and needle valve (ERHARD)

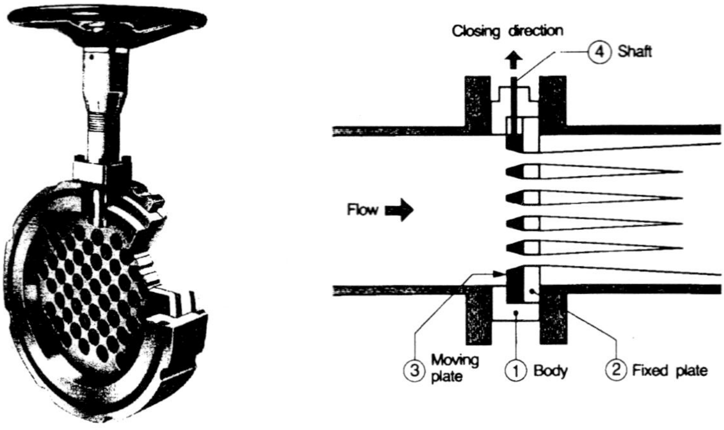

Figure 7. Scheme of a control valve with favorable control characteristics (KUBOTA-NEYRTEC)

In Figures 8. and 9. are shown two different valves, based on the same principle. The needle valve in Figure 8. has a fixed annular streamlined internal body with control piston which can be moved in the flow direction. Depending on the duty required, there are two models, with a vaned ring in order to reduce energy loss, and with a slotted cylinder, in order to increase energy loss and to improve control characteristics in systems with high friction losses. "Y" pattern control valve (BERMAD) in Figure 9., can be equipped with V-port throttling plug. In that way the accuracy, stability and reliability have been improved, and cavitation noise and risk reduced, especially for high pressure reduction and/or low flow cases.

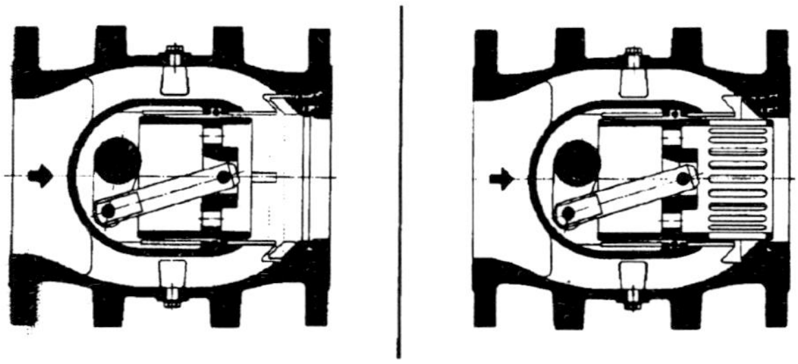

Figure 8. Two variants of the same valve a) for low head losses, vaned, b) for high head losses, slotted cylinder (ERHARD)

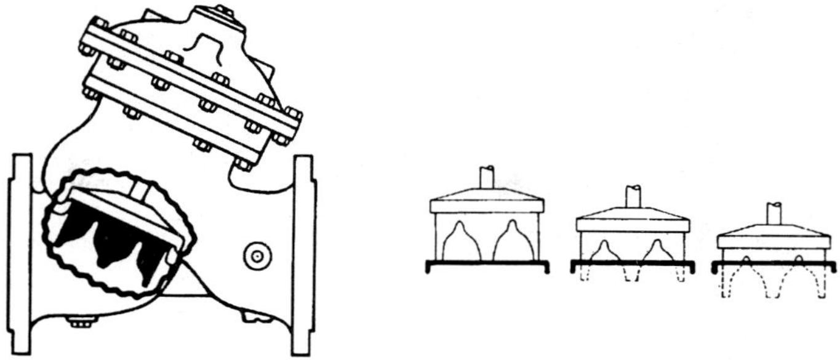

Figure 9. A way to increase head loss in the control valve for high pressure difference services (BERMAD)

In Figure 6. the characteristics of two valves, butterfly and needle (slotted cylinder type, Figure 8.b.) are shown. It can be seen that the butterfly valve can not be used as a control valve in systems with considerable friction losses. In a hypothetical pipeline 20 km long, with diameter equal 0.8 m, with estimated pipe wall roughness 1 mm, at available difference, $\Pi_1 - \Pi_2 = 70m$, the performances of the two valves from Figure 6. has been examined. Control characteristics, flow rate vs. valve position, are given in Figure 10. The movement of butterfly valve from $\tau = 1.0$, completely open position, to $\tau = 0.3$, has not been felt by the system (change of flow less than 5 %). Usage with opening less than 10 % is not recommended, so the valve can control flow between 75 and 100 % of maximum flow, $Q_{\max}$. On the other hand, the needle valve equipped with slotted cylinder to promote energy dissipation, can be used for smooth control over a wide range between 20 and 100 % of maximum flow.

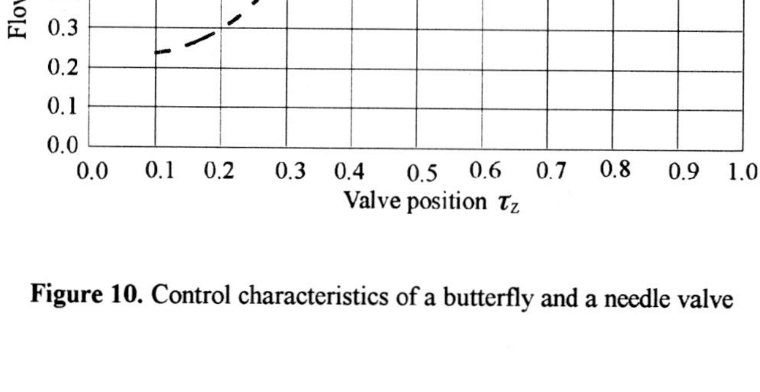

Figure 10. Control characteristics of a butterfly and a needle valve

In order to manipulate a valve in steady and transient operating regimes, one has to overcome significant forces (hydrodynamic, friction etc.) acting on the control and isolating elements of a valve. It can be done either manually, using a gearbox to increase the torque (as for the valve in Figure 7.), or by specially designed devices called *actuators*. There are several types of actuators: electric, pneumatic and hydraulic. Sometimes, actuators are equipped with manual devices, which can override the action of the actuator and which can be used in case of emergency or actuator malfunction.

The principle of operation of a control valve with hydraulic actuator is given in Figure 11. With a help of a small three-way pilot control valve, hydrostatic forces on the diaphragm and washer are used to move the stem and change the valve opening. Pressure difference on the diaphragm can be supplied from some external system (pneumatic - for pneumatic actuators, oil), or it can be provided internally from the valve body and working fluid. Apart from the one-chambered diaphragm unit shown in Figure 11., there are two-chambered units, where the diaphragm is either completely separated from the working fluid, or communication between the inner chamber is only inhibited (Figure 8.) in order to achieve damping effects during closing.

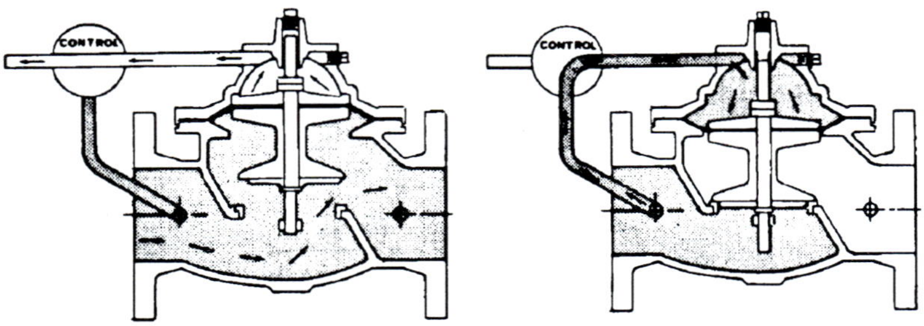

Figure 11. Operation of a diaphragm actuated control valve equipped with a three-way control

4.4. EXAMPLES OF APPLICATION

The list of possible applications of control valves is very long. Usually the same basic valve, with small modifications, can be used for different purposes, depending on the control loop and additional control devices implemented. The most typical applications are explained in the next subsections, using data and recommendations of valve manufacturers (KUBOTA-NEYRTEC, BERMAD, MUESCO, ERHARD).

4.4.1. Water Level Control

Two simple cases of water level control in a reservoir are given in Figure 12. They look similar but they are not.

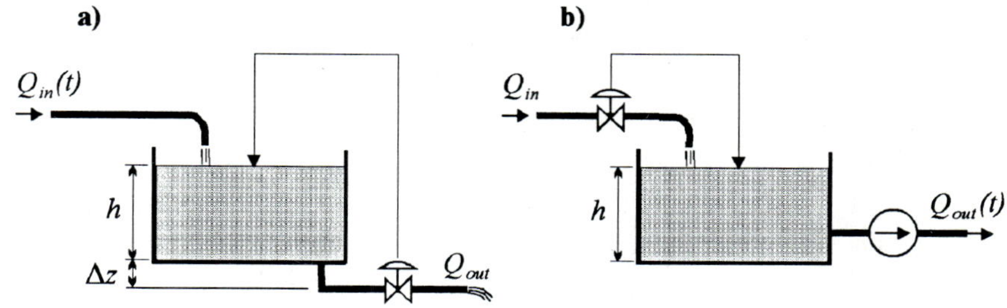

Figure 12. Control valves used for water level control

Following the simplest rules of control (control action is initiated when there is an intolerable difference between measured value and set point), control valve in a) case will have *increase-increase* action, i.e., increase of flow, Q_{in} will result in increase of water level and, as a countermeasure, an increase of valve opening will follow. In the second case, uncertainties come from the downstream side, from Q_{out}. Increase in Q_{out} will cause a *decrease* of level in the tank and it will require an increase of valve opening and vice versa (*increase-decrease action*). There is one more difference between the two cases. In the second case every discrepancy, no matter how small, between Q_{in} and Q_{out} will cause emptying or over spilling of the tank. This process can not be left unattended and uncontrolled. The first process, however, possesses a property of auto-regulation. Small differences between Q_{in} and Q_{out} will be compensated by small changes of water level. (There is a relationship between Q_{out} and h, i.e., $Q_{out} \propto \sqrt{h + \Delta z}$.)

The first group of processes is easier for control, but efficient control is an imperative for the second group of processes.

4.4.2. Control Valves in Pumping Stations

A standard application of control valves in pumping stations is given in Figure 13. When used with centrifugal pumps, the check valve (CV) has to be closed during pump start-up and to open gradually to prevent motor overloading. A relief valve (RV) is used as a protective device against water hammer and to prevent pressure lock-in. It should open immediately after the system pressure surpasses a set point, and closes slowly in a programmed manner.

However, these two valves can have complementary roles of a permanent type. They can be used as flow control valves. Valve (CV), with additional control devices, can reduce the flow by increasing the energy loss (*flow throttling*). Valve (RV) can do the same job by diverting some amount of water to the upstream tank. Energy consumption is higher, but the pump is operating closer to the rated point, with higher efficiency.

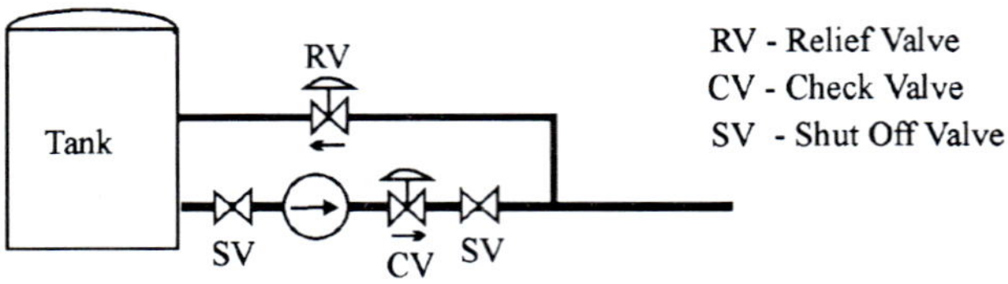

Figure 13. Control valves in a standard scheme pumping station

A typical arrangement of valves with a deep well pump is given in Figure 14. Relief and check valves have the same roles as in standard pumping stations. An additional valve (PCV) is typical for deep well pumps, because the vertical pump column is only partially filled with water. The pump is started and stopped with pump control valve (PCV) opened. The first rush of air and water is discharged to atmosphere, preventing water from being forced too rapidly into the pipeline.

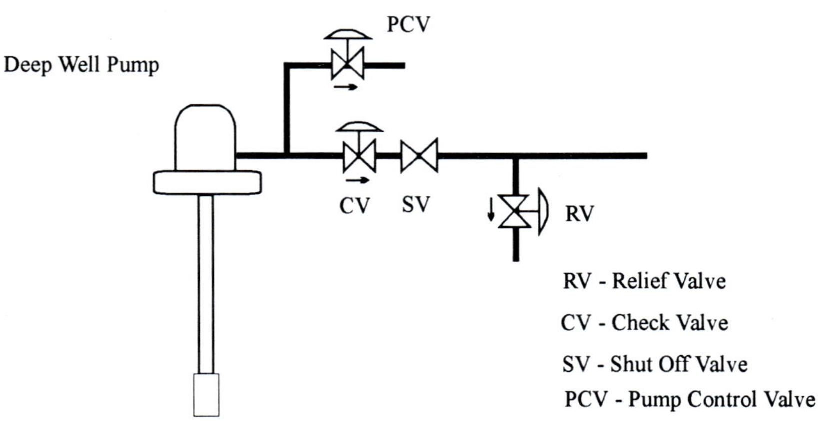

Figure 14. Control valves with deep well pumps

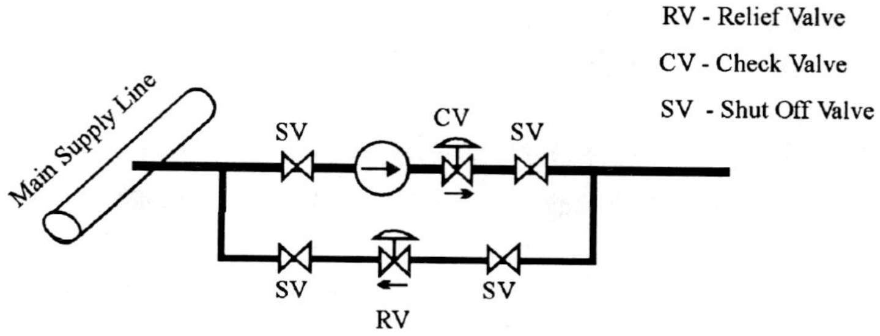

Figure 15. Control valves in a booster pumping station

An arrangement of control valves, very similar to the standard one, is used with booster pumping stations (Figure 15.).

4.4.3. Pressure Reducing and Sustaining Valves

One of the most frequent services of a control valve is to keep pressure at a downstream location (called *the control node*). If the pressure at the control node is above the set level, the control valve will close, if it is below, the control valve will open.

The control node can be a point immediately downstream of the control valve, or at a distant place in a network. In the former case it can operate locally (controlled by local control loop), while in the latter case, the pressure at control node (CN) has to converted into digital form and transmitted to the pressure reducing valve (PRV).

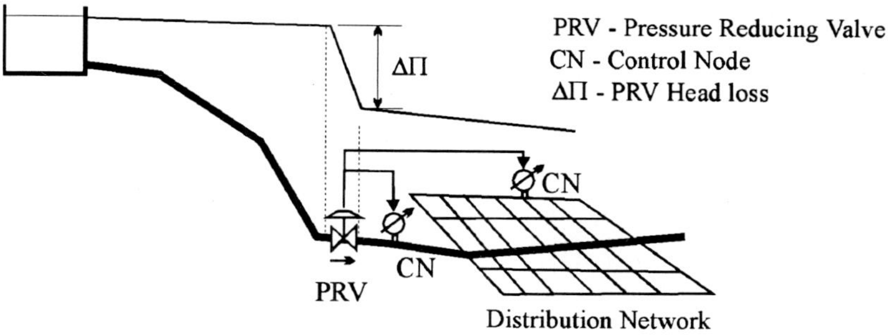

Figure 16. Pressure reducing control valve

A pressure reducing valve can operate as a control valve only if the upstream pressure is above the set pressure. If it is below, the PRV cannot add the energy to water. Also, if the PRV is completely closed, it can not influence the pressure at the control node; the system pressure is controlled by some other elements of the system (pumps, control devices etc.).

A control valve can be used to maintain a set pressure at a location upstream of it (Figure 17.). Comparing to the PRV, this is the opposite role. With this service they are called the pressure sustaining valves. This type of control has similar limitations to a PRV. It can not raise pressure at the control node after it is completely closed, and it cannot lower it more after it is completely opened.

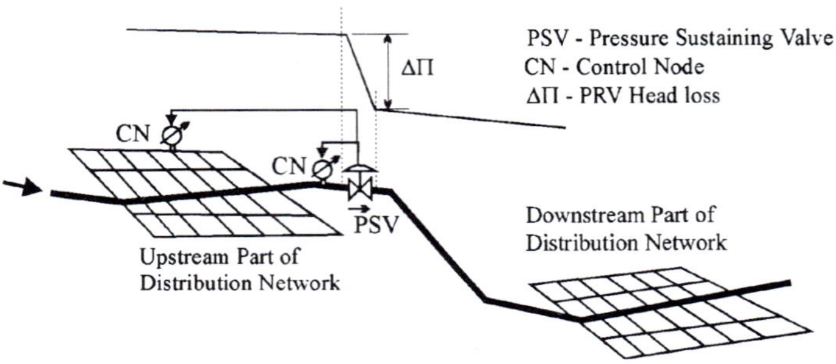

Figure 17. Pressure sustaining valve

Pressure reducing and pressure sustaining valves can fulfill their tasks only if their operating range extends over the greater part of valve stem travel. If the control valve characteristics are like those of the butterfly valve in Figure 10., the overall performance of the control valve will be poor.

4.5. CONCLUDING REMARKS

Control valves in large water supply systems are very important for safe and reliable operation. Design criteria for control valves differ from criteria for other valves used in water supply systems. For example, considerable energy loss is required for flow control in long gravity pipelines and so on.

The most important characteristics of control valves are durability, high reliability, cavitation free operation, ease of manipulation, accurate positioning, full range sensitivity, etc.

The implementation of control valves in overall system control is much easier if their hydraulic characteristics are satisfactory. Otherwise their operation can be an additional source of trouble in the system.

4.6. REFERENCES

BERMAD Control Valves (1994) Catalogue. Evron, Israel.

Coulbeck B., Orr C.H. (Ed.) (1988) Computer Applications in Water Supply, Volume 2 - Systems Optimization and Control. Research Studies Press Ltd. and John Wiley & Sons Inc.

ERHARD Armaturen (1987) Catalogue. Johannes Erhard, H. Waldenmaier Erben, Germany.

KUBOTA-NEYRTEC Valves (1986) Catalogue. Kubota Ltd.Osaka, Japan.

MUESCO Inc. (1986) Regulations and Control of Liquids. Catalogue. Houston, Texas.

Obradović D. (1991) Control Valves and Booster Pumping Stations. Lecture Notes, IHE Delft.

Shinskey F.G. (1988) Process Control Systems, Application, Design and Tuning. Third Edition, McGraw-Hill Publishing Company.

Chapter III

New Technologies for Water Treatment

Chapter III

New Technologies for Water Treatment

INTRODUCTORY STATEMENT

New technologies in drinking water supply are needed basically for three reasons:

1. To obtain excellent drinking water quality for public health,
2. There is a growing concern for nature and conservation of the environment,
3. For a major part of the world population, costs for sufficient and safe drinking water are a dominating factor.

During the last few decades new insights have emerged between, on the one hand, quality of food in general and of drinking water in particular, and public health on the other.

Consenquently an increasing number of drinking water quality parameters are covered by legislation, for example for lead, aluminium, chlorinated hydrocarbons, pesticides, solvents, etc. In addition the limits are becoming more stringent, and more compounds can be detected at lower concentrations.

New philosophies are being developed concerning microbiological, toxicological (mutagenity, carcinogenity, teratogenity), organoleptic and operational quality parameters, and also for new treatment technologies.

However, there is also a requirement that these new treatment technologies do not themselves introduce undesirable compounds into the water.

In general the pollution of nature and the environment is still continuing. Raw water sources, nature and the environment are being polluted more and more.

Hence, more and more compounds have to be removed. On the other hand, from the point of view of the general responsibility for the natural environment, drinking water companies have to treat and produce drinking water in such a manner that no new polluting compounds are accumulated in the environment during production, distribution and usage (liquid and solid waste, lead, copper, soap, detergent, energy, etc.).

Safe drinking water is a first priority for human life, while a major part of the world population has very limited finances to pay the costs of a safe drinking water supply.

Therefore new technologies have to be developed in such a way that theses technologies have costs during production and usage of drinking water which are as low as possible.

The lowest costs during usage mean for example, that no bottled water is consumed and minimum costs for soap, detergents, energy, maintenance etc. are realised.

These financial purposes must be consistent with the environmental goals.

1. New Technologies for Water Treatment: General Aspects

J.P. van der Hoek
A. Graveland
Amsterdam Water Supply
Vogelenzangseweg 21
2114 BA Vogelenzang
THE NETHERLANDS

1.1. INTRODUCTION

Raw water for drinking water production can originate from ground water, surface water, or river bank infiltrated water. The potable water produced from these raw water sources has to meet certain standards, described in legislation and in drinking water rules. These standards can be divided into four groups: (1) microbiological standards, (2) toxicological standards, (3) aesthetic/organoleptic standards, and (4) operational standards. Pollutants in the raw water can be divided into six groups. For every group, relevant treatment processes are available.

Nowadays, the quality of ground water and surface water is deteriorating and new opinions and views are being developed concerning optimal treatment of water. Ground water is threatened by acidification and overfertilization, disposal of waste and agricultural practice (pesticides). New issues in surface water treatment are the awareness of the formation of harmful disinfection by-products during chlorination, the presence of pesticides and organic micropollutants in the raw water, and the tendency to tighten the limits for drinking water and environmental standards. All these aspects require the modification of existing treatment plants and the incorporation of new processes. In surface water treatment, the focus is on the removal of organic matter, such as nutrients for biological growth, pesticides and other organic toxic micropollutants, and the production of hygienically safe and biologically stable drinking water without the use of chlorine. For the removal of organic matter such as nutrients, pesticides and other organic micropollutants, three approaches are discussed: (1) the removal of these compounds by activated carbon filtration or membrane filtration, (2) the chemical oxidation of these compounds by advanced oxidation processes, and (3) the biodegradation of these compounds by biological activated carbon filtration.

NATO ASI Series, Partnership Sub-Series, 2. Environment – Vol. 15
Water Supply Systems – New Technologies
Edited by Č. Maksimović, F. Calomino, and J. Snoxell

To produce hygienically safe and biologically stable drinking water, slow sand filtration or activated carbon filtration in combination with UV-disinfection are promising techniques. For the removal of organic matter in general, and pesticides and other organic micropollutants in particular, three concepts are presented for drinking water treatment in the future: (1) chemical treatment, (2) physical treatment and (3) biological/chemical treatment.

1.2. SOURCES OF RAW WATER: GROUND WATER, SURFACE WATER, RIVER BANK INFILTRATED WATER

For production of drinking water, in principle three sources are available: surface water from rivers or lakes, ground water and water after river bank infiltration. In the latter case, water can be extracted from wells located near the surface water source. Traditionally many drinking water companies preferred ground water above surface water as a source for potable water production. Arguments for this preference are the relatively high quality of the ground water, the low treatment costs and the ease of operation of the relatively simple ground water treatment plants. The most important general characteristics of ground water and surface water as were valid until recently are summarized in Table 1. and in Table 2.

Table 1. Characteristics of ground water

Advantages:	Disadvantages:
- hygienically safe	- often limited availability
- constant temperature	
- constant quality	
- not vulnerable to accidents	

Table 2. Characteristics of surface water

Advantages:	Disadvantages:
- large quantities available most of the time	- disinfection necessary
	- fluctuating temperature
	- biologically and chemically not stable
	- fluctuating quality
	- vulnerable to accidents

By bank infiltration, the advantages of ground water can be achieved. Fluctuating concentrations and temperatures are balanced, and thus the water becomes less sensitive to accidents when compared with surface water, although pollution is spread easily through the catchment. During bank infiltration also the quality of the water is improved: suspended solids are removed from the water, and many contaminants are biologically degraded. The biological and chemical stability of the water is increased.

Due to environmental problems like overfertilization, resulting in increasing nitrate and phosphate concentrations, and the use of pesticides in agriculture, the quality of ground water is deteriorating and the preference for ground water is becoming less pronounced from the point of view of quality.

1.3. STANDARDS AND GUIDELINES FOR DRINKING WATER QUALITY

The raw water originating from surface water or ground water mostly contains pollutants in such a concentration that it is not immediately suitable as drinking water. Therefore, it has to be treated in order to obtain drinking water. However, in legislation and in drinking water rules, the concept of "sound drinking water" is not well defined. On the other hand, the World Health Organization and the European Community describe guidelines, guide levels and maximum admissible concentrations that should be met by the drinking water. These standards can be divided into four groups:

1. microbiological standards;
2. toxicological standards;
3. aesthetic/organoleptic standards;
4. operational standards.

According to the WHO the most important standards concern the microbiological standards. The first target of drinking water treatment has to be the production of water that meets the microbiological standards. As soon as these standards are met the other standards have to be aimed at. Thus the order of standards mentioned above can be considered as a functional sequence.

Microbiological standards and toxicological standards are health related. Many microbiological parameters are used as an indicator for the presence of pathogenic organisms in the water. Toxicological standards are calculated based on risk-assessment models. For a long-life exposure, the WHO allows a risk-factor of 10^{-5}. In the calculation of the standards, an ADI (Acceptable Daily Intake) or TDI (Tolerable Daily Intake) is used. Drinking water only contributes partly to this ADI or TDI, also food and air contribute to the total exposure.

Examples of toxicological standards are amongst others nitrate, nitrite, (heavy) metals and many organic compounds.

Aesthetic/organoleptic standards are related to the fact that the water should have a high aesthetic value, and use of drinking water must be attractive and enjoyable. This means that the water should be bright and free of colour, odour and taste.

Operational standards do not concern health related aspects, but are guidelines to obtain a good technical operation of the treatment plant and the distribution system, and to guarantee a good water quality at the consumers' tap. An example of an operational standard is the aggressivity of the water. Aggressivity may affect the distribution system. Indirectly these standards are also related to microbiological, toxicological and aesthetic aspects.

A complete overview of the standards cannot be given here, but an indication of the way in which the standards affect the normal constituents of the water, and so affect the treatment scheme necessary to produce drinking water from surface water or ground water, can be summarized as follows:

- the presence of phosphorus, nitrogen and carbon containing compounds interfere with the bacteriological standards and toxicological standards. P, N and C may act as nutrients for microbial growth, while carbon containing compounds also include pesticides and other organic micro pollutants which are toxic,

- calcium and hydrogencarbonate determine the hardness of water. Hardness is related to the copper and lead uptake of the water in the distribution system, and thus the toxicological aspects of the water,

- chloride and sulphate result in corrosion of the distribution system, especially when cast iron pipes are used. This may result in increased turbidity and colour of the water, leading to a drop in aesthetic quality.

1.4. CATEGORIES OF POLLUTANTS AND UNIT OPERATIONS

In the guide lines of the World Health Organization (WHO, 1993) and in the European Council Directive relating to the quality of water intended for human consumption (European Community, 1980) a large list of impurities is described that are not allowed to exceed a certain guideline (GL) or maximum admissible concentration level (MAC). Several processes are available to treat the raw water in such a way, that the GL-values or MAC-values can be met. In principle, these contaminants can be divided into six groups, and for each group the most suitable treatment processes can be chosen. Table 3. summarizes the contaminants and relevant treatment processes.

Table 3. Groups of pollutants with suitable treatment processes

Pollutants		Unit operations
1	Suspended solids: clay, algae, bacteria, viruses, microfauna	Coagulation, rapid filtration
2	Heavy metals: lead, copper, cadmium, chromium, mercury, etc.	Coagulation, rapid filtration, softening, hyperfiltration
3	Salts: Cations: Na^+, K^+, NH_4^+, Ca^{2+}, Mg^{2+}, Fe^{3+}, Al^{3+}, etc.	Rapid filtration, softening, hyperfiltration
	Anions: Cl^-, NO_3^-, HCO_3^-, SO_4^{2-}, PO_4^{3-}	Coagulation, softening, hyperfiltration
4	Organic compounds: - nutrients - toxicological compounds - smell, colour, taste - anti-corrosion/scaling	(Biological) activated carbon filtration, hyperfiltration
5	(Micro)organisms: bacteria, bacteriophages, viruses, cysts, etc.	Coagulation, rapid filtration, ozonation, hyperfiltration
6	Radio-active compounds	Coagulation, rapid filtration, (biological) activated carbon filtration, hyperfiltration

The first group concerns suspended solids like clay particles, algae, bacteria, viruses and microfauna. In practice, suspended solids are removed by a conventional treatment scheme including coagulation and flocculation using iron or aluminium salts, followed by sedimentation or flotation to allow much of the suspended flocs and other matter to settle out of the water, and rapid filtration to remove most of the remaining suspended matter from the water. With these treatment processes, suspended solids can be removed from the water to below 5 g/m^3 (Boller, 1994). Removal of bacteria and viruses is limited with these processes. In the Surface Water Treatment Rules of the United States, a log reduction of 2 is given for viruses and a log reduction of 2.5 for *Giardia* cysts by this conventional treatment process.

The second group of pollutants concerns the heavy metals like lead, copper, cadmium, chromium, mercury, nickel etc. By the traditional treatment scheme as described above part of these pollutants can be removed from the water by adsorption on and occlusion in the flocs. However, precipitation of these heavy metals and adsorption/absorption on flocs is rather pH-sensitive, and often rather extreme pH-values are necessary to achieve a good removal efficiency. Another possibility is hyperfiltration (reverse osmosis) by which in theory all ions can be removed from the water, and thus also these heavy metals.

The third group of pollutants concerns the dissolved salts. These can be divided into the possitively charged cations (Na^+, K^+, NH_4^+, Ca^{2+}, Mg^{2+}, Fe^{2+}/Fe^{3+}, Al^{3+}, etc.) and the negatively charged anions (Cl^-, NO_3^-, HCO_3^-, SO_4^{2-}, PO_4^{3-}). Removal of hardness (Ca^{2+} and Mg^{2+}) is possible by softening processes by means of precipitation or crystallization of $CaCO_3$ (Graveland et al., 1983). Removal of ammonia, iron and manganese can be achieved by rapid filtration, as is normally practised in ground water treatment. For removal of sodium and potassium, hyperfiltration is a good alternative. The anions may be removed by coagulation (PO_4^{3-}), softening processes (HCO_3^-), or membrane processes (Cl^-, SO_4^{2-}, NO_3^-). For nitrate removal, specific denitrification processes may be considered.

The fourth group concerns the organic compounds. The total number of different organic compounds probably amounts to a few thousands. Depending on their nature and amount in the raw water they can have the following functions:

- as nutrients to stimulate (micro)biological growth and cause an accumulation of detritus (DOC, Dissolved Organic Carbon; AOC, Assimilable Organic Carbon) in the distribution system,

- to cause a toxicity such as mutagenity, carcinogenity or teratogenity: THM (trihalomethanes), AOX (adsorbable halogenated compounds), pesticides, solvents etc.,

- to cause smell, colour and taste,

- to act as a means of anti-corrosion and/or anti-scaling.

Processes to remove organic compounds from water include:

- physical processes such as adsorption on adsorbants (activated carbon),
- chemical processes such as oxidation by means of ozone, hydrogen peroxide or UV (Advanced Oxidation Processes),
- membrane processes such as hyperfiltration and nanofiltration,
- biological processes like biological activated carbon filtration. In this process a small dose of ozone is used prior to the activated carbon filtration to increase the AOC-content and stimulate biodegradation of the organic matter during activated carbon filtration.

The fifth group comprises micro-organisms like bacteria, viruses and cysts (e.g. *Giardia*). These parameters are all related to the hygienic quality of the water and health aspects. With the conventional treatment scheme of coagulation and rapid filtration a log reduction of 2-2.5 can be achieved. With chemical disinfection (e.g. chlorine, chlorine dioxide, chloramine, ozone, UV radiation) inactivation depends on the concentration used and the contact time (CT-value), the temperature and the pH. Recent experiments have shown that disinfection can also be achieved by membrane processes like hyperfiltration, nanofiltration and ultrafiltration. Log-reductions of at least 5-7 have been reported (Boller, 1994, Anselme et al., 1993).

The last group concerns radio active compounds. These compounds may be removed by either coagulation and rapid filtration, (biological) activated carbon filtration or hyperfiltration.

1.5. NEW ISSUES IN GROUND WATER TREATMENT AND SURFACE WATER TREATMENT

For over a century, both ground water and surface water have been used for drinking water purposes, usually ground water where available, and surface water always had to be treated to achieve drinking water quality.

Traditionally, key issues in ground water treatment were iron, manganese and ammonia removal, achieved by aeration and filtration often with pH correction and hardness reduction. An example of a conventional ground water treatment scheme is shown in Figure 1.

Nowadays, where the quality of ground water is threatened, extension of the treatment scheme is required. The most important threats are:

- acidification and overfertilization: this results in an increase in nitrate, phosphate, sulphate and hardness in the water, and an increase in the concentration of aluminium, nickel, cadmium and other heavy metals,

- disposal of waste: this may lead to pollution of ground water by volatile organic compounds,

- agriculture: the use of pesticides may result in deterioration of the ground water quality.

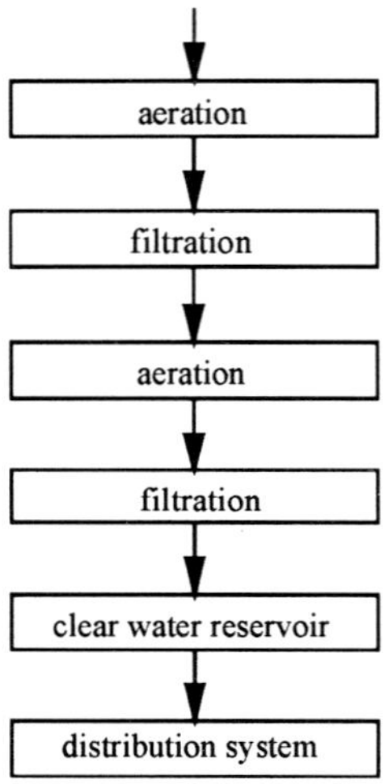

Figure 1. Conventional ground water treatment based on dual aeration and filtration

Most of these contaminants require additional treatment steps, or modification of existing treatment schemes. For example, nitrate requires a nitrate removal process, while volatile organic compounds may require air-stripping. For pesticides, additional processes like (biological) activated carbon filtration, advanced oxidation or membrane filtration may be needed.

The main issues in surface water treatment so far have been:

- clarity of the water, as most raw river water is high in turbidity,

- an acceptable taste and odour, which is not the case for surface water that is highly polluted with domestic, agricultural and industrial waste,
- hygienic reliability,
- storage of water in open storage reservoirs or in the subsoil, to cover periods of insufficient quality or quantity and to enable selective intake of water.

An example of a conventional treatment scheme for surface water is shown in Figure 2.

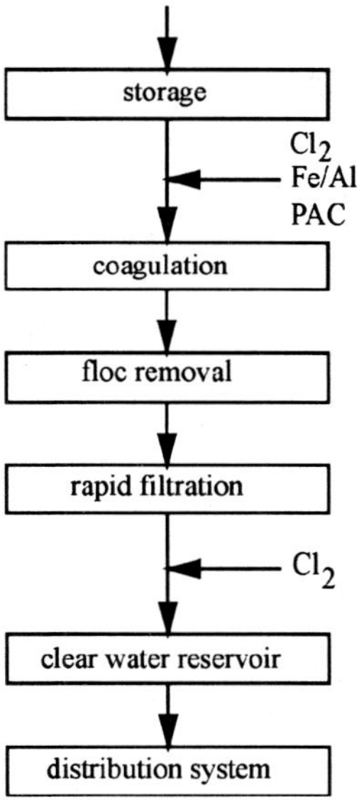

Figure 2. Conventional surface water treatment based on coagulation, rapid filtration and multiple chlorination (1960-1970)

Turbidity was reduced by coagulation using iron or aluminium salts, followed by sedimentation or flotation and rapid filtration. Taste and odour problems could be solved through the addition of powdered activated carbon.

Hygienic reliability was achieved by means of disinfection with chlorine and post chlorination guaranteed the required bacteriological quality. Selective intake and quality improvement of the raw water were achieved by storage reservoirs at the beginning of the treatment process. Dune water supply companies in the Netherlands started infiltration of pre-treated river water in the subsoil to enlarge the storage and capacity.

Also slow sand filtration has been used in this process as a final treatment step to reduce or to avoid the use of chlorine.

Nowadays this conventional surface water treatment scheme is changing as a result of new threats and changing opinions concerning robust, safe and reliable treatment schemes:

- harmful disinfection by-products are produced during the reaction of chlorine with humic acids (Rook, 1974). Both chlorinated compounds and brominated compounds might be produced. Although chloramine is used as an alternative, most attention in Europe is given to slow sand filtration, artificial recharge, UV-disinfection, ozonation and chlorine dioxide as disinfection barriers,

- in order to avoid the use of a persistent disinfectant like chlorine, it is necessary to produce a biologically stable drinking water. This can only be achieved when the content of organic compounds (DOC), N-compounds and P-compounds which all act as nutrients and result in regrowth in the distribution system, is very low. Biological stability of the water is related to the AOC concentration: the concentration of assimilable organic carbon. When the AOC concentration is below 10 µg/l, no regrowth in the distribution system will occur (Van der Kooij et al., 1990) and no persistent disinfectant has to be dosed. To reach biological stability of the water, appropriate treatment steps have to be used (dual filtration, slow sand filtration) and good management of the treatment plant is required. Besides, a reliable distribution system is required in which no recontamination takes place,

- pesticides and organic micropollutants have been detected in surface water. They originate from discharges into rivers by industries, and from surface run-off from agricultural land. In Table 4., the presence of pesticides in the river Rhine and the river Meuse is summarized.

Table 4. Presence of pesticides in river Rhine and river Meuse

	Number of pesticides			
	River Meuse		River Rhine	
µg/l	till 1990	1991	till 1990	1991
> 1	0	0	12	2
> 0.1	6	9	43	11

Table 5. Current and future standards for drinking water according to the EC and the Netherlands Waterworks Association

Parameter	Current standard	New standard
Na	120 mg/l	100 mg/l
Total hardness	1.5 mmol/l	1.0 mmol/l
pH	8.3	8.6
Pb	50 µg/l	10 µg/l
Cu	3 mg/l	1.5 mg/l
DOC	5 mg/l	3 mg/l
Colour	20 mg Pt/l	10 mg Pt/l
AOX		20 µg/l
Pesticides	0.1 µg/l	0.1 µg/l
Chloropesticides	0.1 µg/l	0.01 µg/l
THM	70 µg/l	30 µg/l
AOC		10 µg/l
Modified fouling index		$3\ s/l^2$
Aeromonads		50/ml leaving plant

Additional treatment steps such as (biological) activated carbon filtration, advanced oxidation or membrane filtration is necessary to remove these substances from the raw water:

- in addition to the presence of pesticides and other organic micropollutants as mentioned above, it must be realised that many more pesticides and micropollutants may be present in the water. For example, of a total of 320 different pesticides applied in the Netherlands, 225 are significant in drinking water supply, based on use, production, their presence in surrounding countries, mobility and persistence in the environment. Analytical methods are only available for 94 of these pesticides (Hopman et al., 1989). Polar pesticides and their metabolites in particular cause serious analytical problems. This makes it very likely that a large number of pesticides occur in surface water and bank-infiltrated water that are not expected to be removed by customary treatment processes, including activated carbon. It is not yet clear whether advanced oxidation processes, ozonation and membrane filtration will be a sufficiently effective method for these components. As analytical techniques are being developed and make it possible to detect their presence in water, it must be realised that the standards that the treatment processes will have to meet are getting tighter,

- there is a tendency to tighten the limits for drinking water and environmental standards. This will affect parameters that relate to organic matter (DOC, AOX, pesticides, other micropollutants), and to conditioning (softening) of the water. Moreover, it will also result in standards for the quality of recharge water, for the use of chemicals and for the production of sludge.

Table 5. summarizes some quality standards derived from current and future EC drinking water regulation and from the Netherlands Waterworks Association.

1.6. NEW TECHNOLOGIES IN SURFACE WATER TREATMENT IN RELATION TO THE NEW ISSUES

Looking at the new issues in surface water treatment mentioned in the preceding paragraph, new technological developments will have to focus on the removal of organic compounds from the water (DOC, N-compounds, P-compounds, and especially pesticides and organic micropollutants) and on the production of hygienically safe and biologically stable drinking water without the use of chlorine.

1.6.1. Removal of Organic Matter as Nutrients, Pesticides and Other Organic Micropollutants

With respect to the removal of organic matter (nutrients), pesticides and other organic micropollutants, in principle three approaches are possible:

- in the first approach the main objective is to **remove** organic matter, pesticides and other organic micropollutants from the water phase by activated carbon filtration or by hyper/nanofiltration. Those compounds that can not be removed from the water by these processes for economic or technical reasons have to be oxidised (see second approach),

- in the second approach the main objective is to **oxidise** (chemical oxidation) the organic matter, pesticides and other organic micropollutants with an Advanced Oxidation Process (AOP). AOC that is produced in the AOP is removed by one or more filtration processes,

- in the third approach organic matter, pesticides and other organic micropollutants are **biodegraded**. By means of a low ozone dose the biodegradation capacity is increased and organic matter, pesticides and other organic micropollutants are partially oxidised to yield more biodegradable products. In fact AOC is created by ozonation and in the subsequent activated carbon filtration the biodegradation capacity is increased. Hence, the organic matter can be biodegraded or adsorbed, depending on the nature of the material. Also the more persistent organic compounds may be biodegraded, as a result of the high biological activity in the carbon filter.

The main characteristics of the three approaches will be discussed below.

Activated carbon filtration has been applied in the past for removal of smell, taste and odour, trihalomethanes and also pesticides. It can also be used for the removal of organic matter in general and micropollutants in particular. However, the extent of removal largely depends on the polarity of the compounds. Highly nonpolar compounds (e.g. lindane) adsorb strongly on carbon, but polar compounds like trichloro-acetic acid and dikegulac have a low adsorption capacity, resulting in short running times of the carbon filters, a high reactivation frequency and high reactivation costs. Other factors that determine the adsorption capacity for pesticides and other organic micropollutants are:

- the presence of natural organic matter (humic substances) in the water: due to a relatively high concentration, they are adsorbed preferentially and reduce the adsorption capacity for pesticides and organic micropollutants,

- kinetic factors like surface loading, empty bed contact time, etc.

As membrane filtration processes, hyperfiltration or nanofiltration can be used. In these processes, water is driven by pressure through a semi-permeable membrane that (partially) rejects dissolved compounds. In this way, the feed water is divided into a permeate (product water) and a concentrate (brine). The principle of this process is shown in Figure 3.

Hyperfiltration membranes have already been used a long time for desalination of brackish and sea water and retain almost 100% of monovalent ions. Nanofiltration membranes retain approximately 50% of these compounds. Both types of membranes retain more than 90% of nearly all other components (bivalent ions, micro organisms, DOC). This is also the case for pesticides and organic micropollutants with a molecular weight of 300 or more. No final conclusion can yet be drawn about organic substances with a lower molecular weight than 300, as research into this subject is being carried out at the moment. Besides, removal of pesticides and organic micropollutants is strongly affected by the size and structure of the compounds, and the characteristics of the membrane material. In addition it must be mentioned that hyper- and nanofiltration need an extensive pre-treatment to avoid fouling of the membranes, and post-treatment to adjust the pH and hardness after membrane filtration. Also the concentrated waste stream, the brine, has to be disposed of.

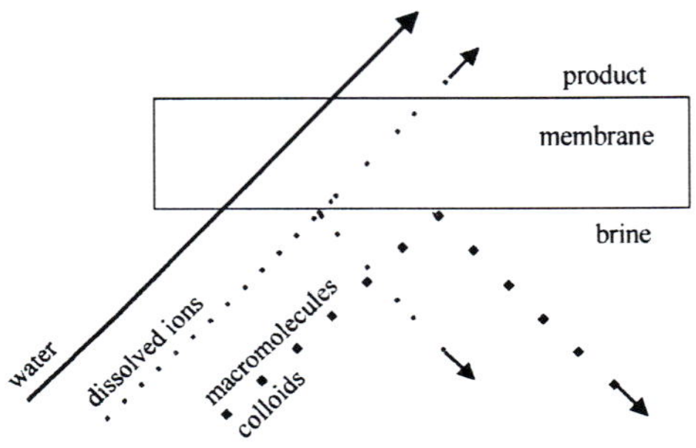

Figure 3. The principle of hyperfiltration and nanofiltration

From the characteristics of activated carbon filtration and membrane filtration it is clear that these processes are not an absolute barrier against pesticides and organic micropollutants. In particular, polar compounds (in activated carbon filtration) and low molecular weight compounds (in membrane filtration) are only partially removed. So, additional processes might be necessary to form a complete barrier.

Chemical oxidation can be achieved by ozonation or advanced oxidation. Some of these processes are summarized in Table 6. The effect of AOP is based on the formation of hydroxyl radicals, an extremely reactive and unstable form of the hydroxide ion lacking one electron and thus uncharged. Hydroxyl radicals might also be formed as a result of the reaction between O_3 and H_2O.

Ozone itself reacts selectively and incompletely with organic material. Reaction rate with aromatics and unsaturated chains is very high, while ozone does not react with saturated aliphatics and chlorinated compounds under practical circumstances. The AOPs are much more reactive. Reaction rates for all organic pollutants are very high, so in theory it is possible to convert all these compounds to CO_2 and H_2O, but then a high dose and long contact time are necessary (40-80 mg O_3/l, 20-400 mg H_2O_2/l, t = 60-300 min). In practice, the objective is selective oxidation of individual contaminants at doses and contact times normally applied in drinking water production (2-8 mg O_3/l, 1-3 mg H_2O_2/l, t = 15 min) (Kruithof et al., 1994). The required dose and contact time depend on the nature of the compound, the oxidation degree required, the amount and nature of other organic material present (DOC) and the amount and nature of inorganic material present (HCO_3^-, CO_3^{2-}).

Table 6. Advanced oxidation opocesses and hydroxyl radical formation

O_3	$3\ O_3 + H_2O$	$\Rightarrow$	$2\ OH^- + 4\ O_2$
O_3/H_2O_2	$2\ O_3 + H_2O_2$	$\Rightarrow$	$2\ OH^- + 3\ O_2$
O_3/UV	$O_3 + H_2O$ + UV	$\Rightarrow$	$2\ OH^- + O_2$
UV/H_2O_2	H_2O_2 + UV	$\Rightarrow$	$2\ OH^-$

An important drawback of ozonation with high ozone dose and AOP is the formation of by-products like brominated compounds and bromate. Only by applying a high H_2O_2/O_3 ratio or a low pH can this be restricted. Removal of the pre-cursor bromide prior to ozonation or advanced oxidation is an alternative. Also AOC is produced, which has to be removed by subsequent biological filtration steps. It can be concluded that ozonation and AOP can be used as barriers against pesticides and organic micropollutants. However, the applicability is restricted as a result of the formation of by-products like brominated compounds, especially bromate. AOP and ozonation have to be followed by filtration for AOC removal.

An alternative for AOP and ozonation is biological activated carbon filtration (BACF) (Graveland, 1994). In this process a small part of the organic material, pesticides and other organic micropollutants are converted to AOC by a low dose of ozone (1-2 mg/l), causing production of 100-150 µg AOC/l and a higher biodegradation capacity in the subsequent carbon filtration due to a higher bacteria density, resulting from the increased AOC concentration. During the activated carbon filtration both biodegradation and adsorption take place. So, a combination of biological oxidation and adsorption is achieved, resulting in a prolongation of running times of the carbon filters with a factor of 10. An additional advantage of ozonation in the BACF process is the disinfection of bacteria, viruses and cysts. Due to the lower ozone dosage as compared with the chemical oxidation in the advanced oxidation processes, the risk of bromate formation is much lower.

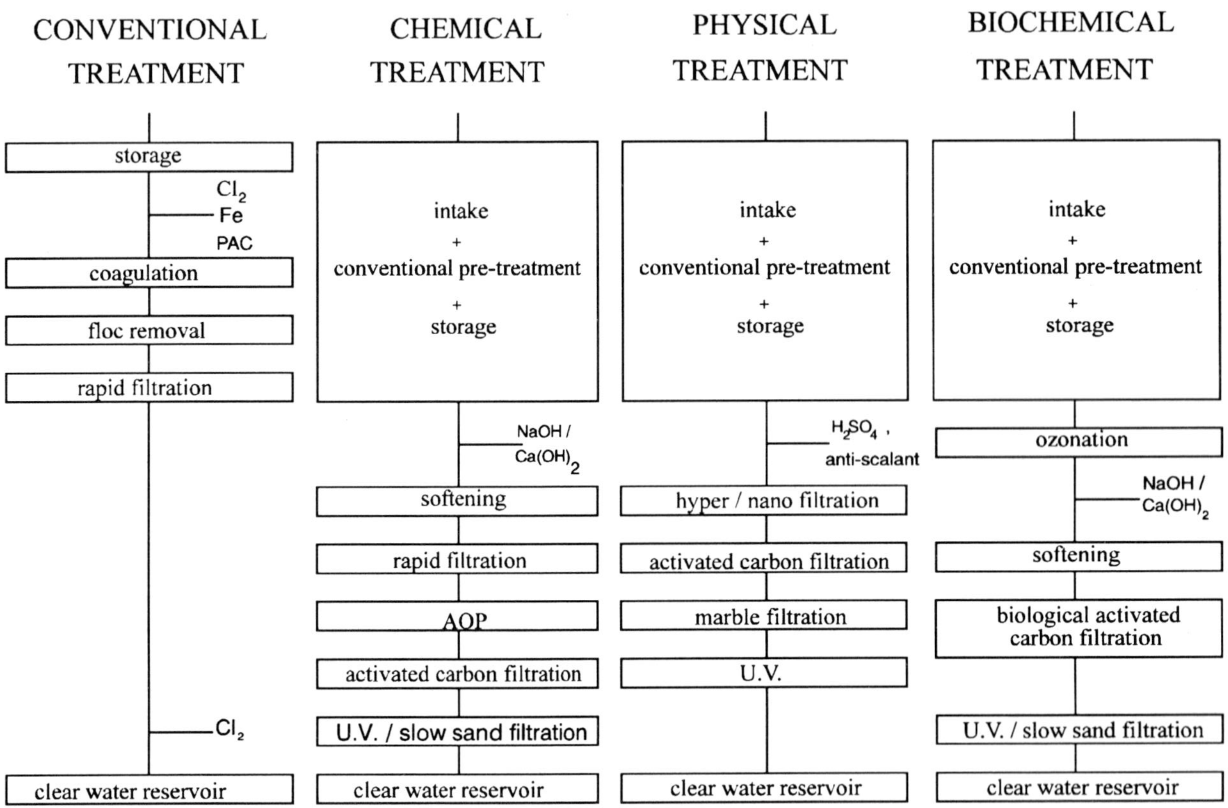

Figure 4. New process schemes in surface water treatment

1.6.2. Production of Hygienically Safe and Biologically Stable Water

With respect to the production of hygienically safe and biologically stable drinking water, the use of chlorine as disinfectant is avoided if possible.

As the main disinfection barrier, the application of slow sand filtration, artificial recharge, UV-radiation, ozone and chlorine dioxide is considered. Also the disinfection capacity of membrane filtration is currently being studied (Boller, 1994, Anselme et al., 1993).

To abandon the use of chlorine as postdisinfectant, the production of a biologically stable water is the objective. When the AOC concentration of the water can be reduced to 10 µg/l or lower, no regrowth will occur in the distribution system and thus no postdisinfection is necessary. As final treatment steps, slow sand filtration or activated carbon filtration can be used. With slow sand filtration, it is possible to produce both biologically stable and hygienically safe drinking water. With activated carbon filtration, it is possible to produce biologically stable drinking water. To achieve hygienically safe drinking water (colony counts below the admissible concentrations) it might be necessary to use UV for final disinfection.

1.7. EVALUATION

As a result of pollution of raw water sources and more stringent limits for drinking water, new processes have to be applied, both in ground water treatment and in surface water treatment.

Focusing on surface water treatment, the traditional process scheme will have to be modified to meet the new standards and to remove contaminants like organic matter (nutrients), pesticides and other organic micropollutants. In fact three approaches can be used: the chemical approach, the physical approach and the biological/chemical approach. Figure 4. shows these approaches, together with the conventional treatment scheme for surface water.

In all three options pre-treatment is similar to the conventional treatment of surface water: coagulation, rapid sand filtration and storage. Then the new processes are incorporated for softening, removal of organic matter, pesticides and other organic micropollutants, disinfection and biological stability.

In **chemical treatment,** softening is achieved by pellet softening or precipitation, and if necessary followed by direct filtration for removal of carry-over. The AOP (O_3/H_2O_2) results in both oxidation and disinfection. Activated carbon filtration removes AOC from the water, any remaining nonpolar pesticides and organic micropollutants. Postdisinfection to produce hygienically safe drinking water is achieved by UV-radiation or slow sand filtration.

The **biological/chemical approach** resembles the chemical approach, but in contrast to chemical treatment a relatively low ozone dose is applied to restrict formation of by-products. After ozonation, the water is first softened by pellet softening, and then treated by biological activated carbon filtration. Contact times of 30-40 min are needed to stimulate and maximize biodegradation in this carbon filter, and to reach the lowest costs per cubic meter and to meet the legal standards. Compounds which are less biodegradable will be removed by adsorption. As with chemical treatment, biological stability and hygienic safety is achieved by slow sand filtration and/or UV-disinfection.

In the **physical treatment process**, hyper- or nanofiltration is used for removal of pesticides and other organic micropollutants, softening and disinfection (retention of bacteria, viruses and cysts by the membranes). To avoid precipitation of salts on the membranes, H_2SO_4 and if necessary an anti-scalant is dosed. After membrane filtration the hardness has to be adjusted, for example by marble filtration or by mixing with water, conventionally treated. Carbon filtration acts as an additional barrier for low molecular weight compounds that might pass the membranes. Post-disinfection is achieved by UV-radiation.

Although these schemes concern direct treatment of surface water, they are also suitable for indirect treatment. In these cases, the water is stored in the subsoil following treatment. Reasons for subsoil storage are balancing and maintaining a strategic supply, and improvement in the bacteriological quality of the water.

It is not possible to give a general preference for one of the treatment schemes. In practice, existing surface water treatment plants have to be upgraded. So, the choice for one of these schemes depends strongly on the set-up of the existing plant (can process-units, already present, be incorporated in the new treatment scheme?), the composition of the raw water to be treated (what are the most important contaminants, and what is their concentration?), and environmental aspects such as the possibility to dispose of the brine in cases of membrane filtration all have to be considered.

Because all three schemes have advantages and disadvantages, in practice also combinations of the schemes are being investigated and constructed. In the next section examples of such combinations as applied by the Amsterdam Water Supply will be discussed in detail.

1.8. REFERENCES

Anselme C., Mandra V., Baudin I., Mallevialle J. (1993) Optimum Use of Membrane Processes in Drinking Water Treatment. Proceedings of the 19th International Water Supply Congress, Budapest, 2-8 October 1993, Special Subject 2 "Membrane Technology": SS2.1-SS2.11.

Boller M. (1994) Trends in Water Filtration Technology. Aqua 40: 65-75.

European Community (1980) Council Directive of 15 July 1980 Relating to the Quality of Water Intended for Human Consumption. Official Journal of the European Community L229 23: 11-29.

Graveland A. (1994) Application of Biological Activated Carbon Filtration at Amsterdam Water Supply. Paper Presented at the IWSA Specialised Conference on Activated Carbon in Drinking Water Treatment, Amsterdam.

Graveland A., Dijk J.C. van, Moel P.J. de, Oomen J.H.C.M. (1983) Developments in Water Softening by Means of Pellet Reactors. Journal American Water Works Association December 1983: 619-625.

Hopman R., Beek C.G.E.M. van, Janssen H.M.J., Puijker L.M. (1989) Pesticides and Drinking Water Supply in the Netherlands. KIWA Communication 113 (in Dutch).

Kooij D. van der, Hijnen W.A.M. (1990) Criteria for Defining the Biological Stability of Drinking Water as Determined With AOC Measurements. Proceedings of the AWWA Water Quality Technology Conference, San Diego, California, 11-15 November 1990.

Kruithof J.C., Schippers J.C., Dijk J.C. van. (1994) Drinking-Water Production From Surface Water in the 1990s. Aqua 43: 47-57.

Rook J.J. (1974) Formation of Haloforms During Chlorination of Natural Water. Water Treatment Examination 23: 234-245.

World Health Organisation (1993) Guidelines for Drinking Water Quality.

2. New Technologies for Water Treatment: Case Studies

J.P. van der Hoek
A. Graveland
Amsterdam Water Supply
Vogelenzangseweg 21
2114 BA Vogelenzang
THE NETHERLANDS

2.1. INTRODUCTION

Amsterdam Water Supply operates two production plants: the River-Dune Waterworks (RDWW, 70.10^6 m^3/year) and the River-Lake Waterworks (RLWW, 31.10^6 m^3/year).

Both are for the main part surface water treatment plants. The basic process scheme for both plants consists of pre-treatment, storage for quality improvement, levelling, selective intake and strategic supply (RDWW: dune infiltration; RLWW: self purification basin), and post-treatment.

The historical development of the RDWW and the RLWW is discussed over the last 20 years, and the need for introduction of new processes in the treatment schemes is explained.

New processes concern softening by means of pellet reactors, removal of organic matter (nutrients for biological growth, pesticides and other organic micropollutants) by biological activated carbon filtration, and hyperfiltration for desalination, an disinfection as an additional final safety barrier for the removal of pesticides and other organic micropollutants. Optimal incorporation of these processes in the existing treatment schemes is discussed.

Finally, the new processes are discussed in more detail, with emphasis on the principles of the process, the merits and limitations of the process, and some technical and operational aspects of the process.

As it is said the Amsterdam Water Supply operates two water production plants: the River-Dune Waterworks (RDWW) and the River-Lake Waterworks (RLWW). The fundamental process schemes of both waterworks are shown in Figure 1. and are more or less comparable.

NATO ASI Series, Partnership Sub-Series, 2. Environment – Vol. 15
Water Supply Systems – New Technologies
Edited by Č. Maksimović, F. Calomino, and J. Snoxell

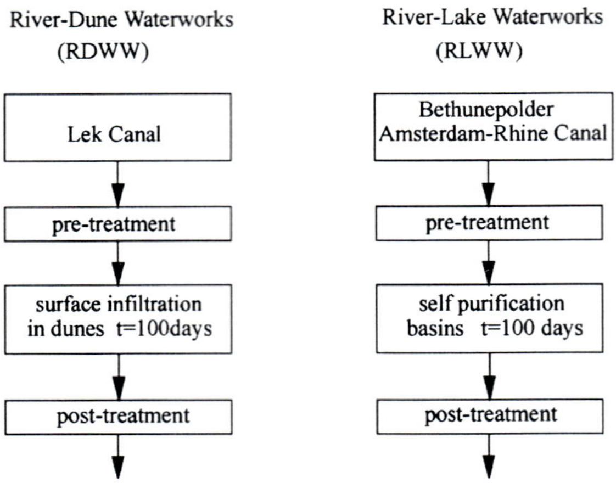

Figure 1. Basic process scheme of the River-Dune Waterworks (RDWW) and the River-Lake Waterworks (RLWW)

In the RDWW water from the Lek Canal is pre-treated in Nieuwegein by coagulation and rapid sand filtration, and then transported to the dunes west of Amsterdam (Leiduin). There the pre-treated water is infiltrated in the dunes for quality improvement, to enable selective intake of water and to cover periods of insufficient intake (strategic supply). After a residence time of approximately 100 days the water is abstracted from an open collecting basin and after rapid sand filtration for removal of suspended solids, algae and plant components and nitrification, the water is further treated to achieve drinking water quality.

In the RLWW raw water from the Bethune Polder and the Amsterdam-Rhine Canal is treated by coagulation before it enters a self purification basin with a residence time of approximately 100 days. These treatment facilities are located in Loenen. Before the water is transported to Weesperkarspel for post-treatment, rapid sand filtration is carried out in order to remove suspended solids, algae, plant components and the cells of *Dreissena polymorpha* as well as to allow ammonia to be converted to nitrate. In this way the rapid sand filtration functions also as a transport treatment instead of a transport chlorination. In Weesperkarspel post-treatment is carried out.

The raw water quality of the Lek Canal and the Amsterdam-Rhine Canal may be characterized as river Rhine water with a high salt and phosphate content, a low organic matter content and many xenobiotic material (AOX, pesticides). The water from the Bethune Polder reclamation area may be characterized as having a low salt content and a high content of phosphate, ammonia and humic acids. Total hardness of both water types is almost equal. Most of the present time only water from the Bethune Polder is used as raw water source in the RLWW.

The average composition of the water, after pre-treatment and storage in the dunes or basin, which has to be treated in Leiduin (RDWW) or Weesperkarspel (RLWW) is summarized in Table 1. In this table also the requirements set by Amsterdam Water Supply for the final water quality are included.

Table 1. Water quality after dune infiltration (RDWW) and after storage and rapid sand filtration (RLWW), and treatment targets

	RDWW	RLWW	treatment targets
Cl^- (mg/l)	130	75	80
SO_4^{2-} (mg/l)	60	10	as low as possible
DOC (mg/l)	2	5-6	< 2.5
AOC (μg/l)	5-6	20	< 10
AOX (μg/l)	15	10-15	< 5
pesticides (μg/l)	≥ 0.1	0	< 0.1
total hardness (mmol/l)	2.7	2.5	1.5
Br^- (μg/l)	200	100	

2.2. HISTORICAL DEVELOPMENT OF THE PROCESS SCHEME OF THE RLWW

Figure 2. shows the historical development of the RLWW over the last 20 years. In 1983, disinfection using chlorine has been stopped. The main reason was the formation of harmful halogenated compounds during chlorination. Depending on the content of organic matter expressed in mg DOC/l, bromoform and non-volatile brominated hydrocarbons and/or chloroform and non-volatile chlorinated hydrocarbons are formed.

The elimination of chlorination was possible as slow sand filtration was already available as the final purification step. With slow sand filtration it appeared to be possible to produce hygienically safe and biologically stable water (AOC < 10 μg/l), thus no postdisinfection was necessary.

In 1987, softening was introduced in the treatment scheme with pellet softening as the chosen technique. The process itself will be discussed in one of the next paragraphs. The reasons for introducing central softening are presented in Table 2. (Van Ammers et al., 1986, Kostense, 1988).

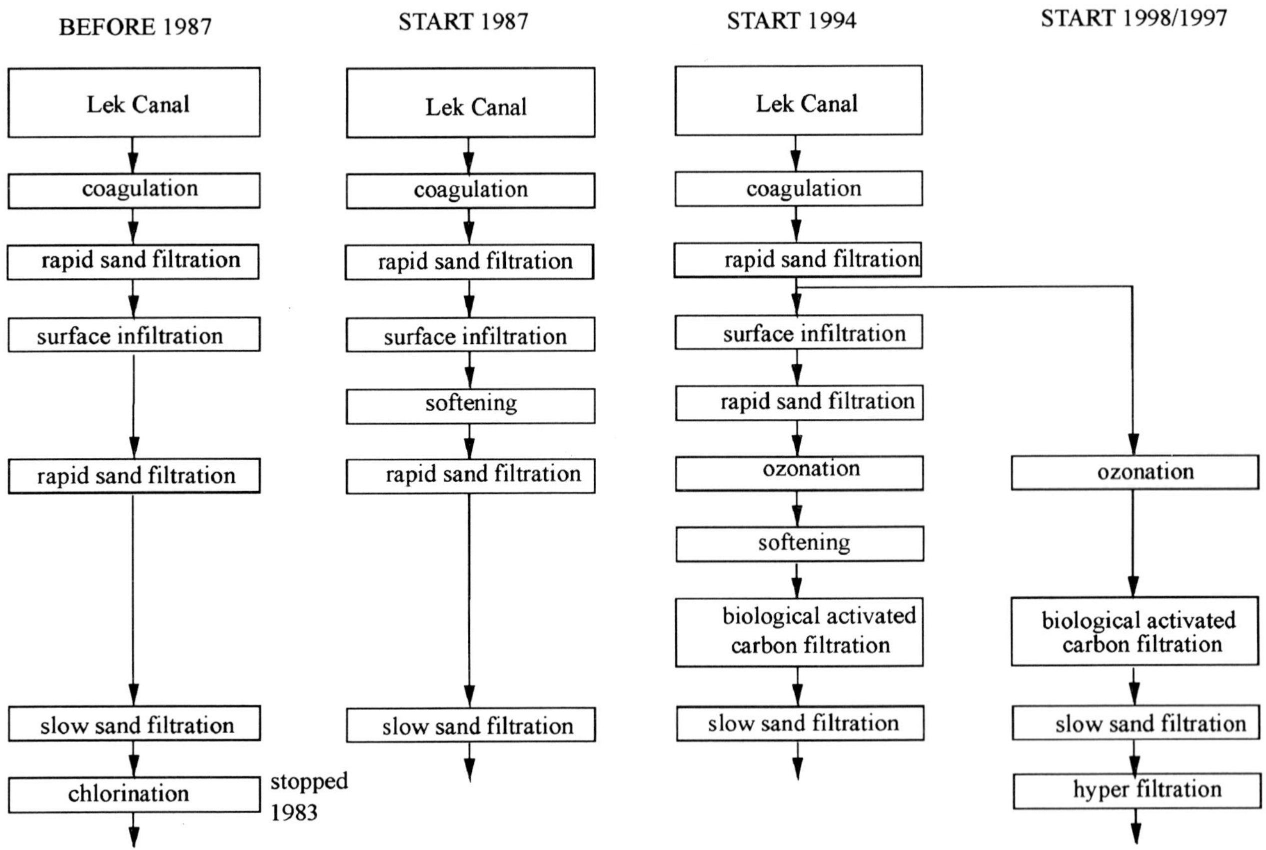

Figure 3. Historical development of the River-Dune Waterworks

Table 2. Arguments for central softening

Public health aspects:

- reduction of the uptake of heavy metals (Pb, Cu, Zn) in the distribution system
- exclusion of private softeners, which may lead to corrosion and bacteriological problems

Economic aspects:

- reduced consumption of detergents
- reduced scaling and wear of kitchen apparatus
- reduced energy consumption of heating apparatus
- reduced wear of clothing

Environmental aspects:

- reduced heavy metals (Pb, Cu) in sludge and effluent from sewage treatment plants
- reduced detergent use and phosphate content of sewage
- reduced brine discharge from private softeners

Aesthetic aspects:

- prevention of stains
- consumer comfort

In pellet softening the pH is increased by addition of NaOH, Na_2CO_3 or $Ca(OH)_2$ to crystallize $CaCO_3$. Because a high pH during ozonation results in a fast disintegration of ozone to oxygen, softening is carried out after ozonation.

In 1992 biological activated carbon filtration (BACF) was introduced to replace coagulation/rapid sand filtration. Coagulation/rapid sand filtration had been applied for the removal of organic matter (especially DOC). However, removal of organic matter by this process is limited. Thus, the load of the subsequent slow sand filters with organic matter acting as nutrient for microbial growth might be relatively high, which complicates the production of biologically stable drinking water and stimulates the development of *Aeromonads*. Besides, the process is costly and generates large amounts of sludge. Moreover, the presence of pesticides and other organic micropollutants was becoming more important, while limits for these compounds were becoming more stringent. With coagulation/rapid sand filtration, removal of these compounds was insufficient. Therefore, biological activated carbon filtration was introduced, which means light oxidation by means of ozone, followed by carbon filtration. In the process AOC is increased to 100-150 µg/l and adsorption as well as biodegradation occurs in the carbon filter (Graveland, 1994). For the reduction of DOC, an important target in the RLWW, biodegradation plays the dominant role. Thus, in the BACF process the carbon filter acts like a bioreactor. The process will be described in more detail in one of the next paragraphs.

Carbon filtration is placed after the softening process for the following reasons:

- it also serves as post-filtration after softening if sludge (carry-over) is escaping from the pellet reactors. If this sludge should reach the slow sand filters, clogging may occur and frequent cleaning of the slow sand filters by scraping might be necessary. However, this approach necessitates the use of activated carbon that resists backwashing, preferably with water and air,

- due to biodegradation in the biological activated carbon filters, lowering of the pH takes place, making a small dose of lye necessary of up to SI = + 0.6 (Langelier Saturation Index) after the carbon filtration. In case of carbon filtration preceding the softening, first more lye has to be dosed in the softening process and later some hydrochloric acid again would be necessary. Minimal use of chemicals has been a major objective.

For introduction of the BACF in the existing treatment scheme, the already present pre-ozonation has been used as the ozonation step in the BACF, while the rapid sand filters, present after the coagulation process, were converted into carbon filters. Also new carbon filters have been built.

2.3. HISTORICAL DEVELOPMENT OF THE PROCESS SCHEME OF THE RDWW

Figure 3. shows the development of the process scheme of the RDWW over the last ten years, and a proposed modification in the near future. As in the RLWW, chlorination was stopped in 1983, and for the same reasons.

Softening was introduced in the treatment scheme in 1987. In contrast to the RLWW, softening is carried out directly on the raw water abstracted from the collecting basin, before rapid sand filtration. To avoid clogging of the nozzles in the pellet reactors the water first passes a sieve (3 mm) to remove larger particles (aquaguards).

After the discovery that certain pesticides and other organic micropollutants (amongst others bentazone, metolachlorine, atrazine) passed through the layer of dune sand, in 1989 it was

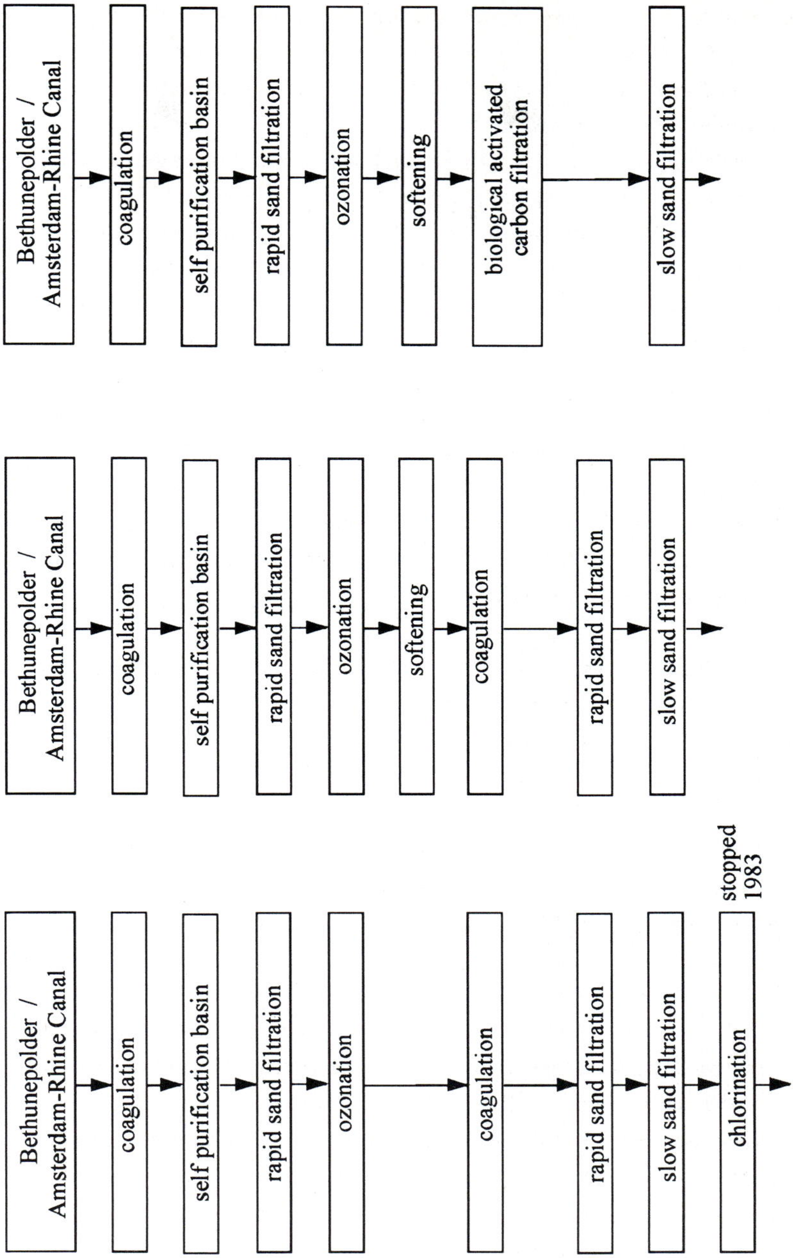

Figure 2. Historical development of the River-Lake Waterworks

decided to extend the treatment scheme with biological activated carbon filtration for the removal of these compounds. As in the RLWW the main objective of BACF is the removal of organic matter that acts as a nutrient (DOC) in general and AOX and pesticides in particular, in the RDWW the main objective of BACF is the removal of pesticides and organic micropollutants (AOX etc.), albeit that also in the RDWW the removal of nutrients by BACF is important to increase the biological stability of the produced water and to extend the running time of the slow sand filters. The sequence of the process units will be ozonation - softening - activated carbon filtration - slow sand filtration. Rapid sand filtration will be replaced in the process scheme and will be the first process step in the post-treatment after dune infiltration.

The main reasons for the new location of the rapid sand filtration are:

- pollution of the pellet reactors with material from the collecting basin will be avoided,

- by two filtration steps (rapid sand filtration and activated carbon filtration) a drastic reduction of organic material (nutrients, DOC) is reached, which results in prolonged running times of the slow sand filters and an increased biological stability of the produced water.

Treatment of the water according to the extended and modified process scheme started in November-December 1994.

In the near future the capacity of the RDWW will have to be increased from 70 to 83.10^6 m^3/year. Extension of the dune infiltration by surface infiltration is not allowed. Therefore, alternatives have been developed and are being studied currently. One of these alternatives concerns the direct treatment of surface water without dune passage. Pre-treated river Rhine water (taken from the Lek Canal and pre-treated by coagulation and rapid sand filtration) will be further treated by ozonation, biological activated carbon filtration, slow sand filtration and finally by hyperfiltration (reverse osmosis). Hyperfiltration is necessary (1) to lower the relatively high salt content of the Rhine water, (2) to soften the water, (3) to remove pesticides and other organic micropollutants or reaction products of ozonation that might pass the biological activated carbon filtration, and (4) to act as an additional disinfection barrier. Extensive treatment preceding hyperfiltration (BACF and slow sand filtration) is necessary to avoid mechanical damage and organic, biological and colloidal fouling of the membranes, and to reduce the cleaning frequency of the membrane units. Besides, as in this alternative no storage reservoir or dune passage is used which balances fluctuating concentrations, enables selective intake and serves as an strategic supply, a robust, reliable treatment scheme is required with several barriers to guarantee continuous drinking water production, independant of the raw water quality. Hyperfiltration will be discussed in more detail in one of the next paragraphs.

2.4. THE NEW TREATMENT PROCESSES

The new treatment processes that have been introduced in the RLWW and RDWW, or that will be introduced in the short term, concern pellet softening, biological activated carbon filtration and hyperfiltration. In this paragraph, these processes will be discussed in more detail. Special attention will be paid to the fundamentals of the processes, the merits, advantages and limits of the processes, and finally some operational aspects.

2.5. WATER SOFTENING BY MEANS OF PELLET REACTORS

2.5.1. Principles of Pellet Softening

Softening basically involves the dosing of an alkaline reagent (NaOH, caustic soda; $Ca(OH)_2$, lime or Na_2CO_3, soda ash) to increase the pH in the case of NaOH and $Ca(OH)_2$, and to increase the CO_3^{2-} content in the case of Na_2CO_3.

By increasing the pH, hydrogencarbonate is converted into carbonate, the solubility product of calcium carbonate will be exceeded and nucleation takes place:

$$HCO_3^- + OH^- \Rightarrow H_2O + CO_3^{2-}$$

$$CO_3^{2-} + Ca^{2+} \Rightarrow CaCO_3\downarrow$$

A difference can be distinguished between homogeneous nucleation (precipitation) and heterogeneous nucleation (crystallization). In the presence of a large crystallization area spontaneous homogeneous nucleation in the water phase is avoided and crystallyzation of $CaCO_3$ on seed crystals proceeds at a lower pH compared with precipitation.

Precipitation takes place when the pH, after addition of the alkaline reagent, is about 1-1.5 above the equilibrium pH. Precipitation:

- takes a relatively long time (1-4 hours),
- produces voluminous sludge flocs in the water phase.

Crystallization takes place when the pH is about 0.3-0.4 above the equilibrium pH. Crystallization:

- proceeds very fast (10 s),
- produces a solid waste when crystallization takes place on the surface of sand grains, providing the required crystallization surface. Dry, hard and durable marble grains are produced.

Figure 4. shows the difference between crystallization and precipitation.

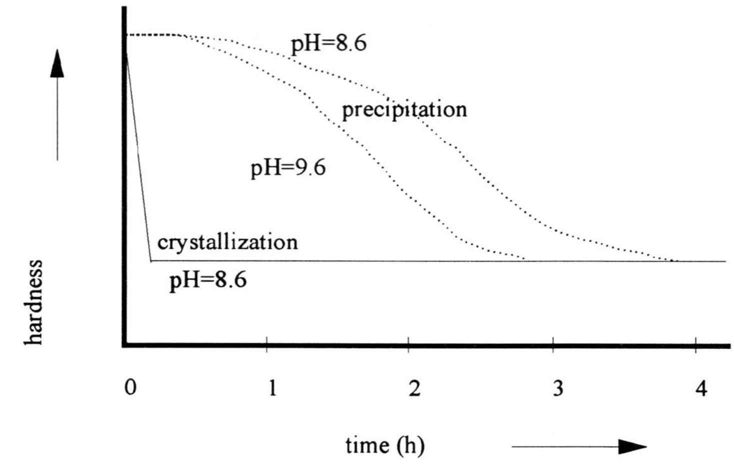

Figure 4. Time required for softening by crystallization and precipitation

The mechanism that determines the rate of the process concerns micro-nucleation. In the case of precipitation (no crystallization surface) this can be quickened, but then an even higher pH is required (in Figure 4. pH = 9.6 instead of pH = 8.6).

In a pellet reactor sand grains are fluidized homogeneously by the upflow velocity of the water and the fluidized bed provides the crystallization surface. The grains are graded over the height of the reactor and move freely in the upward flow of the water so that cementing of the grains is prevented. The principle design of the pellet reactor is shown in Figure 5.

The Amsterdam reactor type ("Reductor") (Graveland et al., 1983) is cylindrical over the entire height of 5 m. The top consists of a cone to prevent wash-out of grains. The bottom of the reactor is equipped with specially designed nozzles that allow a separate injection of water and caustic soda solution into the reactor. With enough of these nozzles per square meter proper initial mixing is obtained while the turbulence in the reactor can be kept relatively low in the higher region to enable a homogeneously fluidized bed.

This, in combination with the cylindrical shape, ensures a stable flow pattern and a complete classification of the fluidized bed; i.e., the larger grains are found at the bottom and the smaller grains at the top of the bed, indicating a good plug flow. The Amsterdam reactor can be operated with caustic soda and soda ash.

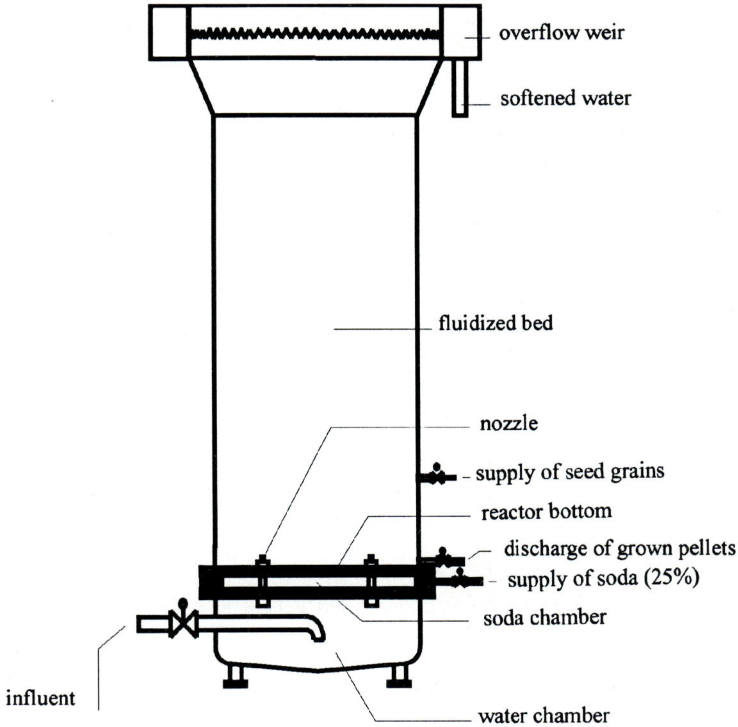

Figure 5. The Amsterdam reactor: softening of water by means of crystallization

Due to the crystallization of $CaCO_3$ on the grains, they grow into pellets. The larger and heavier pellets hydraulically grade towards the base of the reactor where they can be removed periodically and are replaced by the same amount of smaller diameter seeding grains. As only a part of the pellets is removed, this procedure can take place during full operation.

Because the sedimentation velocity of the grains is much higher than that of $CaCO_3$ flocs, the surface load of pellet softeners is about 20-100 times that of a sludge process (precipitation process). The high rate of crystallization allows a short detention time.

Some important process conditions are:

- a minimum flow through the reactor is required to prevent cementing of the grains. This is called the minimum fluidization velocity and is governed by the grain size of the pellets and the temperature. The minimum fluidization velocity for grains of 1 mm (the size, at which grains are normally removed from the reactor) is approximately 50 m/h at 20 °C,

- the dosage of the alkaline reagent must be constructed in such a way that mixing is fast and homogeneous over the whole cross-section of the reactor in order not to exceed the pH value at which precipitation takes place and sludge flocs are formed,

- the water and chemical must be mixed in the presence of seed grains with a high specific surface area to obtain immediate crystallization.

2.5.2. Merits of the Process

The most important advantages of pellet reactors (crystallization) above the sludge process (precipitation) are:

- the crystallization process produces a relatively small quantity (approximately 10 percent by volume compared with the sludge process) of dry, hard, and durable marble grains (pellets), that can be used for other purposes, including treatment of aggressive water, neutralization of acid wastewater, and utilization for road construction and cement manufacturing,

- the difference between effluent pH and equilibrium pH for the crystallization process is less than it is for a precipitation process, indicating a lower required supersaturation. This involves a higher chemical efficiency and fewer, if any, scaling problems,

- treatment plants that use the crystallization process are reliable and relatively simple in operation as compared with the precipitation process. In this process, sludge processing itself requires operational action,

- investment costs for the crystallization process are much lower than those for the sludge process. Due to the high surface load that can be applied (100 m/h and higher) the required surface area of the reactors is low. Also costs for operation and maintenance are lower owing to higher reliability and the form of $CaCO_3$ produced.

With softening, reduction of the Cu- and Pb-dissolving potential is the objective.

Observations in the Netherlands show that by pellet softening the Cu- and Pb-dissolving potential is reduced by 60-80% and 40-50% respectively (Van Ammers et al., 1986).

2.5.3. Limitations

Softening with NaOH and $Ca(OH)_2$ affects the concentration of hydrogencarbonate, while softening with NaOH and Na_2CO_3 affects the concentration of sodium, as can be seen from the chemical reactions below:

$$Ca^{2+} + HCO_3^- \quad + \quad \underbrace{Na^+ + OH^-}_{\text{caustic soda}} \Rightarrow CaCO_3\downarrow + H_2O + Na^+$$

$$Ca^{2+} + 2HCO_3^- \quad + \quad \underbrace{Ca^{2+} + 2OH^-}_{\text{lime}} \Rightarrow 2CaCO_3\downarrow + 2H_2O$$

$$Ca^{2+} \quad + \quad \underbrace{2Na^+ + CO_3^{2-}}_{\text{soda ash}} \Rightarrow CaCO_3\downarrow + 2Na^+$$

The use of lime results in twice as much reduction in HCO_3^- as compared with caustic soda. Soda ash does not change the HCO_3^- concentration. Caustic soda and soda ash both result in an increase in the Na^+ concentration.

In order to distribute water with enough pH buffering capacity, and to avoid corrosion of cast iron or copper pipes, it is recommended to maintain the bicarbonate concentration above 2 mmol/l (Van der Hoven and Van Eekeren, 1988). In the current standards, the standard for sodium is 120 mg/l. Hence, the choice of chemicals is determined by the raw water quality.

Where the HCO_3^- concentration of the raw water is already too low, only soda ash can be used.

Where the HCO_3^- concentration is high, caustic soda or lime can be used. With moderate HCO_3^- concentrations, only caustic soda may be used.

When the Na^+ concentration in the raw water is high, the sodium standard may exclude the use of soda ash and caustic soda.

In practice, post-filtration is necessary to remove carry-over from the pellet reactor. Especially in the case of softening with lime, carry-over can be high (50-70 mg/l, turbidity up to 40 NTU), although with the use of special methods for the preparation of the lime milk (e.g. application of lime with high chemical purity, or use of mixing water free of carbon dioxide and hydrogencarbonate) a reduction of 60-80% in carry-over can be obtained (Van Eekeren et al., 1994).

When softening is carried out with caustic soda, the turbidity increase is generally less than 1 NTU. This leads to filter runs of the post-filters of one week or longer (Graveland et al., 1983).

2.5.4. Technical and Operational Aspects

The dimensions of pellet reactors vary and depend on the construction type (Kostense, 1988). In practice, diameters of 1-3 m are applied and heights of 5-14 m. The upflow velocity ranges from 60-100 m/h in case of lime, and from 50-125 m/h in case of caustic soda. The fixed bed height is approximately 2 m, while after fluidization the bed height is approximately 1.5-2.5 times higher.

The grains, used as crystallization area, can be either ordinary sand (silicate) or garnetsand. Advantages of garnetsand are:

- a high specific weight, so the seed grains can be small at the same upflow velocity without the risc of wash-out. The small particle diameter results in a high specific surface area on which crystallization can take place,

- garnet sand has a relatively rough surface structure, promoting nucleation of $CaCO_3$ on the grains.

2.6. REMOVAL OF ORGANIC MATTER BY MEANS OF BIOLOGICAL ACTIVATED CARBON FILTRATION

2.6.1. Principles of Biological Activated Carbon Filtration

Biological activated carbon filtration (BACF) is based on the combination of ozonation and activated carbon filtration. As already discussed in the preceding paragraphs, BACF is applied in order to reduce the content of organic matter for (micro)biological growth (nutrients) in general, and that of pesticides and other organic micropollutants in particular. In BACF the water is first ozonated with a low ozone dose (< 1-2 mg O_3/l) to produce AOC (assimilable organic carbon) up to a concentration of 100-150 μg/l. As a result, the biodegradation capacity in the subsequent carbon filters is increased, because the increase in AOC yields a higher bacterial density in these filters.

Hence, the organic matter can be biodegraded or adsorbed, depending on the nature of the material. Compared with activated carbon filtration without preceding ozonation, the biodegradation of the organic matter is strongly stimulated, and also organic matter, already adsorbed on the carbon, might be biodegraded (bio-regeneration). The biodegradation of natural organic matter (NOM) favours adsorption of pesticides and other organic micropollutants, because it lowers NOM-preloading onto the carbon. Also the more persistent organic compounds may be biodegraded as a result of the higher biological activity in the carbon filters.

2.6.2. Merits of the Process

Ozone preceding activated carbon filtration has two important positive effects:

1. By ozonation the AOC content of the water is increased, stimulating the biological activity in the subsequent carbon filters. Because both adsorption and biodegradation of organic matter takes place in the carbon filters, the running time of the carbon filters is prolonged with a factor of 10. Due to the longer running times between reactivations, costs for reactivation of the carbon are substantially reduced. Figure 6. shows an example of the ozonation on the running time of carbon filters. From this figure it can be seen that for a decrease of DOC, biodegradation plays a dominant role. Thus the carbon filter acts like a bioreactor.

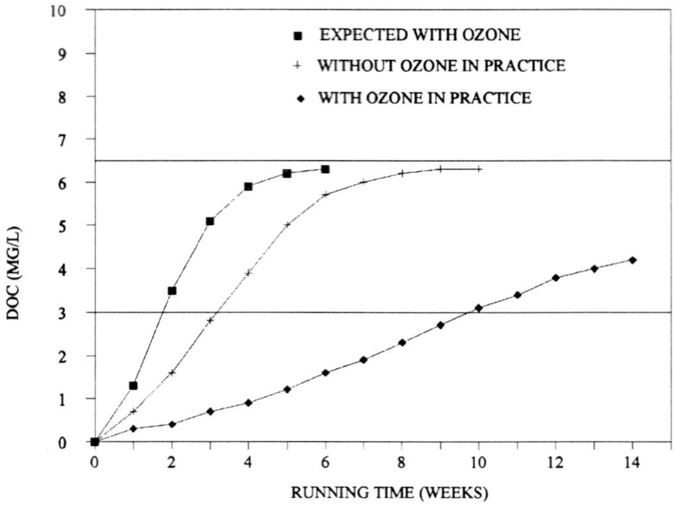

Figure 6. The DOC values in the influent (6.5 mg/l) and effluent of the carbon filters as a function of the running time, with and without preceding ozonation

Figure 7. shows the production of AOC as a function of the applied ozone dose (DOC content of the raw water 2 mg/l).

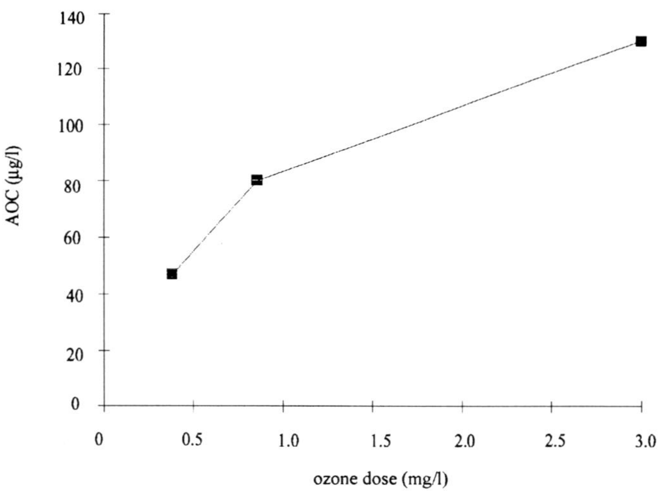

Figure 7. AOC production as function of the ozone dose

2. Ozonation has a disinfection capacity (CT = 1-2). Especially in the winter the number of viruses in the river Rhine water, used in the RDWW, can be high (40.000/m^3), and also in the initial filtration steps. The still present micro-organisms can be sufficiently inactivated by ozonation with CT = 1-2 mg/l/min.

2.6.3. Limitations

Ozonation can have two negative effects:

1. Ozonation may result in the oxidation of halogens, and bromate (BrO_3^-) can be formed as a result of the presence of bromide (Br^-) in the raw water. Bromate is known to be carcinogenic (Kurokawa et al., 1990). The World Health Organization recommends a provisional guideline value of 25 µg/l with an excess lifetime cancer risk of 7×10^{-5}. In the Netherlands a maximum admissible concentration of 0.5 µg/l is proposed with an excess lifetime cancer risk of 10^{-6}.

   The amount of bromate that is produced is affected by the DOC content and bromide content of the raw water, and varies with the ozone dose. For the RDWW and the RLWW the extent of bromate production has been measured at several ozone doses. Figure 8. shows that with an ozone dose below 1 mg/l, as is applied in the BACF process, bromate production is below 1 µg/l, both for the RDWW and for the RLWW.

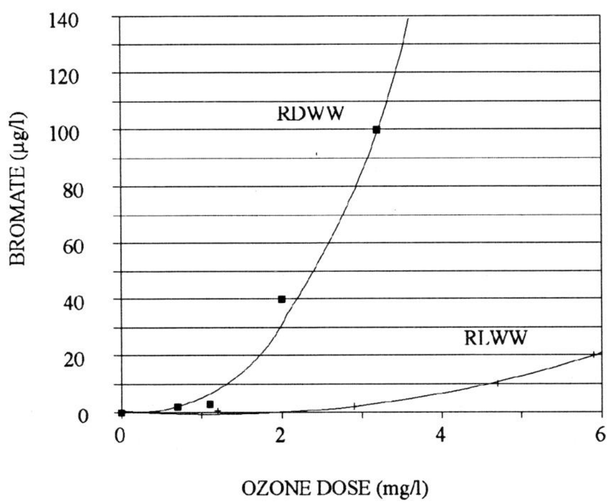

	RDWW	RLWW
DOC (mg/l)	2.0	6.5
Cl (mg/l)	150	70
Br (µg/l)	200	100

Figure 8. The amount of bromate, produced by various doses of ozone for the RDWW and the RLWW

2. The BACF process may result in too high a content of AOC in the effluent of the slow sand filters (AOC > 10 µg/l). However, with adaquate doses of O_3, AOC production can be restricted to a maximum of 100-150 µg/l, which can be removed from the water by good operation of the subsequent carbon filters and slow sand filters. Figure 9. shows the AOC concentration in the water after ozonation, and after activated carbon filtration + slow sand filtration by three ozone dosages, as measured in a pilot plant treating the raw water of the RDWW.

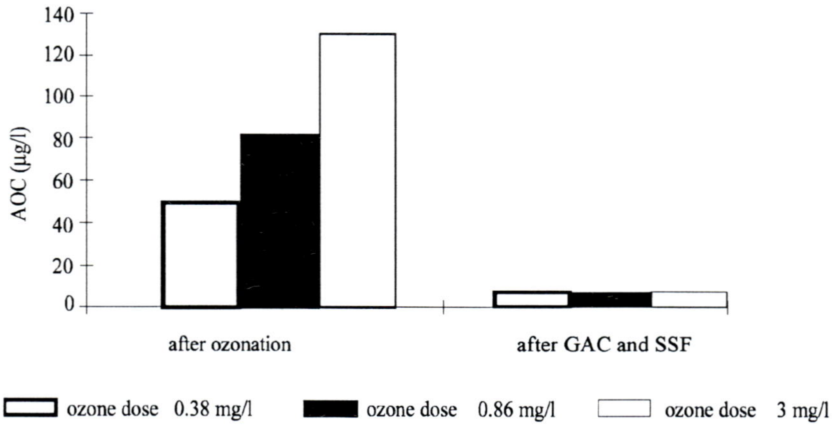

Figure 9. Removal of AOC by activated carbon filtration (GAC) and slow sand filtration (SSF) after ozonation with varying ozone doses

2.6.4. Technical and Operational Aspects

The empty bed contact time of the carbon filters in the BACF process amounts to 40 min. This contact time is used to stimulate and maximize biodegradation in the filters, while with this contact time also the lowest total costs per cubic meter are reached. Figure 10. shows the capital costs, reactivation costs and total costs as a function of the empty bed contact time.

As an alternative for one-stage filtration with a contact time of 40 min, two-stage carbon filtration can be applied, with in each step a contact time of 20 min. When a filter in the first stage is loaded and shows break-through above a certain value, it will be reactivated. A filter that has operated in the second stage subsequently can be placed in the first stage. With this approach a filter can be loaded more heavily before it is taken out of operation for reactivation. Reactivation costs are decreased with this set-up (15%), but investment costs will be higher. In the RLWW one-stage BACF is used, in the RDWW two-stage BACF is used.

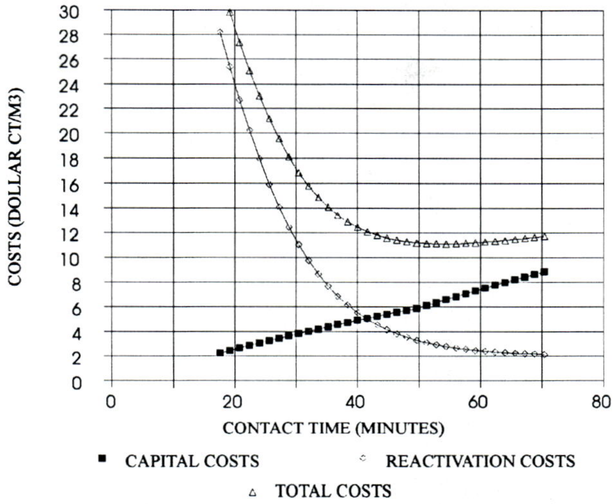

Figure 10. Costs per cubic meter of drinking water as a function of the empty bed contact time: capital costs, reactivation costs and total costs

As a result of biological activity in the carbon filters, biomass accumulates and head loss increases. For this reason, the filters have to be backwashed periodically (every 1-4 weeks). To avoid formation of mudballs that move down into the filter bed during backwash with only water, backwash has to be carried out with water in combination with air. Due to the biological activity during carbon filtration, the effluent of the carbon filters might contain a low oxygen content. Therefore, dosing of oxygen into the effluent of the carbon filters might be necessary.

2.7. REMOVAL OF CONTAMINANTS BY MEANS OF MEMBRANE FILTRATION

2.7.1. Principles of Membrane Filtration

Membrane filtration is the general name for processes that use a semi-permeable membrane for the separation of water and contaminants. In pressure driven membrane processes, the water is driven through the membrane while the contaminants are retained by the membrane. The membrane acts as a selective barrier, permitting the passage of certain components and retaining other components, as shown in Figure 11.

The feed water is split in a product stream (the low concentrated permeate) and a waste stream (the concentrate or brine, a high concentrated waste stream).

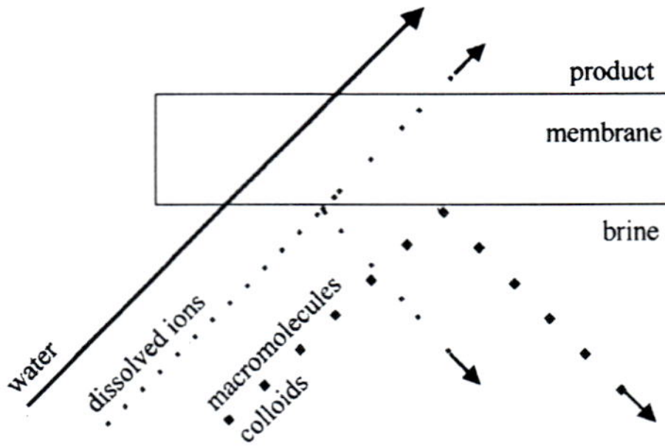

Figure. 11. Principle of membrane filtration

A distinction can be made between microfiltration, ultrafiltration, nanofiltration and hyperfiltration, the last one also named reverse osmosis. Figure 12. defines the different membrane technologies as a function of the membrane pore size, related to the size of particles occuring in the water.

Microfiltration membranes are used for the removal of colloidal and suspended matter, which are in the pore size of the membrane. Ultrafiltration membranes retain colloids and macromolecules such as proteins, which have a molecular weight larger than 1000-100,000. The selectivity of microfiltration and ultrafiltration membranes is primarily based on sieving. Nanofiltration membranes and hyperfiltration membranes both retain dissolved ions and organic compounds, the rejection by hyperfiltration being superior to that by nanofiltration: hyperfiltration retains monovalent and multivalent ions, while the retention of monovalent ions by nanofiltration is limited. Nanofiltration therefore is a suitable technique for the removal of hardness and higher molecular weight compounds. Hyperfiltration is a desalination technique, also removing low molecular weight compounds, and has been used widely for desalination of brackish water and sea water. Nanofiltration and hyperfiltration are diffusion controlled processes.

As can be seen from Figure 12., in principle disinfection can be achieved by ultrafiltration, nanofiltration and hyperfiltration, as these processes retain bacteria and viruses.

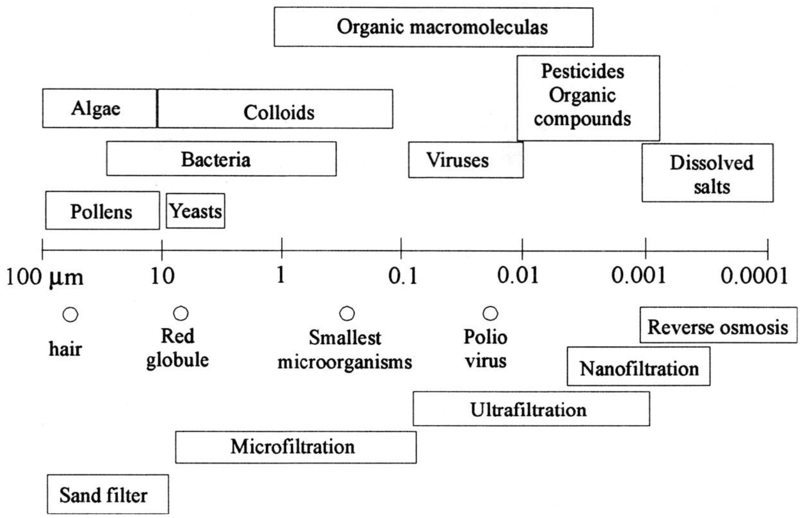

Figure 12. Relation between membrane pore size and the size of the "particles" retained by the membranes

The membrane processes are operated at different pressures. Table 3. summarizes the main characteristics.

Table 3. Main characteristics of pressure driven membrane processes

Filtration process	Particle size (nm)	Pressure (bar)	MWCO [1]	Specific flux ($l/m^2/d/bar$)
micro	> 100	< 1		1000-5000
ultra	10-100	5	100-100000	400-1000
nano	1-2	5-10	200-500	100-400
hyper	0.2-1.0	15-60	100	50-100 (BW [2]) 1-5 (SW [3])

[1] Molecular Weight Cut Off (90% of compound is rejected)
[2] Brackish Water
[3] Sea Water

The membrane structure can be either homogeneous, asymetric or composite. In hyperfiltration and nanofiltration nowadays the Thin Film Composite membranes (TFC membranes: 0.1 μm separating material, usually polyamide; 50-75 μm microporous base, usually polysulphon; reinforced by a polyester support fabric, 125-150 μm) in particular are gaining application. Compared with the traditionally used asymetric Cellulose Acetate membranes (CA membranes) they have superior qualities with respect to stability (no compaction, not biodegradable, no hydrolysis), retention and flux.

2.7.2. Merits of the Process

Membrane filtration is a strictly physical, absolute separative technique whose quality performance depends only on the membrane pore size. Conventional filtration processes used in water treatment can only provide relative effluent quality, determined by both the raw water quality and the process operating conditions. In membrane filtration processes the treated water quality is no longer dependent on the raw water quality, but on the pore size occuring in the implemented membrane. Raw water quality variations and operational conditions influence only the process throughput, this might require extensive post-treatment (see "Limitations").

Depending on the type of particles that have to be removed, the most appropriate type of membrane filtration process can be chosen, and several treatment stages can be combined in one process unit (e.g. desalination, softening and disinfection by hyperfiltration).

Since the driving force of the process is pressure, automation is relatively simple. The membrane processes are compact as compared with conventional processes (e.g. microfiltration as alternative for coagulation), resulting in relatively small treatment plants. Moreover, it is a modular process and extension of treatment plants can be quite straight foreward.

2.7.3. Limitations

Although membrane filtration in drinking water treatment is a promising technique that can be used for many purposes, it has also limitations and disadvantages. The most important are:

1. In the process a concentrated waste is produced, the brine, that has to be disposed of,

2. Energy consumption, especially in the case of nanofiltratiom and hyperfiltration, is somewhat higher than that of conventional processes (approximately + 0.5 kWh/m^3). On the other hand, with these new processes a higher drinking water quality can be obtained as

compared with conventional treatment, and the membrane processes can save costs, e.g. costs for softening in the case of hyperfiltration and nanofiltration,

3. The performance of nanofiltration and hyperfiltration for the removal of organic low molecular weight compounds is not yet fully known. Most of the membranes, available at the moment, have been designed for desalination of sea water and brackish water (hyperfiltration), and for softening (nanofiltration). Research into the rejection of these compounds by the membranes is now being carried out. Development of new membrane materials might be required to reach the high drinking water standards for these compounds,

4. In principle, disinfection is possible with ultra-, nano- and hyperfiltration. A log-reduction of at least 5-7 has been reported (Boller, 1994, Anselme et al., 1993). Two important questions however remain:

   - can the integrity of the membranes and membrane modules be guaranteed, what is the chance and risk of leakages (connections of the elements (O-rings), application of glue, etc.),

   - how can the integrity of the membranes be monitored in full-scale plants in a relatively simple manner. In research pilot-plants, dosing of bacteriophages has been applied to measure high log-reductions. In full-scale plants this is not possible. With a relatively simple analytical technique such as conductivity, only a leakage in the range of log 2 - log 3 can be measured. So, new analytical techniques have to be developed, or a good strategy applying, for example, conductivity has to be developed to monitor the integrity of a full-scale plant.

5. The membranes might release undesirable, toxic substances into the water, especially new membranes. In many countries, approval for toxicological parameters is required for the use of membranes in drinking water treatment, stressing the need for migration tests.

2.7.4. Technical and Operational Aspects

In practice, membrane systems consist of a high pressure pump and a number of modules, with in every module a number of membranes (modules or pressure vessels). The capacity of the modules is limited, so modules in parallel are required to obtain the desired product flow. In order to obtain proper flow velocities along the membranes (important to prevent fouling) the recovery (this is the permeate flow as percentage of the feed flow) per module is limited to 40-50%, otherwise the brine flow would become too low. When a higher total recovery is required, modules must also be arranged in series. An often applied arrangement is the so called "X-mas tree" arrangement, as shown in Figure 13. After the first stage, the brine of this stage (50-60% of the original feed) is sent to the second stage. So, the number of modules in this second stage is about 50% of that of the first stage. In the same way a third and fourth stage can be used to achieve the desired overall recovery of the plant.

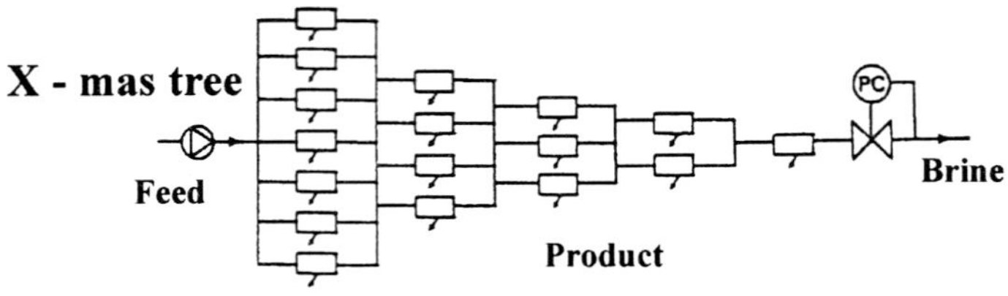

Figure 13. Membrane staging

In the case of nanofiltration and hyperfiltration, attention must be paid to pre-treatment of the water. In order to prevent colloidal and biological fouling, which would result in damage of the membranes and require frequent cleaning of the membranes, the water to be treated should have a MFI (Modified Filtration Index, a parameter determined by means of filtering a water sample through a 0.45 μm membrane filter) below 0.5-1 s/l^2 and an AOC concentration (Assimilable Organic Carbon) below 10 μg/l (biologically stable water to prevent bacterial growth). Possibilities to reach this high quality of the water are the use of microfiltration or ultrafiltration preceding the hyper/nanofiltration, or the use of ozonation-activated carbon filtration-slow sand filtration preceding the hyperfiltration, as applied by Amsterdam Water Supply. An advantage of the second alternative concerns the removal of organics from the water, including pesticides and other organic micropollutants, before the hyperfiltration. As a consequence, the brine of the hyperfiltration will not contain organic pollutants and disposal of the brine will be easier.

In the concentrate, high concentrations occur and the solubility of certain inorganic salts can be exceeded, resulting in scaling of these salts on the membranes (e.g. $CaCO_3$, $BaSO_4$, $CaSO_4$, $SrSO_4$, CaF_2). In order to prevent the precipitation of $CaCO_3$ the pH of the feed is adjusted by acid dosing to the feed. In order to prevent scaling of the other compounds, an anti-scalant can be dosed which inhibits precipitation and crystallization of salts. Then the solubility product of the compounds may be exceeded by a factor up to 50-100 without problems.

In the case of hyperfiltration, the pH and hardness of the product water will be very low and the CO_2 content will be high (CO_2 produced by the acid dose to the feed and passing the membrane). Therefore, a pH correction is necessary and the hardness has to be corrected, for example by filtration over $CaCO_3$ grains, which will simultaneously neutralize CO_2 and increase hardness. An attractive alternative set-up is mixing of the product water with hard water in a split-treatment mode. In this way the desired hardness and pH may be obtained without additional treatment, depending on the water composition and split-treatment ratio.

As a summary, the main key-design aspects which have to be taken into account in membrane filtration processes are shown in Figure 14.

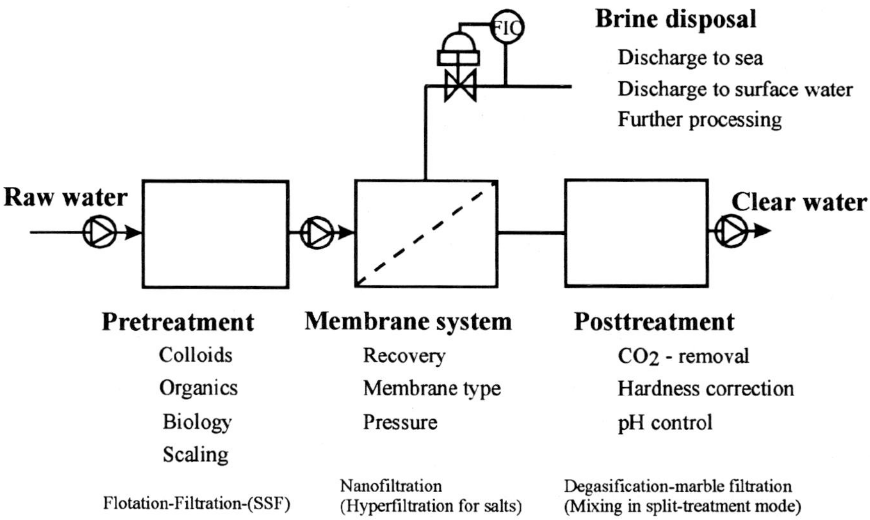

Figure 14. Key design aspects in membrane filtration

2.8. REFERENCES

Anselme C., Mandra V., Baudin I., Mallevialle J. (1993) Optimum Use of Membrane Processes in Drinking Water Treatment. Proceedings of the 19th International Water Supply Congres, Budapest, 2-8 October 1993, Special Subject 2 "Membrane Technology": SS2.1-SS2.11.

Ammers M. van, Dijk J.C. van, Graveland A., Nuhn P.A.N.M. (1986) State of the Art of Pellet Softening in the Netherlands. Water Supply 4: 223-235.

Boller M. (1994) Trends in Water Filtration Technology. Aqua 40: 65-75.

Eekeren M.W.M. van, Paassen J.A.M. van, Merks W.A.M. (1994) Improved Milk-of-Lime for Softening of Drinking Water - The Answer to the Carry-Over Problem. Aqua 43: 1-10.

Graveland A. (1994) Application of Biological Activated Carbon Filtration at Amsterdam Water Supply. Paper presented at the IWSA Specialised Conference on Activated Carbon in Drinking Water Treatment, Amsterdam, 27-28 September 1994.

Graveland A., Dijk J.C. van, Moel P.J. de, Oomen J.H.C.M. (1983) Developments in Water Softening by Means of Pellet Reactors. Journal American Water Works Association December 1983: 619-624.

Hoven Th.J.J. van den, Eekeren M.W.M. van (1988) Optimal Composition of Drinking Water. KIWA Communication No 100, Nieuwegein, The Netherlands, October 1988 (in Dutch).

Kostense A. (1988) Softening in Pellet Reactors. KIWA Communication No 102, Nieuwegein, The Netherlands, October 1988 (in Dutch).

Kurokawa Y. et al. (1990) Toxicity and Carcinogenicity of Potassium Bromate - A New Renal Carcinogen. Environ. Health Persp. 87: 309-335.

3. Treatment of Drinking Water in the United States on the Threshold of the 21st Century

Thomas G. Sanders
Tory L. Champlin
Department of Civil Engineering
Colorado State University
Fort Collins, Colorado 80523
USA

3.1. INTRODUCTION

Three basic objectives of water treatment are (ASCE and AWWA, 1990):

- Production of water that is safe for human consumption,
- Production of water that is appealing to the public, and
- Production of water using facilities which can be constructed and operated at a reasonable cost.

The production of biologically and chemical safe water is the primary goal in the design and operation of water treatment plants. A properly designed treatment plant, however, does not guarantee safety. Skilled and properly trained operators that are attentive to sanitary requirements of water supplies and distribution systems are also equally important.

Today, water treatment professionals are facing multiple concerns including the elimination of contaminants which cause acute health effects such as pathogenic organisms along with the removal of trace contaminants which have long-term or chronic health effects. While water supplies have remained relatively constant, the demand for these resources continues to increase. Surface waters and ground waters are primarily the sources used for drinking water. Contamination of surface waters through dilution with treated wastewaters has continued to degrade surface water quality. Despite dramatic improvements in wastewater

NATO ASI Series, Partnership Sub-Series, 2. Environment – Vol. 15
Water Supply Systems – New Technologies
Edited by Č. Maksimović, F. Calomino, and J. Snoxell

treatment since the 1960s, further degradation of surface waters is still occurring. An additional challenge is the growing awareness of increased contamination of ground water sources. Ground water aquifers are threatened by agricultural contamination from nitrates and other inorganics and infiltration by manufactured synthetic organic chemicals. There is still a need to improve water quality. With continued economic development and population growth, demands on water supplies will require innovative approaches to treat and protect water quality. The demand for water professionals will be extensive in the upcoming century.

3.1.1. Definitions of Public Water Systems in the United States

A public water system in the United States (US) is defined as a system that provides piped water for human consumption to at least 15 service connections, or serving an average of at least 25 people for at least 60 days each year (USEPA, 1993). In 1992, 197,060 water systems in the US including Indian lands and US territories were classified as public water systems (Pontius, 1994a).

Due to the number of different types and sizes of water systems, public water systems are further subdivided into three types of water systems:

- **Community water systems** - provide water to the same population year-round,

- **Non-transient, non-community water systems** - regularly serve at least 25 of the same people at least six months of the year (e.g., schools, factories, and hospitals with their own water supply,

- **Transient, non-community water systems** - serve transitory customers in non-residential areas (e.g., campgrounds, motels, and gas stations).

Reasons for the different classifications of public water systems relate to length of exposure to a particular drinking water. For example, persons who drink from a transient, non-community water system have only a brief exposure to contaminants in the water. In contrast, persons who drink daily from a community water system are exposed to the contaminants on a continual basis. Due to these differences in exposure, community, non-transient non-community, transient non-community water systems are regulated differently and have different water quality standards.

Community water systems are the main focus of this paper. Non-community systems are mentioned briefly regarding waterborne disease outbreaks discussed later.

3.1.2. Growth of Community Water Systems

Water treatment began in the US in the late 1800s. In 1887, an experimental station consisting of rapid sand filters was constructed in Lawrence, Massachusetts by the Massachusetts State Board of Health, after an outbreak of typhoid fever (Culp et al., 1986). Soon after slow sand filters were introduced in Albany, New York (Craun, 1986). Other filtration experiments were also conducted in Louisville, Kentucky, and in New Orleans, Louisiana in the 1890s. Early experiments, however were not completely successful. Attempts to filter muddy waters resulted in clogging and short operational cycles. This problem was not alleviated until aluminum sulfate (alum) was introduced as a pretreatment. Although the use of alum had been proposed in the 1757, it was not effectively demonstrated in the United States until 1885 (Vesilind and Peirce, 1982).

In 1908, the introduction of chlorine disinfection was introduced in the United States at Jersey City, New Jersey. The dramatic effect of chlorination was marked in the years that followed by dramatic decreases in deaths due to typhoid fever and other waterborne diseases.

Table 1. Growth of community water systems in the United States from 1800 to 1958 (Craun, 1986)

Year	Number of Water Facilities	Number of Purification Facilities	Number of Disinfection Facilities
1800	16		
1850	83		
1870	243		
1880	598	1	
1890	1,878	NA	
1896	3,196	NA	
1924	9,850	638	NA
1930	NA	1,531	2,917
1934	10,790	NA	NA
1938	12,760	NA	4,054
1940	NA	1,855	4,590
1948	NA	2,054	6,137
1958	17,808	2,517	8,845

As shown in Table 1., community water systems were developed for most major cities in the US during the late 1800s and early 1900s. Disinfection of water to prevent waterborne disease was used increasingly in the years prior to 1940.

As shown in Table 2., by 1904, only 10% of the urban population in the US received filtered water. By 1958, that number had increases to 58% of the urban population.

The growth of rapid sand filtration has been more dramatic than slow sand filtration. By 1958, rapid sand filters supplied water 95% of the population receiving filtered water.

Table 2. Growth of sand filtration in the United States (Craun, 1986)

Year	Slow Sand Filtration (population)	Rapid Sand Filtration (population)	Percent of Urban Population Served by Filtered Water
1800			
1850			
1870	0	0	0
1880	30,000	0	<1
1890	35,000	275,000	2
1900	360,000	1,500,000	6
1904	560,000	2,600,000	10
1910	3,883,000	6,922,000	28
1914	5,398,000	11,983,000	36
1924	5,054,000	18,610,000	38
1940	3,579,200	26,751,700	41
1948	2,014,635	39,731,460	47
1958	3,177,380	55,606,245	48

3.1.3. Water Sources

There are approximately 58,000 community water systems in the US serving approximately 245 million people. The remainder of the population receives its residential water from private wells or other non-federal regulated systems (i.e., those serving fewer than 15 service connections or 25 people). As Table 3. shows, 10,768 community water systems (18% of the total number of systems) providing water to 140 million people (57% of the total population) receive surface water, and 47,898 community water systems (82% of the total number of systems) providing water to 105 million people (43% of the total population) receive water from ground water sources. Based on the total number of systems, 62% are considered very small systems (servicing 25 to 500 people) providing water to only 2% of the population. In contrast, only 1% of the total systems are considered very large (more than 100,000 people) providing water to 45% of the population. A breakdown of other system sizes, their number and population served is shown in Table 3.

Table 3. Surface and ground water sources used by Community Water Systems (CWSs) in the United States at the end of 1992 (Pontius, 1994a)

System Size	Number of CWSs served by Surface and Ground Water Sources				Total Population Served by Surface and Ground Water Sources			
	Surface Water	Ground -Water	Total Number of CWSs	Percent of CWSs	Surface Water $x10^6$	Ground -Water $x10^6$	Population Served by CWSs $x10^6$	Percent of Population Served by CWSs
Very small (25-500)	3,526	32,985	36,511	62	0.72	4.85	5.57	2
Small (501-3,300)	3,678	10,838	14,516	25	5.70	14.35	20.05	8
Medium (3,301-10,000)	1,746	2,505	4,251	7	10.35	14.38	24.73	10
Large (10,001-100,000)	1,598	1,464	3,062	5	45.99	39.04	85.04	35
Very Large (> 100,000)	220	106	326	1	76.95	32.85	109.80	45
Total	**10,768**	**47,898**	**58,666**	**100**	**139.71**	**105.47**	**245.19**	**100**

3.2. THE PROCESS TO IMPROVING WATER QUALITY

As shown in Figure 1., there are three basic steps followed for improving drinking water quality: (1) the development of criteria based on identified health effects, (2) the creation of laws and standards regulating the quality of treated water, and (3) the selection of treatment goals by community water systems to achieve regulated levels.

The quality of drinking water throughout the US is determined by federally enforced regulations. These regulations establish water quality standards concerning contaminants which have a direct effect on public health. The criteria for these standards are developed by various groups which define contaminants and their associated concentrations which should not be exceeded in treated drinking waters.

Figure 1. Process of improving water quality in the US

Once determined, these criteria serve as a basis for establishing water quality standards. These standards are promulgated into federal laws and regulations. Community water systems are then required to meet these standards which may involve modifying existing treatment practices or the design and construction of new facilities. Water quality goals are contaminant concentrations chosen by community water system to ensure a quality of water that meets regulated standards.

3.2.1. Health Effects - Identifying Risk

Water quality criteria are developed solely on the basis of data and scientific judgment without consideration of technical or economical feasibility. These criteria may be developed for different uses. For a single contaminant, separate criteria can be applied for drinking water based on either health effects or appearance.

3.2.2. Regulations - Driving Water Quality

There are two major regulations which govern the quality of water in the US: the Clean Water Act (1948 through 1972) and the Safe Drinking Water Act (1974). The Clean Water Act, originally enacted in 1948, regulates the discharge of pollutants into US waterways. This law directly effects the discharge of wastewater from municipal and industrial treatment facilities into rivers, lakes and surrounding oceans. The Safe Drinking Water Act, enacted in 1974, regulates harmful contaminants in public drinking water systems.

Setting water quality standards is important and many times a controversial aspect of water treatment. Since the passage of the Safe Drinking Water Act in 1974, emphasis has been placed on linking low levels of contaminants to higher incidence of disease such as cancer. The ability to measure quantities of contaminants improve and our knowledge of their health effects increases, water professionals are faced with increasingly complex water quality regulations. Understanding and forecasting the promulgation of these regulations are important to water professionals for a number of reasons. Standards affect the selection of raw water sources, choice of treatment processes and design criteria, ability to modify existing treatment processes and treatment costs.

Unlike the development of water quality criteria, water quality standards must consider factors such as economic impact, technical feasibility, and political realities. Establishing standards is a careful balance between the benefits of protecting public health and the costs of meeting various water quality levels.

3.2.2.1. Clean Water Act. The objective of the Clean Water Act is the restoration and maintenance of the "chemical, physical and biological integrity" of the waters in the US. Two goals were included in the 1972 legislation: zero discharge of pollutants by 1985 and an interim goal and where possible, water quality that is both "fishable" and "swimable" by 1983 (USEPA, 1993). While these dates have passed and the established goals have not been met, the goals still remain, and efforts to attain these goals continue. There is still, however, some question as to the motive of zero discharge. Is it the complete elimination of pollutants in the discharging effluent, or an effluent having a concentration of pollutants less than or equal to the concentrations in the receiving stream? Prohibiting the discharge of any pollutant at any concentration is one possible interpretation.

3.2.2.2. Safe Drinking Water Act. The Safe Drinking Water Act (SDWA) directs the US Environmental Protection Agency's (EPA's) Administrator to develop national primary drinking water regulations that incorporated maximum contaminant level goals (MCLGs), maximum contaminant levels (MCLs) and treatment techniques. Under this act, EPA set maximum contaminant levels for 10 inorganics, 6 pesticides and one set of disinfection by-products (i.e., trihalomethanes). Since 1974, the SDWA has been amended four times. Each time, new contaminants have been added. In January 1993, the SDWA was amended again and now includes a total of 117 regulated contaminants, 14 inorganics, 6 microbiological contaminants and 6 radiological contaminants (USEPA, 1993). A listing of the regulated contaminants is provided in Tables A., B., C. and D. shown in the Appendix. Included in these lists are both regulated and non-regulated contaminants. The status of regulated contaminants are "final" and are identified in the fourth column of each table. Non-regulated contaminants are cited, because either EPA requires more data to determine a contaminant's significance or to determine any additional limits that may be required. In cases where water systems are especially vulnerable to a particular contaminant (regulated or non-regulated), a specific monitoring program may be requested by EPA.

In addition to primary standards, EPA also sets secondary drinking water standards which specify secondary maximum contaminant levels (SMCLs). These standards are primarily to protect public welfare and deal primarily with contaminants affecting drinking water odor and appearance. A list of secondary standards is provided in Table E. shown in the Appendix. These standards are intended as guidelines and are not federally enforced (Montgomery, 1985).

3.2.3. Selecting Water Treatment Goals

Water quality goals are set by local governments and community water systems. These goals are typically more stringent than standards and may include contaminants and other measurements not identified in the regulations. For example, a local government or community water system may decide to set a turbidity goal of 0.2 nephelometric turbidity units (NTUs) even though the standard is 0.5 NTUs. Setting treatment goals below standards ensure satisfactory performance. Even though many decision makers question the value (and additional costs) to exceed federal standards. The American Water Works Association (AWWA) is also an example of a professional organization that sets goals for drinking water treatment. Alternatively, a community water system may elect to provide a water quality better than required by standards or for contaminants not regulated. Examples include goals for trihalomethanes (THMs) that are below standards or based on other surrogate parameters such as total organic carbon (TOC) or ultraviolet light absorbance at a wavelength of 254 nanometers. The decisions on setting goals involve determination of costs, benefits, and political philosophy of the local government or community water system.

3.3. CHALLENGES FACING DRINKING WATER TREATMENT

Community water system's concerns are derived from impending regulations. The impetus of the majority of the research is driven by the following regulations:

- **Surface Water and Ground Water Treatment Rules** which will require disinfection and, in the case of surface waters, filtration of drinking water for communities serving more than 25 persons,

- **Amendments to the Surface Water Treatment Rule** which requires EPA to promulgate MCLs for and additional 25 contaminants every three years,

- **Disinfectant/Disinfection By-Product Rule** which will regulate disinfection by-products that have been identified or are being identified as carcinogens, and

- **Lead and Copper Rule** which limits the levels of lead and copper at the consumers tap by corrosion within the distribution system.

Table 4. Utility Research Efforts as identified by AWWARF Survey (AWWA, 1992)

Research Areas	Number of CWSs Responding	Percent of Total Response (a)
Corrosion Control	447	57.7
Water Conservation	445	57.4
Wellhead Protection	298	38.5
Filtration	297	38.3
Watershed Management	264	34.1
Chlorine	263	33.9
Ozone	197	25.4
Permanganate	172	22.2
Chloramines	159	20.5
Granular Activated Carbon	150	19.4
Chlorine Dioxide	145	18.7
Microbiological Treatment	99	12.8
Air Stripping	91	11.7
Powdered Activated Carbon	84	10.8
Reverse Osmosis	43	5.5
Ion Exchange	31	4.0
Ultrafiltration	29	3.7

(a) 775 community water systems responded to this survey

Table 4. summarizes the results from a recent survey conducted by AWWA concerning ongoing research activities by 775 community water systems (AWWA, 1992). According to this survey, nearly 60% of the respondents are involved in controlling corrosion and water conservation. Other areas of research include disinfection with chlorine and the use of alternative disinfectants (i.e., ozone, chloramines, and chlorine dioxide), filtration including sand, reverse osmosis and ultrafiltration, and organic removal technologies such as granular activated carbon (GAC), powdered activated carbon (PAC), and air stripping. Microbiological treatment is also becoming increasingly important regarding biofilms and biofilm regrowth in distributions systems. Concerns with biofilms include colonization of pathogenic organisms and increased levels of disinfection by-products.

3.3.1. Waterborne Disease

3.3.3.1. General Consideration Since the germ theory of disease was developed by Pasteur and Koch in the 1870s (ASCE & AWWA, 1990), the issue of waterborne disease from contaminated water has been a concern. During the period between 1920 to 1992, 1,683 waterborne outbreaks were reported in the US resulting in 466,486 cases of illness and more than a 1,000 deaths. Figures 2. and 3. show the number of outbreaks and cases of illness by etiology during this period. Of these outbreaks, 43% occurred from community water systems resulting in 83% of the reported illnesses and 71% of the reported deaths (Centers of Disease Control, 1985, Craun, 1986, St. Louis, 1988, Levine et al, 1990, Herwaldt et al, 1992, and Moore et al, 1993).

During 1920 through 1936, waterborne outbreaks were reported most frequently for community water systems. After 1936, outbreaks were generally more frequent for non-community water systems (i.e., systems which do not serve residents but serve tourists). In general, the number of outbreaks reported by community water systems have decreased since the late 1920s through the mid-1950s (Craun, 1986). Since 1965, however, the number of outbreaks have been increasing steadily from 56 outbreaks during 1961-1965 to 196 outbreaks during 1976-1980. The number of outbreaks reported during 1979 through 1980 by community water systems were the greatest since 1942. The decline in the number of reported waterborne outbreaks during the 1940s through the 1960s was not considered associated with any new developments in water treatment or widespread use of a treatment technology (Craun, 1986). Disinfection and filtration which showed dramatic decreases in waterborne outbreaks in the early 1900s, were developed and applied in the 1940s (Craun, 1986).

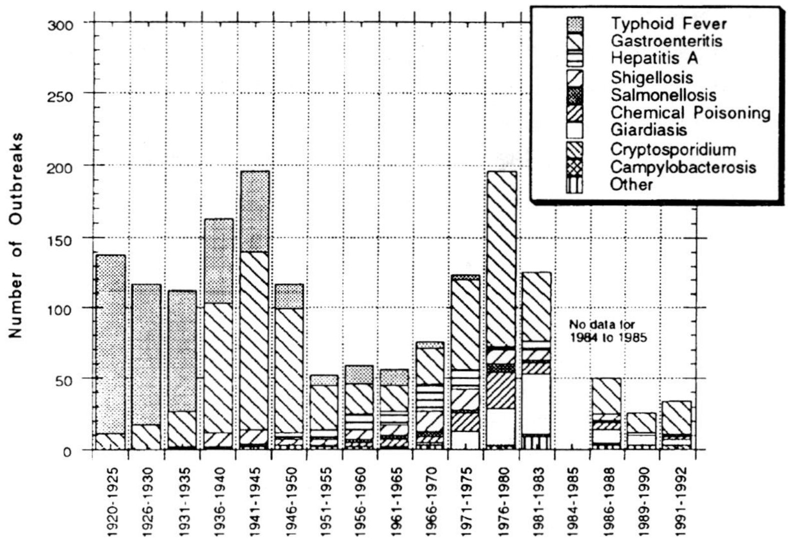

Figure 2. Total number of waterborne outbreaks reported by etiology from community, non-community and individual systems

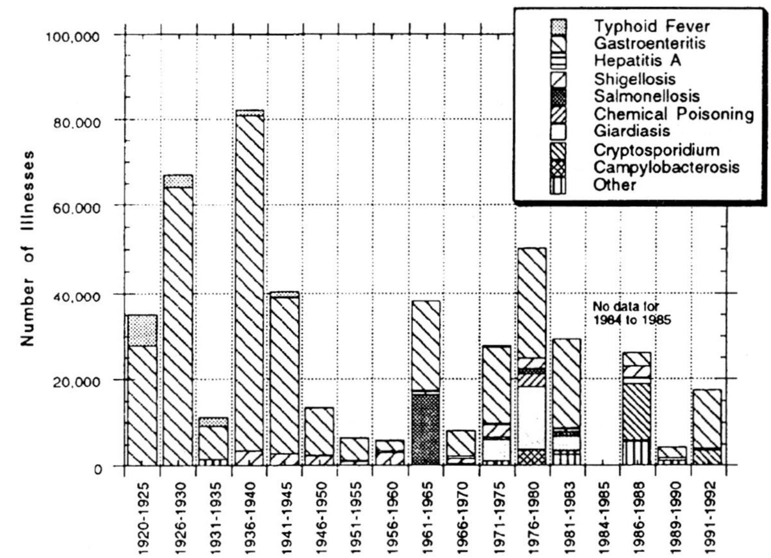

Figure 3. Total number of waterborne illnesses reported by etiology from community, non-community and individual systems

3.3.1.2. Identification and Surveillance Techniques Giardia lamblia, which was considered a benign microorganism for many years, has been related to numerous waterborne disease outbreaks in the mid-1970s. The identification of this microorganism's pathogenic nature coupled with recent research at Colorado State University which showed that filtration without proper coagulation, flocculation, and sedimentation is ineffective in removing *Giardia* (Hendricks et al., 1988a, 1988b, 1986a and 1986b) has promulgated the Surface Water Treatment Rule requiring disinfection and filtration for all surface water sources.

More recently, *cryptosporidium* has been linked with a number of waterborne outbreaks in the US (AWWA Satellite Teleconference, 1994). In the well-publicized outbreak in Milwaukee in April, 1993, 370,000 people were affected by *cryptosporidium* (Fox and Lytle, 1993). Inadequate dosing of polyaluminum chloride, used as a coagulant, was determined as the primary cause. Other treatment deficiencies related to waterborne outbreaks during the period between 1920 to 1980 are listed in Table 5. One major cause for outbreaks during this period was inadequate filtration and other allied treatments.

Table 5. Treatment deficiencies related to waterborne outbreaks and illnesses in community, non-community and individual water systems in the United States between 1920 to 1980 (Craun, 1986)

Deficiency	Number of Outbreaks	Number of Illnesses
Inadequate filtration/allied treatment	37	77,062
Inadequate disinfection when only treatment	104	45,823
Interrupted disinfection when only treatment	102	16,965
Inadequate disinfection with other treatment	4	1,785
Interruption of disinfection with other treatment	8	8,715
Inadequate control of chemical feed	9	700
Contamination of distribution system	187	72,254
No treatment of surface water	152	10,799
No treatment of well water	522	80,145
Source contamination	20	51,283
Contaminated storage facilities	30	8,359

The approach used to prevent waterborne outbreaks is the "system of multiple barriers." As opposed to relying on only one treatment process for removing pathogenic microorganisms (i.e., disinfection), a series of processes or barriers are considered (i.e., coagulation-flocculation-sedimentation, filtration, and disinfection). This system of barriers begins with the protection of surface water supplies from human and animal contact.

The removal of pathogenic microorganisms is usually expressed using a logarithmic scale. For example, 90% removal equals 1-log, 99% removal equals 2 logs, 99.9 removal equals 3 logs, and 99.99% removal equals 4 logs. Under the SDWA, community water systems are required to achieve 3-log removal for *Giardia* cysts and 4 log removal for viruses. Table 6. shows the removal credits expressed as logs for different filtration technologies. Conventional filtration (i.e., coagulation, flocculation, sedimentation, and filtration) receives the highest log-credit of 2.5. New technologies such as cartridge filters, bag filters and membranes may also be used, but need to be demonstrated under normal operating conditions.

Log credits are also given to other processes. These credits are additive, resulting in a total credit for a series of treatment processes. The log credit equation for filtration and disinfection is expressed in the following equation:

Removal		Inactivation		3-log (99.9%)
Credit	+	Credit for	$\geq$	for *Giardia*
Filtration		Disinfection		4-log (99.99%) for viruses

In other words, the log removal achieved by filtration, plus the log inactivation achieved by disinfection, must be greater or equal to 3 logs for *Giardia* and 4 logs for viruses.

Table 6. Removal credits for various filtration technologies (Pontius, 1993)

Technology	Turbidity Limit[(a)] (NTU)	Log Removal Credit
Conventional treatment[(b)]	0.5	2.5 *Giardia* 2 Viruses
Direct filtration[(c)]	0.5	2 *Giardia*
Slow sand filtration	1	2 *Giardia* 2 Viruses
Diatomaceous earth filtration	1	2 *Giardia* 1 Viruses
Other technologies	1	Must be demonstrated

(a) Filtered-water turbidity must be less than or equal to the level indicted in 95 percent of the measurements taken every month.

(b) Coagulation, flocculation, sedimentation, and filtration.

(c) Coagulation, with or without flocculation, and filtration.

Cryptosporidium is not currently regulated, but is on EPA's Drinking Water Priority List for possible regulation (Pontius, 1994b). *Cryptosporidium* oocysts are resistant to chlorination. Chlorine and monochloramine at typical concentrations have little effect (Pontius, 1994b). Chlorine dioxide has been shown somewhat effective (Pontius, 1994b). Ozone is the most effective and can completely inactivate *Cryptosporidium* oocyst. Filtration and watershed protection are the only conventional control practices available for most community water systems. Greater than 97% removal (1.5 log removal) of *Cryptosporidium* oocysts can be achieved with proper coagulation and filtration (Pontius, 1994b).

3.3.2. Disinfection By-Products

Based on a study published in the *American Journal of Public Health* [82, 955 (1992)], people who drink chlorinated surface waters are more likely to develop bladder and rectal cancer than those drinking un-chlorinated water (Morris et al., 1992).

According to study findings, the risk of getting bladder cancer was 9% and rectal cancer was 15% greater for people drinking chlorinated water (Hileman, 1992). It is estimated that of the 47,000 cases of bladder cancer reported each year in the US, 4,200 cases are associated with drinking chlorinated water (Hileman, 1992; and Water Quality Association, 1992). Of the 44,000 cases of rectal cancer reported each year in the US, 6,500 cases are associated with drinking chlorinated water (Hileman, 1992; and Water Quality Association, 1992). Other study results have suggested that the cancer risk from drinking chlorinated water is 3 times greater than drinking un-chlorinated water (Cantor, 1992).

Termed disinfection by-products (DBPs), these contaminants in treated drinking water has been the main focus of research in the US in the past two decades. DBPs are the result of chlorinating surface waters containing levels of natural organic matter (e.g., humic and fulvic acids). Halides from either chlorine or naturally present bromide are substituted onto the organic molecules.

The chemical structures of trihalomethane and haloacetic acid species are shown in Figures 4. and 5., respectively. A complete list of identified disinfection by-products is provided in Table 7. Important to note is that a significant number of DBPs have not yet been identified. Identified DBPs represent only 50% of the total organic halide (TOX) concentrations measured in most chlorinated surface waters (AWWA Satellite Teleconference, 1993).

Factors which influence DBP formation include contact time, chlorine dose, temperature, pH, the level of natural organic matter in the water and the bromide concentration (AWWA Satellite Teleconference, 1993).

For example:

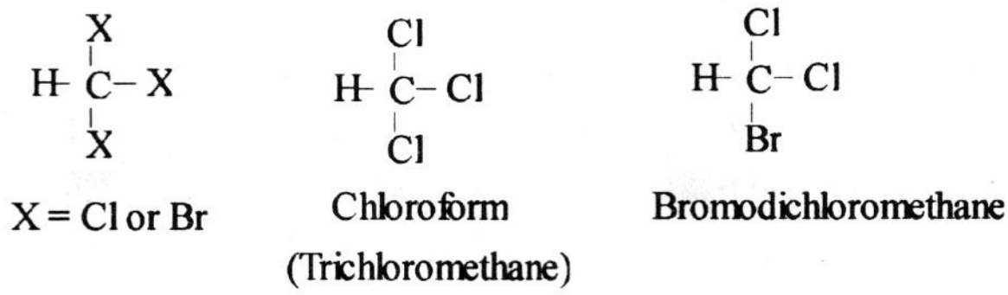

Figure 4. Chemical structure of trihalomethane (THM) species (AWWA Satellite Teleconference,1993)

For example:

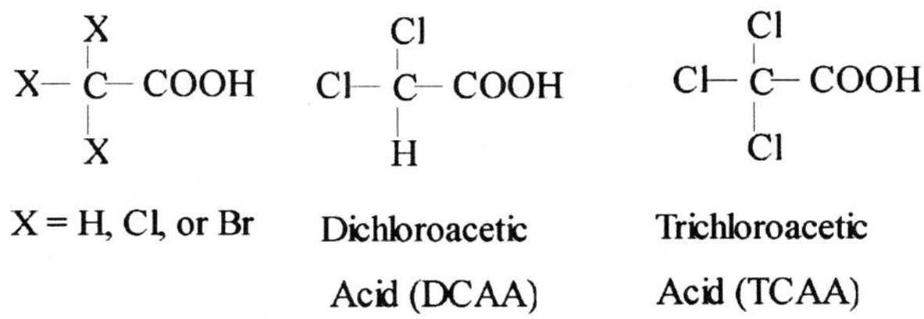

Figure 5. Chemical structure of haloacetic acid species (AWWA Satellite Teleconference, 1993)

A balance between the need to disinfect water to prevent waterborne disease and to minimize the creation of DBPs that are linked with causing cancer is essential. Figure 6. shows graphically the risks of waterborne disease and cancer caused by DBPs with increasing concentrations of chlorine.

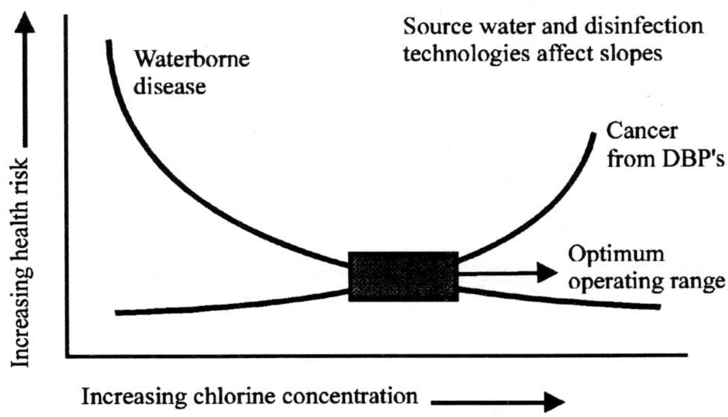

Figure 6. Risk trade-offs between waterborne disease and cancer caused by disinfection by-products (adapted from AWWA Satellite Teleconference, 1993)

At low concentrations of chlorine, the risk of waterborne disease is high and the risk of cancer from formed DBPs is low. As chlorine levels are increased, the risk of waterborne disease is lowered and the risk of cancer from formed DBPs is increased.

Best available technologies for controlling DBPs include: moving the point of chlorination to the end of the treatment plant, using alternative disinfection (e.g., chloramines, ozone), or removing the natural organic matter or precursors to DBPs prior to chlorination (e.g., enhanced coagulation, granular activated carbon, or nanofiltration). Moving points of pre-chlorination to later stages in treatment allows for removal of DBP precursors by conventional processes. Historically, water treatment plants have pre-chlorinated prior to the treatment plant and prior to filtration as a means of preventing biological growth. Moving these points of chlorination allows for the coagulation/sedimentation and filtration processes to reduce DBP precursors prior to chlorination. Research has also shown that the addition of ammonia to chlorinated water stops DBPs from continuing to form in the distribution system. Adding ammonia prior to the distribution system stops the formation of DBPs (AWWA Satellite Teleconference, 1993).

Table 7. Disinfection by-products (DBPs) formed by chlorination (AWWA Satellite Teleconference, 1993)

Disinfection By-Products	
Trihalomethanes (THMs)	**Haloacetic Acids (HAAs)**
Chloroform	Monochloroacetic Acid
Bromodichloromethane	Dichloroacetic Acid
Dibromochloromethane	Trichloroacetic Acid
Bromoform	Monobromoacetic Acid
	Dibromoacetic Acid
Cyanogen Halides	Bromochloroacetic Acid
Cyanogen Chloride	
Cyanogen Bromide	**Haloacetonitriles (HANs)**
	Dichloroacetonitrile
Halopicrins	Trichloroacetonitrile
Chloropicrin	Dibromoacetonitrile
Bromopicrin	Tribromoacetonitrile
	Bromochloroacetonitrile
Chlorol Hydrate	
	Haloketones, Haloaldehydes, Halophenols
MX [3-chloro-4-(dichloromethyl)-5-hydroxy-2(5H)-furanone]	

While substituting chlorination with ozone eliminates the DBPs listed in Table 7., other DBPs with unknown health effects are formed. Ozone also does not leave a residual needed to disinfect the distribution system. A secondary disinfectant which is capable of leaving a residual in the distribution system is still required. Enhancing or increasing coagulation concentrations has shown to increase DBP precursor removals effectively reducing the formation of DBPs. Granular activated carbon and nanofiltration (a membrane technology) have also been shown to reduce DBP precursors.

3.3.3. Corrosion Control

A total of 11 chemical poisonings have occurred from corrosive waters leaching copper and lead from plumbing materials during the period between 1971 to 1983 (Craun, 1986). Six acute copper poisonings with measured copper levels of 4, 12.5 38.5 and 80 mg/l occurred during 1971 through 1980. Five copper and lead poisonings occurred during 1981 to 1983.

The concern of copper and lead leaching from piping and their acute and chronic health effects has promulgated the Lead and Copper Rule. This rule sets what are known as "action levels" for lead and copper of 15 μg/l and 1,300 μg/l, respectively, measured in water discharging from residential taps.

Treatment approaches for reducing copper and lead levels has included: pH adjustment to achieve a more positive Langlier Index (i.e., promote calcium carbonate saturation and deposition), feeding various phosphate-based corrosion inhibitors (e.g., zinc orthophosphate), and replacing lead service lines. Results of corrosion control studies have been positive showing compliance with the action levels (Ramaley, 1993). Another benefit of controlling corrosion has been the reduction of metal levels in biosolids from associated wastewater treatment plants.

3.3.4. Biofilms and Biological Regrowth

Biofilms and biological regrowth within the pipes is a recent and new concern in water treatment. Biofilms refer to microbial cells that have attached to pipe surfaces or on particles (suspended in the water or stationary within a filter). Biofilms may consist of numerous species, sometimes in relatively high numbers. Residual chlorine concentrations in distribution systems have an important impact on the diversity of these biofilms (AWWA, 1988).

Investigations of distribution systems have shown that biofilms occur and may be responsible for the release of bacteria into treated drinking water. Biofilms have been linked with coliform outbreaks in New Haven, Connecticut (AWWA, 1988).

Conditions conducive to biofilm growth include (AWWA, 1988):

- High pH (pH range 6.8 to 7.5) in raw, treated effluent and distribution water simultaneously, coupled with low concentrations of free chlorine,
- High turbidity in raw, treated effluent, and distribution water simultaneously, coupled with low concentrations of free chlorine, and
- High coliform counts in the treated effluent, coupled with low concentrations of free chlorine.

The problems of biofilms has sparked research into biological treatment at drinking water plants. Historically, biological treatment was prevented using pre-chlorination in front of treatment plants and ahead of filters. With the problems of disinfection by-products, pre-chlorination has been almost eliminated. Now, biological treatment within filters is being considered as a method for reducing organics in water, resulting in lowering disinfection by-products and in removing microbial nutrients needed for biofilm growth. Biologically active filters have resulted in less biological activity in distribution systems (AWWA, 1988). Assimilable organic carbon (AOC) is a new analytical measurement used to detect the level of organics in water that would contribute to biofilm growth. This measurement is currently being explored in research and by water utilities (Reid, 1991).

3.4. ESTABLISHED AND EMERGING TECHNOLOGIES IN THE UNITED STATES

3.4.1. Overview of Available Technologies

Table 8. provides a listing of the available treatment technologies used in water treatment. The technologies are organized according to type of treatment needed to remove key contaminants. Summary data is provided for each technology and alternatives are compared on the basis of performance, suitability to treatment plant size, degree of acceptance, conditions required for effective treatment, operational and maintenance requirements, and compatibility with other treatment processes.

Table 8. (p1 of 2) Experimental, emerging and established water treatment technologies in the United States (EPA, 1989)

Technologies	Stage of Acceptability	Size Suitability	Comments
FILTRATION - MICROBIAL CONTAMINANTS			
Conventional	Established	All	Most common: adapted for adding other processes
Direct	Established	All	Lower cost alternative to conventional filtration
Slow sand	Established	Mostly small	Operationally simple: low cost, but requires large land areas
Package plant	Established	Mostly small	Compact, variety of process combinations available
Diatomaceous earth	Established	Mostly small	Limited applicability; potentially expensive for small systems
Membrane	Emerging	Mostly small	Experimental, expensive
Cartridge	Emerging	Small	Experimental; expensive
DISINFECTION - MICROBIAL CONTAMINANTS			
Chlorine	Established	All	Most widely used method; concerns about health effects of by-products
Chlorine dioxide	Established	All	Relatively new to the US; concerns about inorganic by-products
Monochloramine	Established	All	Secondary disinfectant only; some by-product concerns
Ozone	Established	All	Very effective; requires a secondary disinfectant
Ultraviolet radiation	Established	All	Simple, no established harmful by-products; requires a secondary disinfectant
Advanced oxidation (ozone plus hydrogen peroxide plus ultraviolet radiation)	Emerging	All	Limited information concerning disinfection aspects of this process

Table 8. (p2 of 2) Experimental, emerging and established water treatment technologies in the United States (EPA, 1989)

ORGANIC CONTAMINANTS			
Granular activated carbon	Best available technology	All	Highly effective; potential waste disposal problem
Packed column aeration	Best available technology	All	Highly effective for volatile compounds; potential air emissions issues
Powdered activated carbon	Established	Large	Requires conventional treatment processes
Diffused aeration	Established	All	Variable removal effective
Multiple tray aeration	Established	All	Variable removal effectiveness
Oxidation	Experimental	All	By-product concerns
Reverse osmosis	Emerging	Medium	Variable removal effectiveness; expensive
Mechanical aeration	Experimental	All	Mostly for wastewater treatment; high energy requirements, easy to operate
Catenary grid	Experimental	All	Performance data are scarce; potential air emissions issues
Higee aeration	Experimental	Small	Compact, high energy requirements; potential air emissions problems
Resins	Experimental	Small	Data scarce
Ultrafiltration	Emerging	Small	Primarily for turbidity; data for organics removal is limited
INORGANIC CONTAMINANTS			
Reverse osmosis	Established	Medium	Highly effective; expensive; potential waste disposal problems
Ion exchange	Established	Medium	Highly effective; expensive; potential waste disposal problems
Activated alumina	Established	Small	Highly effective; expensive; potential waste disposal problem
Granular activated carbon	Experimental	Small	Experimental for radionuclide removal; potential waste disposal problem
CORROSION CONTROLS			
pH control	Established	All	Potential conflict with other treatments
Corrosion inhibitors	Established	All	Variable effectiveness depending on type of inhibitor

3.4.2. Established Technologies

Table 9. lists the most commonly used treatment technologies for treating water sources (FRDS, 1994). These technologies were identified in the Federal Reporting Data System (FRDS). Information in the FRDS is supplied by State Agencies in compliance with the quarterly reporting requirements under the Safe Drinking Water Act. The accuracy of this information is dependent upon the quality of information provided by State Agencies. The entire list of treatment technologies used and their significance are listed in Table F. in the Appendix.

Table 9 . Common used treatment technologies used for treating drinking water (FRDS, 1994)

Water Treatment Practices	
Adsorption	**Disinfection**
Ion Exchange	Gaseous chlorination, Post
	Gaseous chlorination, Pre
Aeration	Hypochlorination, Post
Slate Tray	
	Filtration
Conventional processes	Rapid Sand
Rapid mix	Pressure Sand
Coagulation	
Flocculation	**Fluoridation**
Sedimentation	
	Lime-soda ash softening
Corrosion Control	
Polyphosphate inhibator	

3.4.3. Emerging Technologies

Although many of the so-called emerging technologies listed in Table 8. have been used either in research or demonstration plants, their use by community water systems has been limited. For a technology to emerge into practice lies with the effect of new regulatory requirements. The direction of water treatment research is often driven by regulations. The more pertinent technologies having a strong potential of emerging in the next century include:

Alternative Disinfection	*Membrane Filtration*
Ozone disinfection	Ultrafiltration
Ultraviolet disinfection	Nanofiltration
Biological Treatment	*Granular Activated Carbon*
Biofilters	*Computer Control Systems*
Granular activated carbon	

3.4.3.1. Alternative Disinfection. Alternative disinfection technologies such as ozone and ultraviolet light will emerge as a result of regulations targeting harmful disinfection by-products and pathogenic organisms such as *cryptosporidium* which is more resistant to chlorination.

3.4.3.2. Biological Treatment. Allowing microbial activity in conventional filters has shown to reduce microbial nutrients in treated water resulting in lowering the potential for biofilm growth within distribution systems (AWWA Research Foundation, 1988). With concerns about increased levels of coliforms due to biofilms, biological treatment in conventional filters and on granular activated carbon may emerge as promising technologies. Further research is needed.

3.4.3.3. Membrane Filtration. Membrane filtration is gaining wider spread acceptance due its ability to remove a variety of contaminants including organic and inorganic contaminants and pathogenic organisms. Ultrafiltration has shown to be effective in disinfection and removal of large weight organics. In conjunction with either coagulation or granular activated carbon, ultrafiltration has been effective in reducing natural organics that are precursors to disinfection by-products. Nanofiltration, introduced in the early 1980s, is very effective in reducing disinfection by-product precursors without any other treatment.

3.4.3.4. Granular Activated Carbon. Granular activated carbon will emerge as a technology used to reduce levels of disinfection by-product precursors as well as regulated chlorinated hydrocarbons.

3.4.3.5. Computer Control Systems. Computerization continues to be one of the fastest areas for emerging technologies. Computer control systems allow for real time monitoring of water characteristics and process performance. Sensors on influent waters to treatment plants can detect quickly abnormal water characteristics which may lead to process upsets or reduced treatment efficiency. Plant operations such as filter backwashing can be automated by computer controls.

3.5. TECHNOLOGY COST COMPARISONS

The capital costs, annual operational and maintenance (O & M) costs, and the total costs per 10 cubic meters of treated water (capital costs have been amortized over a 20 year period at a 10% fixed interest) for different treatment technologies are listed in Table 10. (USEPA, 1990). These costs are based on a small treatment plant size of 400 m^3 per day (100,000 gallons per day).

As indicated by Table 10., each treatment technology has different mix of capital and operating and maintenance costs. Technologies with high capital costs often have lower operating and maintenance costs (e.g., slow sand filters); those with lower capital costs often have higher operational and maintenance costs (e.g., hypochlorite solution - disinfection). Thus, communities which can not afford the initial capital costs usually become burdened with higher operational and maintenance costs later.

Based on the costs listed in Table 10., reverse osmosis and covered slow sand filters are the most expensive technologies with costs of \$6.40 and \$6.60 per 10 m^3 of water treated, respectively. Reverse osmosis capital and operational and maintenance costs are expensive relative to other technologies. Reverse osmosis equipment which includes high pressure pumps, pressure vessels and computer automation is expensive. Operational and maintenance costs include membrane replacement and also high energy requirements. Costs for covering slow sand filters is a major capital cost. Land requirements are normally large resulting in large covers. In comparison, the capital costs for uncovered slow sand filters is nearly half that for covered filters (Table 10.).

Of the disinfection technologies listed, ozone is the most expensive. Capital costs which include the cost of an ozone generator are significantly higher than the other disinfection technologies. Operational and maintenance costs, however, are less for ozone compared with hypochlorite solution disinfection.

Table 10. Estimated costs of drinking water technologies for a 400 m^3/day (100,000 gal/day) plant (USEPA, 1990)

Treatment Technologies	Capital Cost (a,b) $1994	Annual O & M (a,b) $1994	Total Cost per 10 m^3 (c)
FILTRATION			
Slow sand filters; covered	$740,000	$9,800	$6.60
Slow sand filters; uncovered	$427,000	$9,100	$4.10
Diatomaceous earth vacuum filters	$131,000	$14,200	$2.00
Diatomaceous earth pressure filters	$135,000	$13,500	$2.00
PACKAGE PLANT FILTERS			
Coagulation/Filtration with tube settlers	$225,000	$14,000	$2.80
Pressure depth clarifiers/Pressure filter	$263,000	$13,300	$3.00
Pressure depth clarifier/Pressure filter with GAC adsorber	$314,000	$20,800	$4.00
DISINFECTION			
Gas feed chlorination	$13,400	$4,500	$0.40
Hypochlorite solution	$5,200	$7,100	$0.50
Pellet feed chlorinators	$2,100	$5,100	$0.40
Ultraviolet light	$33,200	$2,700	$0.50
Ozonation-high pressure	$50,100	$6,500	$0.80
ORGANIC CONTAMINANTS			
Granular activated carbon; pressure vessel	$223,000	$18,400 (6 mo carbon)	$3.10
Packed tower aerator	$58,000	$3,700	$0.70
INORGANIC CONTAMINANTS			
High pressure reverse osmosis	$351,000	$52,700	$6.40
Low pressure reverse osmosis	$351,000	$38,000	$5.40
Cation exchange	$193,000	$10,900	$2.30
Anion exchange	$147,000	$13,100	$2.10
Activated alumina	$133,000	$18,700	$2.40

(a) Construction costs generally include manufactured equipment, concrete, steel, labor, pipes and valves, electrical equipment and instrumentation, housing, site evacuation, some other site work, general contractor's overhead and profit, engineering costs, financial and administrative costs, and interest costs during construction. Construction costs do not include land costs, legal fees, interface piping, roads, and certain other site work. O & M costs generally include annual energy, labor, and chemical costs. Construction costs can vary depending on specific data characteristics. O & M costs can vary, up to plus or minus 100 percent for some technologies, depending on such variables as feed water characteristics, flow rate, and chemical dosage requirements.
(b) Capital costs and O & M costs have been inflated at an annual percentage rate of 5% from originally reported 1989 cost figures reported in EPA, 1990 (F/P, 5%, 5 years = 1.276).
(c) Costs include capital costs annualized at 10 percent interest over 20 years (A/P, 10%, 20 years = 0.1175) plus annual O & M costs.

Granular activated carbon is more expensive than packed tower aeration due to significant capital costs and the expense of regenerating carbon. Packed tower aeration, however, is only effective for volatile organic compounds. Non-volatile organics would require granular activated carbon or some other technology.

For removing inorganic contaminants, ion exchange resins (i.e., cation and anion) are the least expensive technologies both with respect to capital and operational and maintenance costs. Reverse osmosis (high and low pressures) have significantly higher capital and operational and maintenance costs.

3.6. RESEARCH AT COLORADO STATE UNIVERSITY

3.6.1. Pilot-Plant Studies

A Water Treatment Research Facility (WTRF) was constructed at Colorado State University in November, 1992. This facility includes two pilot-plants: (1) an advance water treatment pilot-plant and (2) a membrane filtration pilot-plant. The advanced water treatment pilot-plant was funded by the US Army Corps of Engineers Construction Engineering Research Laboratory in Champaign, Illinois (Hendricks et al., 1993).

Processes include low and high G rapid mix, flocculation and coagulation, plate settling, multi-media filtration, ozone, granular activated carbon, ion exchange and air stripping. Rated at a flow of 75 liters per minute (20 gal/min), this pilot-plant is at a scale large enough to simulate full-scale operation. The membrane filtration pilot-plant was funded by the USEPA Risk Reduction Engineering Laboratory (RREL) in Cincinnati, Ohio (Champlin and Hendricks, 1994). This pilot-plant consists of six (6) 4"x 40" pressure vessels, each provided with its own high pressure multi-stage centrifugal pump.

The WTRF occupies an area of about 10 m x 20 m (30 ft X 60 ft) at the south end of the Hydraulics Laboratory in the Engineering Research Center, located at the Foothills Campus of Colorado State University in Fort Collins, Colorado. Figure 7. shows the layout of the different processes and supporting equipment. Tables 11. and 12. list some of the features of each pilot plant.

The purpose of the pilot plant is to permit testing of alternative treatment trains for a given set of contaminants and water quality. It has also been used to evaluate individual processes. Specific waters can be synthesized by adding chemical spikes, pathogenic organisms or particles. The pilot plant is fitted with sensors and sampling ports to allow determination of various water contaminant profiles as they change through treatment. Thus, the effects of different variables on processes can be assessed. The selection of an appropriate treatment train for a given contaminated water results in effective treatment at lower cost. At the same time, optimization of a unit process design can result in significant cost savings. Being able to conduct research adds to the knowledge base on the behavior of the respective process which translates into lower cost designs.

The pilot plant is computer controlled, in that valves can automatically be opened and closed along with the collection of sensor data. Flow, temperature, pressure, conductivity, turbidity and particle sensors send data directly to a computer. The computer is programmed to detect problems in the operation of the pilot plant.

For example, flow sensors on influent and effluent lines are used to detect leaks. Differences in flows in and out of the pilot plant greater than 10% indicate a leak to the computer, shutting the pilot plant down. Sensor data also provides a means to trouble shoot and to explain unexpected events.

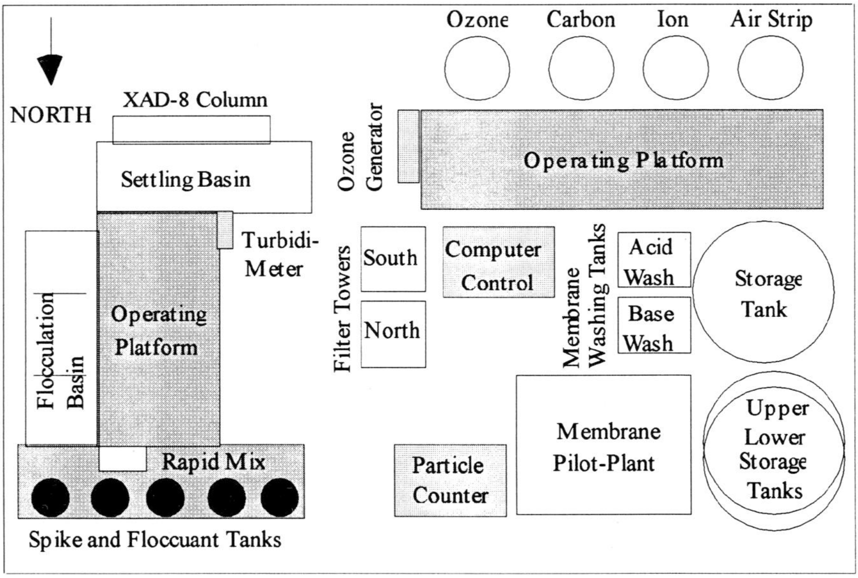

Figure 7. Layout of water treatment research facility (WTRF) at Colorado State University (Hendricks et al., 1993)

Table 11. Available unit processes at the water treatment research facility (Hendricks et al., 1993)

Unit Process	Objective
Advanced Water Treatment Pilot-Plant	
Low G rapid mix	Disperses coagulant
High G rapid mix	Disperses coagulant more effectively
Coagulation	Charge neutralization of colloidal particles
Flocculation	Promote collisions between particles increasing particle size
Lamella plate settling	Permit settling of settlable particles
Dual-media filtration	Removal of small charged particles
Ozone (generation, gas transfer, reaction)	Oxidize organic molecules and disinfection
Granular activated adsorption	Adsorb organic molecules
Clino-thiolite ion exchange	Adsorbs cations, particularly metal ions
Air stripping	Transfer of volatile organic compounds to the gas phase
Membrane Filtration Pilot-Plant	
Six Membrane Pressure Vessels	Removal of particles, molecules, ions in accordance with pore size of membranes tested

Table 12. Appurtenant equipment to water treatment research pilot plant (Hendricks et al., 1993)

Appurtenant Equipment	
Components	
Head boxes	Motors
Pumps	Stirrers
Metering pumps	Blowers
Chemical storage tanks	Computers
Air compressor	Storage tanks
Control System	
Sensors; flow, water levels, pressure, temperature, conductivity, pH, turbidity, particle counts	
Automated on-off valves	
Animated analogs	
Plot generation	
Sampling	
Taps at each node point between processes	
Taps along column towers	
Laboratory Support	
Gas chromatograph	Zeta meter
Total organic carbon	Particle counter
UV analyzer	Spectrophotometer
Fluorometer	Turbidimeter
Safety Plan	
Containment area to capture spills	Approvals for discharge to city sewer
Carbon vent traps	Secondary containment
Power back-up for computers	Personnel health and safety training
Computer shut-down from sensor data	Computer alarms

3.6.2. Survey of Treatment Technologies

Treatment of contaminated waters is a major issue in the United States and around the world. Problems vary due to differences between municipal and industrial needs, the varied types of contaminated sources, the wide range of contaminants, and the increasingly more stringent regulations. Treatment trains, comprised of an array of unit processes, are selected typically for the type of contaminants present and the degree of treatment required. With the wide range of contaminants involved and the need for economical treatment, the selection of unit processes for treatment trains is uncertain.

As one part of a two year project with the US Army Corps of Engineers Research Laboratory, Colorado State University was involved in research to delineate performance of varying treatment technologies for a variety of water contaminants. One delineation has been the development of a technology/performance matrix, intended to assist engineers in the preliminary selection of treatment unit processes (Champlin and Hendricks, 1993).

Based upon reviews of data bases and journal articles, a technology/performance matrix which indicates expected removals for nearly 650 different chemicals and water contaminants versus 50 different treatment technologies was developed (Champlin and Hendricks, 1993). Listed contaminants were defined by federal laws and regulations.

The matrix shows clearly the limited availability of treatability information. For the 46,000 cells that comprise the matrix, about 80% show the symbol "NA" (not available) with respect to treatability information (Champlin and Hendricks, 1993). Thus the matrix shows that for most of the contaminants listed, there is a dearth of knowledge on treatment efficiency, design criteria, operational guidelines.

The matrix shows also the reliance that has been placed primarily on the activated sludge and activated carbon technologies. About 25% of the contaminants listed in the matrix are effectively treated (50% removal or better) using activated sludge (Champlin and Hendricks, 1993). By the same token, about 10% of the contaminants listed in the matrix are effectively treated using activated carbon technologies (i.e., GAC and PAC). A smaller percentage of contaminants are treated effectively by the other listed technologies.

3.7. REFERENCES

ASCE and AWWA (1990) Water Treatment Plant Design. Second Edition, McGraw-Hill Publishing Co. New York, New York. ISBN 0-07-001542-2.

AWWA Research Foundation (1992) WIDB Water Industry Data Base: Utility Profiles. American Water Works Association, Denver, Colorado. ISBN 0-89867-643-6.

AWWA Research Foundation (1988) Bacterial Regrowth in Distribution Systems, Research Report. Subject Area: Distribution Systems. American Water Works Association, Denver, Colorado. ISBN 0-89867-418-2.

AWWA Satellite Teleconference (1994) Preventing Waterborne Disease: Is Your System at Risk? Participant Guide. Broadcast April 8, 1994. American Water Works Association, Denver, Colorado. ISBN 0-89867-742-4.

AWWA Satellite Teleconference (1993) Disinfectants and Disinfection By-Products: Understanding the Proposed D/DBP Rule. Participant Guide. Broadcast October 15,1993. American Water Works Association, Denver, Colorado. ISBN 0-89867-708-4.

Cantor K.P. (1992) Cancer Risks Associated with Chlorination By-Products. WaterReview Technical Report, A Publication of the Water Quality Research Council. Vol. 10, No. 2, 1992.

Centers for Disease Control (1985) Water-Related Disease Outbreaks Surveillance. Annual Summary 1984, US Department of Health and Human Services, Public Health Service. Issued November, 1985.

Champlin and Hendricks (1994) Nanofiltration For Treatment of Low-Turbidity - Waters Membrane Fouling and Removals of Disinfection By-Product Precursors. Report prepared for USEPA Risk Reduction Engineering Laboratory, Cincinnati, Ohio. Prepared by Colorado State University, Department of Civil Engineering, Fort Collins, Colorado.

Champlin and Hendricks (1993) Treatment Train Modeling for Aqueous Contaminants. Volume 2 of 5, Matrix of Contaminants and Treatment Technologies, Report prepared for the US Army Construction Engineering Research Laboratory, Champaign, Illinois. Prepared by Colorado State University, Department of Civil Engineering, Fort Collins, Colorado.

Craun G.F. (1986) Waterborne Diseases in the United States. CRC Press, Inc., Boca Raton, Florida.

Culp G.L., Wesner G.M., Culp R.L. (1986) Handbook of Public Water Systems. Van Nostrand Reinhold Co. Inc. New York, New York.

FRDS (1994) Federal Reporting Data System. Information supplied by Primary Agents in Compliance With Quarterly Reporting Equirments Under the Safe Drinking Water Act. Search of Reporting Treatment Plant Processes, July 15, 1994.

Fox K.R., Lytle D.A. (1993) The Milwaukee Cryptosporidiosis Outbreak: Investigation and Recommendations. Proceedings to the AWWA Water Quality Technology Conference, Part II - Session 3D through ST6. November 7-11, 1993, Miami, Florida.

Hendricks D.W. et al. (1993) Treatment Train Modeling for Aqueous Contaminants. Volume 1 of 5, Summary Report, Report prepared for the US Army Construction Engineering Research Laboratory, Champaign, Illinois. Prepared by Colorado State University, Department of Civil Engineering, Fort Collins, Colorado.

Hendricks D.W. et al. (1986a) Removing Giardia Cysts from Low Turbidity Waters by Rapid Rate Filtration. Journal of AWWA., May, 1986.

Hendricks D.W. et al. (1986b) Rapid Rate Filtration of Low Turbidity Water Using Field-Scale Pilot Filters. Journal of AWWA., December, 1986.

Hendricks D.W. et al. (1988a) Removing Giardia Cysts and Other Particles from Low Turbidity Waters Using Dual-Stage Filtration. Journal of AWWA. Vol. 80, No. 2, February, 1988.

Hendricks D.W. et al. (1988b) Bench-Scale Evaluations of Coagulants for Low Turbidity Water. Journal of AWWA, April, 1988.

Herwaldt B.L., Craun G.F., Stokes S.L., Juranek D.D. (1992) Waterborne Disease Outbreaks. 1989-1990, Centers for Disease Control Morbidity and Mortality Weekly Report, 1992, Vol. 40, No. SS-3, p1-21.

Hileman B. (1992) Cancer Risk Found From Water Chlorination. Chemical & Engineering News. July 13, 1992.

Lay T. (1993) SDWA Reauthorization - Good Time to Plan for Future. AWWA Opflow, Vol. 19, No. 1, ISSN 0149-8029, January, 1993.

Levine W.C., Craun G.F. (1990) Waterborne Disease Outbreaks. 1986-1988, Centers for Disease Control Morbidity and Mortality Weekly Report. March, 1990, Vol. 39, No. SS-1, p1-13.

Montgomery J.M. (1985) Water Treatment - Principals & Design. 1985. John Wiley & Sons, Inc. New York, New York.

Moore A.C., Herwaldt B.L., Craun G.F., Calderon R.L., Highsmith A.K., Juranek D.D. (1993) Surveillance for Waterborne Disease Outbreaks - United States. 1991-1992, Centers for Disease Control Morbidity and Mortality Weekly Report. November 19, 1993, Vol. 42, No. SS-5, p1-22.

Morris R.D., Audet A.M., Angelillo I.F., Chalmers T.C., Mosteller F. (1992) Chlorination, Chlorination By-products and Cancer: A Meta-Analysis. American Journal of Public Health. July 1992, Vol. 82, No. 7, p. 955-963.

Pontius F.W. (1994a) SDWA Advisor, Regulatory Update Service. American Water Works Association Publication. Denver, Colorado. ISBN 0-89867-736-X.

Pontius F.W. (1994b) Part II - What You Should Know About Cryptosporidium. AWWA Opflow, Vol. 21, No. 7, ISSN 0149-8029, June, 1994.

Pontius F.W. (1993) Surface Water Treatment Rule Targets Microbials. AWWA Opflow, Vol. 19, No. 5, ISSN 0149-8029, May, 1993.

Pontius F.W. (1992) Plan, Prepare Now for Impact of Phase V Regs. AWWA Opflow, Vol. 18, No. 11, ISSN 0149-8029, November, 1992.

Ramaley B.L. (1993) Monitoring and Control Experience Under the Lead and Copper Rule. Journal of AWWA. Vol. 85, No. 2, February, 1993.

Reid J. (1991) Emerging Technologies - Some Are Years Away, But Others are Coming to a Plant Near You. AWWA MainStream, July 1991.

St. Louis M.E. (1988) Water-Related Disease Outbreaks, 1985. Centers for Disease Control Morbidity and Mortality Weekly Report, CDC Surveillance Summaries. June, 1988, Vol. 37, No. SS-2, p15-24.

US Environmental Protection Agency (1978) Estimating Costs for Water Treatment as a Function of Size and Treatment Plant Efficiency. Municipal Environmental Research Laboratory, Cincinnati, Ohio. EPA-600/2-78-182.

US Environmental Protection Agency (1989) Technologies for Upgrading Existing or Designing New Drinking Water Treatment Facilities. Office of Drinking Water, Cincinnati, Ohio. EPA/625/4-89/023.

US Environmental Protection Agency (1990) Environmental Pollution Control Alternatives: Drinking Water Treatment for Small Communities. Center for Environmental Research Information, Cincinnati, Ohio. EPA/625/5-90/025.

US Environmental Protection Agency (1992) Drinking Water Seminar Handbook. US Environmental Protection Agency, Region VIII, Denver Colorado and Department of Civil Engineering, University of Wyoming, Laramie, Wyoming.

US Environmental Protection Agency (1993) Drinking Water Regulations and Health Advisors. Office of Water, Washington D. C. EPA 822/F-93-011.

Vesilind P.A., Peirce J.J. (1982) Environmental Engineering. Butterworth Publishers, An Ann Arbor Science Book. Woburn, Massachusetts.

Water Quality Association (1992) Findings Link Chlorination with Bladder and Rectal Cancer. WaterReview Technical Report, A Publication of the Water Quality Research Council. Vol. 7, No. 2, 1992.

3.8. APPENDIX OF DATA TABLES

Table A. (p1 of 6) United States drinking water standards, health effects and best available technologies (BAT) for treatment of organic contaminants

Contaminant	Health / Advisories		Standards [(b)]			
	Effect	Refer. Dose [(a)] (mg/kg/d)	Status	MCLG (mg/l)	MCL (mg/l)	BAT [(c)]
ORGANICS						
Acenaphthene		0.06	-	-	-	
Acifluorfen		0.013	F	zero	-	
Acrylamide		0.001	F	zero	TT	PA
Acrylonitrile		-	D	zero	-	
Adipate (diethyhexyl)	Carcinogen, liver and reproductive system effects	0.6	-	0.4	0.4	
Alachlor	Carcinogen	0.01	F	zero	0.002	GAC
Aldicarb		0.001	D	0.007	0.007	GAC
Aldicarb sulfone		0.001	D	0.007	0.007	GAC
Aldicarb sulfoxide		0.001	D	0.007	0.007	GAC
Aldrin		0.00003	D	-	-	
Ametryn		0.009	F	-	-	
Ammonium sulfamate		0.28	F	-	-	
Anthracene (PAH)		0.3	-	-	-	
Atrazine	Effects offspring	0.035	F	0.003	0.003	GAC
Baygon		0.004	F	-	-	
Bentazon		0.0025	F	0.02	-	
Bemz(a) anthracene (PAH)		-	-	zero	0.0001	
Benzene	Carcinogen	-	F	zero	0.005	GAC, PTA
Benzo(a) pyrene (PAH)	Carcinogen	-	-	zero	0.0002	GAC
Benzo(b) fluoranthene (PAH)		-	-	zero	0.0002	
Benzo(g,h,i) perylene		-	-	-	-	
Benzo(k) fluoranthene		-	-	zero	0.0002	
bis-2-Chloroisopropyl ether		0.04	F	-	-	
Bromacil		0.13	F	-	-	
Bromobenzene		-	D	-	-	
Bromochloroacetonitrile		-	D	-	-	
Bromochloromethane		0.013	F	-	-	
Bromodichloromethane (THM)		0.02	D	zero	0.1	
Bromoform (THM)		0.02	D	zero	0.1	
Bromomethane		0.001	F	-	-	
Butyl benzyl phthalate (PAE)		0.2	-	zero	0.1	
Butylate		0.05	F	-	-	
Butylbenzene n-		-	D	-	-	
Buthylbenzene sec-		-	D	-	-	
Butylbenzene tert-		-	D	-	-	
Carbaryl		0.1	F	-	-	

Footnotes are listed at the end of the table

Table A. (p2 of 6) United States drinking water standards, health effects and best available technologies (BAT) for treatment of organic contaminants

	Health / Advisories		Standards [b]			
Contaminant	**Effect**	**Refer. Dose [a] (mg/kg/d)**	**Status**	**MCLG (mg/l)**	**MCL (mg/l)**	**BAT [c]**
Carbofuran	Nervous and reproductive system	0.005	F	0.04	0.04	GAC
Carbon tetrachloride	Carcinogen	0.0007	F	zero	0.005	GAC, PTA
Carboxin		0.1	F	-	-	
Chloral hydrate		0.0002	D	0.06	-	
Chloramben		0.015	F	-	-	
Chlordane	Carcinogen	0.00006	F	zero	0.002	GAC
Chlorodibromomethane (THM)		0.02	D	zero	0.1	
Chloroethene		-	D	-	-	
Chloroform (THM)		0.01	D	zero	0.1	
Chloromethane		0.004	F	-	-	
Chlorophenol (2-)		0.005	D	-	-	
p-Chlorophenyl methyl sulfide/ sulfone/sulfoxide		-	-	-	-	
Chloropicrin		-	-	-	-	
Chlorothalonil		0.015	F	-	-	
Chlorotoluene o-		0.02	F	-	-	
Chlorotoluene p-		0.02	F	-	-	
Chlorpyrifos		0.003	F	-	-	
Chrysene (PAH)		-	-	zero	0.0002	
Cyanazine		0.002	D	0.001	-	
Cyanogen chloride		-	-	-	-	
Cymene -p		-	D	-	-	
2,4-D	Nervous system, liver and kidney	0.01	F	0.07	0.07	GAC
DCPA (Dacthal)		0.5	F	-	-	
Dalapon	Kidney and liver effects	0.026	F	0.2	0.2	GAC
Di[2-ethylhexyl] adipate	Carcinogen, liver and reprod. syst. ef.	0.6	-	0.4	0.4	GAC, PTA
Diazinon		0.00009	F	-	-	
Dibenz(a,h) anthracene (PAH)		-	-	zero	0.0003	
Dibromoacetonitrile		0.02	D	-	-	
Dibromochloropropane (DBCP)	Carcinogen	-	F	zero	0.0002	GAC, PTA
Dibromomethane		-	-	-	-	
Dibutyl phthalate		0.01	-	-	-	
Dicamba		0.03	F	-	-	
Dichloroacetaldehyde		-	D	-	-	
Dichloroacetic acid		0.04	D	zero	-	
Dichloroacetonitrile		0.008	D	-	-	

Footnotes are listed at the end of the table

Table A. (p3 of 6) United States drinking water standards, health effects and best available technologies (BAT) for treatment of organic contaminants

Contaminant	Health / Advisories Effect	Refer. Dose [(a)] (mg/kg/d)	Standards [(b)] Status	MCLG (mg/l)	MCL (mg/l)	BAT [(c)]
Dichlorobenzene o-	Nervous system, lung, liver and kidney	0.09	F	0.6	0.6	GAC, PTA
Dichorobenzene m-		0.09	F	0.6	0.6	
Dichlorobenzene p-	Liver and kidney	0.1	F	0.075	0.075	GAC, PTA
Dichlrodifluoromethane		0.2	F	-	-	
Dichloroethane (1,1-)		-	D	-	-	
Dichloroethane (1,2-)	Carcinogen	-	F	zero	0.005	GAC, PTA
Dichloroethylene (1,1-)	Liver and kidney	0.009	F	0.007	0.007	GAC, PTA
Dichlroethylene (cis-1,2-)	Nervous system, liver, circulatory system and carcinogen	0.01	FF	0.07	0.07	GAC, PTA
Dichloroethylene (trans-1,2-)	Nervous system, liver and circulatory	0.02	F	0.1	0.1	GAC, PTA
Dichloromethane	Carcinogen	0.06	F	zero	0.005	PTA
Dichlorophenol (2,4-)		0.003	D	-	-	
Dichloropropane (1,1-)		-	D	-	-	
Dichloropropane (1,2-)	Carcinogen, liver, lungs and kidneys	-	F	zero	0.005	GAC, PTA
Dichloropropane (1,3-)		-	D	-	-	
Dichloropropane (2,2-)		-	D	-	-	
Dichloropropene (1,1-)		-	D	-	-	
Dichloropropene (1,3-)		0.0003	F	zero	-	
Dieldrin		0.00005	F	-	-	
Diethyl phthalate (PAE)		0.8	D	-	-	
Diethylene glycol dinitrate		-	-	-	-	
Diethylhexyl phthalate (PAE)		0.02	D	zero	0.006	
Diisopropyl methylphosphonate		0.08	F	-	-	
Dimethrin		0.3	F	-	-	
Dimethyl methylphosphonate		0.2	F	-	-	
Dimethyl phthalate (PAE)		-	-	-	-	
1,3-Dinitrobenzene		0.0001	F	-	-	
Dinitrotoluene (2,4-)		0.002	F	-	-	
Dinitrotoluene (2,6-)		0.001	F	-	-	
tg 2,6 & 2,4 dinitrotoluene		-	-	-	-	
Dinoseb	Thyroid and reproductive organ effects	0.001	F	0.007	0.007	GAC
Dioxane p-		-	F	-	-	
Diphenamid		0.03	F	-	-	
Diphenylamine		0.03	F	-	-	

Footnotes are listed at the end of the table

Table A. (p4 of 6) United States drinking water standards, health effects and best available technologies (BAT) for treatment of organic contaminants

Contaminant	Health / Advisories: Effect	Health / Advisories: Refer. Dose (a) (mg/kg/d)	Standards (b): Status	Standards (b): MCLG (mg/l)	Standards (b): MCL (mg/l)	BAT (c)
Diquat	Liver, kidney, GI tract effects; cataract formation	0.0022	-	0.02	0.02	GAC
Disulfoton		0.00004	F	-	-	
Dithiane (1,4-)		0.01	F	-	-	
Diuron		0.002	F	-	-	
Endothall	Liver, kidney, GI tract, reprod. system effects	0.02	F	0.1	0.1	GAC
Endrin	Liver, kidney and heart effects	0.0003	F	0.002	0.002	GAC
Epichlorohydrin		0.002	F	zero	TT	
Ethylbenzene	Kidney, liver, nervous system	0.1	F	0.7	0.7	GAC, PTA
Ethylene dibromide (EDB)	Carcinogen	-	F	zero	5E-5	GAC, PTA
Ethylene glycol		2	F	-	-	
ETU		0.00008	F	-	-	
Fenamiphos		0.00025	F	-	-	
Fluometron		0.013	F	-	-	
Fluorene (PAH)		0.04	-	-	-	
Fluorotrichloromethane		0.3	F	-	-	
Fog Oil		-	D	-	-	
Fonofos		0.002	F	-	-	
Formaldehyde		0.15	D	-	-	
Gasoline, unleaded (benzene)		-	D	-	-	
Glyphosate	Liver and kidney effects	0.1	F	0.7	0.7	OX
Heptachlor	Carcinogen	0.0005	F	zero	0.0004	GAC
Heptachlor epoxide	Carcinogen	1E-05	F	zero	0.0002	GAC
Hexachlorobenzene	Carcinogen	0.0008	F	zero	0.001	GAC
Hexachlorobutadiene		0.002	F	0.001	-	
Hexachlorocyclopentadiene	Carcinogen	0.007	-	0.05	0.05	GAC, PTA
Hexachloroethane		0.001	F	-	-	
Hexane (n-)		-	F	-	-	
Hexazinone		0.033	F	-	-	
HMX		0.05	F	-	-	
Indeno(1,23,-c,d) pyrne (PAH)		-	D	zero	0.0004	
Isophorone		0.2	F	-	-	
Isopropyl methylphosphonate		0.1	D	-	-	
Isopropylbenzene		-	D	-	-	
Lindane		0.0003	F	0.0002	0.0002	GAC
Malathion		0.02	F	-	-	

Footnotes are listed at the end of the table.

Table A. (p5 of 6) United States drinking water standards, health effects and best available technologies (BAT) for treatment of organic contaminants

	Health / Advisories		Standards [(b)]			
Contaminant	**Effect**	**Refer. Dose [(a)] (mg/kg/d)**	**Status**	**MCLG (mg/l)**	**MCL (mg/l)**	**BAT [(c)]**
Maleic hydrazide		0.5	F	-	-	
MCPA		0.0015	F	-	-	
Methomyl		0.025	F	-	-	
Methoxychlor	Liver, kidney, nervous and reprod. systems	0.005	F	0.04	0.04	GAC
Methyl ethyl ketone		-	F	-	-	
Methyl parathon		0.00025	F	-	-	
Methyl tert butyl ether		0.005	D	-	-	
Metolachlor		0.15	F	-	-	
Metribuzin		0.025	F	-	-	
Monochloroacetic acid		-	D	-	-	
Monochlobenzene	Kidney, liver, nervous system	0.02	F	0.1	0.1	GAC, PTA
Naphthalene		0.004	F	-	-	
Nitrocellulose (non-toxic)		-	F	-	-	
Nitroguanidine		0.1	F	-	-	
Nitrophenol p-		0.008	F	-	-	
Oxamyl (Vydate)	Kidney	0.025	F	0.2	0.2	GAC
Paraquat		0.0045	F	-	-	
Pentachloroethane		-	D	-	-	
Pentachlorphenol		0.03	F	zero	0.001	GAC
Phenanthrene (PAH)		-	-	-	-	
Phenol		0.6	D	-	-	
Picloram	Kidney and liver	0.07	F	0.5	0.5	GAC
Polychlorinated biphenyls (PCBs)	Carcinogen	-	P	zero	0.0005	GAC
Prometon		0.015	F	-	-	
Pronamide		0.075	F	-	-	
Propachlor		0.013	F	-	-	
Propazine		0.02	F	-	-	
Propham		0.02	F	-	-	
Propylbenzene n-		-	D	-	-	
Pyrene (PAH)		0.03	-	-	-	
RDX		0.003	F	-	-	
Simazine	Carcinogen, circulatory system	0.005	F	0.004	0.004	GAC
Styrene	Liver and nervous system	0.2	F	0.1	0.1	GAC, PTA
2,4,5,-TP (Silvex)		0.01	F	-	-	GAC
2,3,7,8-TCDD (Dioxin)	Carcinogen	1E-09	F	zero	3E-8	GAC
Tebuthiuron		0.07	F	-	-	
Terbacil		0.013	F	-	-	
Terbufos		0.00013	F	-	-	
Tetrachloroethane (1,1,1,2-)		0.03	F	-	-	

Footnotes are listed at the end of the table.

Table A. (p6 of 6) United States drinking water standards, health effects and best available technologies (BAT) for treatment of organic contaminants

Contaminant	Health / Advisories: Effect	Refer. Dose (a) (mg/kg/d)	Standards (b): Status	MCLG (mg/l)	MCL (mg/l)	BAT (c)
Tetrachloroethane (1,1,2,2-)		-	D	-	-	
Tetrachloroethylene	Carcinogen	0.01	F	zero	0.005	GAC, PTA
Tetranitromethane		-	-	-	-	
Toluene	Kidney, nervous system and lung	0.2	F	1	1	GAC
Toxaphene	Carcinogen	0.1	F	zero	0.003	GAC, PTA
2, 4,5-TP	Nervous system, liver and kidney	0.0075	F	0.05	0.05	
1, 1, 2-Trichloro-1, 2, 2-trifluoroethane		-	-	-	-	
Trichloroacetic acid		0.1	D	0.1	-	
Trichloroacetonitrile		-	D	-	-	
Trichlorobenzene (1,2,4-)	Kidney and liver effects	0.01	F	0.07	0.07	GAC, PTA
Trichlorobenzene (1,3,5-)		0.006	F		-	
Trichloroethane (1,1,1-)	Liver, nervous system, circ. system	0.035	F	0.2	0.2	GAC, PTA
Trichloroethane (1,1,2-)	Carcinogen, kidney and liver effects	0.004	F	0.003	0.005	GAC, PTA
Trichloroethanol (2,2,2-)		-	-	-	-	
Trichloroethylene	Carcinogen	-	F	zero	0.005	GAC, PTA
Trichlorophenol (2,4,6-)		-	D	-	-	
Trichloropropane (1,1,1-)		-	D	-	-	
Trichloropropane (1,2,3-)		0.006	F	-	-	
Trifluralin		0.0075	F	-	-	
Trimethylbenzene (1,2,4-)		-	D	-	-	
Trimethylbenzene (1,3,5-)		-	D	-	-	
Trinitroglycerol		-	F	-	-	
Trinitrotuluene		0.0005	F	-	-	
Vinyl chloride	Carcinogen	-	F	zero	0.002	PTA
Xylenes	Liver, kidney, nervous system	2	F	10	10	GAC, PTA

(a) Reference dose is an estimate of a daily exposure level to the human population that is likely to be without appreciable risk of deleterious effects over a lifetime (EPA, 1993). Units are mg contaminant per kg mass of human per day.

(b) Status refers to the current state of regulations pertaining to the particular contaminant (F refers to final, D refers to draft and P refers to proposed). MCLG refers to maximum contaminant level goal. A non-enforceable concentration that is protective of adverse human health effects (EPA, 1993). MCL refers to the maximum contaminant level. This concentration is the maximum permissible level which is delivered to any public water system. TT refers to the treatment technique required, no MCL are stated (EPA,1992).

(c) BAT refers to the best available technology for treatment (EPA, 1990; EPA, 1992; and Opflow, Nov. 1992). Abbreviations: AA-Activated Alumina; AER-Aeration; CC-Corrosion Control; C/F Coagulation/flocculation; DDF Direct and Diatomite Filtration; DIS - Disinfection; ED Electrodialysis; FIL - Filtration; GAC - Granular Activated Carbon; IE - Ion Exchange; LS - Lime Softening; PA - Polymer Addition; PAC - Powdered Activated Carbon; PTA - Packed Tower Aeration; and RO - Reverse Osmosis.

Table B. (p1 of 2) United States drinking water standards, health effects and best available technologies (BAT) for treatment of inorganic contaminants

Contaminant	Health / Advisories Effect	Refer. Dose [(a)] (mg/kg/d)	Standards [(b)] Status	MCLG (mg/l)	MCL (mg/l)	BAT [(c)]
INORGANICS						
Aluminum		-	D	-	-	
Ammonia		-	D	-	-	
Antimony	Decreases growth and longevity	0.0004	F	0.006	0.006	C/F, RO, GAC
Arsenic		-	D	-	0.05	
Asbestos (fibers/L >10μm length)		-	-	7 MFL	7 MFL	C/F, DD, CC
Barium		0.07	F	2	2	IE,LS, RO,EI
Beryllium	Carcinogen, damage to bones and lungs	0.005	D	0.004	0.004	AA,IE,RO, LS, C/F
Boron		0.09	D	-	-	
Cadmium		0.0005	F	0.005	0.005	C/F, IE, LS, RO
Chloramine		0.1	D	4	-	
Chlorate		-	D	-	-	
Chlorine		0.08	D	4	-	
Chlorine dioxide		0.003	D	0.08	-	
Chlorite		-	D	-	-	
Chromium (Total)		0.005	F	0.1	0.1	C/F, IE, LS, RO
Copper		-	-	1.3	TT	
Cyanide	Spleen, brain, and liver effects	0.022	F	0.2	0.2	IE, RO, CH
Fluoride		0.12	-	4	4	
Hypochlorite		-	-	4	-	
Hypochlorus acid		-	-	4	-	
Lead (at tap)		-	-	zero	TT	
Manganese		0.005	D	-	-	
Mercury (inorganic)		0.0003	F	0.002	0.002	C/F, GAC, LS, RO, PAC
Molybedenum		0.005	D	-	-	
Nickel	Heart and liver effects	0.02	F	0.1	0.1	IE, RO, LS
Nitrate (as N)		1.6	F	10	10	IE, RO, ED
Nitrite (as N)		0.16	F	1	1	IE, RO
Nitrate + Nitrite (both as N)		-	F	10	10	
Selenium		0.005	-	0.05	0.5	AA, C/F, LS, RO, ED

Footnotes are listed at the end of the table.

Table B. (p2 of 2) United States drinking water standards, health effects and best available technologies (BAT) for treatment of inorganic contaminants

Contaminant	Health / Advisories		Standards (b)			
	Effect	Refer. Dose (a) (mg/kg/d)	Status	MCLG (mg/l)	MCL (mg/l)	BAT (c)
Silver		0.005	D	-	-	
Sodium		-	D	-	-	
Strontium		0.6	D	-	-	
Sulfate	Gastronteritis	-	-	-	-	
Thallium	Kidney, liver and brain effects	0.00007	F	0.0005	0.002	AA, IE
Vanadium		-	D	-	-	
White phosphorus		0.00002	F	-	-	
Zinc		0.3	F	-	-	
Zinc chloride (as Zn)		0.3	F	-	-	

(a) Reference dose is an estimate of a daily exposure level to the human population that is likely to be without appreciable risk of deleterious effects over a lifetime (EPA, 1993). Units are mg contaminant per kg mass of human per day.
(b) Status refers to the current state of regulations pertaining to the particular contaminant (F refers to final, D refers to draft and P refers to proposed). MCLG refers to maximum contaminant level goal. A non-enforceable concentration that is protective of adverse human health effects (EPA, 1993). MCL refers to the maximum contaminant level. This concentration is the maximum permissible level which is delivered to any public water system. TT refers to the treatment technique required, no MCL are stated (EPA,1992).
(c) BAT refers to the best available technology for treatment (EPA, 1990; EPA, 1992; and Opflow, Nov. 1992). Abbreviations: AA - Activated Alumina; AER - Aeration; CC - Corrosion Control; C/F Coagulation/flocculation; DDF Direct and Diatomite Filtration; DIS - Disinfection; ED Electrodialysis; FIL - Filtration; GAC - Granular Activated Carbon; IE - Ion Exchange; LS - Lime Softening; PA - Polymer Addition; PAC - Powdered Activated Carbon; PTA - Packed Tower Aeration; and RO - Reverse Osmosis.

Table C. United States drinking water standards and health effects for treatment of radionuclides

Contaminant	Health / Advisories		Standards (b)		
	Effect	Refer. Dose (a) (mg/kg/d)	Status	MCLG (mg/l)	MCL (mg/l)
RADIONUCLIDES					
Beta particle and photon acitivity		-	-	zero	4 mrem
Gross alpha particle activity		-	-	zero	15 pCi/L
Radium 226		-	-	zero	20 pCi/L
Radium 228		-	-	zero	20 pCi/L
Radon		-	-	zero	300 pCi/L
Uranium		0.003	-	zero	20 μg/L

(a) Reference dose is an estimate of a daily exposure level to the human population that is likely to be without appreciable risk of deleterious effects over a lifetime (EPA, 1993). Units are mg contaminant per kg mass of human per day.
(b) Status refers to the current state of regulations pertaining to the particular contaminant (F refers to final, D refers to draft and P refers to proposed). MCLG refers to maximum contaminant level goal. A non-enforceable concentration that is protective of adverse human health effects (EPA, 1993). MCL refers to the maximum contaminant level. This concentration is the maximum permissible level which is delivered to any public water system. TT refers to the treatment technique required, no MCL are stated (EPA,1992).

Table D. United States drinking water standards, health effects and best available technologies (BAT) for treatment of microorganisms and surrogate contaminants

Contaminant	Health / Advisories		Standards (b)			
	Effect	Refer. Dose (a) (mg/kg/d)	Status	MCLG (mg/L)	MCL (mg/L)	BAT (c)
Crptosporidium			L	-	-	
Giardia lamblia			F	zero	TT	FIL, DIS
Legionella			F	zero	TT	FIL, DIS
Standard Plate Count			F	NA	TT	FIL, DIS
Total Coliforms			F	zero	-	FIL, DIS
Turbidity			F	NA	PS	FIL
Viruses			F	zero	TT	FIL, DIS

(a) Reference dose is an estimate of a daily exposure level to the human population that is likely to be without appreciable risk of deleterious effects over a lifetime (EPA, 1993). Units are mg contaminant per kg mass of human per day.

(b) Status refers to the current state of regulations pertaining to the particular contaminant (F refers to final, D refers to draft and P refers to proposed). MCLG refers to maximum contaminant level goal. A non-enforceable concentration that is protective of adverse human health effects (EPA, 1993). MCL refers to the maximum contaminant level. This concentration is the maximum permissible level which is delivered to any public water system. TT refers to the treatment technique required, no MCL are stated (EPA,1992).

(c) BAT refers to the best available technology for treatment (EPA, 1990; EPA, 1992; and Opflow, Nov. 1992). Abbreviations: AA - Activated Alumina; AER - Aeration; CC - Corrosion Control; C/F Coagulation/flocculation; DDF Direct and Diatomite Filtration; DIS - Disinfection; ED Electrodialysis; FIL - Filtration; GAC - Granular Activated Carbon; IE - Ion Exchange; LS - Lime Softening; PA - Polymer Addition; PAC - Powdered Activated Carbon; PTA - Packed Tower Aeration; and RO - Reverse Osmosis.

Table E. United States drinking secondary maximum contaminant level (SMCL) water standards (USEPA, 1993)

Contaminant	Status (b)	Standard(a) (SMCL)
Aluminum	F	0.05 to 0.2
Chloride	F	250
Color	F	15 C.U.
Copper	F	1.0
Corrosivity	F	non-corrosive
Fluoride	F	2.0
Foaming Agents	F	0.5
Iron	F	0.3
Manganese	F	0.05
Odor	F	3 TON
pH	F	6.5 to 8.5
Silver	F	0.1
Sulfate	F	250
Total Dissolved Solids	F	500
Zinc	F	5

(a) Secondary standards provide guidance relating to aesthetic quality. These standards are only guidelines and are not enforced.

(b) Status refers to the current state of regulations pertaining to the particular contaminant (F refers to final, D refers to draft and P refers to proposed). MCLG refers to maximum contaminant level goal. A non-enforceable concentration that is protective of adverse human health effects (EPA, 1993). MCL refers to the maximum contaminant level.

Table F. (p1 of 2) Surface water treatment practices in the United States (FRDS 1994)

Water Treatment Practices	Number of Plants [a]	Percent of Total Plants
Treatment Facilities		
Public water systems	52,567	100.00
Treatment applied by seller	1,101	2.09
Treatment applied by plant	47,990	91.29
Treatment applied at entry point	55	0.10
No treatment/not applicable	9,004	17.13
Treatment not reported	413	0.79
Other treatment	84	0.16
Adsorption		
Activated alumina	19	0.04
Granular activated carbon	687	1.31
Powdered activated carbon	412	0.78
Ion Exchange	3,008	5.72
Aeration		
Cascade	445	0.85
Diffused	125	0.24
Packed Tower	236	0.45
Slate Tray	1,666	3.17
Spray	134	0.25
Algae Control	219	0.42
Conventional Processes		
Rapid mix	817	1.55
Coagulation	2,615	4.97
Flocculation	1,075	2.05
Sedimentation	1,774	3.37
Corrosion Control		
Biometallic phosphate inhibitor	120	0.23
Hexametaphosphate inhibitor	130	0.25
Orthophosphate inhibitor	161	0.31
Polyphosphate inhibitor	838	1.59
Silicate inhibitor	20	0.04
Disinfection		
Ammoniation	106	0.20
Chloramines	271	0.52
Chlorine Dioxide	83	0.16
Gaseous chlorination, Post	6,647	12.64
Gaseous chlorination, Pre	2,145	4.08
Hypochlorination, Post	11,758	22.37
Ozonation, Post	18	0.03
Ozonation, Pre	47	0.09
Permanganate	780	1.48
Peroxide	7	0.01
Ultraviolet Radiation	659	1.25

Footnotes at end of table

Table F. (p2 of 2) Surface water treatment practices in the United States (FRDS 1994)

Water Treatment Practices	Number of Plants [a]	Percent of Total Plants
Distillation	5	0.01
Filtration		
Unfiltered, avoiding filtration	45	0.09
Unfiltered, must install filtration	661	1.26
Not subject to SWTR	5,668	10.78
Cartridge	316	0.60
Diatomaceous Earth	67	0.13
Greensand	484	0.92
Microscreening	49	0.09
Pressure Sand	1,073	2.04
Rapid Sand	2,385	4.54
Slow Sand	263	0.50
Ultrafiltration	67	0.13
Reverse osmosis	256	0.49
Electrodialysis	2	0.00
Fluoridation	2,955	5.62
Iron Removal	203	0.39
pH adjustment	2,141	4.07
pH adjustment, post	597	1.14
pH adjustment, pre	638	1.21
Reducing Agents		
Sodium bisulfate	3	0.01
Sodium sulfite	2	0.00
Sulfur dioxide	10	0.02
Sequestration	915	1.74
Softening	186	0.35
Lime - soda ash	1,669	3.17
Taste and odor	136	0.26

(a) Information provided by data search of Federal Reporting Data System (FRDS). FDRS contains information supplied by Primary Agents in compliance with the quarterly reporting requirments under the Safe Drinking Water Act. The accuarcy of all the information in this database is dependant on the quality of the information provided by the Primary Agents.

Table G. Ground water treatment practices in the United States (AWWA Research Foundation, 1992)

Ground Water Treatment Practices	Number of Plants (a)	Percent of Plants Surveyed
Aeration	155	26.3
Predisinfection/Oxidation		
Chlorine	437	74.1
Chlorine dioxide	3	0.5
Chloramines	31	5.3
Ozone	1	0.2
Permanganate	31	5.3
Reverse osmosis	8	1.4
Ion exchange	21	3.6
Lime/Soda Ash Softening	92	15.6
Recarbonation with CO2	33	5.6
Iron & Manganese Removal	137	23.2
Filtration	176	29.8
Organics Removal		
PAC addition	2	0.3
GAC adsorption	22	3.7
Air Stripping	35	5.9
Off gas treatment	2	0.3
Resin Adsorption	1	0.2
Radionuclide Removal		
Radon	2	0.3
Radium/Uranium	0	0.0
Fluoridation	256	43.4
Defluoridation	0	0.0
Corrosion Control		
pH adjustment	146	24.7
Alkalinity adjustment	33	5.6
Corrosion inhibitors	124	21.0
Post-Disinfection		
Chlorine	182	30.8
Chlorine dioxide	1	0.2
Chloramines	26	4.4
Any Disinfection (Pre and/or Post)	513	86.9
No Treatment	53	9.0

(a) 590 utilities responded to this survey

Chapter IV

Emerging Problems and Technologies in Water Treatment

Chapter IV

Emerging Problems and Technologies in Water Treatment

INTRODUCTORY STATEMENT

Almost from the introduction of public water supply, the emphasis on water treatment shifted from point of use to centralised treatment. Due to the development of more stringent water - quality standards, potable water is now being treated in centralised facilities to very high standards at substantially increased costs. In approaching minimum risk from waterborne disease, environmental engineers have begun to focus more on subtle risks from cancer. As we exhaust more resources in the pursuit of a longer list of chemical contaminants regulated at lower concentrations, we must not loose sight of fundamental requirements for public health. A long standing philosophy of producing water that "tastes good" has resulted in design practices that frequently emphasise the use of granular activated carbon and ozone. At the same time, water disinfection practices have also evolved. First, a concern for halogenated by-products of disinfection resulted in reduction of chlorine doses, substitution by alternative disinfectants, and better application of chlorinating technology. More recently, it has been found that treatment - resistant pathogens (viruses and pathogenic protozoa, especially Giardia and Cryptosporidum) may limit the degree to which reduction in chlorine dosage may be used to control disinfection by product formation without posing unacceptably high risk of infectious disease.

Changes in drinking water regulations are causing a stronger coupling between the level of drinking water treatment and the quality of the water supply. Lower quality supplies may require such capital intensive processes as filtration, high - level disinfection (such as ozonation) and perhaps membrane treatment. In some cases, this provides strong incentive for starting and / or strengthening watershed protection programs. In other cases, it may provide an incentive to locate and reduce major sources of contaminant loading that contribute to a water supply used for drinking water. This may be particularly important in the case of microbial loading.

Development of new technologies and approaches to water treatment can not entirely eliminate the need to further improve currently available treatment options. As a matter of fact, enhanced coagulation is designated as a possible alternative to meet more stringent disinfection by-product regulations. In addition optimisation of conventional settling and filtration processes would certainly have major benefits for water utilities.

Intermediate transfer of risks in water treatment must also be avoided, since we are merely sweeping risk from potable water to the air we breathe, and to our landfills. Risk and benefit - cost analysis must be extended to consider life - cycle impacts on the water treatment industry. These actions are needed not only to improve drinking water quality but to provide information needed by decision makers in balancing environmental quality with other valuable objectives of our society (schools, hospitals, etc.). We must also carefully examine our human health ethic to be better prepared for water recycling that is already being implemented in many arid areas.

1. Control of Disinfection By-Products in Drinking Water: Regulations and Costs

Radisav D. Vidić
Department of Civil and Environmental Engineering
University of Pittsburgh
949 Benedum Hall, Pittsburgh PA 15261
USA

1.1. HISTORY OF DBP FORMATION AND REGULATIONS

The regulation of disinfection by-products (DBPs) in the United States was based on the discovery of the presence of chloroform as a by-product of chlorination in drinking water in the early 1970s. This suspected human carcinogen was indirectly regulated as one of the four trihalomethanes (THMs) under the interim total THM (TTHM) regulation established by the US Environmental Protection Agency (USEPA) in 1979. An interim maximum contaminant level (MCL) of 0.10 mg/l was established for water utility systems serving more than 10,000 people, based on the running annual average of four quarterly samples.

It is significant that the standard was established based on then-available health effects data on chloroform ($CHCl_3$). The National Academy of Sciences and the USEPA's Carcinogen Assessment Group estimated the incremental risks associate with exposure to chloroform in drinking water. Assuming that the population consumes 2 l of water (containing 0.10 mg/l of chloroform) daily, for 70 years, they estimated a risk of 3-4 excess cases of cancer per 10,000 people. Adoption of 0.10 mg/l MCL for TTHMs was understood to be a compromise position; public health considerations had to be balanced with technological and economic considerations. In the development of regulations in general, two categories of risk targets are established. The first is the Maximum Contaminant Level Goal (MCLG). The second is the Maximum Contaminant Level (MCL). The MCLG is a health-based target, usually established at a 10^{-6} risk level. The MCL is an enforceable standard and might be as high as 10^{-4} because of limitations in technology for removing or measuring the contaminant. The intent in establishing the MCL for TTHMs, however, was to revise the MCL in future regulations "based upon an updated assessment of technological and economic feasibility, implementation of experience and additional toxicological information". Few data existed at the time on the

NATO ASI Series, Partnership Sub-Series, 2. Environment – Vol. 15
Water Supply Systems – New Technologies
Edited by Č. Maksimović, F. Calomino, and J. Snoxell

other three THMs: bromodichloromethane ($CHCl_2Br$), dibromochloromethane ($CHClBr_2$), and bromoform ($CHBr_3$), but their structural similarities to chloroform led the USEPA to conclude to be inappropriate to exclude them from regulation.

Besides THMs, di- and trichloroacetic acids were identified as a second major class of DBPs in chlorinated water (Quimby et al., 1980, Miller and Uden, 1983). Other frequently identified halogenated DBPs include haloacetonitriles, haloketones, chloropicrin, cyanogen chloride, and chloral hydrate. A comprehensive study by the National Academy of Sciences in 1987 (National Academy of Sciences, 1987) found that a number of these other halogenated DBPs also have adverse effects. It is believed that dichloroacetic acid is a more potent carcinogen than any of the THMs (Bull and Kopfler, 1991) while a halogenated furanone (3-chloro-4-(dichloromethyl)-5-hydroxy-2(5H)-furanone, also known as MX) has been found to be extremely mutagenic, even at concentrations as low as 0.05 μg/l.

In the mid-1980s, Congress became disenchanted with the pace at which the USEPA was setting drinking water regulations. Spurred by the perception of a substantial public health risk being posed by the US drinking water, Congress passed a Safe Drinking Water Act (SDWA) amendments in 1986. Congress prescribed, with limited flexibility, the first 83 compounds to be regulated and the future rate at which the USEPA would establish standards (25 new standards every three years).

The most common treatment modification implemented by the water supply industry to comply with the new MCL for TTHMs include moving the point of chlorination downstream in the treatment process, decreasing chlorine doses, and utilizing chloramines as alternative primary or secondary disinfectant instead of free chlorine (McGuire and Meadow, 1988). These modifications generally lead to the compliance with the new rule but the questions were raised about the microbiological quality of the finished water. To ensure that the microbial quality of finished drinking water was not compromised, the USEPA promulgated the Surface Water Treatment Rule (SWTR) and the Total Coliform Rule (TCR) in 1989 (USEPA, 1989a.,b.). The SWTR states that all public water systems using any surface water or ground water under direct influence of surface water must disinfect and may be required to implement filtration as well. It further requires that the treatment must reliably achieve at least 99.9 percent (3-log) removal and/or inactivation of *Giardia Lamblia* cyst and 99.99 percent (4-log) removal and/or inactivation of viruses. The key concept in the SWTR is the requirement that the water must be in contact with a disinfectant concentration for a sufficient period of time in order to provide specified degrees of inactivation (C·t concept). Furthermore, a measurable disinfectant residual must be achieved in at least 95% of distribution system samples analyzed.

The regulation of DBPs has been complicated by the discoveries that alternative disinfectants to free chlorine also produce by-products that have adverse health effects. Ozonation of bromide-containing waters will produce bromate which has been identified as a probable human carcinogen while the addition of chlorine dioxide to water results in the formation of chlorite which causes hepatotoxicity in animals and possible hemolytic anemia (Haag and Hoigne, 1983, Couri et al., 1982).

Balancing between the risk associated with DBPs and the risk associated with microbial disease is complicated by the following factors: lack of occurrence data for various DBPs; insufficient information concerning the nature of unidentified (50%) TOX; lack of information regarding the by-products formed by using the alternative disinfectants to free chlorine (chloramines, ozone, chlorine dioxide, etc.); uncertainty regarding the health effects of various DBPs; and the uncertainty regarding the C·t values required for the inactivation of *Giardia* and *Cryptosporidium* cysts (Singer, 1994).

Because of the lack of sufficient technical data on DBPs, the USEPA could not, within the prescribed time frame, establish defensible standards for all of the DBPs included in the initial set of contaminants. They did issue a "strawman" regulation in 1989 for public comment (Pontius, 1990). This rule was to serve as basis for discussion and to seek input from the affected parties early in the rule-making process. Because of the technical complexity and uncertainty that exists in various aspects of this rule, the agency instead chose to develop the proposed rule using the negotiated rule-making process - referred to as regulatory negotiation, or "Reg-Neg". In this approach, all interested parties that have a stake in the regulation, including the USEPA, meet as a group and, with the assistance of a third party that acts as a facilitator, collectively develop the regulation. This Reg-Neg process was initiated in September 1992 and was completed in mid-1993. The current status of negotiated D-DBP is presented at the end of this manuscript.

1.2. OCCURRENCE OF DISINFECTION BY-PRODUCTS

THMs and other halogenated DBPs result from the interaction of the disinfectants with organic precursor (in most cases natural organic material (NOM)) as shown by:

$$\begin{bmatrix}\text{Precursor}\\ \text{Material}\end{bmatrix} + [\text{Disinfectant}] \rightarrow \begin{bmatrix}\text{Disinfection}\\ \text{By-Products}\end{bmatrix}$$

The major halogenated DBPs identified in finished waters are:

- Trihalomethanes (THMs)
  - Chloroform
  - Bromodichloromethane
  - Dibromochloromethane
  - Bromoform

- Haloacetic Acids (HAAs)
  - Monochloroacetic acid
  - Dichloroacetic acid
  - Trichloroacetic acid
  - Monobromoacetic acid
  - Dibromoacetic acid
  - Tribromoacetic acid
  - Bromochloroacetic acid
  - Bromodichloroacetic acid
  - Dibromochloroacetic acid

- Aldehydes
  - formaledehyde
  - acetaldehyde
  - dichloroacetaldehyde
  - trichloroacetaldehyde

- Haloacetonitriles (HANs)
  - Dichloroacetonitrile
  - Trichloroacetonitrile
  - Dibromoacetonitrile
  - Tribromoacetonitrile
  - Bromochloroacetonitrile

- Cyanogen Halides
  - Cyanogen chloride
  - Cyanogen bromide

- Haloketones
  - 1,1-dichloropropanone
  - 1,1,1-trichloropropanone
  - 1,1-dichloro-2-butanone
  - 3,3-dichloro-2-butanone
  - 1,1,1-trichloro-butanone

- Halopicrins
  - Chloropicrin
  - Bromopicrin

- MX

A recent survey of DBPs in finished drinking water in the United States, which included 35 water-treatment facilities utilizing a broad range of source water qualities and treatment processes, revealed that THMs were the predominant class of DBPs present, with the HAAs being the second most dominant class (Krasner et al., 1989). The median TTHM concentration was 39 μg/l, which compared very well with the values reported in previous surveys (AWWARF survey of 727 utilities in 1988). The median HAA concentration was 19 mg/l, approximately half of the TTHM concentration. Formaldehyde and acetaldehyde, by-products of ozonation, can also be produced by chlorination. Cyanogen chloride was produced preferentially in chloraminated waters.

1.3. FACTORS INFLUENCING THE FORMATION OF DBPs

The formation of DBPs occurs primarily through the substitution reactions of chlorine with the potential reactive sites in the organic molecule. The factors influencing these reactions include: pH, contact time, temperature and season, nature and concentration of organic precursors, disinfectant type and dose, and bromide concentration.

pH. Historical data confirmed that the increase in pH leads to higher concentration of THMs in finished water (Oliver, 1978, Stevens et al., 1989). On the other hand, higher pH leads to overall reduction in TOX formation since many of the halogenated DBPs other than THMs tend to hydrolyze at alkaline pH-values (Johnson and Jensen, 1986, Stevens et al., 1989). The most obvious implication for water treatment is the direct trade-off between TTHM control (lower pH) and the control of most of the other DBPs (higher pH).

Contact time. Concentrations of THMs and HAAs in chlorinated generally increase with increasing contact times (Stevens et al., 1989) while some of the halogenated DBPs (haloacetonitriles and haloketones) form rapidly upon chlorination but then decay in the distribution system due to hydrolysis and continuing reaction with residual chlorine. This behavior has major implications on the exposure of consumers to DBPs depending upon their proximity to the treatment plant or the point of rechlorination.

Temperature. Formation of DBPs is highly influenced by the seasonal changes in raw water quality. Higher temperatures in summer months lead to faster kinetics and therefore, increased chlorine demand which causes more chlorine to be applied and higher DBP concentrations are observed. Such behavior is very well documented for THM (White, 1986). It is reasonable to assume that most of the other DBPs will exhibit similar behavior. On the other hand, hydrolysis rates for chloral hydrate, dihaloacetonitriles, and possibly chloropicrin are also likely to increase, possibly reducing the presence of these species. Another aspect of seasonal changes are the changes in the nature of organic precursors and the concentrations of bromide.

Chlorine dose. As the chlorine dose increases, TOX increase in proportion to the available organic precursor. Higher chlorine doses favor the formation of HAAs over THMs and the formation of trihalogenated HAAs over di- and monohalogenated HAAs (Pourmoghaddas et al., 1993).

Bromide concentration. In the absence of bromide ion, only chlorinated by-products are formed. In the presence of bromide, free chlorine (HOCl) rapidly oxidizes bromide to hypobromous acid (HOBr), which then also reacts with organic precursors to produce mixed chloro-bromo substitution products. The extent of bromine incorporation depends on the ratio of bromide ion concentration to the applied free chlorine. As that ratio increases, the extent of the formation of brominated DBPs increases (Pourmoghaddas et al., 1993). This is true for both THMs and HAAs.

A number of these factors have been incorporated into DBP predictive equations that are being used by the USEPA and by the Reg-Neg group to asses impact of various candidate MCLs on the water supply industry and the extent to which utilities treating waters of different quality would be able to comply with these MCLs (Malcolm Pirnie, 1992).

1.4. CONTROL OF DISINFECTION BY-PRODUCTS

There are several options for DBP control. As discussed earlier, DBPs tend to increase with time so that moving point-of-disinfection to the end of the treatment process will minimize the time for by-product formation. This approach however, does not have any impact on the formation of DBPs in the distribution system which is of increasing concern in the US.

Other options for the control of DBPs include: source water quality control; removal of precursor material or NOM before it interacts with the disinfectant; removal of the DBPs once they are found; and use of a disinfectant that minimizes the formation of by-products. The following section provides some insight into the treatment options, effects of process variables, and the treatment technology costs associated with some of these options for controlling DBPs. Since the impact of control of THMs is similar to the control of other halogenated by-products, the costs developed will be those associated with THM control.

1.4.1. Source Water Quality Control

Many studies have documented that the concentrations of DBPs tend to directly correlate with the increases in algal metabolites present in the source water. Consequently, one of the approaches in preserving the quality of source waters with respect to algal blooms is to implement an adequate nutrient management program that may include: storm-water detention basins to trap nutrients, land-use controls (limit developments on watersheds used for water supply), installation of aeration systems and implementation of harvesting programs to control nutrient cycling in reservoirs.

Since the presence of bromide ions was shown to have a pronounced impact on the formation DBPs (drives the formation of more brominated species that are believed to be more harmful to human health), it is also advisable to implement programs that would limit the presence of bromide ions in the source water. Such programs could include the control of salt water or brine intrusion into the water source by installing structural or hydrodynamic barriers.

Recent studies showed that the water management concept known as aquifer storage and recovery (ASR) has a positive impact on halogenated DBPs in finished water (Singer et al., 1993). This concept involves underground storage of treated drinking water by injection into a suitable storage zone during those periods of the year when available capacity of treatment facilities exceed system demand. The stored water is later recovered to meet system demands that exceed treatment capacity, usually without the need for additional treatment other than disinfection. The same well is used for both recharge and recovery. Such practice was shown to result in the elimination of THMs and HAAs produced during treatment and in decrease in the concentration of residual DBP precursors.

1.4.2. Precursor Removal

There are three ways to effectively remove DBP precursors: enhanced coagulation, activated carbon adsorption, and membrane processes.

Enhanced coagulation. Both aluminum and ferric salts have effectively been used to coagulate NOM (Kavanaugh, 1978). The optimal pH range for alum ($Al_2SO_4{\cdot}xH_2O$) coagulation is between 5.5 and 6.0. This pH range is easily achieved in water of low alkalinity (less than 50 mg/l as $CaCO_3$) by the addition of alum. However, in waters of very low or very high (above 100 mg/l as $CaCO_3$) alkalinity, it might be necessary to add an appropriate amount of a base (lime or caustic soda) or an acid (sulfuric acid), respectively, to maintain the pH in a desired range. Alum and ferric chloride dosages depend quite heavily on the TOC content and alkalinity of raw water, the hydrophobic/hydrophilic distribution of the TOC (hydrophobic fraction of TOC is believed to constitute 30-70% of TOC and is more susceptible to coagulation), and the pH of coagulation an have to be optimized for each raw water source thereby, precluding general conclusions with respect to design and cost parameters.

GAC Adsorption. Most applications of GAC for the control of precursor (NOM) would require a separate postfiltration adsorber bed to accomplish appropriate removals. Empty-bed contact times (EBCT) in excess of 15 minutes are required with regeneration frequency on the order of three to six months.

USEPA Drinking Water Research Division developed cost estimates for the utilization of GAC adsorption for the removal of DBP precursor based on the field studies conducted in Cincinnati (OH), Jefferson Parish (LA), and Manchester (NH) (Clark et al., 1994). The cost estimates given in Table 1. were established based on the assumption of 20 min EBCT and for two different scenarios of MCLs for TTHMs (Ce < 100 μg/l and Ce < 50 μg/l). For systems of 380, 3,800, and 38,000 m^3/day, pressure contactors were assumed and for systems of 95,000, 190,000, and 380,000 m^3/day, concrete gravity contactors were assumed. For systems of 380 and 3,800 m^3/day replacement of spent carbon with virgin carbon was used in calculations, while the on-site reactivation was assumed for the other four systems.

Membrane Filtration. To achieve TOC removal in excess of 75%, membrane filtration generally requires the use of nanofilters, with membranes having a molecular weight cutoffs (MWC) of 200-500 daltons. For most applications, pretreatment is required to prevent fouling of the membrane. For the cost analysis, it is assumed that a 20.3x100.6 cm element removed organics having the molecular weights greater than 200 daltons (Clark et al., 1994). Based on field experience, the removal of DOC, TOX and DBP can be assumed to be in the excess of 95%. Figure 1. shows the impact of MWC on TOX formation potential (TOXFP)

Table 1. GAC cost summary for THM precursor removal in ¢/m^3

Plant Design Flow (m^3/day)	Plant Average Flow (m^3/day)	Ce < 100 μg/l	Ce < 50 μg/l
380	190	64.3 - 76.0	66.5 - 85.5
3,800	1,900	24.4 - 35.2	26.5 - 43.8
38,000	26,600	14.1 - 18.2	15.0 - 21.1
95,000	64,600	9.4 - 12.8	10.1 - 15.3
180,000	126,600	7.6 - 10.6	8.2 - 12.8
380,000	266,600	6.4 - 9.8	7.0 - 12.1

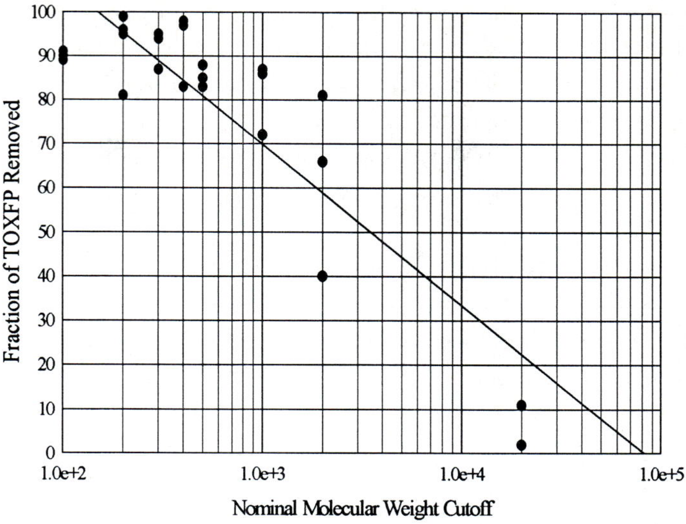

Figure 1. Impact of MWC on TOXFP removal in membrane processes

Table 2. Nanofiltration cost summary (¢/m^3)

	Ground-Water System		Surface-Water System
System permeate capacity (m^3/day)	Surface Concentrate Disposal	Deep Well Injection	Pretreatment with alum, settling, filtration, deep well injection
380	53.9	63.1	120.9
3,800	35.1	40.7	79.7
19,000	27.7	31.9	63.1
38,000	25.6	29.6	58.3
95,000	24.6	28.2	55.4
190,000	22.2	25.6	50.7
380,000	20.3	23.2	45.9

Two types of systems were considered in cost analysis: (1) groundwater with an average feed pressure of 1 MPa, an average flux of 0.61 m/day, and 10 membrane skids with nine on line and one on standby; and (2) surface water with an average pressure of 0.7 MPa, an average flux of 0.41 m/day, and 13 membrane skids with 12 on line and one on standby. Both systems are operated using a three-stage membrane configuration at 85% recovery. Assumptions for the cost analysis of groundwater systems included both brine disposal to and existing surface pond or stream as well as deep well injection. Surface water treatment by nanofiltration included pretreatment using alum coagulation, solids contact clarification and rapid sand filtration. Table 2. summarizes the costs for all three options.

As can be seen from Figure 2., which compares the costs of the three alternatives, nanofiltration is more expensive for the removal of DBP precursors from surface water than GAC adsorption. On the other hand, small systems utilizing ground water (below 38,000 m^3/day), could find nanofiltration an attractive option to GAC adsorption.

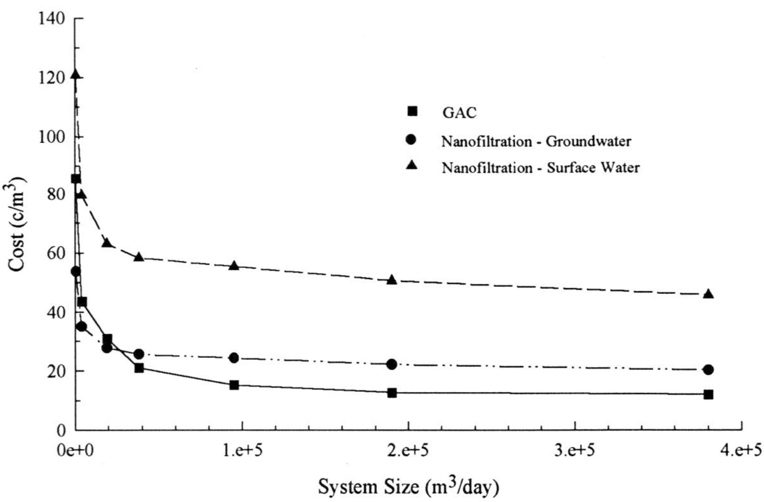

Figure 2. Cost for removing precursors (150 μg/l to 50 μg/l)

1.4.3. Removal of DBPs

Once the DBPs are formed, it is quite a task to remove them from the finished drinking water prior to distribution into the system. There are basically two options for the removal of low levels of THMs: (1) GAC adsorption and (2) air stripping. To illustrate the cost impact of removing DBPs (in this case THMs), costs were calculated for these treatment options assuming the initial THM concentration of 150 μg/l (Clark et al., 1994). It should be noted that the composition of THMs in the finished water may vary according to source water characteristics. Cost calculations were performed using the design and operational parameters given in Table 3.

Chloroform was selected to represent THMs that are difficult to remove by GAC adsorption (low adsorptive capacity) but are easily stripped using a packed column aeration (Henry's Law constant = 0.087). On the other hand, dibromochloromethane is much less susceptible to removal by air stripping (Henry's Law constant = 0.030). Two design options were considered for cost analyses: reducing the influent concentration of TTHMs of 150 μg/l (75 μg/l of chloroform and 75 μg/l DBCM) to levels below 100 μg/l (Option 1.) and below 50 μg/l (Option 2). The assumptions in calculating the costs for GAC adsorption are the same as those used in the analysis of GAC adsorbers for precursor removal. Table 4. summarizes the costs for GAC adsorption, air stripping and air stripping with off-gas control. As is apparent from this table, removal of THMs from finished water is quite expensive with GAC adsorption being the most expensive alternative.

Table 3. Design options for packed-bed air stripping tower

Parameter	Option 1.	Option 2.
Air/Water ratio	40	40
Depth (m)	3	4.5
Diameter (m)	2.1	2.1
Influent (μg/l)		
Chloroform	75	75
DBCM	75	75
Effluent (μg/l)		
Chloroform	15	6
DBCM	23	12

1.4.4. Use of Alternative Disinfectants

Performance and costs associated with the use of alternative disinfectants to free chlorine is described in the following Section.

Table 4. Cost for removal of THMs from an influent level of 150 μg/l (¢/m^3)

	Ce < 100 μg/l			Ce < 50 μg/l		
Plant Size (m^3/day)	GAC	Air Stripping	Air Stripping with offgas control	GAC	Air Stripping	Air Stripping with offgas control
380	72.3	14.6	36.5	72.8	16.1	40.0
3,800	31.9	4.0	12.5	32.2	5.1	15.1
38,000	16.9	1.9	5.6	17.2	2.4	6.9
95,000	11.7	1.7	5.2	11.9	2.2	6.5
180,000	9.6	1.6	5.0	9.8	2.1	6.3
380,000	8.9	1.6	5.0	9.1	2.0	6.2

1.5. CURRENT STATUS OF DBP REGULATIONS

The negotiating process among the parties involved (Reg-Neg) resulted in the development of essentially three rules (Pontius, 1993):

- An Information Collection Rule (ICR),
- An Enhanced Surface Water Treatment Rule (ESWTR),
- The Disinfectant-Disinfection By Product (D-DBP) Rule itself.

The ICR is intended to provide information concerning raw-water quality, treatment practices, and DBP concentrations in finished drinking waters across the United States for the surface water systems serving more than 10,000 people. These systems will also be required to collect information concerning the occurrence of pathogenic microorganisms (i.e., *Giardia, Cryptosporidium,* and viruses) in raw-water supplies.

The ESWTR is aimed at potentially increasing the requirements for removal and/or inactivation of *Giardia* and viruses above those specified in the existing SWTR. Such prospect came as a result of several waterborne outbreaks of cryptosporiodiosis and the fact that the current SWTR does not address *Cryptosporidium* and the degree of protection against this organism. Subsequently, there is a need to include the required C·t values for *Cryptosporidium* in the newly developed regulations. One response to this issue could come through the requirement for a 3- to 6-log removal of *Giardia* based on the source water cyst concentration. This would however, require greater doses of disinfectants which would increase the level of DBPs in the finished water. The ESWTR provisions will be based on the findings of the ICR monitoring.

Table 5. Disinfection/disinfection by-product rule: proposed MCLs and MRDLs

Compound	MCL/MRDL
Total trihalomethanes	80 μg/l
Haloacetic acids (total of five)	60 μg/l
Bromate	10 μg/l
Chlorite	1.0 mg/l
Chlorine	4.0 mg/l
Chloramines	4.0 mg/l
Chlorine Dioxide	0.8 mg/l

The D-DBP Rule will be developed in two stages. Stage I contains compromise MCLs for TTHMs, total HAAs, bromate, and chlorite along with maximum residual disinfectant levels (MRDLs) for chlorine, chloramines and chlorine dioxide based on the running annual averages computed quarterly (Table 5.). The MCL for TTHMs and HAAs in the Stage II are now proposed at 40 and 30 μg/l, respectively but will be renegotiated following collection of additional information under the ICR.

Best available technology (BAT) for meeting the Stage I MCLs for THMs and HAAs is enhanced coagulation (or enhanced softening) or GAC adsorption with a 10-min EBCT and 180-day reactivation cycle. If the TOC concentration (used as a surrogate for DBP precursors) exceeds 2 mg/l at the point of first continuous disinfection, enhanced coagulation must be implemented. Water systems that serve more than 10,000 customers and whose water contains TOC above 4 mg/l level at the point of first continuous disinfection will be required to conduct pilot-scale studies to evaluate GAC or membrane processes for precursor removal. Ground water systems that serve more than 50,00 people will have to conduct similar studies if the treated water TOC exceeds 2 mg/l.

The bottom line for DBP regulation is that promulgation of a long-term DBP rule must await additional scientific findings and that the ICR will provide the key data to support this rule-making process.

1.6. REFERENCES

Bull R.J., Kopfler F.C. (1991) Health Effects of Disinfectants and Disinfection By-Products. American Water Works Association Research Foundation, Denver, CO.

Clark R.M., Adams J.Q., Lykins B.W. (1994) DBP Control in Drinking Water: Cost and Performance. J. Envir. Engrg. Div., ASCE, 120(4), 759.

Couri D., Abdel-Rahman M.S., Bull R.J. (1982) Toxicological Effects of Chlorine Dioxide, Chlorite, and Chlorate. Envir. Health Perspectives, 46, 13.

Haag W.R., Hoigne J. (1983) Ozonation of Bromide-Containing Waters: Kinetics of Formation of Hypobromous Acid and Bromate. Environ. Sci. Technol., 17(5), 261.

Johnson J.D., Jensen J.N. (1986) THM and TOX Formation; Routes, Rates and Precursors. J. AWWA, 78(4), 156.

Kavanaugh M.C. (1978) Modified Coagulation for Improved Removal of Trihalomethane Precursors. J. AWWA, 70(11), 613.

Krasner S.W., McGuire M.J., Jacangelo J.G., Patania M.L., Reagan K.M., Aieta E.M. (1989) The occurrence of disinfection by-products in U.S. drinking water. J. AWWA, 81(8), 41.

Malcolm Pirnie Inc. (1992) Water Treatment Plant Simulation Program. Version 1.21. User's Manual. USEPA, Washington, D.C.

McGuire M.J., Meadow R.G. (1988) AWWARF Trihalomethane Survey. J. AWWA, 80(1), 61.

McGuire M.J., Marshall D.K., Tate C.H., Aieta E.M., Howe E.W., Crittenden J.C. (1991) Evaluating GAC for Trihalomethane Control. J. AWWA, 83(1), 38.

Miller J.W., Uden P.C. (1983) Characterization of Non-Volatile Aqueous Chlorination Products of Humic Substances. Environ. Sci. Technol., 17(3), 150.

National Academy of Sciences (1987) Drinking Water and Health: Disinfectants and Disinfection By-Products, Vol. 7, National Academy Press, Washington, D.C.

Pontius F.W. (1990) Complying With the New Drinking Water Quality Regulations. J. AWWA, 82(2), 32.

Pontius F.W. (1993) Reg-Neg Process Draws to a Close,. J. AWWA, 85(9), 18.

Pourmoghaddas H., Stevens A.A., Kinman R.K., Dressman R.C., Moore L.A., Ireland J.C. (1993) Effect of Bromide Ion on Formation of HAAs During Chlorination. J. AWWA, 85(1), 82.

Oliver B.G. (1978) Chlorinated No-Volatile Organics Produced by the Reaction of Chlorine With Humic Materials. Canadian Research, 11(6), 21.

Quimby B.D., Delaney M.F., Uden P.C., Barnes R.M. (1980) Determination of the Aqueous Chlorination Products of Humic Substances by Gas Chromatography With Microwave Emission Detection. Anal. Chem., 52, 259.

Singer P.C., Pyne R.D.G., Avs M., Miller C.T., Mojonier C. (1993) Examining the Impact of Aquifer Storage and Recovery on DBPs. J. AWWA, 85(11), 85.

Singer P.C. (1994) Control of Disinfection By-Products in Drinking Water. J. Envir. Engrg. Div., ASCE, 120(4), 727.

Stevens A.A., Moore L.A., Miltner R.J. (1989) Formation and Control of Non-Trihalomethane Disinfection By-Products. J. AWWA, 81(8), 54.

US Environmental Protection Agency (1979) National Interim Primary Drinking Water Regulations; Control of Trihalomethanes in Drinking Water. Federal Register, 44(231; Nov.29), 68624.

US Environmental Protection Agency (1989a) National Primary Drinking Water Regulations; Filtration and Disinfection; Turbidity, Giardia Lamblia, Viruses, Legionella, and Heterotrophic Bacteria. Federal Register, 54(124; Jun.29), 27468.

US Environmental Protection Agency (1989b) Drinking Water; National Primary Drinking Water Regulations; Total Coliforms. Federal Register, 54(212; Jun.29), 27544.

White G.C. (1986) Handbook of Chlorination. 2nd edition, Van Nostrand Reinhold Co., Inc., New York.

2. Control of Disinfection By-Products in Drinking Water: Case Studies of Alternative Disinfection Technologies

Radisav D. Vidić
Department of Civil and Environmental Engineering
University of Pittsburgh
949 Benedum Hall, Pittsburgh PA 15261
USA

2.1. CHLORAMINES

Chloramines have been used in water treatment since the early 1900s (Wolfe et al., 1984) and renewed interest in their use occurred in recent years since they represent a viable alternative for the control of trihalomethanes (THMs). Other advantages of using chloramines include increased residual stability, reduction in odor problems caused by chlorination, and the lowest cost among main alternatives (Wolfe et al., 1984).

Chloramines are formed when hypochlorous acid (HOCl) reacts with ammonia or organic nitrogen in the water according to following reactions:

$$HOCl + NH_3 \rightarrow NH_2Cl + H_2O$$

$$HOCl + NH_2Cl \rightarrow NHCl_2 + H_2O$$

$$HOCl + NH_2Cl \rightarrow NCl_3 + H_2O$$

The species of chloramines present in water are dependent on pH, amount of chlorine relative to ammonia nitrogen (Table 1.), and temperature (formation of chloramines is significantly retarded below 10 ^{o}C).

NATO ASI Series, Partnership Sub-Series, 2. Environment – Vol. 15
Water Supply Systems – New Technologies
Edited by Č. Maksimović, F. Calomino, and J. Snoxell

Table 1. Existence of chloramine species in water

Predominant species	pH range	Chlorine : Ammonia-N (mg Cl2/l:mg NH3-N/l)
monochloramine	> 6.5	< 5:1
dichloramine	4.3 - 6.5	5:1 to 7.6:1
trichloramine	< 4.3	after breakpoint chlorination

Normally, product water from treatment plants has a pH ranging from 6 to 9 with monochloramine as the predominant species. Furthermore, a large majority of utilities operate at low chlorine to ammonia nitrogen ratio (3:1 to 4:1) which also favors the formation of monochloramines. Some utilities however, choose to operate at higher ratios of 6:1 to 7:1, where dichloramine is the dominant chloramine form, thereby increasing disinfection efficiency and reducing ammonia usage (Trussell and Kreft, 1984). Dichloramine and trichloramine in drinking water, however, may cause taste and odor problems at concentrations of 0.8 and 0.02 mg/l, respectively (White, 1986).

Controversy exists regarding the effectiveness of chloramines as a microbiocidal agent. In 1987 EPA labeled chloramines as a secondary disinfectant (White, 1986). However, the effectiveness of chloramines under laboratory experimental conditions is apparently lower than in drinking water treatment plants where results show that coliform bacterial counts have been maintained within the MCL of 1 colony/100 ml using the chloramines as disinfectants (Wolfe et al., 1984, Thompson and Ameno, 1987).

Water utilities using chloramines have employed the following patterns for the addition of ammonia and chlorine in the treatment process: (a) prechlorination upstream a substantial distance from ammonia application, (b) preammoniation and mixing of ammonia before application of chlorine, and (c) concurrent addition of ammonia and chlorine. Option (a) allows for primary disinfection by chlorine prior to the addition of ammonia to form chloramines that would provide secondary disinfection in distribution systems. Additional advantages of this option include reduced chlorine demand of the finished water and longer lasting chloramine residuals. Some utilities were capable of maintaining THM levels below 0.10 mg/l even with 2 hours of free chlorine contact time. Several utilities have added ammonia and chlorine at the same location and have maintained residual throughout the entire treatment process and distribution system (Trussel and Kreft, 1984, White, 1986). Most utilities using monochloramine as a secondary disinfectant switch back to free chlorine periodically in order to eliminate biological growths that may have colonized various equipment in the treatment plant as well as the distribution system.

Care should be exercised to prevent the appearance of excess ammonia in the finished water since that might cause nitrification (biological conversion of ammonia to nitrite (NO_2^-) and nitrate (NO_3^-)) to occur yielding a decrease in the total chlorine and ammonia concentrations in the storage reservoirs and distribution systems. This is especially important in warm waters. One possible method to prevent this problem is to adjust the Cl_2:NH_3-N ratio (4:1 to 4.5:1 is recommended) and maintain adequate combined residual (0.5 to 1.0 mg/l) (White, 1986, Lieu et al., 1993).

Another advantage of monochloramine over free chlorine is that monochloramine produces appreciably lower amounts of disinfection by-products (DBPs). Some dichloroacetic acid can be formed, however, and cyanogen chloride formation is greater than with free chlorine (Jacangelo et al., 1989, Krasner et al., 1989).

2.1.1. Ann Arbor Water Treatment Plant (George et al., 1990)

System Description. The Ann Arbor Water Treatment Plant is a split treatment lime-softening plant which uses water from two sources: 85% was surface water from the Huron River, and 15% was from groundwater extraction wells. The overall capacity of the plant is 131 m^3/min divided between two plants. The combined raw water contained high hardness (279 mg/l as $CaCO_3$) and low turbidity (5 NTU). Well water was used for split treatment and for pH control together with recarbonation. The treatment process incorporated the addition of polymer in the secondary rapid mix chamber while phosphate and sodium hypochlorite were added immediately prior to filtration and sodium silicofluoride was added after filtration along with postchlorination. The final total chlorine residual at the plant was 4.0 mg/l.

After 1978, chlorine was added at the raw water pumping station for disinfection and 1.5 to 3.5 mg/l potassium permanganate was used for taste and odor control. TTHM levels in the finished water averaged 105 μg/l while the maximum trihalomethane potential (MTP) varied from 250 to 600 μg/l. These high THM levels may have been enhanced by the addition of chlorine and $KMnO_4$ at the same location while high pH of the finished water (9 to 10.8) certainly promoted THM formation.

In September 1981, ammonium sulfate was introduced to a primary rapid mix chamber to react with free chlorine and thereby reduce THM concentrations. However, being that ammonium sulfate is highly hygroscopic and therefore, extremely difficult to feed, this practice was quickly replaced by the addition of liquid anhydrous ammonia through the existing chlorinator at the raw water pumping station. Chlorine to ammonia nitrogen ratio was maintained at 3:1 which provided approximately 0.25 mg/l residual free ammonia.

System Performance. With the plant expansion and continuous use of coagulant the plant was producing good quality water with respect to turbidity (average below 0.2 NTU). Since the conversion to chloramines, TTHM concentrations in the finished water were reduced to approximately 15 μg/l while still providing product water with <1 total coliform bacteria/100 ml at the final combined chlorine concentration in the plant effluent of 4.0 mg/l. Heterotrophic plate counts (HPCs) within the distribution system were generally less than 10 counts/ml (when the residual concentration of 3.5 mg/l was maintained during the initial period, HPC averaged 20 counts/ml). A slight increase (from 2.4 to 4.1 pcu) in the color of the finished water was observed since chloramines are ineffective in oxidizing coloring agents.

Problem Areas. The introduction of gaseous ammonia into the raw water stream softens the water, resulting in the precipitation of the salts that caused a scale to form around the injector. Fish loses were reported by students attending the local university since they were not aware of chloramine levels in water and did not take appropriate precautions to remove it prior to use in their aquariums like most of the commercial fish industry in the area.

Economic Analysis. For the implementation of anhydrous ammonia, a 908 kg/day chlorinator was converted to ammonia applicator together with a purchase of a pump and a daily storage tank. Table 2. at the end of this section presents capital costs together with an increase in operation and maintenance (O&M) costs incurred as a result of conversion to chloramines. The labor requirements increased by 2.3 hours per week and the time required for laboratory analyses increased by 1.5 hours per week. In addition, an ammonia dosage of 1 mg/l combined with a chlorine dosage increase of 1.1 mg/l comprised all increases in chemical costs. The total annual capital and O&M costs for the conversion were $20,712 (amortization of capital expenses is based on 20 year life span and six percent compound interest) which translated to an increase in treatment costs of $0.99/1,000 m^3.

2.1.2. Contra Costa Water District, Ralph D. Bollman Water Treatment Plant (George et al., 1990)

System Description. This is a conventional surface water treatment plant. The raw water source for this plant is the Sacramento-San Joaquin River delta. Water from the river delta is pumped into Mallard Reservoir (total volume of $3.8 \cdot 10^6$ m^3) where it is held until treatment. This water is moderately hard (94 mg CaCO3/l) with an average pH and turbidity of 8.1 and 29 NTU, respectively. In dry years, chloride ion concentration can be as high as 300 mg/l resulting in high THM levels, particularly the bromide-containing compounds.

During this period, the raw water contained nearly 1 mg bromide/l and the treated water had THM levels of 275 µg/l, 75% of which was bromofrom.

To address this problem, a chloramination system was added to the plant in September 1981. A small amount of chlorine was added to the raw water as a coagulant aid (0.75 mg/l), which produced no residual after mixing. Chlorine was also added prior to the filters in the amount sufficient to produce a 0.75 mg/l residual in the filter effluent. Ammonia was then applied at a rate of approximately 0.2 mg/l or at a 3.75:1 chlorine to ammonia weight ratio. The location of ammonia addition was selected to reduce a free chlorine contact time to only 10 min. Since the conversion to chloramination, it has also been a standard practice to add enough chlorine, for about one hour each night, to the influent water to produce a 0.5-0.7 mg/l residual in the filters. This practice was used to reduce the amount of algal growth in the sedimentation basin and filters.

System Performance. THM concentrations were reduced to an average of 47 µg/l by the conversion to chloramination. During a six-months period in 1985, TTHM levels in drinking water increased when state of California required the plant to stop using the chloramination because of some problems experienced by dialysis patients in southern California. Bacteriological quality of chloraminated water showed some improvement over chlorinated water. Total coliform levels remained nearly undetectable while HPCs were lowered from an average of 39 counts/ml to an average of 21 counts/ml. 95 percent of distribution system samples contained a combined chlorine residual. Chlorine-like odor complaints were eliminated and while some other utilities using the same raw water source experienced taste and odor problems caused by algae, this was not a problem at Ralph D. Bollman treatment plant due to significant free chlorine contact time at the plant.

Problem Areas. Hemodialysis patients are very susceptible to health problems that might be caused by chloramines. On one occasion, a local hospital removed a water purification process from a patient treatment step and chloramines reached the dialysis equipment. Several patients suffered anemia for a short time until the hospital personnel corrected the problem. Change to chloramination has increased the amount of chlorine required to clean storage reservoirs in the distribution system from 4 to 12 m^3 of 12% chlorine solution per 10^6 m^3 of reservoir capacity.

Economic Analysis. The process changes required the purchase and installation of an ammoniator, diffusers, and piping while the 2.37 m^3 anhydrous ammonia storage tank was provided at no cost. O&M costs included the chemical costs for the addition of ammonia (0.26 mg/l), ammoniation system maintenance, and savings realized through the reduction in the chlorine dosage of 1.1 mg/l. The conversion to chloramination at this plant resulted in the overall savings of \$0.14/1,000 m^3 (Table 2.).

Table 2. Increase in annual (1988) costs for conversion to chloramines (\$)

Item	Ann Arbor	Contra Costa
Capital Costs		
Chlorinator conversion kit	847	--
Ammoniators	--	5,240
Storage tank	678	0
Pump	460	--
Hose, valves, fittings	862	10,480
Engineering	--	--
Total	2,847	15,720
Annual amortized capital cost	248	1,370
Operation & Maintenance Cost Increase		
Labor	1,555	480
Laboratory time	438	--
Chlorine	12,710	-17,932
Ammonia	5,761	10,000
Potassium permanganate	--	--
Annual O&M Costs	20,464	-7,452
Total Annual Costs	20,712	-6,082
TOTAL UNIT COSTS (\$/1,000 m^3)	**0.99**	**-0.14**

2.2. CHLORINE DIOXIDE

The application of chlorine dioxide (ClO_2) has been implemented as an alternative to meet THM regulation. In addition, ClO_2 is effective for taste and odor control, decolorization, iron and manganese oxidation, organic oxidation, enhanced coagulation-filtration, in-plant control of biofouling, and disinfection (Lykins and Griese, 1986, Aieta and Berg, 1986). The use of ClO_2 as the primary disinfectant followed by free chlorine or chloramines for distribution residual is a common and effective alternative treatment practice for THM reduction.

Chlorine dioxide is an unstable gas that, like ozone, must be generated on site. It does not react with water to the same extent as chlorine; however, it is volatile and can easily be removed by aeration. Therefore, facilities for the application of ClO_2 should be designed to reduce volatilization of the gas out of the solution. Chlorine dioxide is considered to be a powerful oxidant, but at the pH typical in drinking water treatment, the full oxidation strength of ClO_2 is not utilized. Chlorine dioxide will not hydrolyze in water like chlorine and it exists as dissolved gas in the 2 to 10 pH range. The following reactions define reduction of ClO_2 in aqueous solution:

$$ClO_2 + 5e^- + 2H_2O \rightarrow Cl^- + 4OH^- \quad (1)$$

$$ClO_2 + e^- \rightarrow ClO_2^- \quad (2)$$

Equation (1) is the complete reaction used to describe reduction of ClO_2, but the reaction that often occurs in drinking water application is described by Equation (2) since the oxidizing power of this agent is pH-dependent and the Equation (2) is predominant reaction at pH above 5. Chlorine dioxide can also dissociate in alkaline solutions to chlorite (ClO_2^-) and chlorate (ClO_3^-) ions:

$$2ClO_2 + 2OH^- \rightarrow ClO_2^- + ClO_3^- + H_2O$$

ClO_2^- is also an effective oxidizing agent and will be consumed in oxidation-reduction reactions, although at a much slower rate than chlorine dioxide. This reduction half-reaction is given by:

$$ClO_2^- + 4H^+ + 4e^- \rightarrow Cl^- + 2H_2O$$

During water treatment, approximately 50-70 percent of chlorine dioxide reacted will immediately appear as chlorite and reminder as chlorine. The residual chlorite continues to degrade in the water distribution system via the above reaction, presumably in reactions with oxidizable material in the finished water.

There is some concern about the by-products of chlorine dioxide decomposition. ClO_2, ClO_2^- and ClO_3^- have been implicated as a cause for hemolytic anemia, methemoglobinemia and other hematological effects, with ClO_2^- identified as the most potent. However, if the recommended total residual oxidant level of less than 1.0 mg/l is maintained, no adverse health effects are anticipated (Lykins and Griese, 1986).

Reactions of ClO_2 with NOM do not produce THMs (Aieta and Berg, 1986) and the total production of total organic halides are from 1 to 25% of those produced by chlorine under similar conditions. Recent research has shown that some of these by-product are most likely similar to those produced by ozone (Richardson et al., 1994; see discussion about ozone by-products). However, ClO_2 reacts with THM precursors to render them unreactive to THM formation by as much as 50%, thereby reducing the amount of THMs that could be formed by post chlorination. Postchlorination dosages required to maintain a residual however, are generally higher than before the change to ClO_2 (Aieta and Berg, 1986).

Chlorine dioxide is superior to free chlorine as a biocide. However, during the period of high turbidity (>100 NTU), an increase in taste and odor and a decrease in disinfection efficiency may occur. This behavior can be attributed to an increase in organic matter rather than suspended matter. In general, most studies found chlorine dioxide to be an effective and rapid bactericide (unaffected by pH in the 6.5 to 8 range) and equal or superior to chlorine on a mass-dosage basis (Aieta and Berg, 1986).

Generation of chlorine dioxide in water treatment is most commonly performed through the reaction of sodium chlorite and chlorine (gas or liquid) (Miller et al., 1978):

$$2NaClO_2 + Cl_{2(g)} \rightarrow 2NaCl + 2ClO_2$$

$$5NaClO_2 + 4HCl \rightarrow 4ClO_2 + 5NaCl + 2H_2O$$

The objective is to produce a solution with highest purity and the greatest yield (purity is the ratio of ClO_2 to the total of all chlorine oxidative compounds). If a gas chlorinator is used, the chlorinator effluent must be at pH 3.5 or lower and contain at least 500 mg/l chlorine to accomplish the above objective. This condition is typically achieved by setting the chlorine-to-sodium chlorite ratio at 1:1 in order to provide excess chlorine. It is also possible to add acid to achieve the desired pH. The most application nowadays use the reaction of gaseous chlorine with sodium chlorite under vacuum (Aieta and Berg, 1986).

2.2.1. Bristol County Water Authority's Child Street Treatment Plant (George et al., 1990)

System Description. This treatment plant obtained its raw water from a $1.76 \cdot 10^6$ m^3 reservoir that was fed from three other reservoirs. Raw water was treated in a conventional alum coagulation, rapid sand filtration system rated at 11 m^3/min with an average water production of 8.9 m^3/min. The reservoir was subjected to salt water intrusion for a few days each year when high winds, high tide, and low water levels in the reservoir occurred simultaneously. Raw water had high color (87 pcu) with low alkalinity (17 mg CaCO3/l), low turbidity (2 NTU), and significant amounts of iron (0.6 mg/l) and manganese (0.19 mg/l). The maximum trihalomethane potential of raw water occasionally exceeded 3000 μg/l which forced this utility to seek alternative treatment method. One of the criteria for selecting the treatment scheme was the requirement for the oxidation of iron and manganese. Typically, chlorine, alum, coagulant aid and lime were added at the raw water intake. Adding potassium permanganate at the low-lift pump station and moving the point of chlorination to just before the rapid sand filters reduced the TTHM levels only from 227 to 125 μg/l. Subsequently, a new treatment approach had to be adopted.

Based on the results of trial runs, the modifications included the addition of chlorine dioxide before the pretreatment clarifier (detention time of 3.75 hours at average flow) and moving the point of chlorine addition to clear well. Chlorine dioxide was applied at 1 to 2 mg/l to provide approximately 0.1 mg ClO_2/l residual after the pretreatment clarifier. Chlorine dioxide was produced from a 25% sodium chlorite solution and chlorine gas in stoichiometric quantities under vacuum. PAC was added into the settling basin for taste and odor control. Lime was added to adjust pH and alkalinity just before filtration. Chlorine and fluoride were added at the clear well which provided 25 min of detention time.

System Performance. The THM levels in finished water were reduced significantly from an average of 112 to 53 μg/l. Chlorine dioxide also oxidized iron and manganese to average finished water levels of 0.043 and 0.037 mg/l, respectively. With the use of chlorine dioxide, the average chlorine dosage requirements were reduced from 4.1 to 3.2 mg/l. Correspondingly, the chlorine residual was reduced from 1.1 to 0.8 mg/l. Even though the raw water color increased from an average of 74 pcu during the period of chlorine usage to an average of 101 pcu after the change in the treatment process, the new system was capable of producing a product water with a slightly less color (4.7 pcu) than achieved before chlorine dioxide treatment (5.1 pcu). In addition, the plant continued to produce water with an average turbidity of 0.57 NT and odor of 1.8 TON.

Problem Areas. During the summer of 1986, the system experienced taste and odor problem which was not solved even with the addition of PAC and the cause was never identified. The only other concern was with the analytical procedure for residual ClO_2 measurements (low confidence levels in the results obtained by the procedure).

Economic Analyses. Capital expenditures included a chlorine dioxide generator, sodium chloride handling equipment (mixing tank, barrel luft, mixer) and a block building. The use of 0.97 mg/l potassium permanganate was discontinued and chlorine dosage was reduced by 0.9 mg/l. Sodium chlorite was added to the water at a concentration of 1.72 mg/l. Other O&M costs included 2 hours/week labor for the operation of generation facilities and 16 hours per year for equipment maintenance together with some replacement parts (needle valves, o-rings, etc.) The combined annual capital costs and O&M costs equaled $13,556 resulting in the overall increase in treatment costs of $2.8/1,000 m^3 (Table 3.).

2.2.2. Kentucky American Water Company (George et al., 1990)

System Description. Raw water for the plant was withdrawn from the Kentucky River. The plant had a design capacity of 105 m^3/min with an average production of 74 m^3/min. Prior to conversion to chlorine dioxide, a dosage of 5-7 mg/l of chlorine was added to the influent water to maintain a residual throughout the treatment process. A cationic polymer was added as the primary coagulant and alum was used as a coagulant aid when needed. Optimal pH for coagulation was achieved by the addition of lime. Water treatment was accomplished in 10 Aldrich units (Hydrotreators). THM analyses revealed average concentrations of 125 µg/l with only 37% of the total being chlorinated species.

Once the utility recognized that bromide in the Kentucky River was aggravating the THM levels in the finished water, an alternative treatment method had to be implemented. A chlorine dioxide generating system capable of generating 304 kg of chlorine dioxide per day was installed at the plant and 25% sodium chlorite solution was used to feed the generator. Choice of solution over a dry powder was made because the dry powder is a strong oxidizing agent that poses a significant hazard if spilled and the mixing of a dry powder would require an intensive labor. A 2-month trial run was conducted by feeding chlorine dioxide at a dosage from 0.6 to 1.25 mg/l to provide a residual prior to filtration of 0.1 mg/l. During this period, THM levels were reduced by about 60% but a significant number of complaints from customers about chlorinous- or petroleum-like odors were reported. There was no success in identifying source of the problem and it was decided to use chlorine dioxide only during periods when excessive THM production occurred. After the addition of potassium permanganate at the river intake to control ClO_2 odors, a number of complaints significantly decreased. The point of chlorination was then moved from the rapid mix to the top of the filters and the addition of chlorine dioxide was reduced only to two quarters throughout the year when THM levels became excessive.

System Performance. A major concern during the trial run was consistently higher bacteriological counts throughout the treatment system even though they were always below state regulations. After portions of distribution system were flushed to control odors, distribution system bacteriological quality improved tremendously indicating a requirement for postchlorination. THM levels decreased from 140 to 90 μg/l while the average amount of chlorine added to the system remained practically unchanged (516 vs. 477 kg/day) yielding an appreciable decrease in chlorine dosage (from 5.1 to 3.9 mg/l) due to the increase in average flow.

Problem Areas. Due to an observed decrease in the efficiency of chlorine dioxide generator from 95 to 90%, it was necessary to clean the media once a week with a mild acid and replace it twice in four years. The cause of odor problems instigated with the use of chlorine dioxide were never identified and the problems occurred even though no chlorine dioxide could be identified in the effluent (pretreatment with potassium permanganate seemed to eliminate many of the problems). In conjunction with poor biological control capabilities of ClO_2 observed at this plant, this problem forced the utility to investigate other disinfection alternatives. In addition, analytical methods for chlorine dioxide, chlorate, and chlorite that were capable of providing accurate results were too time consuming and complex for everyday use by a water treatment plant operators.

Economic Analysis. The capital cost consisted of the purchase of a chlorine dioxide generator, two mixing tanks and mixers, metering pump, jet pump, two filament-wound tanks, and two stainless-steel centrifugal pumps totaling $20,340. The average sodium chlorate dosage was 1.55 mg/l while the chlorine dosage was reduced by 1.2 mg/l. An estimated additional labor hours associated with laboratory analyses, maintenance, and operation of the generator was 31 hours per week for a 6-month period. The combined annual amortized capital cost and O&M costs totaled $55,299 which translated in the increase in treatment costs of $1.43/1,000 m^3 (Table 3.).

Table 3. Increase in annual (1988) costs for conversion to chlorine dioxide ($)

Item	Bristol County	KY-American Water Co.
Capital Costs		
ClO_2 generator	12,180	
Handling Equipment	3,606	
Block Building	23,189	
Pump	--	
Hose, valves, fittings	--	
Engineering	--	
Total	39,975	20,340
Annual amortized capital cost	3,398	1,773
Operation & Maintenance Cost Increase		
$NaClO_2$	22,907	69,124
Chlorine	-2,397	-26,758
Potassium permanganate	-12,204	--
Labor	220	11,160
Laboratory time	1,432	
Replacement parts	200	
Annual O&M Costs	10,158	53,526
Total Annual Costs	13,556	55,299
TOTAL UNIT COSTS ($/1,000 m^3)	**2.80**	**1.43**

2.3. OZONE

Ozone is one of the most powerful oxidizing agents available for practical applications in water treatment. It does not impart tastes and odors to water and has beneficial effects on microflocculation, enhanced suspended solids removal, and pretreatment for biological activated carbon. Ozone, however, does not provide a disinfection residual and must, in most cases, be followed with secondary disinfection to maintain residual in the distribution system. Ozone is a highly reactive gas that follows two basic modes of reaction:

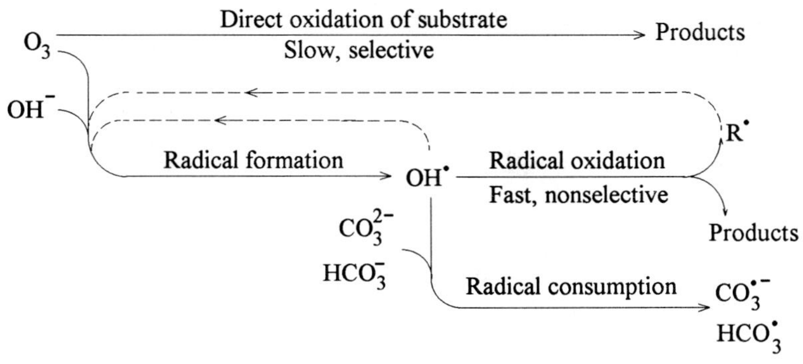

The rate of ozone decomposition is a complex function of temperature, pH, ozone concentration, type and concentration of organic compounds, and type and concentration of inorganic constituents. Conditions of low pH favor direct oxidation reactions while high pH or high concentrations of organic matter favor the auto-decomposition of ozone via the hydroxyl radical (Hoigne and Bader, 1976, Gurol, 1985).

Ozone has received considerable attention for THM control since it is the most effective oxidant and disinfectant used in water treatment practice and a number of studies showed that in combination with monochloramine provides effective treatment while minimizing halogenated DBP formation. Ozone, however, reacts with NOM to produce a variety of oxidation by-products like aldehydes (e.g., formaldehyde, acetaldehyde, glyoxal, methyl glyoxal), aldo- and ketoacids (e.g., pyruvic acid), acids (e.g., oxalic, succinic, formic, acetic), and hydrogen peroxide (Glaze et al., 1989, Weinberg et al., 1993). Of these, the aldehydes appear to be of greatest concern from a public-health standpoint at this time (Bull and Kopfler, 1991), but insufficient information is currently available to ascertain the health risks due to exposure to aldehydes. All of the oxidation products listed are for the most part biodegradable and contribute to the assimilable organic carbon (AOC) content of the water which can lead to nuisance biological growths in flocculation basins, sedimentation basins, filters, and the distribution system if not properly controlled. One effective way of controlling these growths is through the use of biologically active filtration (Reckow et al., 1992, Weinberg et al., 1993).

Although ozone itself does not produce halogenated DBPs, it can result in the production of brominated DBPs (bromate, bromoform, brominated acetic acids, bromopicrin, brominated acetonitriles) when bromide-containing waters are ozonated. Ozone will oxidize bromide (Br^-) to hypobromous acid (HOBr), which in turn can react with NOM to produce the fully brominated analogs of the chlorination by-products (Siddiqui and Amy, 1993). In addition, hypobromite (OBr^-) in equilibrium with HOBr can be oxidized further by ozone to produce bromate (BrO_3^-) which has been classified by EPA as a B2 carcinogen (probable human carcinogen) and is likely to be regulated with an MCL equal to its practical quantification level of 10 μg/l. Bromate formation can be controlled by ozonation at acidic pH-values, at which hypobromous acid dominates over hypobromous ion (Siddiqui and Amy, 1993).

Virtually all-large scale ozone generators incorporate the corona discharge method for the production of ozone with varying cell configuration, operation voltage, operation frequency, cooling method, and air preparation. Ozone production depends mainly on the geometry of the generator, the dielectric properties, the current frequency, the dryness and content of the feed gas, the pressure, and process temperature. Since the ozone yield is decreased with increasing temperature and dew point, it is necessary to provide process cooling and feed-gas (air or oxygen) preparation system. The use of oxygen as feed gas allows higher ozone production (by a factor of 2), larger ozone concentrations, smaller generators, and lower energy consumption. Nonetheless, ozone is produced from air in the majority of cases.

2.3.1. Bell Glade Water Treatment Plant (George et al., 1990)

System Description. Raw water for the Bell Glade plant was taken from Lake Okeechobee (large, shallow lake) and had average color between 59 and 74 pcu (peak value as high as 500 pcu), high TOC content of 20 to 30 mg/l (up to 75 mg/l during summer agricultural drainage periods), and MTP average of 808 μg/l. In addition, the underlying limestone combined with natural acids in the water caused high hardness (average monthly concentration of 242.4 mg $CaCO_3$/l with the average maximum concentration of 512 mg $CaCO_3$/l) and high mineral content of the raw water. Treatment process utilized softening, coagulation, flocculation, filtration and disinfection. Prior to treatment change, 30 to 60 mg/l alum and 15 to 50 mg/l of chlorine (breakpoint chlorination) were added to water as it entered the plant. Alum was added for color removal but did not work well because of the high pH (10.5) utilized in the excess lime softening process. Treatment process modification incorporated replacing the chlorine addition to the raw water with ozone to control the color, biological growth in the plant, musty and obnoxious taste and odor compounds, and to oxidize organic THM precursors. The average monthly dosage of ozone was 9.5 mg/l (range from 4.9 to 15 mg/l) while postchlorination was used to maintain residual in the distribution system.

System Performance. One of the most obvious improvements of finished water was in the aesthetic quality with a significant decrease in color resulting in the elimination of customer complaints concerning color, taste and odor. Ozone addition resulted in significant decrease in the instantaneous TTHM concentration average from 643 to 60 μg/l (prior to change, the actual measured TTHM concentration occasionally exceeded 1,200 μg/l) while no significant changes in TOC concentrations were observed. Water softening operation improved slightly after the change to ozonation. The reduction in the finished water pH from 9.7 to 7.8 after the conversion to ozonation was not to minimize ozone decomposition, but to control the precipitation of calcium in the distribution system. A decrease in water turbidity from 0.4 to 0.2 NTU may have been due several factors, including microflocculation and an increase in polymer dosage from 0.04 to 0.08 mg/l.

Problem Areas. No serious problems associate with the change to ozonation were reported by the principal operator. Minor maintenance (e.g., installation of filters in the seal water system of the air compressor, replacement of the steel sleeves in the dryers with stainless steel, and the development of a few pin holes in the dielectric sleeves) were easily resolved and the system has operated reliably since installation.

Economic Analysis. The capital and operation and maintenance costs for Belle Glade water treatment plant are given in Table 4. The estimated total maintenance costs averaged \$4,000/year while power requirements for the ozone generator were estimated at 89,913 kw·hr/month for an average monthly cost of \$5,700. Additional labor requirements included 2 hr/day for operation and 16 hr/year for maintenance. After converting to ozone, the chlorine dosage was reduced from 19.3 to 9.4 mg/l resulting in savings of \$3,523/month. The estimated total yearly increase in O&M costs was \$17,367 which, together with annual amortized capital costs, resulted in total costs of \$78,506/year or \$14.57/1,000m^3.

2.3.2. Monroe Water Treatment Plant (George et al., 1990)

System Description. This conventional water treatment plant used lake water with relatively low turbidity (average of 10 NTU) and moderate hardness and alkalinity of 121 and 87 mg $CaCO_3$/l, respectively. However, because of the shallow lake depth and uneven bottom, frequent strong winds caused violent wave action that disturbed the bottom sediments and produced high turbidities (1,200 NTU maximum). These conditions also created objectionable odors. Spring run-offs and odorous compounds produced by algae and actinomycetes during Summer also affected the raw water quality. Odor control efforts included application of PAC, clarification with alum and intermediate chlorination with contact time prior to filtration. These efforts were successful until the increase in the flow rate decreased the chlorine contact time and the system was altered to incorporate ozonation for the control of taste- and odor-causing compounds. The two ozone generators, each rated at 102 kg ozone/day at concentrations of 1 wt % using air as the feed gas, were installed at that time. Air treatment prior to ozone generator included passage through an inlet filter and compression which increased its temperature to 149 ^{O}C. The dew point was lowered to 20 ^{O}C by a refrigerant dryer and finally to -50 to -100 ^{O}C by transporting the air through molecular sieves and silica gel. Carryover of desiccant was removed using a 10 μm filter prior to the corona cells. The contact chamber consisted of a countercurrent gas-diffused (353 porous aluminum oxide diffusers), positive-pressure (12.4 kPa), recirculating contactor that consistently yielded inlet-to-outlet efficiency from 93 to greater than 99 percent. The contactor had two passes with three stages in each pass (60-70% distributed in the first stage of pass 1 and the remaining 30-40% distributed in the second stage) with an average contact time of approximately 10 min at a flow rate of 47 m^3/min.

System Performance. The raw water was treated with an average ozone dosage of 1.2 mg/l and a maximum of 3.1 mg/l during the period of March through May of each year when taste and odor problems were traditionally most pronounced. As a result, taste and odor complaints were decreased from a high of 100 in one day to zero. In addition, the chlorine dosage was lowered from 4.1 to 2.9 mg/l to produce the residual of 0.7 mg/l. It was also discovered that the chlorine demand of ozonated water increased presumably due to partial oxidation of complexed organic compounds by ozone which produced by-products that were more susceptible to chlorine oxidation. TTHM concentration in finished water prior to ozonation averaged 60-80 μg/l while the average of 70 μg/l was observed during the periods of ozone addition. Chloroform was the predominant halomethane in the finished water, followed by dichlorobromomethane and traces of dibromochloromethane (< 7 μg/l) while no bromoform was ever detected. During the periods when ozone was not used, and average TTHM concentration of 44 μg/l was observed. However, due to the limited number of data, this difference in average concentrations was not found statistically significant. Ozonation did not significantly improve the product water turbidity (0.2 NT prior and after ozonation) but enabled a reduction in alum dosages from 25 to 19 mg/l and longer filter runs (average of 88 hours compared to 55 hours). Occasional presence of cyanide originated from steel mils was successfully treated by ozonation through oxidation to cyanate and subsequent hydrolysis.

Problem Areas. No major problems with the use of ozone was experienced at Monroe Water Treatment Plant. Several interruptions in ozonation occurred due to malfunction in the dew point hygrometer and a complete renewal of desiccant had to be performed once. During the start-up, all rubber material that came in contact with the ozonated water had to be replaced with silicone rubber.

Economic Analysis. The total capital costs amounted to \$754,500. Amortization and insurance comprised more than half of the annual operating costs for this system with the electrical power being the second most important item. Reduction in chlorine and alum dosages by 1.2 and 6 mg/l, respectively, resulted in chemical costs savings of 6,575 and 3,945 \$/month, respectively. The overall annual cost for this system is \$84,650 which resulted in the increase in treatment costs of \$8.51/1,000 m^3.

Table 4. Increase in annual (1988) costs for conversion to ozone (\$)

Item	Belle Glade	Monroe
Capital Costs		
Ozone generator	620,100	312,700
Installation	58,300	192,700
Piping and valves	--	249,100
Training, prestart-up	4,770	--
Lab test	5,512	--
Start-up	9,434	--
Assistance, poststart-up	3,180	--
Total	701,296	754,500
Annual amortized capital cost	61,139	65,578
Annual O&M Costs	17,367	8,552
Total Annual Costs	78,506	84,650
TOTAL UNIT COSTS (\$/1,000 m^3)	**14.57**	**8.51**

2.4. REFERENCES

Aieta E.M., Berg J.D. (1986) A Review of Chlorine Dioxide in Drinking Water Treatment. Journal AWWA, 78(6), 62.

Bull R.J., Kopfler F.C. (1991) Health Effects of Disinfectants and Disinfection By-Products. American Water Works Association Research Foundation, Denver, CO.

George D.B., Adams V.D., Huddleston S.A., Roberts K.L., Borup M.B. (1990) Case Studies of Modified Disinfection Practice for Trihalomethane Control. AWWA Research Foundation, Denver, CO.

Glaze W.H. (1986) Chemistry of Ozone, By-Products and Their Health Effects. Proc. AWWA Annual Conf., AWWA, 1-15, Denver, CO.

Glaze W.H., Koga M., Cancilla D. (1989) Ozonation By-Products. 2: Improvement of an Aqueous-Phase Derivatization Method for the Detection of Formaldehyde and Other Carbonyl Compounds Formed by the Ozonation of Drinking Water. Environ. Sci. Technol. 23(7), 838.

Gurol M.D. (1985) Factors Controlling the Removal of Organic Pollutants in Ozone Reactors. Journal AWWA, 77(8), 55.

Hoigne J., Bader H. (1976) Role of Hydroxyl Radical Reactions in Ozonation Processes in Aqueous Solutions. Water Research, 10, 377.

Jacangelo J.G., Patania N.L., Reagan K.M., Aieta E.M., Krasner S.W., McGuire M.J. (1989) Ozonation: Assessing its Role in the Formation and Control of Disinfection By-Products. Journal AWWA, 81(8), 74.

Krasner S.W., McGuire M.J., Jacangelo J.G., Patania N.L., Reagan K.M., Aieta E.M. (1989) The Occurrence of Disinfection By-Products in US Drinking Water. Journal AWWA, 81(8), 41.

Lieu N.I., Wolfe R.L., Means E.G. (1993) Optimizing Chloramine Disinfection for the Control of Nitrification. Journal AWWA, 85(2), 84.

Lykins B.W., Griese M.H. (1986) Using Chlorine Dioxide for Trihalomethane Control. Journal AWWA, 78(6), 88.

Miller G.W., Rice R.G., Robson C.M., Scullin R.L., Kuhn W., Wolf H. (1978) An Assessment of Ozone and Chlorine Dioxide Treatment Technologies for Municipal Water Supplies - Executive Summary. EPA report EPA-600/8-78-018, USEPA, Cincinnati, OH.

Reckow D.A., Tobiason J.T., Switzenbaum M.S., et al. (1992) Control of Disinfection By-Products and AOC by Ozonation and Biologically Active In-Line Direct Filtration. Proc. AWWA Annual Conf., AWWA, Denver, CO.

Richardson S.D., Thruston A.D., Colette T.W., et al. (1994) Multispectral Identification of Chlorine Dioxide Disinfection Byproducts in Drinking Water. Env. Sci. Technol., 28(4), 592.

Siddiqui M.S., Amy G.L. (1993) Factors Affecting DBP Formation During Ozone-Bromide Reactions. Journal AWWA, 85(1), 63.

Thompson J.C., Ameno J.J. (1987) Preventing the Formation of Trihalomethanes in Florida Groundwater. Journal AWWA, 79(1), 38.

Trussel R.R., Kreft P. (1984) Engineering Considerations of Chloramine Applications. Proceedings AWWA Seminar on Chloramination for THM Control: Principles and Practices, AWWA Annual Conference, Dallas, TX, p. 47.

White G.C. (1986) Handbook of Chlorination. 2nd edition, Van Nostrand Reinhold Co., Inc., New York.

Weinberg H.S., Glaze W.H., Krasner S.K., Sclimenti M.J. (1993) Formation and Removal of Aldehydes in Plants That Use Ozone. Journal AWWA, 85(5), 72.

Wolfe R.L., Ward N.L., Olson B.H. (1984) Inorganic Chloramines as Drinking Water Disinfectants: A Review. Journal AWWA, 76(5), 74.

3. Technological Aspects of Advanced Processes of Coagulation and Flocculation in Water Treatment

Peter Kalinkov
Jean Louis Monzo
Higher Institute of Architecture
and Civil Engineering
1 Christo Smirnenski , 1421 Sofia
BULGARIA

3.1. INTRODUCTION

Creating the optimal conditions for an effective coagulation-flocculation process in raw water treatment is always a question of current interest for specialists working in that field.

Our previous study (Kalinkov and Monzo, 1992) was devoted to the investigation of the influence of initial rapid mixing duration and the major objective of this paper is to investigate the influence of geometrical characteristics of mixers in these processes.

It is well known that the effectiveness of the coagulation-flocculation process depends on the physico-chemical characteristics of water, the type and dose of the coagulant, as well as the duration and intensity of initial rapid mixing.

Contemporary theoretical models for this process combine the physical aspects, such as the hydrodynamics of mixing, with the mechanism of destabilisation of colloidal particles. Initial rapid mixing at non - effective hydrodynamic conditions leads to a higher coagulant consumption as well as to a smaller degree of particle aggregation at a given coagulant dosage.

Different approaches can be used to estimate the effectiveness of mechanical mixers. Camp and Conklin (1970) use the so called turbulent gross grad coefficient *Ct*.

They consider that at a condition of full turbulence of the stream the necessary relative power to start the impeller in water with regards to the unit volume may be expressed with:

NATO ASI Series, Partnership Sub-Series, 2. Environment – Vol. 15
Water Supply Systems – New Technologies
Edited by Č. Maksimović, F. Calomino, and J. Snoxell

$$W_0 = \frac{(2\Pi)^3}{2} \rho f C_t N = 124 \rho f C_t N \qquad (1)$$

where
- ρ water density (kgm^{-3});
- f projected area of the impeller blades normal to the direction of the tangential velocity (m^2);
- C_t turbulent gross drag coefficient (-);
- N speed of impeller (s^{-1}).

Vrale and Jorden (1971) use apparent aggregation rate Kapp, which according to them, can be defined by the equation:

$$\frac{dn}{dt} = Kapp\, n^2 \qquad (2)$$

where
- $\frac{dn}{dt}$ is the rate of changing water turbidity during aggregation;
- $Kapp$ aggregation rate constant
- n water turbidity

Equation (2) can be integrated in the interval for n (Mo, Mt) and t (0, T).
After integration Equation (2) in the upper interval:

$$\frac{1}{M_t} = \frac{1}{M_0} + KappT \qquad (3)$$

where
- Mo water turbidity after initial mixing;
- Mt water turbidity after T min flocculation (slow mixing);
- T flocculation or slow mixing time.

It is possible, using the theory of mixing presented by Vrale and Jorden, to obtain experimentally the apparent aggregation rate at different conditions of initial rapid mixing and the same (constant) flocculation conditions, and to draw a conclusion for the intensity of destabilisation, e.g. the influence of hydrodynamic conditions of mixing on the effectiveness of the coagulation - flocculation process.

3.2. EXPERIMENTAL SET UP AND PROCEDURE

Laboratory investigation has been carried out in 5.0 l plexiglass vessels and different geometrical configurations (Figure 1.) and impellers (Figure 2.) the dimensions of which are valid under the condition of ratio D/C = 0.56; at a constant area of the cross section *f*, perpendicular to the plane of rotation.

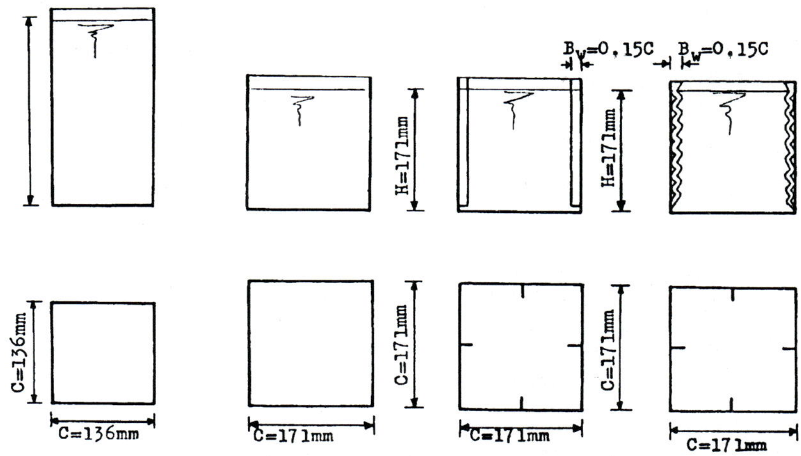

a) without baffles and C/H = 0.5

b) without baffles and C/H = 1

c) with simple baffles and C/H = 1

d) with cogged perforated baffles and C/H = 1

Figure 1. Geometrical configuration of mixers used in this study

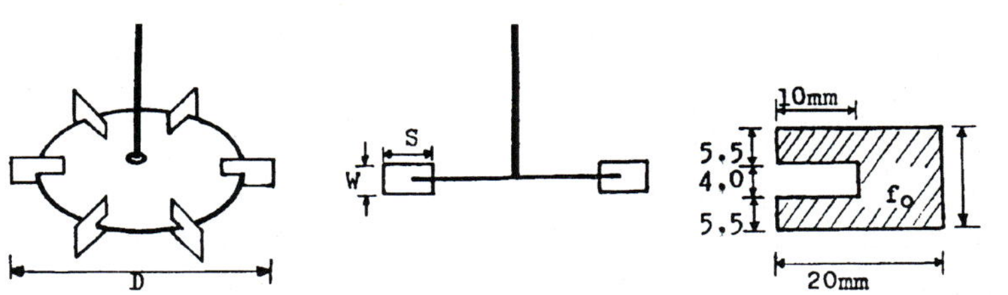

$f = 6fo = 15.60\ cm^2$

Figure 2. Impeller - disc turbine type

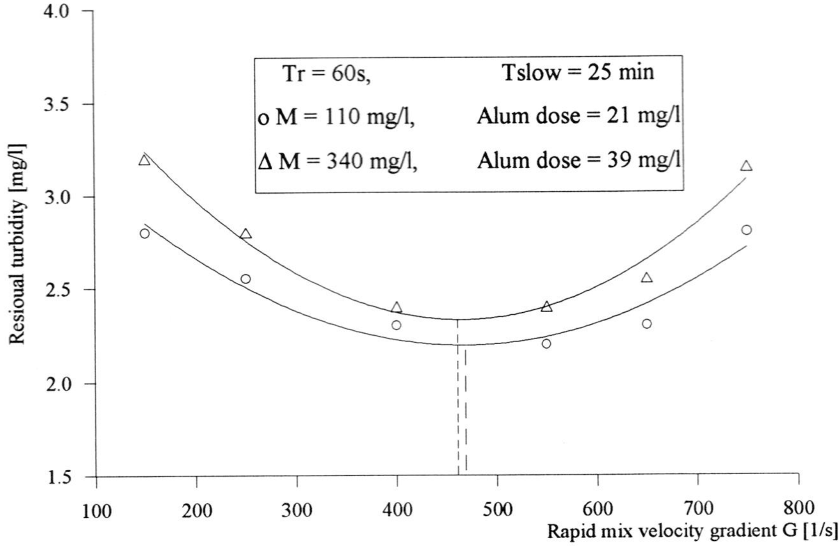

a) Classical coagulation and optimal Alum dose

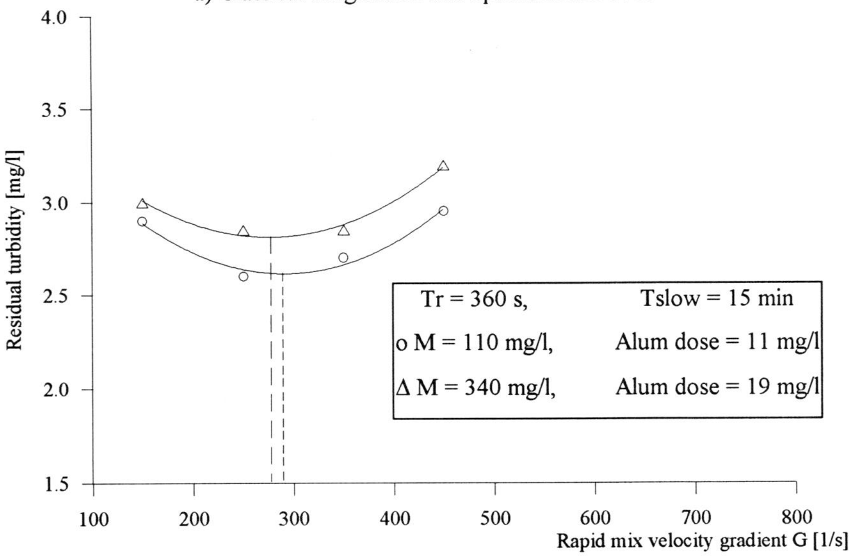

b) Prolonged coagulation and reduced Alum dose

Figure 3. Optimal rapid mix velocity gradient

Four mixer configurations have been investigated with square bottom without baffles, at ratio C/H = 1 and the same vessel with C/H = 0.5; with square bottom with simple baffles built in and ratio C/H = 1; with square bottom with perforated cogged baffles and C/H = 1. For all configurations G = f(N), where G is the velocity gradient (s^{-1}) and N is the impeller rotation speed (min^{-1}).

An artificially obtained turbid water is used, obtained by adding bentonite clay to distilled water. A 5% solution of $Al_2(SO_4)_3 18H_2O$ is used as a coagulant.

Water temperature, pH and alkalinity were held constant and only the physical parameters of rapid mix were varied. pH and alkalinity were adjusted with a buffer made of sodium hydroxide, sodium bicarbonate, sodium chloride and acetic acid as suggested by Packham (1962).

Two series of experiments have been carried out with different water turbidity, as follows M_1 = 26.2 EM, corresponding to 110 mg/l according to SiO_2 scale and M_2 = 80.9 EM, corresponding to 340 mg/l according to SiO_2 scale. Preliminary optimal coagulant doses for these turbidities have been defined by classical method with conventional jar - test equipment. They are as follows: Dc_1 = 21 mg/l and Dc_2 = 39 mg/l Alum.

An experiment has been implemented to determine the optimal values of velocity gradient G for classical water treatment (60 s rapid mixing and 25 min slow mixing) and prolonged initial rapid mixing (360 s rapid mixing and 15 min slow mixing) with 50% reduced dose of coagulant for a vessel without baffles at C/H = 1.

The results obtained are shown on Figure 3a. and Figure 3b. It is obvious that in the first case the optimal values of the velocity gradient are 550 s^{-1} and 500 s^{-1} for water turbidity of 110 mg/l and 340 mg/l respectively according to SiO_2 scale. And in the second case these values are 250 s^{-1} and 230 s^{-1}.

The investigation of all configurations is carried out by two procedures: classical coagulation with optimal coagulant dose and prolonged coagulation (prolonged initial rapid mixing) and coagulant dose reduced with 50%.

The rapid mixing is carried out in the range of optimal velocity gradient values. In the course of every experiment the rapid mixing is followed by a slow one with different duration - from 5 to 25 min and constant impeller speed of about 45 min^{-1} which corresponds to a velocity gradient of 40 s^{-1}.

Seventy two series of experiments have been carried out which give the opportunity for the apparent aggregation rate for the different cases to be defined. The results of the experiments with vessels without baffles and ratio C/H = 1, implemented with the classical procedure are given in Table 1. and in Table 2. the results for prolonged initial rapid mixing are shown. In all experiments after the period of slow mixing follows a 15 min period of settlement and after that is measured the residual water turbidity.

Table 1. Residual turbidity values as a function of slow mix time under classic coagulation

Raw water turbidity: 26.2 EM = 110 mg/l according to SiO_2 scale Alum dose: 21 mg/l Initial Rapid mix time: t_R = 60s; C/H = 1												
Initial Rapid mix Velocity Gradient G (s^{-1})												
slow mix time	150		250		400		550		650		750	
t_s	Mt	1/Mt	Mt	1/Mt	Mt	1/Mt	Mt	1/Mt	Mt	1/Mt	Mt	1/Mt
min	mg/l	1/mg	mg/l	1/mg	mg/l	1/mg	mg/l	1/mg	mg/l	1/mg	mg/l	1/mg
5	10.05	0.099	8.69	0.115	8.40	0.119	7.94	0.126	8.77	0.114	9.98	0.100
10	6.36	0.157	6.25	0.160	4.75	0.210	4.63	0.216	4.69	0.213	6.78	0.148
15	4.25	0.235	4.01	0.249	3.78	0.264	3.34	0.299	3.58	0.279	4.06	0.246
20	3.43	0.291	2.90	0.345	2.86	0.350	2.85	0.351	2.90	0.345	3.40	0.294
25	2.64	0.379	2.62	0.382	2.34	0.428	2.32	0.431	2.32	0.431	2.88	0.347

Table 2. Residual turbidity values as a function of slow mix time under prolonged coagulation

Raw water turbidity: 26.2 EM = 110 mg/l according to SiO_2 scale Alum dose: 21 mg/l Initial Rapid mix time: t_R = 360s; C/H = 1								
Initial Rapid mix Velocity Gradient G (s^{-1})								
slow mix time	150		250		400		550	
t_s	Mt	1/Mt	Mt	1/Mt	Mt	1/Mt	Mt	1/Mt
min	mg/l	1/mg	mg/l	1/mg	mg/l	1/mg	mg/l	1/mg
5	6.25	0.160	6.37	0.157	5.81	0.172	7.52	0.133
10	3.94	0.254	3.69	0.271	3.71	0.269	3.92	0.255
15	2.87	0.348	2.69	0.371	2.75	0.364	2.87	0.348
20	2.35	0.425	2.04	0.489	2.23	0.448	2.34	0.427
25	1.77	0.564	1.76	0.567	1.73	0.577	1.99	0.502

Based on the results of the experiments presented in Table 1., curves for the relation 1/Mt = f ($t_{\text{slow mixing}}$) are shown in Figure 4. and respectively the results from Table 2. are plotted in Figure 5. Similar values like those presented in Table 1. and Table 2. and plotted curves in Figure 4. and Figure 5. are obtained for all investigated configurations at conditions of initial rapid mixing and initial water turbidity. Only part of these results are given as an illustration, for brevity.

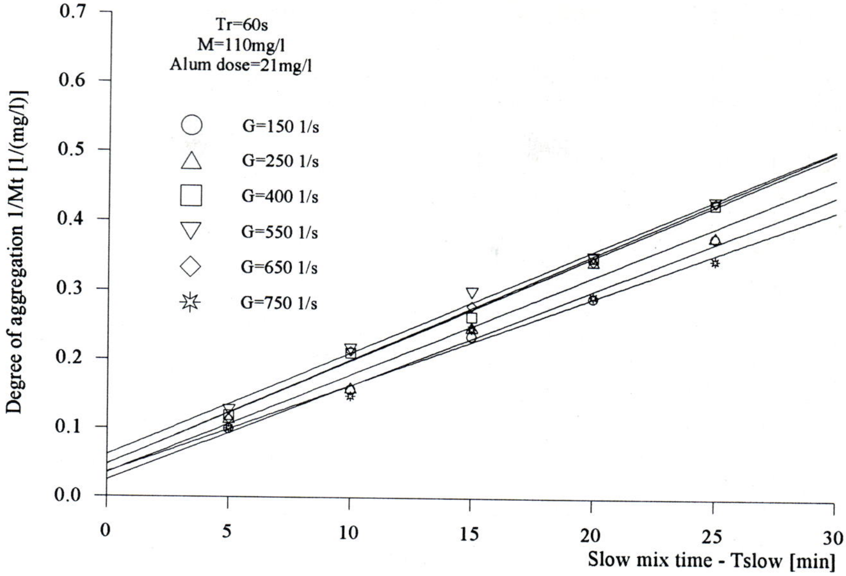

Figure 4. Residual turbidity as a function of slow mix time for various G_{rapid} under classical coagulation

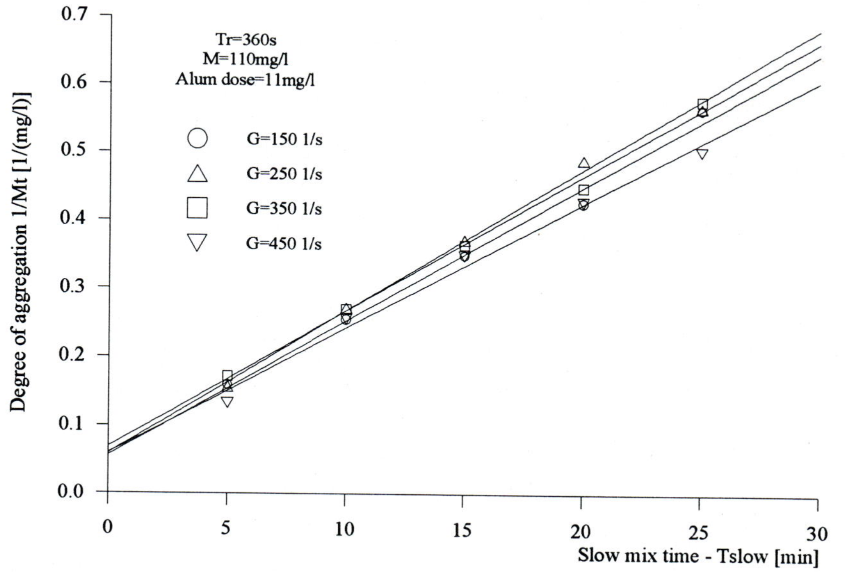

Figure 5. Residual turbidity as a function of slow mix time for various G_{rapid} under prolonged coagulation

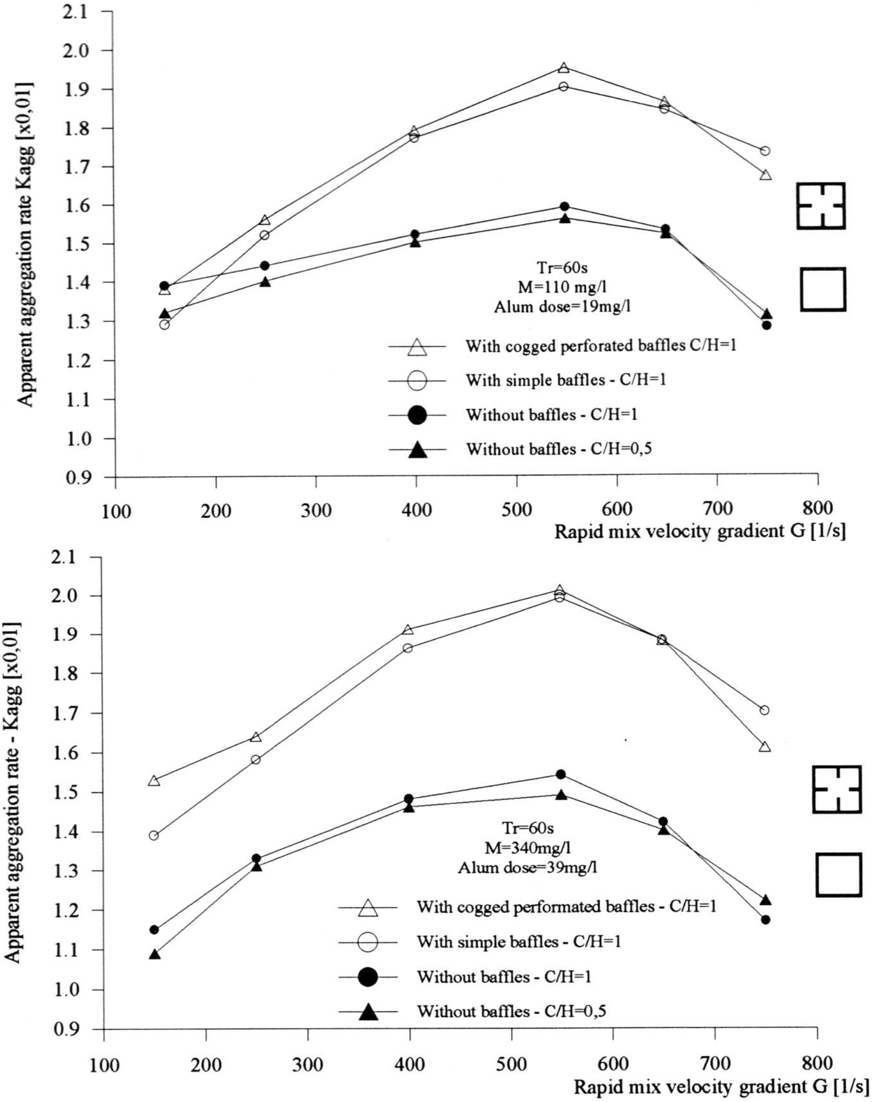

Figure 6. Apparent aggregation rate as a function of G_{rapid} under classical coagulation and optimal Alum dose

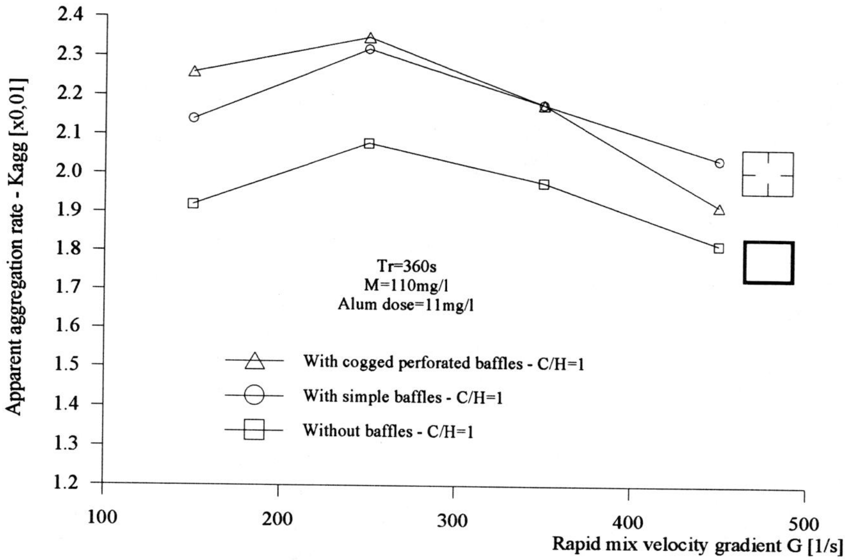

Rapid mix velocity gradient G (s-1)

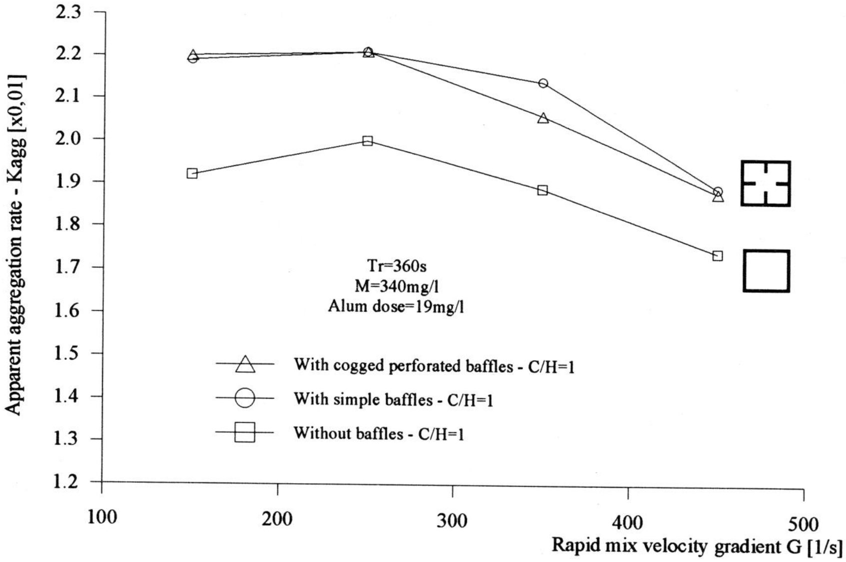

Figure 7. Apparent aggregation rate as a function of G_{rapid} under prolonged coagulation and reduced Alum dose

3.3 RESULTS AND DISCUSSION

Analysing the results of the apparent aggregation rate presented in Figure 6. and Figure 7. definite conclusions can be made. On the first place the curves of Kapp = f (G) increase at the beginning, reach the maximum and after that decrease for all studied configurations. This shows that optimal hydrodynamic conditions exist in all configurations and if they are reached the process of destabilisation and flocculation proceeds in the best way. The fact that the value of this maximum is different means that the different configurations have different abilities concerning insuring the optimal hydrodynamic conditions for the coagulation-flocculation process. In classical treatment the hydrodynamic conditions for the process of destabilisation (coagulation) of colloid particles are improving upwards from a vessel with simple square bottom without baffles to a vessel with simple baffles to a vessel with cogged perforated baffles. While the difference between the results of the vessel with simple and perforated cogged baffles is not very big, the difference between the results of the vessel without baffles and that one with built in baffles is significant and is 27% for the vessel with baffles. The difference between vessels with ratio C/H = 0.5 and C/H = 1 is almost insignificant. A slight priority at C/H = 1 exist, but because of the limited vessel dimensions it is not clearly distinguished.

The above mentioned tendencies can be observed in the experiment with prolonged initial rapid mixing and with 50% reduced coagulant dosages. Here again better results for vessels with baffles are obtained and the difference is 12 - 14%. The most evident feature here is that considerable high values of the apparent aggregation rate are obtained at prolonged initial rapid mixing and reduced coagulant dosages than at the classical coagulation - flocculation. The difference reaches 22% for 110 mg/l and 11% for 340 mg/l according to SiO_2 scale water turbidity.

A possible explanation of the results obtained is that the neutralisation of colloid particles depends on the number of collisions between them and the coagulant. The achievement of efficient collisions depends on one hand on the number of particles participating in the contact and on the other hand on created turbulence vortices and their quality. Thus in a vessel with baffles, because of the microturbulences, the possibilities of collision in the whole volume of the vessel are probably better. In other words, with all other conditions equal, the effectiveness of the process does not depend only on the energy parameter *G* defined by Camp, but also on the character and dimensions of caused turbulences and in that connection on the shape and dimensions of the vessel where mixing is taking place.

The analysis of the obtained results can be done also using the turbulent gross grad coefficient *Ct*. From equation (1) it can be written that:

The apparent aggregation rate Kapp for every particular case of study can be obtained from the plotted curves 1/Mt = Mo + Kapp.T, using the Equation (3). In the present research the apparent aggregation rate value is computed using the least square method. The full study Kapp values as a function of velocity gradient are given in Table 3. and plotted in Figure 6. and Figure 7.

Table 3. Kapp values for various vessel geometric configurations

Water turbidity and Alum dose		Kinetic conditions		Geometric configuration			
				square without baffles		square with baffles C/H=1	
		Rapid mix time t_R	G s^{-1}	C/H = 0.5	C/H = 1	simple	Perf.cogged
			150	0.0132	0.0139	0.0129	0.0138
	M = 110		250	0.0140	0.0144	0.0152	0.0156
	mg/l	60 s	400	0.0150	0.0152	0.0177	0.0179
			550	0.0156	0.0159	0.0190	0.0195
classic	Dc_1 = 21		650	0.0152	0.0153	0.0184	0.0186
	mg/l		750	0.0131	0.0128	0.0173	0.0167
			150	0.0109	0.0115	0.0139	0.0153
coagu-	M = 340		250	0.0131	0.0133	0.0158	0.0164
lation	mg/l	60 s	400	0.0146	0.0148	0.0186	0.0191
			550	0.0149	0.0154	0.0199	0.0201
	Dc_2 = 39		650	0.0140	0.0142	0.0188	0.0188
	mg/l		750	0.0122	0.0117	0.0170	0.0161
	M = 110		150		0.0192	0.0214	0.0226
	mg/l	360 s	250		0.0208	0.0232	0.0235
prolonged	Dc = 11		350		0.0198	0.0218	0.0218
	mg/l		450		0.0182	0.0214	0.0192
coagu-	M = 340		150		0.0192	0.0219	0.0220
lation	mg/l	360 s	250		0.0200	0.0221	0.0221
	Dc = 19		350		0.0189	0.0214	0.0201
	mg/l		450		0.0174	0.0189	0.0188

$$Ct = \frac{Wo}{124.\,\rho.\,f.\,N} \qquad (4)$$

where the relative power *Wo*, which is necessary to operate the impeller, in equation (4) is substituted with the expression $\mu.G^2 = Wo$, which is obtained from the well known Camp's equation $G = (Wo / \mu)^{1/2}$.

Substituting $\mu = 1.01375.10^{-3}$ N.s.m^{-2} for water temperature at tests 19° C, $\rho = 1000$ kg.m^{-3} and $f = 15.6.10^{-4}$ m^2, we receive:

$$Ct = 5.24064.10^{-6}\,\frac{G}{N} \qquad (5)$$

Table 4. Values of turbulent gross drag coefficient C*t* for different vessel geometric configurations

Rapid mix velocity gradient	Geometric configuration of vessel: square without baffles				square with simple baffles		square with perforated cogged baffles	
G (s^{-1})	N	C/H = 0.5	N	C/H = 1	N	C/H = 1	N	C/H = 1
230	196	0.00795	177	0.01080	143	0.02047	149	0.01810
250	208	0.00786	187	0.01082	151	0.02057	158	0.01794
500	331	0.00780	296	0.01080	240	0.02047	251	0.01790
550	352	0.00785	316	0.01085	256	0.02040	266	0.01819

In Table 4. the values of the coefficient *Ct* are presented for the four investigated configurations and optimal *G* values. In this computation we have used the function G = f (N) which we have obtained in our previous studies, the results of which are already published.

It is obvious that at one and the same velocity gradient, the turbulent drag coefficient *Ct* increases in the following sequence: vessel without baffles with C/H = 0.5; vessel without baffles with C/H = 1.0; vessel with perforated cogged baffles and C/H = 1.0; vessel with simple baffles and C/H = 1.0.

Certain contradictions can be noticed at once: according to the Camp and Conklin method the coefficient *Ct* has its maximum value at a vessel with simple baffles, while using the method of Vrale and Jorden, the apparent aggregation rate has its maximum for the vessel with perforated cogged baffles. A contradiction exists also with the results obtained by Mhaisalkar et al. (1991), where the coefficient Kapp is proportional to *Ct*. In our opinion this is probably correct only for a certain correlation between the baffles' width and bottom dimensions, e.g. for Bw/C < 0.10. We consider that the proportion Bw/C must not be increased without restriction and it would be better the baffles to be perforated and cogged at Bw/C ≥ 0.10. In this way dead zone generation behind the baffles is avoided and micro-turbulences are created which favours the process of destabilisation and agglomeration of the particles.

3.4. CONCLUSIONS

The experimental results of this study lead to the following conclusions. The parameters of rapid mixing, velocity gradient, initial mixing duration and geometrical configuration of vessels, influence the effectiveness of coagulation process. Their optimal combination depends on the coagulant dosage at a definite water turbidity. The experimental results confirm once again the conclusion made by us (Kalinkov and Monzo, 1992) that it is possible at prolonged initial mixing to reduce significantly the necessary coagulant dose without decreasing the water clarifying effect.

The most successful parameter for functional assessment of the vessel effectiveness regarding the creation conditions for particle destabilisation turns out to be that proposed by the Vrale and Jorden apparent aggregation rate. Assessing the mixers by the universal turbulent drag coefficient to a certain extent is a wrong step, because the contribution of effective collisions between particles can't be recognised. All results show that vessels with square cross section and perforated cogged baffles have better results than the other configurations which were investigated.

3.5. REFERENCES

Camp T.R., Conklin G.F. (1970) Towards a Rational Jar Test for Coagulation. J.New Engl.Wat. Wks Ass., 84, 325.

Kalinkov P., Monzo J.L. (1992) Decreasing Coagulant Dose by Prolonging Initial Rapid Mixing Time. Annual report of University of Architecture, Civil Engineering and Geodesy, vol. XXXVI, scroll X.

Mhaisalkar V.A., Paramasivam R., Bhole A.G. (1991) Optimising Physical Parameters of Rapid Mix Design for Coagulation - Flocculation of Turbid Waters. Wat.Res., 25, 1, 43 - 52.

Packham R.F. (1962) The Coagulation Process: Effect of pH and the Nature of the Turidity. J.appl.Chem., 12, 556 - 564.

Vrale L., Jorden R.M. (1971) Rapid Mixing in Water Treatment. J.Am.Wat.Wks Ass., 63, 52 - 58.

4. Study of Floc-Blanket Clarifier, Combined with Lamella Settlement under the Condition of Prolonged Coagulation

Peter Kalinkov
Jean Louis Monzo
Higher Institute of Architecture
and Civil Engineering
1 Christo Smirnenski, 1421 Sofia
BULGARIA

4.1. INTRODUCTION

Water clarification working on the principles of contact flocculation is well known from 60 - 70 years ago. Since then different kinds of clarifiers have been constructed and used, where the separation of pre-destabilised colloidal particles from water is effected by the floc-blanket of previously formed flocs. Setting up and the quality of the floc blanket in these facilities working on the principle of hydraulic fluidization, depend on different factors as well as on coagulant dose, quality of formed flocs, upwards stream velocity in the floc - blanket zone, etc. (Азервьер, 1952, Миндц and Кастальский, 1962; Tesarik, 1967, Miller and West, 1968). Recently, lamellas are set up above the floc-blanket. It is considered that the floc-blanket is stabilised and high treatment effect is achieved at a considerable increased upward flow velocity, about 1.6 up to 1.8 mm/s, while in common conditions it is 0.5 up to 0.9 mm/s. According to the data in Marin Galvin (1992), fixing up lamellas gives the opportunity of decreasing the coagulant dose by 14% in comparison with facilities without them.

Using the results of our previous works (Kalinkov and Monzo, 1992, Monzo and Kalinkov, 1989) concerning the possible decrease of the coagulant dose by up to 50% by means of mechanical mixers with prolonged up to 6 min initial intensive mixing, we investigated the behaviour of floc - blanket clarifier and lamella settling, working at the above conditions.

NATO ASI Series, Partnership Sub-Series, 2. Environment – Vol. 15
Water Supply Systems – New Technologies
Edited by Č. Maksimović, F. Calomino, and J. Snoxell

4.1.1. Experimental Set Up and Procedure

For this purpose we used the equipment shown in Figure 1.

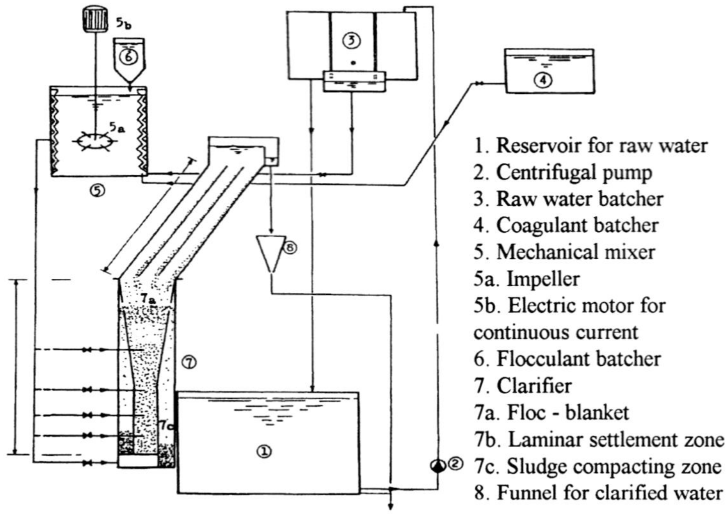

Figure 1. Pilot plant

The investigations are carried out with water, which is turbid with freshly taken sludge from the Pancharevo lake (near Sofia, Bulgaria). Some of the water quality parameters are shown in Table 1., and in Figure 2. settlement curves at natural conditions without coagulant are shown.

Table 1. Physico - chemical parameters of raw water

Turbidity mg/l acc. to SiO_2 scale	Colour degrees at Rubluov's scale	Tempe-rature °C	pH	Alkalinity mgeq/l	Hardness H°	Oxidiza-bility mg/l O_2
50	10	13 ± 1	6.8 ± 0.1	1.38	2.35	1.4 - 1.5
100	10	13 ± 1	6.8 ± 0.1	1.40	2.40	1.4 - 1.5
200	15	13 ± 1	6.8 ± 0.1	1.40	2.40	1.4 - 1.5
300	15	13 ± 1	6.8 ± 0.1	1.40	2.40	1.4 - 1.5

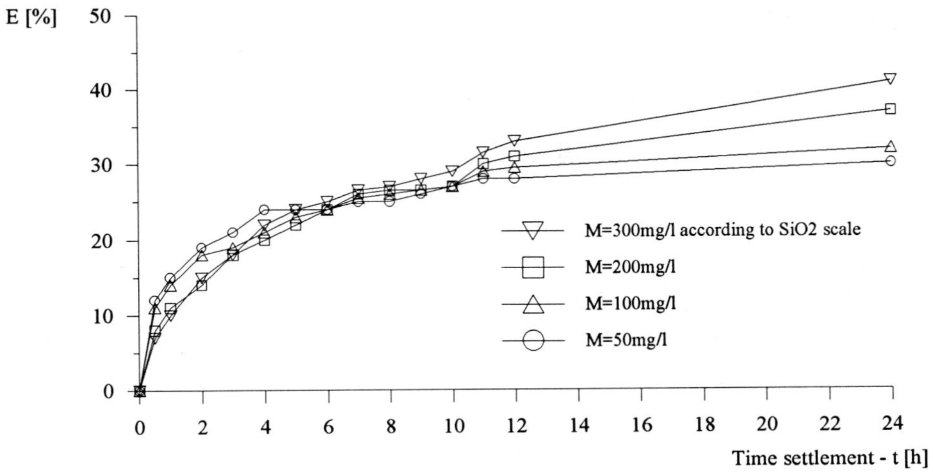

Figure 2. Curves of water settlement at natural conditions without coagulant

At the time of the tests, the water quantity passing through the mechanical mixer was regulated by a proportioning device for raw water 3, and the residence time by the rate of filling up vessel 5. For coagulant, a 0.25% solution of $Al_2(SO_4)_3$ is used, dosed by the proportioning device 4.

A period of about seven hours is necessary for the normal work of the installation after its start. The effect of water clarification and discolouration is defined by the Expression (1):

$$E = \frac{M_0 - M}{M_0} .100\% \qquad (1)$$

$$E = \frac{C_0 - C}{C_0} .100\%$$

where: M_0 and C_0 Turbidity and colour of raw water respectively in mg/l according to SiO_2 scale and in degrees of Rubluov's scale,

M and C Turbidity and colour of clarified water respectively in mg/l according to SiO_2 scale and in degrees of Rubluov's scale.

The optimum values of the necessary coagulant quantity defined by the classical method with jar - test for raw water turbidity shown in Table 1 are respectively 20; 22.5 and 39 mg/l $Al_2(SO_4)_3$.

All tests are carried out at a prolonged coagulation with residence time in the mixer of 6 min, using 50% of the optimum coagulant dose, and velocity gradient in the range 230-300 s^{-1}.

4.2. EXPERIMENTAL DATA PROCEEDING AND RESULT ANALYSIS

Investigations are carried out at different initial raw water turbidity for examining the work of the equipment and defining the behaviour of the floc-blanket in a condition of prolonged initial rapid mixing and reduced coagulant dose. Turbidity and colour of the clarifier water are controlled at different velocity of water flow through lamellas. The residence time in the mixer is 6 min in all tests and velocity gradient in the range of 230 - 300 s^{-1}. The results from the tests are presented in Table 2. and plotted in Figure 3.

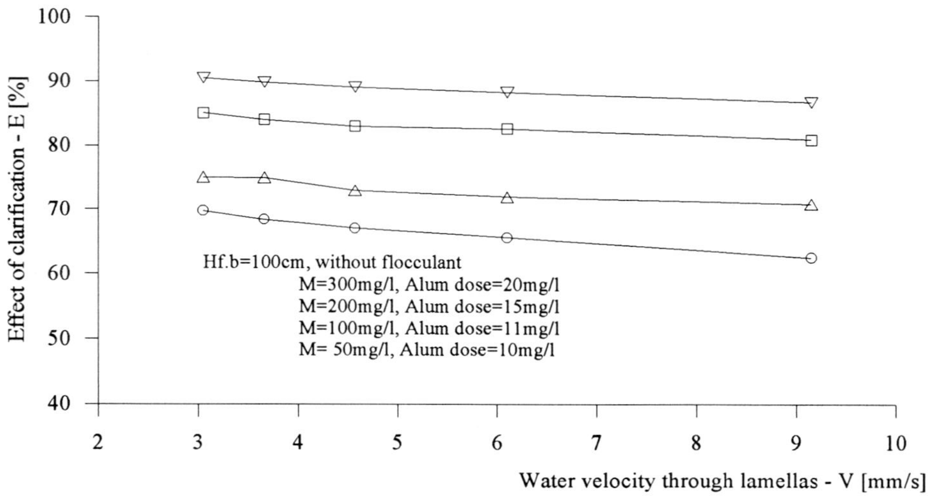

Figure 3. Clarifying effect as a function of water velocity through lamellas

Table 2. Effect of clarification at different water velocity through lamellas and floc-blanket height 1.00 m.

Water velocity through lamellas	Water Turbidity, mg/l according to SiO_2 scale, t° = 13 ± 1°C							
	M0=50 mg/l Alum dose = 10mg/l		M0=100 mg/l Alum dose = 11mg/l		M0=200 mg/l Alum dose = 15mg/l		M0=300 mg/l Alum dose = 20mg/l	
mm/s	M-mg/l	E%	M-mg/l	E%	M-mg/l	E%	M-mg/l	E%
3.05	15.15	69.70	24.95	75.05	30.06	84.97	28.50	90.50
3.66	15.83	68.34	25.02	74.98	32.08	83.96	30.33	89.82
4.57	16.50	67.00	27.05	72.95	34.16	82.98	32.85	89.05
6.10	17.22	65.56	28.05	71.95	34.86	82.57	35.31	88.23
9.15	18.77	62.46	29.15	70.85	38.08	80.96	39.99	86.67

It can be seen that the clarifying effect increases with the increased turbidity of the influent and decreases with velocity increase of water through the lamellas, which is expected. The fact which draws attention is that the effect of clarification is high, the same as at normal conditions in spite of the restricted (50%) coagulant dose. These results can be explained in the following way:

a) Neutralising and following adsorption of destabilised colloidal particles depend on the number of achieved useful collisions between them and the coagulant. On the other hand, useful collisions depend on the number of particles participating in the contact and on the character and quality of occurred turbulences. So in the conditions of decreased coagulant doses it is necessary to prolong the time of initial rapid mixing for reaching the same quality of useful collisions (Kalinkov and Monzo, 1992),

b) The hydrodynamic conditions in the mixer and its influence on the process of coagulation and flocculation has been thoroughly investigated by the authors in a previous study (Monzo and Kalinkov, 1989). The regime of mixing in the present study provides the obtained floc particles to have grain structure, high density and low water consistency,

c) The lamellas provide continuous downward flow of partially compacted large heavy sludge particles, which directly take part in creating and work of the floc - blanket,

d) The contact coagulation, taking part in the floc - blanket is realised on one hand by small floc particles with characteristics described in b) and on the other hand by larger ones, like these in c.).

Due to all this a floc - blanket of thick, compact, heavy, large flocs is formed, which provide effective flocculation regardless of decreased coagulant doses and low influent turbidity.

The test results of the effect of decolouration at a velocity through the lamellas of 3.05 mm/s are shown in Table 3.

Table 3. The effect of decolouration at velocity through lamellas of 3.05 mm/s

RAW WATER					CLARIFIED WATER				
					Colour				
Turbidity, mg/l according to SiO_2 scale	Colour degrees at Rublouv's scale	pH	Alkalinity mgeq/l	Hardness H°	Degrees at Rublouv's scale	E%	pH	Alkalinity mgeq/l	Hardness H°
50	10	6.8 ± 0.1	1.38	2.35	< 5	> 50	6.65	0.88 - 1.05	2.35
100	10	6.8 ± 0.1	1.40	2.40	< 5	> 50	6.64	0.88 - 1.05	2.35
200	15	6.8 ± 0.1	1.40	2.40	< 5	> 60	6.61	0.88 - 1.05	2.35
300	15	6.8 ± 0.1	1.40	2.40	< 5	> 60	6.58	0.88 - 1.05	2.35

4.3. STUDY THE INFLUENCE OF FLOCCULANT ADDITION ON THE CLARIFICATION EFFECT

It is well known that a flocculant is added to improve the coagulation-flocculation process at low temperatures and low turbidity.

A series of tests were carried out to answer the question - what will be the influence of flocculant addition in the specific working conditions of the equipment and at prolonged mixing and decreased coagulant doses.

The studies were carried out at two values of water turbidity, respectively 50 and 100 mg/l according to SiO_2 scale and temperature 14 °C. As a flocculant a 0.05% solution of active silicon acid is used.

It should be noticed without describing the tests in detail, that the floc - blanket could hardly be held in fluidized condition, using doses higher than 0.5 mg/l.

That is why regardless of the optimum flocculant dose of 1.2 mg/l for Mo = 50 mg/l and 0.8 mg/l for Mo = 100 mg/l according to SiO_2 scale at jar-test, the tests were conducted with a dose of 0.25 mg/l. Two working conditions of the equipment are studied:

a) At existence of 1.0 m floc - blanket height,

b) Without floc - blanket.

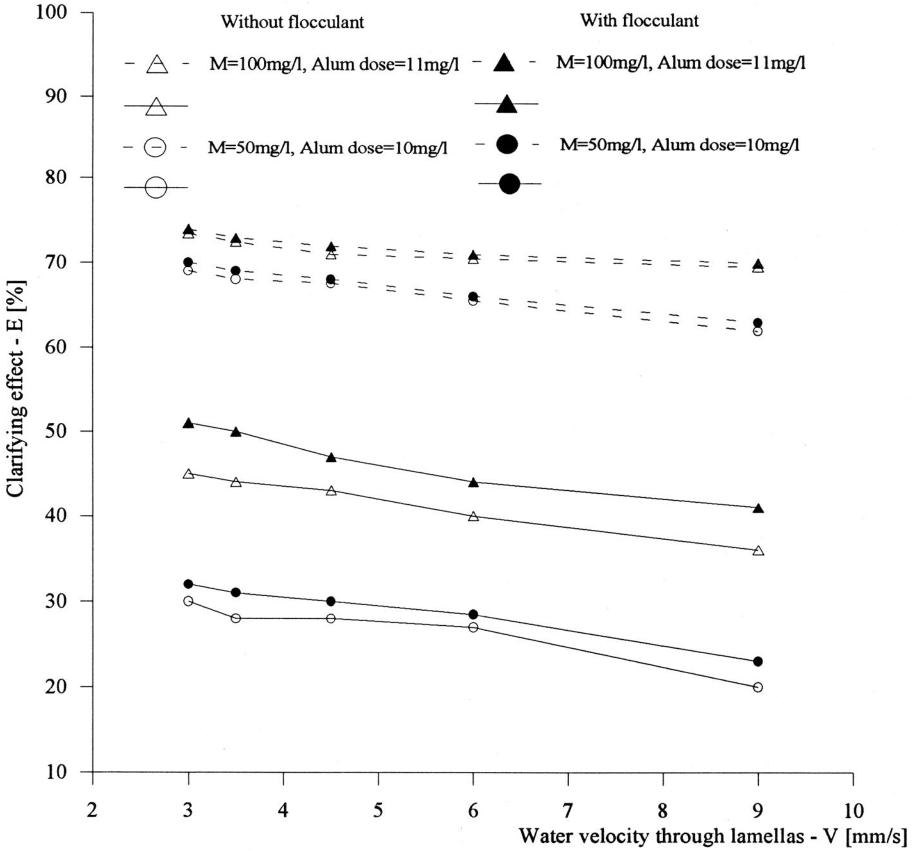

Figure 4. Clarifying effect vs. water velocity through lamellas

The results are shown in Figure 4. It can be seen that the influence of the flocculant is significant when the equipment works without a floc - blanket. It is obvious that the quantities of the floc-blanket formed at the conditions described above don't deserve flocculant addition. There is no effect when adding flocculant with a floc - blanket in existence. It is possible of course that flocculant addition will have significant effect at lower influent water temperatures than these at which the tests were conducted.

4.4. STUDY OF INFLUENCE OF FLOC - BLANKET HEIGHT OVER CLARIFICATION EFFECT

A series of tests were conducted to investigate the influence of floc - blanket height over the clarification effect in concrete conditions and prolonged initial rapid mixing and decreased coagulant doses.

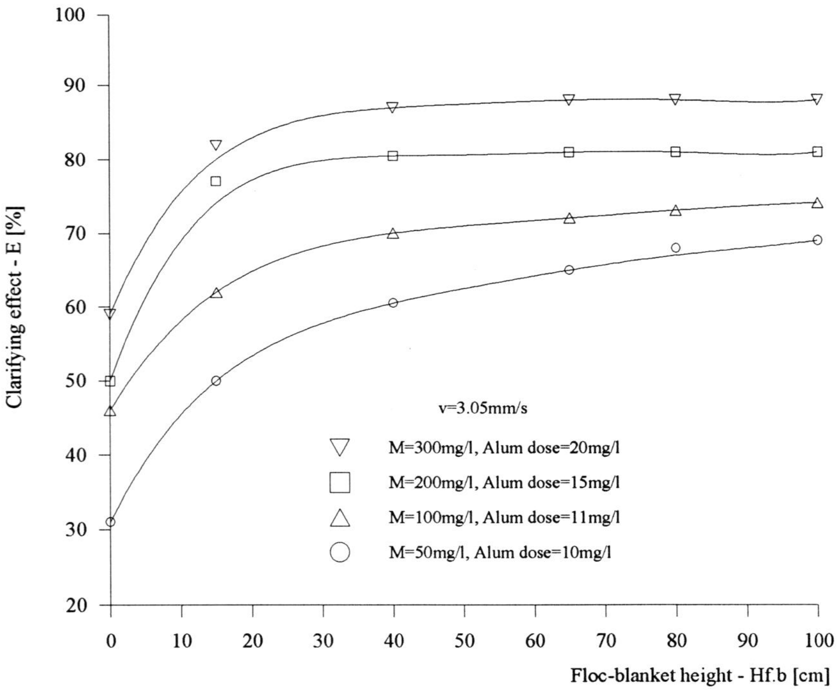

Figure 5. Clarifying effect vs. height of the floc -blanket

The study was carried out at different initial turbidities of raw water, constant velocity in lamellas, equal to 3.05 mm/s and different heights of floc - blanket.

The results are shown in Figure 5. It can be seen that at all values of turbidity the effect of clarification increases with the increase in the floc - blanket height.

This increase is considerable at a height from 0.0 to 0.60 m, and after that the process diminishes. In practice, the increase of the floc - blanket height stops its influence over the clarification effect at 1.0 m for lower turbidity.

This considerable decrease of the necessary floc - blanket height from 1.5 to 2.5 m for ordinary clarifiers to 1.0 m in the concrete case can be explained again by reference to the floc - blancket characteristics described previously.

4.5. COHESION COEFFICIENT ASSESSMENT OF THE FLOC PARTICIPATING IN FLOC - BLANKET SET UP

In order to explain the excellent results obtained so far, the tests were carried on by determining the cohesion coefficient of the floc, which participate in floc - blanket setup, as well as assessment of the concentration of compacted sludge in sthe ludge settler.

In water treatment the cohesion coefficient is defined as the maximum ascending velocity in a zone of floc - blanket, which causes its tearing and dragging out. In our studies it is defined by that described in the Degremont (1989) method and Equation (2):

$$v = K_0\left(\frac{V}{V_0} - 1\right) \qquad (2)$$

where:

- K_0 cohesion coefficient,
- V sludge volume at flogging,
- V_0 initial sludge volume defined from the curve.

The scheme of conducting the test is presented in Figure 6.

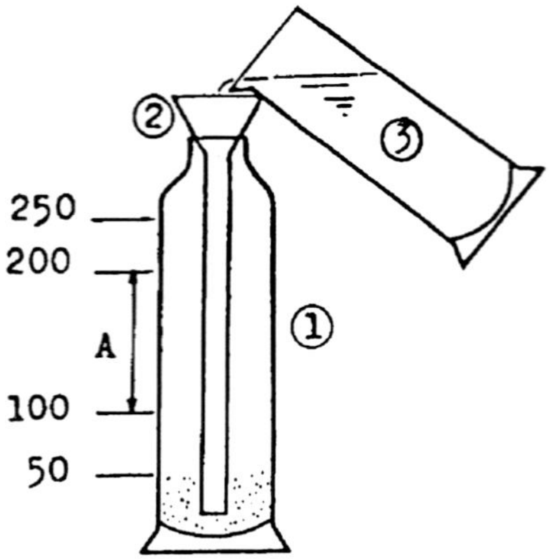

1 - Measuring cylinder

2 - Funnel

3 - Measuring cylinder for adding water

Figure 6. Scheme of conducting the test

Into a measuring cylinder with a volume of 250 ml about 50 ml fresh sludge from the upper third of the floc - blanket is added. As a result of the ascending water velocity, which is increased in vessel 1, the sludge is flogged. This flogging is relatively proportional to the velocity of the upward flow. It is defined by the equation:

$$V_i = \frac{A}{t} \qquad (3)$$

where: A = 100 mm and it is a height corresponding to a volume of 100 ml in the measuring cylinder,

t the time for regular water inflow through the funnel in vessel 1.

By changing the time t different values of flogged volume Vi are assessed. Using the graph, Equation (2) is a straight line, where V_o is a segment from the abscissa axis and the cohesion coefficient is an ordinate of a point from it with abscissa $2V_0$.

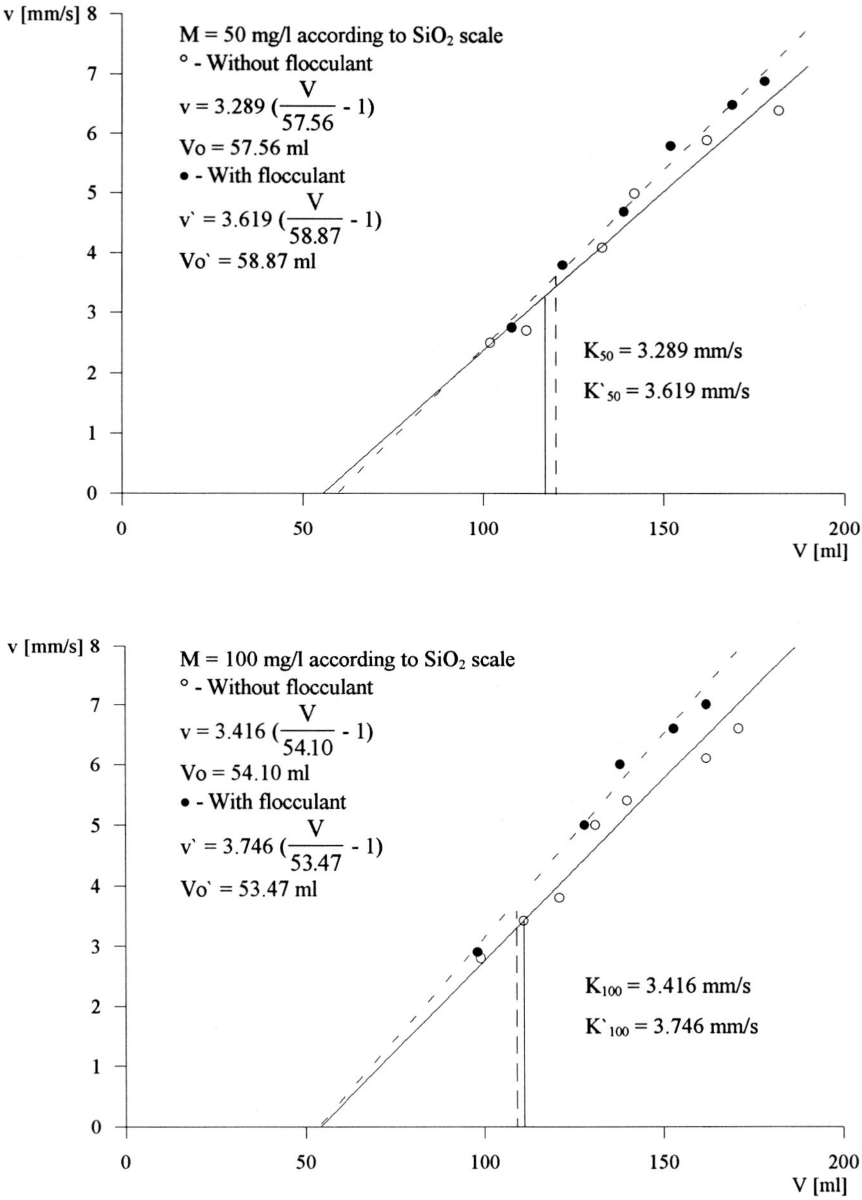

Figure 7. Determining the cohesion coefficient of sludge

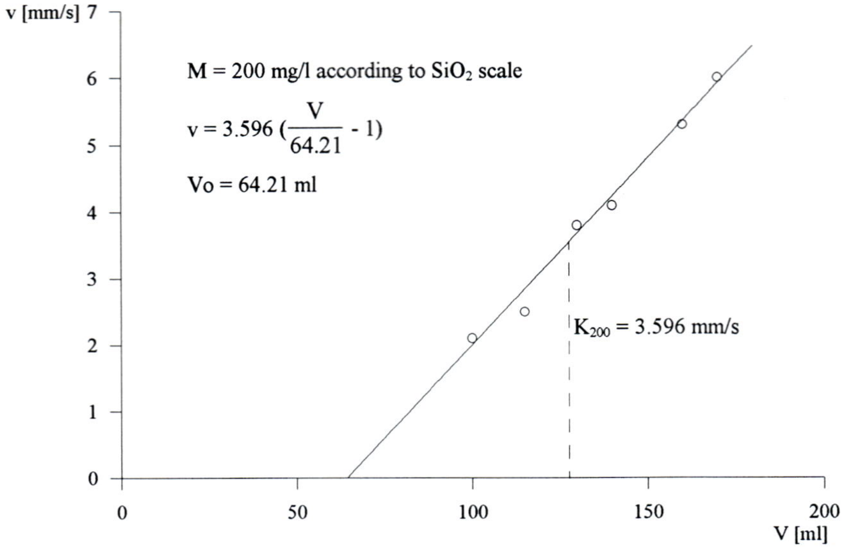

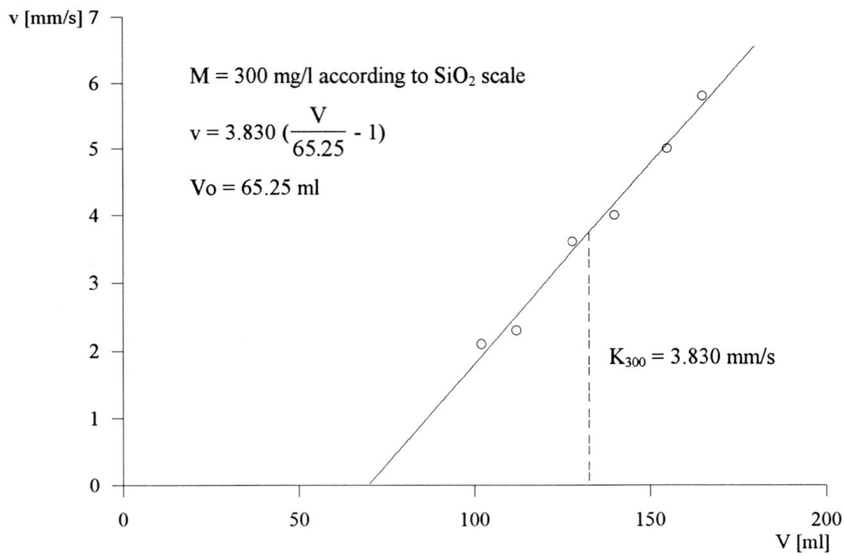

Figure 7. Determining the cohesion coefficient of sludge (continued)

The studies were conducted for water with initial turbidity 50 and 100 mg/l according to SiO_2 scale with and without using flocculant; and 200 mg/l and 300 mg/l without flocculant. On the basis of received data a curve $v = f\,(Vi)$ is plotted, and using regression analysis the equations of the curves are defined, in respect to the cohesion coefficient. The results are presented in Figure 7.

The cohesion coefficient varies from 3.29 for 50 mg/l turbidity without using flocculant up to 3.83 for influent water turbidity 300 mg/l. The values are considerably higher than the cohesion coefficient for a clarifier with a floc - blanket, working at normal conditions. According to Degremont (1989), for equipment operating efficiently, it is in the range of 0.8 to 1.2 mm/s. In practice this means that clarifiers with a floc - blanket working in the conditions of the test could be dimensioned for considerably higher velocities in floc blanket zone, like 3.0 up to 3.5 mm/s.

The investigations confirm once again our conviction, that the flocs obtained after a prolonged phase of intensive initial mixing and contact coagulation in condition of floc-blanket with sludge sliding down from lamellas are extremely compact and heavy, ready for quick and effective separation. The results prove once again, that in these conditions, flocculant addition is not necessary for increasing floc density.

With these tests the minimum velocities of ascending flow have also been investigated, and these still manage to sustain the sludge layer in suspended condition.

4.6. DEFINING THE CONCENTRATION OF COMPACTED SLUDGE

Defining the concentration of the compacted sludge was performed as samples of floc - blanket were left at rest after period *t*, their concentration of suspended particles was defined, using well known methods (Комитет по териториално и селишно устройство при МС КХИПИБКС "Водоканалпроект", 1987). The conducted test results are presented in Figure 8. It can be seen that a residence time of 5 hours is enough for relatively full sludge compaction and in the same time the concentration is several times higher than sludge in traditional clarifiers.

In Table 4. the comparative data of sludge concentration, max. and min. ascending velocities in the floc - blanket zone are presented. It was obtained in this and other studies for equipment working at normal conditions.

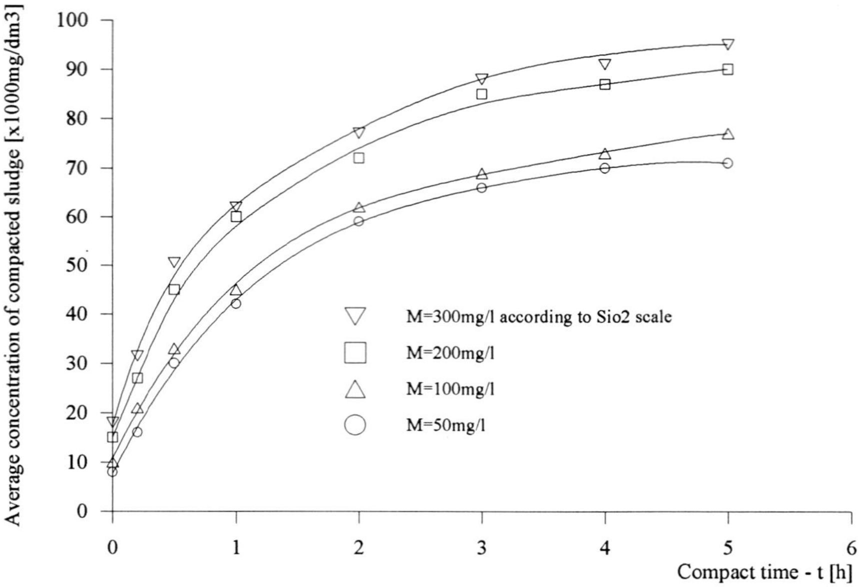

Figure 8. Sludge concentration vs. compact time

Table 4. Comparison of same dimensioning parameters for a floc-blanket clarifier

Water	Floc - blanket clarifier working at normal conditions (according to literature data)			Investigate equipment (test data)		
turbidity mg/l	Dimensioning velocity mm/s		Average concentrat. of compacted sludge after 24 hours	Dimensioning velocity mm/s		Average concentrat. of compacted sludge after 5 hours
	V min	V max		V min	V max	
50 to 100	0.40 to 0.60	1.40 to 1.60	15000 to 40000	1.46 to 1.72	3.24 to 3.42	75000 to 80000
100 to 400	0.60 to 0.80	1.60 to 1.80	40000 to 60000	1.72 to 1.86	3.60 to 3.83	80000 to 100000

4.7. CONCLUSIONS

The investigations confirm the possibility that equipment for water clarification works very well even when the coagulant dose is reduced by 50% of the optimum and the initial intensive mixing is prolonged by about 6 minutes. Especially, clarifiers with a floc-blanket, combined with lamellas settlement and working in process of the above conditions result in the formation of thick compact, large, heavy flocs which enable a quick and effective coagulation-flocculation. This guarantees on one hand a high clarification effect and on the other results in equipment volume, and lower height of floc blanket, with increased water velocity; in addition sludge is more concentrated and the sludge compaction zone consequently smaller.

4.8. REFERENCES

Азервьер С.Х. (1952) Водоснабжение и железодорожном транспорте. Трансжелдориздат. Москва.

Degremont (1989) Metmento Technique de l'eau. Neuvieme edition, Tome 1.

Kalinkov P., Monzo J.L. (1992) Decreasing Coagulant Dose by Prolonging Initial Rapid Mixing Time. Annual report of University of Architecture, Civil Engineering and Geodesy, vol.XXXVI, scroll X.

Комитет по териториално и селишно устройство при МС КХИПИБКС "Водоканалпроект" (1987) Сборник от методики за изследване на отпадъчни води и утайки. София.

Marin Galvin R. (1992) Lamella Clarification in Floc Blanket Decanters - A Case Study. J.Wat. SRT. - Aqua, 41, 1, pp. 28 - 32.

Miller D., West J. (1968) Pilot Plant Studies of Floc Blanket Clarification. J.Am.Wat.Wks.Assoc., 60, 2, pp. 154 - 164.

Миндц М., Кастальский А.А. (1962) Подготовска воды для питиьевого и промишленого водоснабжения. Москва.

Monzo J.L., Kalinkov P. (1995) Influence of Geometrical Configuration of Mixer on the Effectiveness of Initial Rapid Mixing in Water Treatment With Alum. Annual report of University of Architecture, Civil Engineering and Geodesy.

Tesarik I. (1967) Flow in Sludge-Blanket Clarifiers. J.Sanit.Engng Div., Proc, Am.Soc.Civ.Eng., 93, 6, pp. 105 - 120.

Chapter V

Diagnosis and Rehabilitation of Existing Systems

Chapter V

Diagnosis and Rehabilitation of Existing Systems

INTRODUCTORY STATEMENT

Diagnosis, measurement, data acquisition and processing as well as operational management of Water Supply Systems (WSSs) have undergone significant changes during the last two decades under direct influence of information technologies.

The classical sensors and instruments for measurements of flow, levels, velocities and pressure have gained in usefulness when converted from stand alone devices to an element in an integrated system for operational management. In this way a multipurpose use of the data obtained by measurements becomes more feasible for both on-line action and tuning current control devices.

This Chapter consists of two Sections. Some basic principles of flow measurement and their application in Water Supply Systems for monitoring water balance and for quantification of losses are shown in the first Section. General principles of integrating these measurements into system for data acquisition, processing and operational management are also discussed.

The second Section presents these principles in a more detailed way using examples from Wessex Water in the UK. Since the individual consumption on the UK is not metered, the problem of assessment of the key element of water balance is discussed for this advanced system.

1. Measurement Methods for Diagnosis and Rehabilitation of Existing Systems

Čedo Maksimović
IRTCUW, Insitute of Hydraulic Engineering
Faculty of Civil Engineering
University of Belgrade
P.O. Box 895, 11 000 Belgrade
YUGOSLAVIA

1.1. INTRODUCTION

Existing water supply systems are under continuous demand to supply an adequate amount of water of a required quality (health and technical aspect) at the required pressure. These systems are in continuous need of improvment in their operational management in order to meet new regulations, reduce costs and increase levels of service. Modern technology provides the means for such improvements either by the introduction of high level of informatic support after rehabilitation of the existing, aged systems, or by introducing appropriate new technologies into the newly designed and built systems. The computerized operational management systems have to be based on a number of components such as measurement, telemetric transmission, data acquisition, relational data base systems, data processing and hydrodynamic and water quality modeling. It is of upmost importance that the characteristics of the existing systems are known. This can be achieved by a proper combination of field measurements and use of simulation models. Thus, the reliable measurements of certain hydrodynamics quantities such as discharge, velocity, water level, pressure, etc. in addition to the data on status of flow regulation and similar elements in the systems are the major source of information on both the characteristics and conditions in existing systems.

Some of the major tasks of operational management of water supply systems are:- determination of the parameters to be measured, number and placing of sensors and regulating devices, sampling frequency and duration of measurements as well as the level of local and central data processing and/or operational control.

NATO ASI Series, Partnership Sub-Series, 2. Environment – Vol. 15
Water Supply Systems – New Technologies
Edited by Č. Maksimović, F. Calomino, and J. Snoxell

1.2 SCOPE

The scope of the present chapter is to present the tools and methods by which a reliable set of information concerning the existing system can be obtained to serve either for its redesign and rehabilitation or for the operational management. Since the characteristics of elements of a water supply system change over time (the roughness of pipes increases, the capacity of pumps decreases, leakage from pipes increases, etc.) the diagnostic procedure should be applied to the existing systems regularly even if they will not undergo any rehabilitation. Depending on the scope, measurements can be continuous (sensors remain built--in the system) or occasional, ie. sensors are occasionally brought to a particular location for measurements of the "snapshot" type. The types of measurements expressed in terms of data sampling are shown in Figure 1.

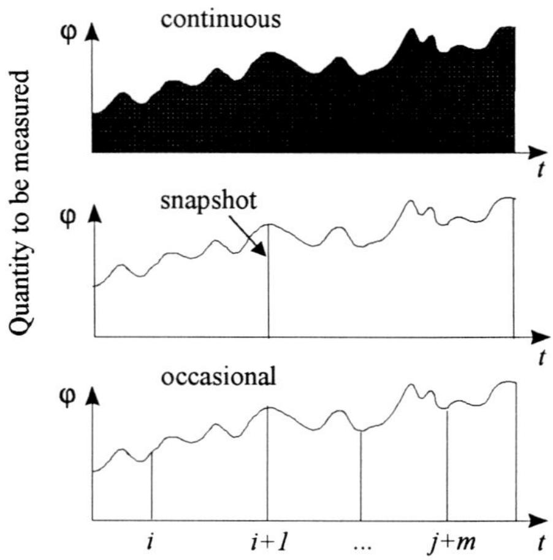

Figure 1. Three major types of data sampling - recording determining three types of measurements

The first type of data sampling is applicable in permanent measuring stations where a sensor (flow meter, level meter, pressure transducer, etc.) is placed permanently in the pipes, reservoir, pumping station, etc. and readings are taken continuously at regular or irregular time intervals.

The second type is usually applied when the single measurement is performed at the particular location. In this case it is usually necessary to prepare the site before that particular measurement (drilling, mounting of a sensor, calibration, etc.).

The third type of measurement (occasional data sampling) is applied at the locations which are accessible and where a portable instrument can be brought and easily mounted for ad hoc measurement and data sampling.

Table 1. Groups of measurements for WSS's depending on the phase in the life span of a particular part of the system

Phase	Group of measurements
Initial field survey	Hydrological, meteorological, hydrodinamic, geologic, etc. (water balance)
Design of a new system	Concrete location, levels, capacity source, etc.
Construction	Pipe roughness (capacity), sealing (tightness)
Commisioning and trial work	Sensor calibration, tightness
Normal operation	Pressure, levels, discharge; for operation management, water losses, water quality
Rehabilitation and expansion	Reduced capacity of mains, leakage, pump characteristics
Shut-of	Final capacity of sources and mains

Regardless to the type of measurement shown in Table 1., one can again distinguish between two groups of measurements serving two different goals ie:

- **indication** ie. the data obtained serve only as an indication of the situation (normal or emergency) in the system, for example main bursts, high night consumption, excessive night pressure etc.; several water companies in the UK follow this concept (described in the next Section),
- **quantification** in which the reliable values of the measured parameter are requested.

In what follows, the preference will be given to the second group of measurements, although they can be used for the first group as well.

In Table 1. the groups of measurements are shown in the function of the place in the life span of water supply system.

Water supply systems have a finite life span. Depending on its age, various types of measurements are performed in the system: before design, in the design phase, during construction and commissioning, during their trial and full application periods, before rehabilitation, and possibly before shut down. Most of the measurements presented here are applicable in all phases of the life span of a WSS.

1.3. MEASUREMENTS FOR WATER BALANCE IN WSS

1.3.1. General Water Balance

Water flows in water supply systems are mostly under pressure. However there are some exceptions such as reservoirs, wells for pumping of ground water, in treatment plants etc. where water flows with a free surface. This fact determines the type of sensors and measuring principles since they normally differ for systems under pressure and those with free surface. Before drawing conclusions about the system's performance, one has to have the access to the reliable results of measurements. The reliability is affected not only by the accuracy of measuring systems (sensors, amplifiers, data transition and recording equipment) but also by the way in which the measurement in performed i.e. where sensors are placed, how to interpret the value of an intermediate quantity, how to average and finally process the signal etc. It also depends on the level of knowledge of the physical characteristics of the system (where pipes are, their characteristics, existence of connections, valves etc.) and upon the status of flow regulation devices (open, closed) etc. One of the major tasks of measurements in the systems is to provide information on water balance (input, own consumption, leakage, consumption by metered consumers, non-accounted for consumption etc.). For this purpose reliable flow measurements are essential (Maksimović, 1993).

The problems of measurement and data acquisition are different in the simple systems which just start to introduce informatic technology and in the complex systems which have a long experience in measurements and data processing such as described by Snoxell in the next Section.

For the beginning, let us consider a very simple water supply system consisting of one source (intake), one treatment plant, one pumping station, network and one reservoir connected by one main conduit with the network (Radojković et al., 1986). .

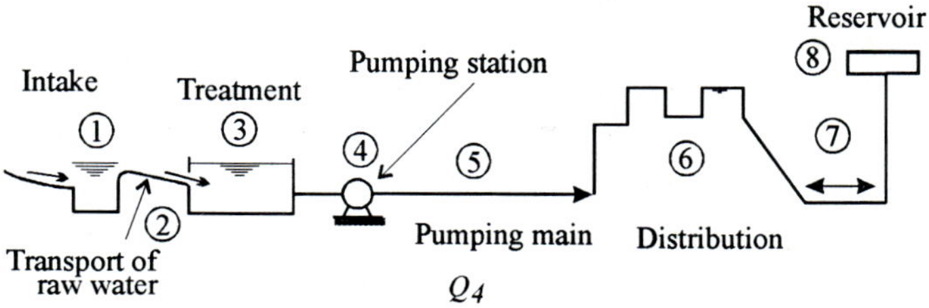

Figure 2. Scheme of a simple water supply system

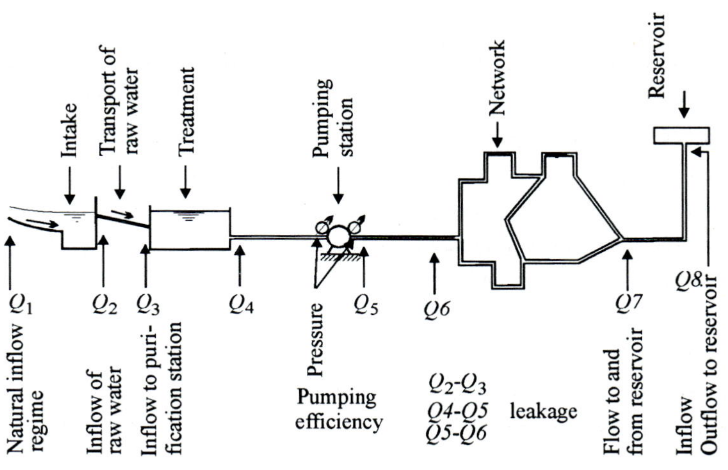

Figure 3. Positioning of flow meters for water balance

In order to keep track of water balance, at least six or seven various positions of discharge measuring equipment has to be considered. Taking, as an example, the system shown in Figure 2. the overall water balance can be made if the following quantities are monitored (Figure 3.).

- natural flow of raw water Q_1 or its integral over a given time periods (day, month, year), through the water course, or inlfow into reservoir, through an underground aquifer etc.,

- withdrawal (abstraction) of raw water Q_2 from a natural resource,

- discharge at the inlet to treatment plant Q_3 which can differ from Q_2 and Q_1 because of raw water losses during transportation, pumping, storage etc.,

- output of treated water Q_4 ($Q_4 < Q_3$),

- discharge of the treated water at the beginning of pumping - trunk main Q_5 and at the end of trunk main - Q_6 ($Q_6 < Q_5 < Q_4$),

- discharge in various pipes in the network or its closed parts (districts),

- discharge to or from the distribution reservoir.

In this case it is relatively easy to measure the amount of water delivered into the system by installing a flow meter (reliable, regularly calibrated and maintained) downstream from the pump - at the beginning of the water main (ie. outlet of the treatment plant). Placing the same meter at the downstream end of the main introduces a higher uncertainty because the information on water losses is lost, that is the amount of water that leaks from the main or is withdrawn by illegal connections (if any), between the treatment plant and the network.

The uncertainty is reduced if flow meters are placed on both ends of the pipe. For example, one can measure both Q_4 and Q_5 provided that both meters are correct. In that case the difference in discharge (or in volume obtained by integration of discharge) represents leakage (losses) and illegal consumption. A similar problem occurs if one tries to determine leakage from a pipe in city networks.

Usually it is very difficult to find out any of these data with a satisfactory reliability. One often finds that certain figures are available by WSS staff, but, if one tries to analyse the source of data (methods of measurement, data recording and analysis) and their accuracy, one is usually faced with the fact that reliable data do not exist. One of the frequently available information is the record on the duration of pumping, presuming that the pump (or pumping station) pumps water with a constant discharge. Simple measurements of the pumping head and power reveals that pumps often operate out of the range of favorable efficiency and that determination of water balance suffers from considerable errors (Maksimović, 1993).

1.4. METHODS OF FLOW MEASUREMENTS IN WSS

There are many books and standards dealing with flow measurements (for example). International (ISO) standard No. ISO/TR 3313 and others also deal with this issue. The discharge measurement methods can be classified in four groups depending on the principle applied as follows:

a) Integral flow measurements by the intermediate measurement of energy loss or drop of piezometric head caused by construction or other obstacle in the flow

Ventury meter is an example of such a principle of flow measurement (Figure 4.).

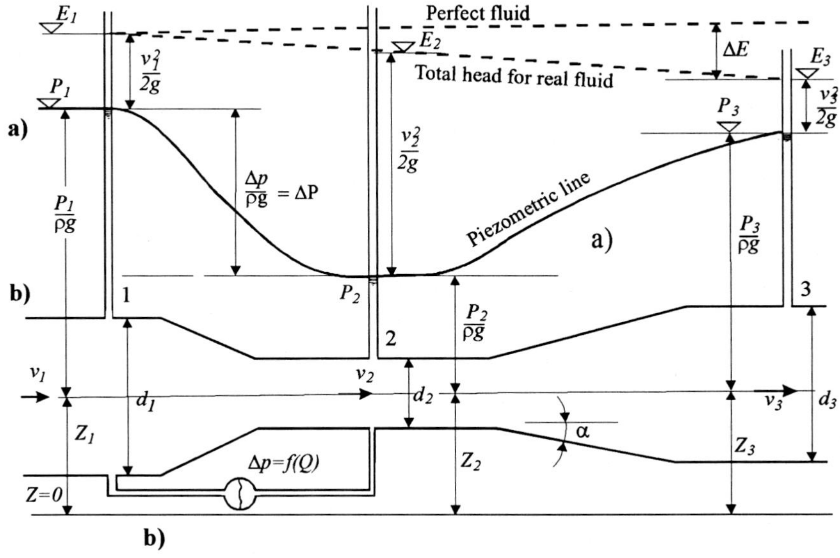

Figure 4. Venturi constriction as a flow measuring device

One of the disadvantages of this type of flow/measurement is that the obstacle in flow causes permanent energy loss in the pipe. The energy loss can be expressed in terms of piezometric head difference ΔP. The flow rate - discharge can be obtained if ΔP is measured.

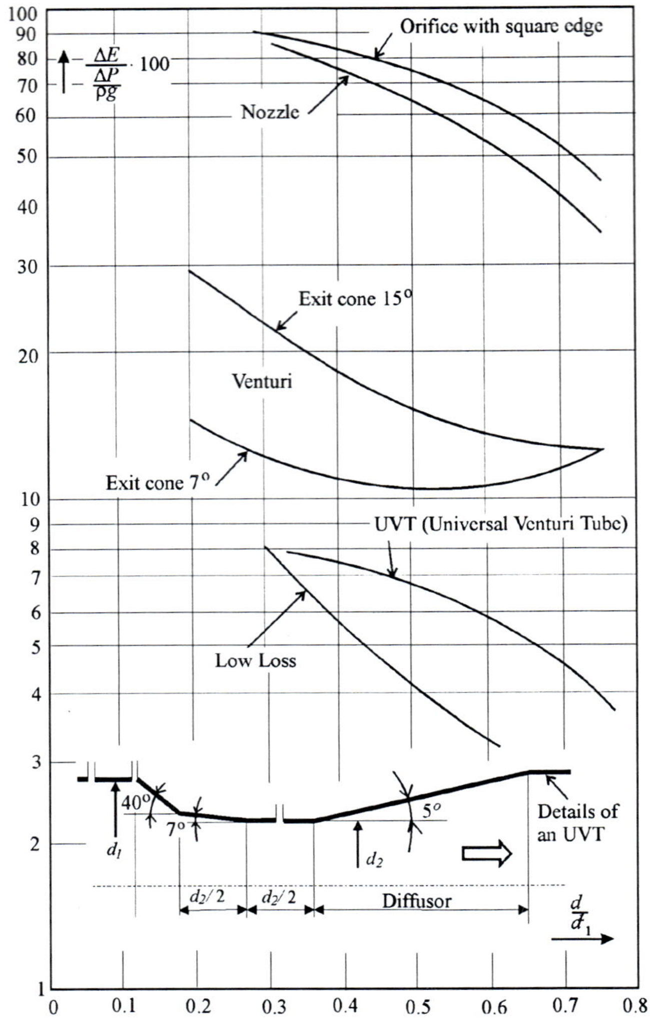

Permanent energy loss for various Venturi type flow meter

Figure 5. Permanent energy loss for various ventury orifice and nozzle types flow meters

The normalized energy loss ($\Delta E / \Delta p/\rho g$) is shown in Figure 5. for several types of Venturi flow meters. For comparison, the same data is shown for orifices with square edge and for a nozzle flow meter. This function is to be taken into account for selection of a new flow meter or upgrade of the exsisting one. In the system in which a big difference in water consumption in various seasons is pronounced it may even be necessary to use different obstacles for different seasons.

b) Integral flow measurement by active elements such as electromagnetic flow meters

The principle of flow measurement by this method is shown in Figure 6. There are several types of this flow meter: with homogeneous and inhomogeneous magnetic fields etc. A disadvantage of this flow meters in the past was that they had to be dismounted and carried to laboratory for calibration purposes. This problem can be solved by using insertion (portable) flow meters for on-line calibration.

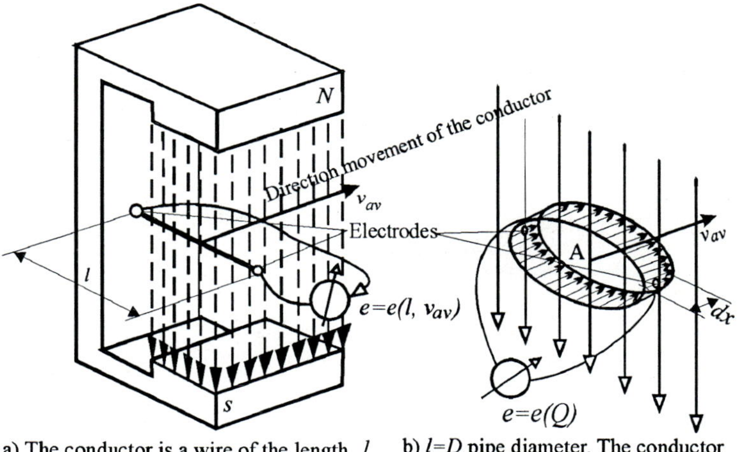

a) The conductor is a wire of the length l
$e = e(l,\ v_{av})$

b) $l=D$ pipe diameter. The conductor is a fluid element of the volume $A \cdot dx$

A = cross section area

The principle of electromagnetic induction applied to electromagnetic flow meter

Figure 6. The principle of electromagnetic induction applied to electromagnetic flow meter

c) Intermediate determination of discharge by correlation with average velocity measurement along a single or multiple path of ultrasound

The principle on which this method is based along with the scheme of an ultrasonic flow meter are shown in Figure 7a. and 7b.

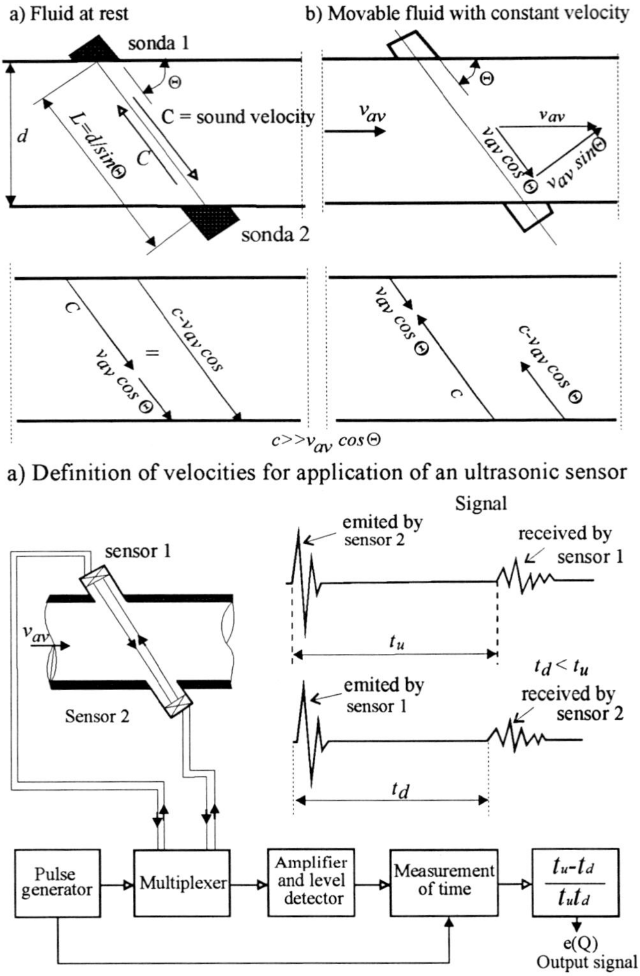

b) Scheme of an flow meter

Figure 7. The principle of an ultrasonic flow meter

The following definitions and abbreviations are used in the basic equations:

- travel time of sound in the upstream direction: t_u,
- travel time of sound in the downstream direction: t_d,
- average velocity along the path of sound v_{av}.

These terms are explained in the Figure 7.a. and 7.b. and are calculated as follows:

upstream time of travel

$$t_u = \frac{L}{c - v_{av}\cos\theta} = \frac{d}{\sin\theta(c + v_{sv}\cos\theta)}$$

downstream time of travel

$$t_d = \frac{L}{c + v_{sv}\cos\theta} = \frac{d}{\sin\theta(c + v_{sv}\cos\theta)}$$

average velocity

$$v_{av} = \frac{t_u - t_d}{t_u t_d} \frac{d}{2\sin\theta\cos\theta}$$

d) Indirect determination of discharge by point velocity measurement (ie. measurement of velocity profile) in a cross section of a pipe

Figure 8. shows the principle based on point velocity measurement by a miniature two velocity component electromagnetic flow meter.

In addition to the above mentioned best known types of principles there are also several other instruments based on various principles of fluid mechanics such as: Corriolis force, vortex, drag force, Anubar, etc. (Bernard, 1988). Detailed presentation of all these principles is beyond the scope of the present text. Although, all of the previously mentioned instruments are well known and well documented, it is surprising how often they are misused in practice.

The major shortcomings and sources errors in practice are caused by the ignorance of basic principles of fluid mechanics and of flow measurements, or by lack of proper calibration or by poor maintenance and by an inadequate or incomplete data recording and processing. For example when a Ventury tube is used for flow measurements, the rating curves are normally provided by the manufacturer.

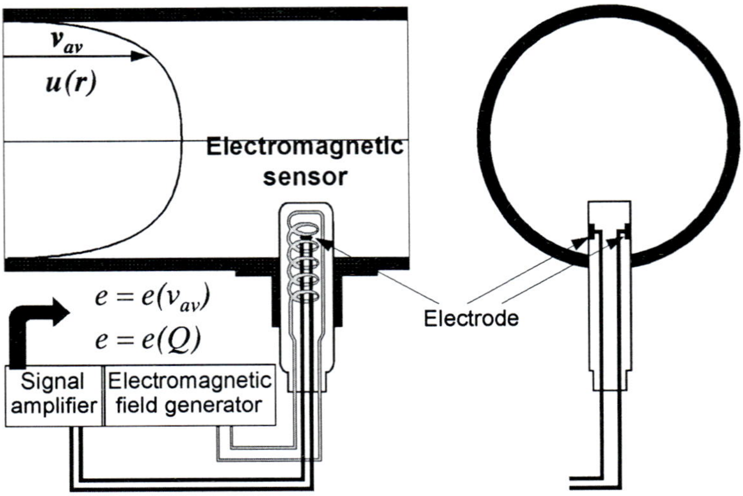

Figure 8. The principle of a two velocity components electromagnetic sensor used for flow measurements

However, application of various types of geometries such as UVT - Universal Ventury Tube, low loss causing permanent energy losses ΔE (Figure 5.), or differences in piezometric heads (ΔP) are obtained. Thus it is necessary to perform calibration both after introduction and in operation.

Traditional techniques for measurement of differences in piezometric head ($P_1 - P_2$ (Figure 5.)) based on mechanical principles can be easily improved by adding a differential pressure transducer (also shown in Figure 4.). After the meter has been installed, occasional (ad hoc) recalibration is rarely done in practice, after the meter has been installed. When the local velocity method is used for flow measurement it is important to determine the distance of the sensitive element of the sensor (electrodes of a miniature electromagnetic flow meter for example) from the wall. The velocity distribution in a circular pipe is known to be dependant on flow regime and pipe roughness as schematized in Figure 9.

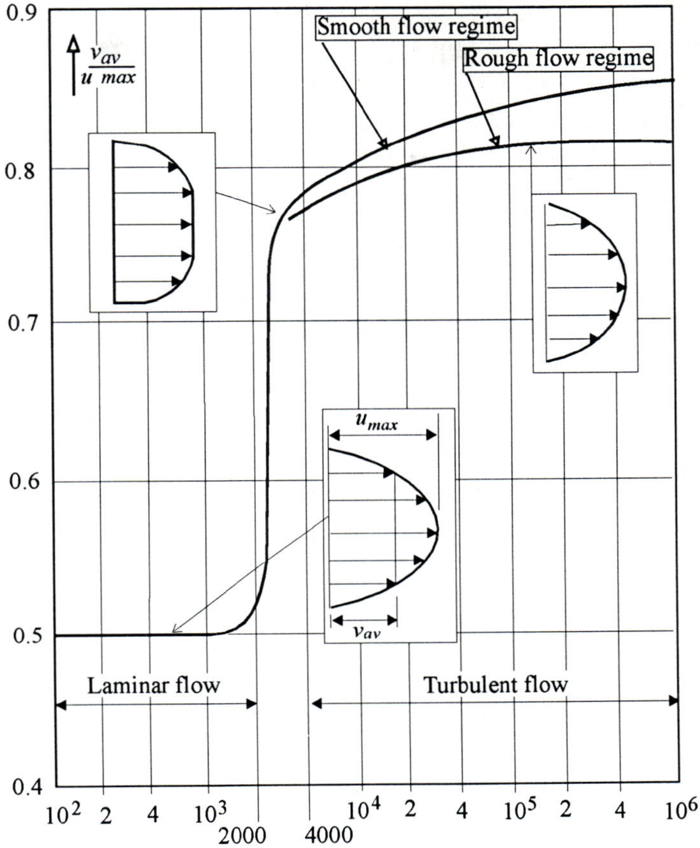

Figure 9. Ratio of average and maximum flow velocity in a circular pipe as a function of flow regime and pipe wall roughness

Similarly, the ratio between average and maximum flow velocity changes accordingly. Positioning of a sensor is normally prescribed or done by the manufacturer.

However, in practice, conditions change during the period of use of a sensor. For example roughness can change and so the rating curve (ie. function of discharge Q v.s. the electrical signal e generated by the point velocity) can be shifted or changed. The problem can be solved by calibration measurements of discharge using an alternative method. Measurement of velocity field in one or more radial positions can be applied to compute the discharge. To perform this kind of measurement it is necessary to have an appropriate access to the pipe. The access to the pipe and a place prepared the calibration is rarely done during the construction.

Thus the drilling and mounting of sensors and auxiliary devices has to be done on the spot before calibration. The modern term "insertion flow meter" is used to denote a portable device that can be inserted in the pipe where needed. In most cases it is a local velocity sensor, used for discharge monitoring, required for water balance or leakage detection. An arrangement for determination of pipe's carrying capacity (ie. hydraulic pipe roughness) is shown in Figure 10. In addition to flow measurement, flow regulation might be needed in order to achieve the desired range of flows. Piezometric heads (or the appropriate differences) have to be measured in two cross sections. Temperature measurements are to be used for determination of viscosity, thus the following can be determined by the sensor used for flow measurements:

- flow (discharge) as a function of flow regime and pipe wall rougness,
- roughness and carrying capacity by measurements of piezometric heads and discharge.

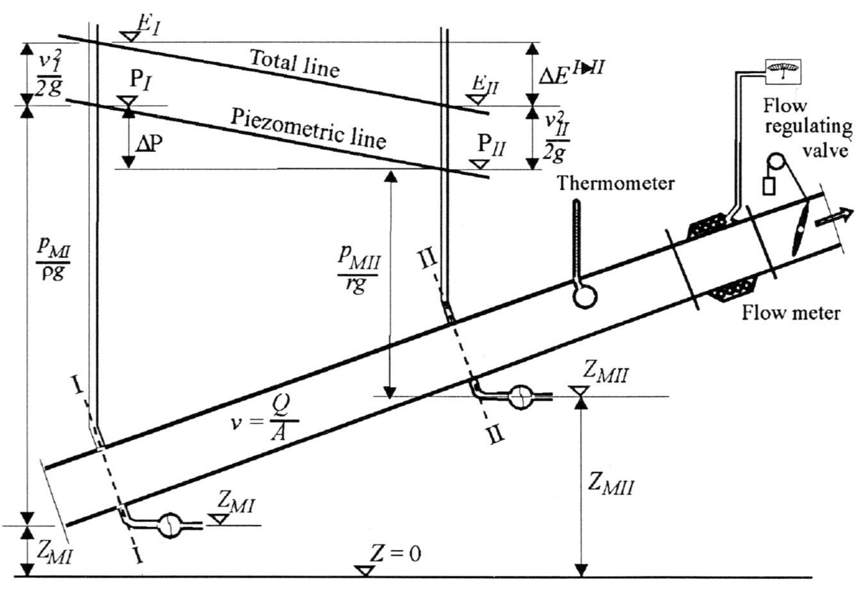

Figure 10. Setup for determination of the pipe's hydraulic roughness and carrying capacity by measurements of piezometric heads and discharge

1.4.1. Measurements for Determination of Water Losses

In determination of water losses there are two major tasks (i) finding--spotting leaks and (ii) quantifying the amount of water that leaks. In a WSS which has a sufficient number of reliable flow meters to keep the control over principal elements of water balance (intake, production, delivery to the network etc.) and in which individual consumers are metered, the information on water losses should be deducible from the data collected regularly. However, many factors make this task difficult:

a) in many systems (for example in the UK) individual consumers (households) are not metered regularly or not at all,

b) where metered, many meters are unreliable, out of order or not functioning properly,

c) consumption for fire fighting, street washing, pipe flushing, is not metered in general,

d) public consumption (fountains for instance) are also not metered.

Thus the determination of leakage is in many cases arbitrary unless special attention is paid to improve the accuracy of data.

Losses are monitored by several methods as follows:

a) passive methods in which the leak is spotted only where water appears on the surface,

b) active method in which the location of leak is determined by tracking the noise generated (leak spotters, correlators, etc.),

c) direct flow measurements in a selected main (or in its reach) by two fixed flow meters (when existing) Figure 11.a. or by two insertion flow meters - (Figure 11.b).

d) Determination of leakage in a part of network by special pumping, monitoring and control equipment mounted in a vehicle. In this case the vehicle with powerful pumps, flow and pressure meters, regulators and other required equipment is driven to the part of the network to be investigated. By closing valves, a defined area (part of network) is isolated from the rest and water is pumped from a small reservoir on the vehicle (outlet discharge) to circulate through the isolated sub network and received back to the vehicle (inlet discharge). All consumptions within the sub network are either disconnected or metered. The difference between the two meters (inlet and outlet) minus the metered consumptions is accounted as leakage or water loss recorded by the change of water level in the reservoir. There are also options without the reservoir. In this case the information on losses are deduced from the changes in pressure or in the pump load.

e) other methods.

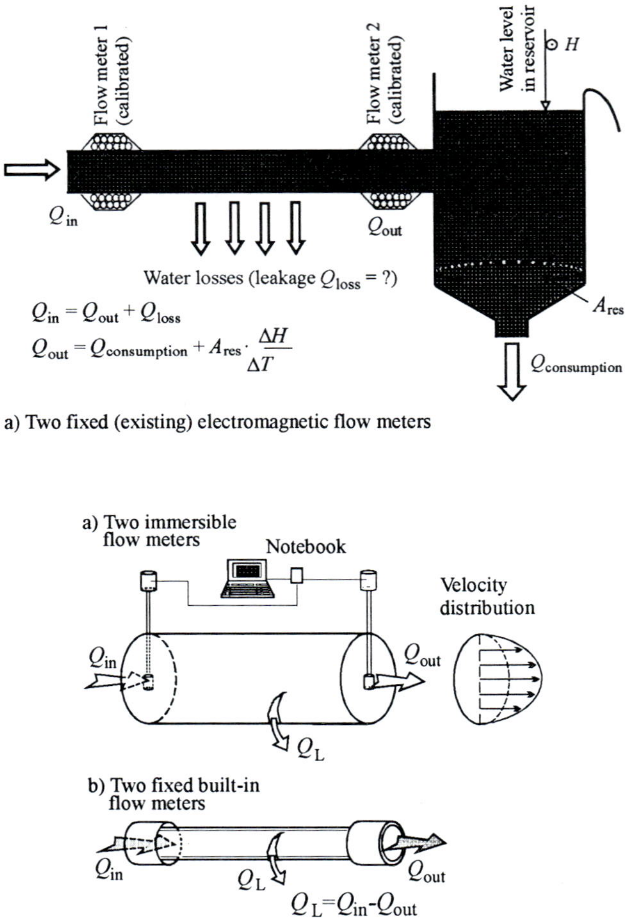

Figure 11. Determination of leakage from a main and quantification of losses by two calibrated flow meters a.) two fixed (existing) flow meters, b) two insertion flow meters - inspection of velocity fields for determination of discharge

In the systems well equipped for leakage detection and reduction, it is argued that it is economically feasible to perform regular leakage spotting annually or even twice a year. The savings in water losses exceeds the costs involved. In the other systems which do not have a regular program of a leakage control the losses of the order between 30 and 45% are considered regular whereas in extreme cases they reach as high as 70%. Irregular delivery of water, rationing, frequent epidemics, and inconvenience caused to the consumers are very frequent in such cases.

Thus, the leakage detection and water losses management should have a high priority in setting up rules for rehabilitation and modernization of the existing systems.

1.5. MEASUREMENTS FOR PRESSURE CONTROL AND OPERATIONAL MEASUREMENT

1.5.1. Pressure Control

One of the goals of operational management in WSSs is to secure continuous water supply with sufficient pressure and of the satisfactory quality to all consumers. Distribution networks are usually devided in several supply zones depending on ground (terrain) elevations. However, the dynamic pressure in the network (in each of the zones) varies in time and in space and it often happens that certain parts of the network are exposed to excessive pressure either continuously or during some periods of a season or of a day. This excessive pressure has adverse effect on home appliances, valves, faucets, etc.; it increases danger of leakage and increases water losses wherever there is a crack. Modern approach in operational management aims at an efficient pressure control within the network.

Pressure can be controlled within network as a whole or an automatic pressure reduction can be applied locally. In any case, for pressure control it is necessary to have an on-line indication of the local or global pressure (Figure 12). Moderately priced pressure reducing valves with actuators are available on the market.

By application of pressure control measures the following improvements can be achieved within WSSs:

- reduction of water losses,

- reduction of frequency of main bursts,
- reduction of energy consumption,
- more stable pressures in the network,
- reduction of damages on the services.

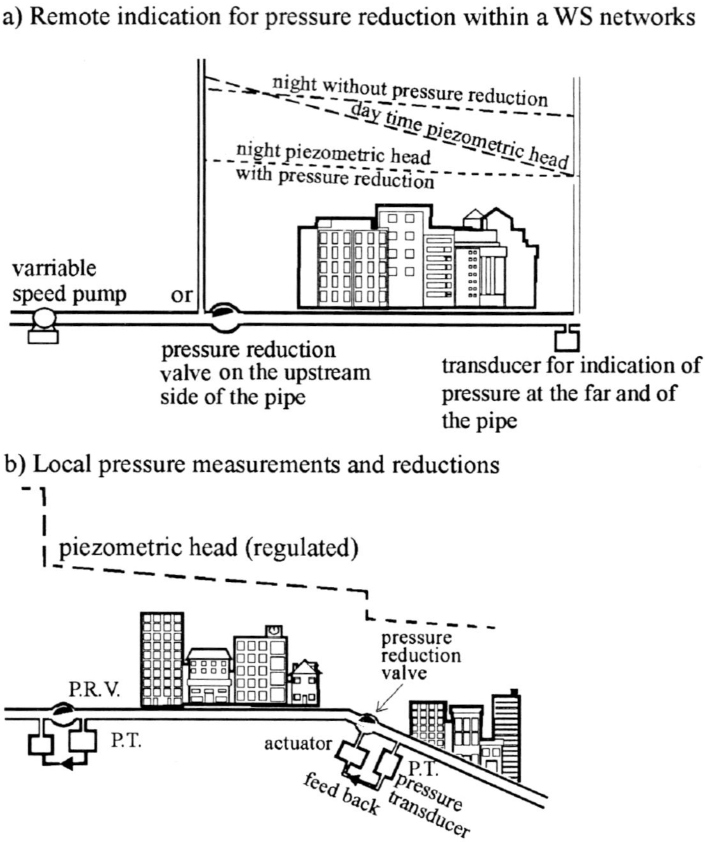

Figure 12. Global and local pressure control

1.5.2. Operational Management

In addition to measurements related to pressure reduction, operational management of WSS (excluding management of water treatment processes) requires other parameters to be monitored such as:

- ground water level in wells and in control holes (piezometers),
- water level and in reservoirs and other facilities with free surfaces,
- status of opening of flow modulating valves,
- states and operating range of pumps (pressure differences--pumping heads),
- flow in water mains,
- pressure at control points,
- parameters of water quality (turbidity, chlorine residual etc.).

Depending on the level of automatic control of a WSS, measurement can be accompanied by a degree of mechanical, automatic or software controlled flow regulation, that can be at local or at the whole network level (Koelle, 1993).

Recent tendencies in development of operation supervision and operational control rely on central data acquisition systems integrated with local instruments and data loggers. Communication system is based on either cables or radio links or public telephone lines.

One of the possible arrangement in handling operational data and commands is shown in Figure 13.

In developing countries the application of the advanced computer-aided operational management system should be planned realistically and implemented in steps. A special attention should be paid to the reliability and accuracy of data being transmitted and stored.

The data obtained in real time or statistically processed time series of particular measuring spots are stored in data files used later for simulation models and operation models. These quantities are of a dynamic nature. For application of the above mentioned models the additional data on network topology are needed. The latter data are almost fixed or change much slower as new pipes are laid or old ones have been rehabilitated.

Sufficient amount of measuring points, positioned in the system, the (reliable) data collected and transferred to the operator's control rooms is one of the prerequisites for introduction of the additional functions to the system's operational management.

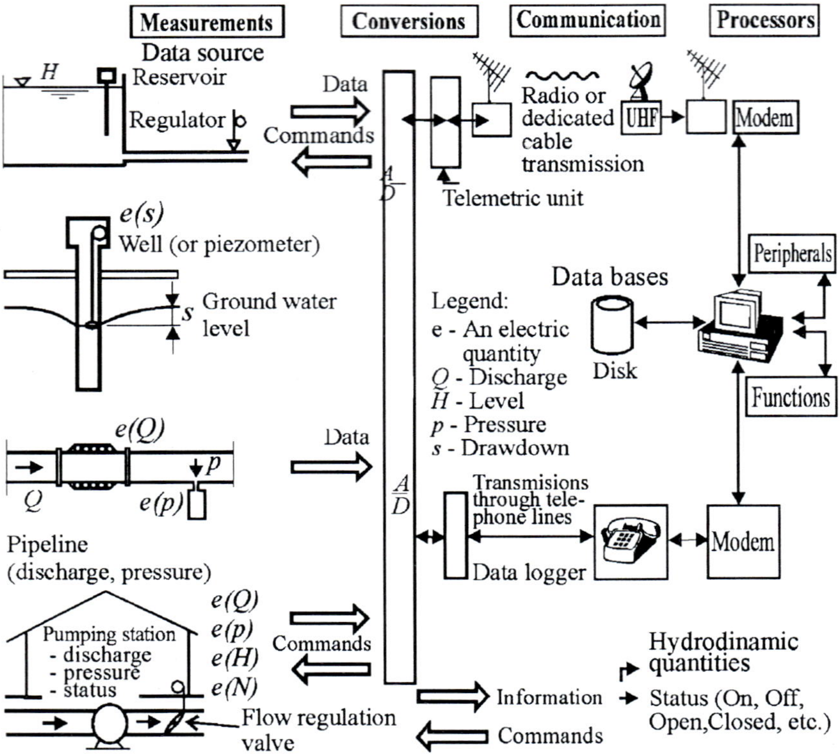

Figure 13. Scheme of a centralized system for operational management of a WSS

Some of the possible functions of integrated operational management are:

- energy consumption optimization,
- leakage and water losses management,
- rehabilitation scheduling,

- resources optimization etc.,
- application of GIS technologies.

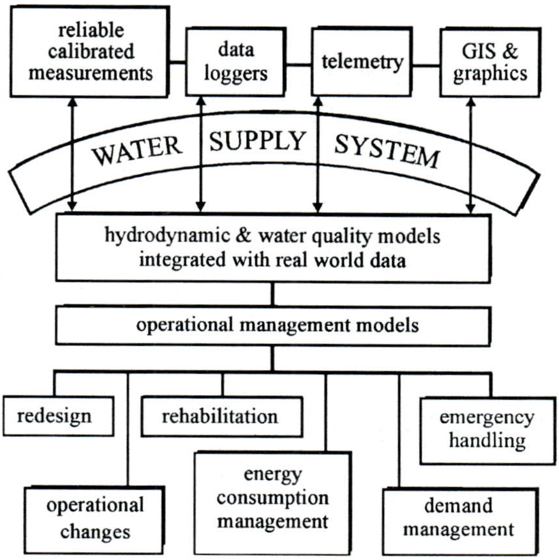

Figure 14. The data sources, techniques and functions of an integrated operational management in a WSS

The scheme of an interaction of measurements, telemetric data transmission, CAD & GIS operational and hydrodynamic modelling with the operational management functions is shown in Figure 14.

One should distinguish between hydrodynamic model and the operational management model. The hydrodynamic models, capable of performing detailed simulations are used for reproduction of measurements and for predictions of future condition on both short term and long term basis, such as presented in the Chapter VI of this book.

The information obtained are essential for planning and redesign of rehabilitations and reconstructions, for development and calibration of operational management models. Operational management models are based on a simplified geometry of network and other parts of the system and are based on the results of modelling of various scenarios simulated by hydrodynamic models. It has to be robust and rugged, reliable and user friendly. It has to have been played with in advance in order to elaborate the and facilitate handling emergency situation as well as regular ones. An example of such a system is given in the next Section by Snoxell.

1.6. CONCLUSIONS

Introduction and widespread of both simulation and operational management models require reliable sets of data obtained by measurements. In planning the upgrade of an existing WSS for introduction of advanced operational management it is necessary to set up priorities. In the systems with limited funding, such as in developing countries, the recommended priorities are as follows:

- placing a minimum number of flow meters in order to monitor a general water balance and for securing their long term observability, detection and quantification of major sources of losses,

- introduction of advanced simulation models, preceded by reliable documenting of network,

- development of a robust operational management model, based on an achievable sets of data telemetered or available through a number of data loggers.

1.7. REFERENCES

Bernard C.J. (1988) Fluid Flowmetering. The Trade & Technical Press Ltd., England.

Koelle E. (1993) Automation Levels in Water Supply Systems. Measurement and Control Variables. Water Supply Systesm, (Ed. Cabrera E., Martinez F.), Computational Mechanics Publications, Southampton Boston, pp. 297-317.

Maksimović Č. (1993) Merenja u hidrotehnici. Measurements in Hydrotechnics (in Serbian). Faculty of Civil Engineering, Belgrade.

Radojković M., Obradović D., Maksimović Č. (1986) Računari u komunalnoj hidrotehnici Computers in Urban Waters Engineering (in Serbian). Gradjevinska knjiga, Belgrade.

2. Diagnosis and Rehabilitation of Existing Systems

John Snoxell
WESSEX WATER
WESSEX House, Passage str.
BS2 0JQ Bristol
UK

2.1. INTRODUCTION

Over a long period of time all water supply networks deteriorate, and losses increase; although this is an obvious and fairly simple fact it presents very complex operational and economic problems for water company management. This Section looks at these questions, how they are dealt with in the UK and how technology is being applied to solve these problems. The Section firstly considers the measurement of inputs to the water supply system by looking at the types of flow metering commonly used and the way in which this data is handled. Calculating a basic water balance on the network is a fundamental element of management and the paper looks at the ways that this is achieved in the UK. Leak detection and quantification is currently undergoing a major review in the UK and the results of that review are outlined in the Section along with the methods that can be employed to reduce leakage.

The following sections go on to look at ways in which existing system can be upgraded using a number of techniques, and suggests that a "total integrated" approach offers the best overall solution where a number of complementary techniques are used to provide the most cost-effective package of measures.

2.2. MEASUREMENT OF INPUT TO WATER SUPPLY SYSTEMS

Firstly, we should look at the basic measuring devices used on typical networks for monitoring inputs to the system at various points.

NATO ASI Series, Partnership Sub-Series, 2. Environment – Vol. 15
Water Supply Systems – New Technologies
Edited by Č. Maksimović, F. Calomino, and J. Snoxell

In the UK all water sources have a daily and annual abstraction rate set down in the licence issued by the UK regulatory body, the National Rivers Authority (NRA). These abstraction rates are measured by the water company, which reports regularly to the NRA and is also audited by the Director General of Water on such matters as meter accuracy, calibration etc. Magnetic flowmeters are now commonly used and these are often telemetered to a central point for data capture, archiving, monitoring and the production of routine output statistics. This basic measurement of water produced forms the basis of all other measurements of consumption, leakage and so on. It is common for modern supply networks to be divided on a geographical basis into demand zones which can be monitored individually or in groups. Each zone will typically have one or more sources, together with imports and exports to or from the zone, and thus has a certain **configuration** of flow meters which must be carefully controlled and monitored.

Generally, major transfers of water are also measured by magnetic flow meters and telemetered back to a central point for continuous monitoring.

Wessex Water has 1million customers, supplied from over 120 sources and the territory is divided into 29 demand zones, monitored by around 450 telemetered flow meters. This monitoring system is used both for source outputs and zonal demands and can be used to check for major losses on the trunk main and reservoir system where, in the UK, it is estimated that around 4% of leakage occurs.

Figure 1. shows part of the Wessex Water territory with zonal boundaries.

Figure 1. Wessex Water - zonal boundaries

The distribution network within each zone is further subdivided into districts, called District Metered Areas (DMA's see Figure 2.), each with between 1000-3000 properties and each metered by one or more waste meters which are generally of a mechanical type (e.g. Kent Helix). Below this level are smaller zones, sometimes called control zones, which are generally used to monitor night flows during step-testing activity (Report No.26,Natioanl Leakage Initiative, Managing Leakage, 1994). The vast majority of domestic customers in the UK are not metered and this means that calculations of "unaccounted" for water are, at best, estimates which include a number of assumptions.

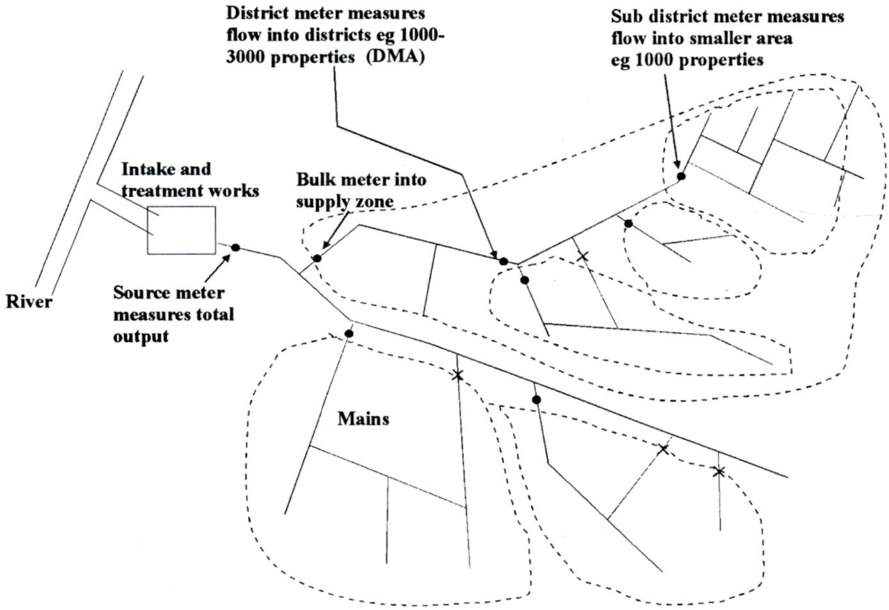

Figure 2. Typical layout of district metered area (DMA) and control zone

Until fairly recently meter readings from DMA's were collected and analysed manually but it is now common that these meters are logged by small, battery-powered data loggers which are collected and downloaded to a central point at regular intervals, thus providing a continuous record of daily flow profiles for each DMA.

Wessex has over 350 DMA's monitored by over 600 data loggers and this presents a major logistical exercise in collecting, collating, analysing and monitoring the data, a task that has to be largely automated if it is to be carried out efficiently. There are a number of trends emerging in the UK to streamline this operation, including "dial-up" loggers which can be interrogated by public telephone line (including cellular phone networks) or logger data can be downloaded into convenient points on the company PC network which can now reach remote points inexpensively via new communication services available (e.g. Integrated Services Digital Network - ISDN).

Thus in the UK we monitor the network at 2 or 3 levels, but do not have the benefit of customer metering as the final element in the water balance.

2.3. CALCULATING THE WATER BALANCE

The mass of data coming in from different levels in the system presents major problems in data handling, archiving, analysis and so on, to produce a complete water balance for the network at regional level. The data is derived from different instruments, is collected in different ways and is transmitted to different points in the organisation - collating, integrating and analysing the data into meaningful information is something which Wessex has solved through the development of its Aquaware software (AquaWare, 1996). This PC-based package reads data from different sources through standard interfaces from telemetry and data loggers and enables the User to configure flow meters in any combination to balance inputs, outputs, transfers etc. The package also reads metered consumption data from corporate billing files and Post Code information on numbers of properties in particular geographical areas - this is useful in the UK, which is predominantly unmetered.

The system allows the User to reconcile production data from a zone with consumption data as measured at DMA level; this gives a good indication of losses on the primary system: pumping mains, reservoir overflows, trunk mains and so on. These losses are generally held to be a small proportion of the total leakage but can be very significant and, most important, can be very cost-effective to repair. The Aquaware system also allows the user to monitor complex configurations of DMA meters and to monitor night flows from DMA's continuously and to build up these zonal night flows into divisional and regional data. This is probably the most reliable form of leakage monitoring in the UK and can be used to monitor the effectiveness of leak detection and repair expenditure and so on. Changes in DMA boundaries and configurations of meters can be handled easily through the software with changes in metered demand allocation automatically adjusted.

In this way the complete water balance for the company can be computed and monitored regularly.

Clearly, to assess leakage rates, some estimate of customer consumption has to be made and a number of studies have been carried out in the UK to this end; while some companies still report relatively high or low consumptions, most now fall in a band between 130-160 l/p/day as the annual average. This can vary considerably with the weather, however, since it is common practice to water the garden with the domestic water supply pushing consumption well above 200-300 l/p/day.

Customer leakage is a further problem - it is currently included in the specific consumption figure and is generally held to be around 8% of total production on average in the UK (Water Metering Trials, 1993).

2.4. LEAK DETECTION AND QUANTIFICATION

Since privatisation of the UK industry in 1989, leakage has become a major concern and the industry has recently reviewed all of its methodologies and procedures in order to produce realistic estimates on which to base major investment decisions; demand management is now probably the most important subject in the UK industry and most companies have embarked on major initiatives to suppress demand. Traditionally, the UK industry has been "resource-led"; it is now definitely "demand driven" and it is quite possible that the UK will not see major new resource developments for the next 5-10 years at least. Consequently, capital expenditure has reduced considerably while the effort and resources put into demand management has increased. Current leakage rates in the UK range between 15-40% expressed as a simple percentage of water produced, which can be very misleading; it is more meaningful to express leakage in terms of loss per Km of water main.

These levels of leakage are high by comparison with some reported continental leakage levels, where 10% losses are often considered to be high. However, leakage levels in the UK are related to the cost-benefits of suppressing leakage by the means now available, and it is the case that on purely economic grounds relatively high leakage levels can be justified, though this may not be politically acceptable; in practice all companies repair small leaks reported by customers purely on PR grounds even though this may be totally uneconomic.

Leakage can be reduced by a variety of means:

- Leak detection, location and repair,
- Rehabilitation (replacement and relining),

- Metering,
- Pressure Reduction.

2.5. LEAK DETECTION, LOCATION AND REPAIR

The business of controlling leakage on a water supply network is inherently complex, especially in the UK where the majority of domestic customers are unmetered. The UK industry has for many years followed the principles and methodologies set out in Report No 26, published by the National Water Council in 1980; this report collated the results of a national survey and recommended "best practices" as perceived at the time. Over the last 10 years this report has been the guiding light for the industry but more recent experience and research has led to the need for a review and revision of some of the guidelines. Thus a national Water Industry body, the National Leakage Initiative, has been set up to carry out this review and the first reports are now starting to appear in draft form.

One major change from previous thinking is related to target levels for leakage. Report 26 set guideline values for urban and rural areas based on best practice in the UK industry. These were expressed in l/property/hour during the night flow period (0200-0600) after due allowance had been made for legitimate domestic night use and legitimate night use by major metered customers. It is now recognised that meaningful targets at zonal level must relate to the basic cost-benefit equation of expenditure on detection and repair and revenue savings on water production or capital savings on new facilities.(National Leakage Initiative, 1994)

Progress has been made in this area in the UK by the use of software which has been developed to understand and quantify annual losses in a water supply area and, specifically, to analyse the components of leakage. The 'Bursts and Background Leakage Estimate' (Babe) software suit has emerged from the UK National Leakage Initiative and can be used to provide an analysis of the components of night flows (IWEM, 1994).

When DMA's with high leakage have been identified the next step is the detection and location of leaks. Leak detection is carried out via a step-test; this is a systematic investigation of a DMA which has been isolated by valve closure and is usually carried out at night when legitimate usage is low (Figure 3.). Valves within the district are closed sequentially and at the input meter the "step" drop in flow is recorded. Traditionally this has been carried out with data loggers which are analysed the next day, but it is now possible to transmit results from loggers on the system back to a central point (using devices such as the Spectrascan Tele-link Step Tester) so that the results are immediately available and visible to the operator - in this way unnecessary valve operations can be avoided. Figure 3. shows a step test and illustrates advances in step-test technology.

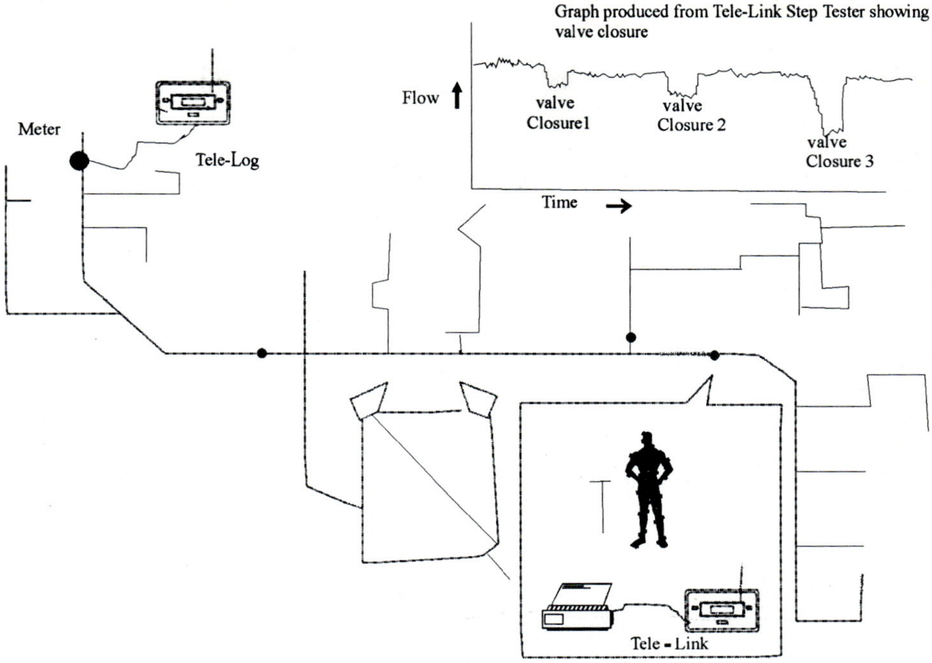

Figure 3. Principle of step testing

Leak detection (ie establishing whether leakage is present and which part of the network it is occurring in) leads to the problem of leak location; here again there have been major advances in technology in recent years.

The most obvious example is in the use of acoustic methods for leak location- so-called leak noise correlation. Figure 4. illustrates the principles involved and shows the basic equipment which is required (Biwater Industries, 1993).

Microphone sensors are placed on each side of the suspected leak at convenient points (eg valve chambers); generally one microphone is connected directly to the correlator while the other transmits its signal to the correlator via radio link. Modern correlators now have sophisticated capabilities for filtering extraneous noise and for accurate operation with relatively small leaks.

These units can also be used with 2 radio units and can be used with non-metallic pipe over relatively long distances.

However, this technology does require proper operation and training is essential - the effectiveness of this equipment can be increased greatly with correct training.

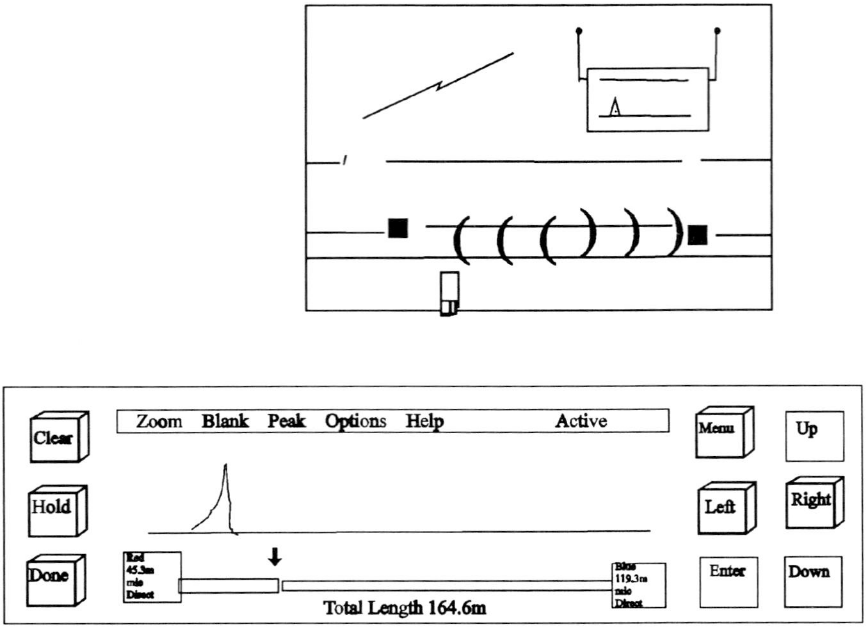

Figure 4. Principle of leak-noise correlation and typical acoustic pattern

2.6. REHABILITATION (REPLACEMENT/RELINING)

Wessex has around 10,000 km of water mains and replaces less than 1% per annum. This is not uncommon in the UK and contrasts with many continental countries where 2% replacement is widespread. Often, higher replacement rates such as these correlate with lower reported leakage rates but there is no clear economic argument for increasing replacement rates to reduce leakage where the marginal cost of water only is taken account of. In addition to replacement, UK companies now have extensive relining programmes generally driven by water quality directives. Wessex now relines more water mains than it replaces, using either cement mortar or epoxy resin, though the effect of relining on leakage is not clear - in some cases an increase in leakage after relining has been identified.

Mains renewal is generally much more expensive than relining, owing to the obvious factors of excavation and so on. However, technology is advancing in all of these areas, making the options more difficult to evaluate.(IWSA, 1994).

The cost of digging a trench, laying a main and reinstating the surface makes open cut renewal an expensive option but this can be overcome by using narrow trenching techniques or guided boring methods which will drill a hole in the ground into which the new pipe is subsequently pulled.

Relining techniques are either structural or non-structural: structural techniques include pipe bursting, slip lining and other insertion techniques; as the name implies they physically reinforce the existing pipeline as well as improving it's hydraulic and water quality characteristics. Non-structural techniques include cement or epoxy lining; these improve water quality characteristics but do not physically reinforce the pipeline. Cement relining obviously reduces the internal diameter significantly which in turn reduces the hydraulic capacity of the pipe. Thus each rehabilitation technology has a variety of functional and cost characteristics which have to be assessed carefully before a choice is finally made.

2.7. METERING

Universal domestic metering is still a contentious subject in the UK and although a new means of charging customers has to be found by the year 2000, it is not certain that metering will be the method adopted. Extensive metering trials have been conducted in the UK, however, and this revealed more fully the extent of domestic leakage i.e. leakage which is primarily the responsibility of the customer. This is now put at around 8% on average and the metering trials showed that demand was suppressed in metered areas both via charging tariffs and through reductions in leakage. It is now considered more cost-effective to install domestic meters than to build new resources in many cases and this argument is now being actively pursued by the UK regulators when capital expenditure programmes are promoted by the companies. Several UK companies are now committed wholly or in part to domestic metering, generally those in the south and east of the country where resource problems are greatest.(Water Metering Trials, 1993)

2.8. PRESSURE CONTROL

Pressure control, when applied in the right situation, is a very effective means of reducing background leakage in distribution networks. It is particularly popular in the UK, which is unmetered and where pressure and flow variations do not affect income directly. Pressure control also reduces the frequency of bursts and consequential damage and generally results in a better service to customers, though pressure has to be managed carefully in the UK in order to meet the conditions of the operating licence issued by the regulator, which is 10 m head at the boundary, at a flow of 9 l/min for a single property. In practice this is generally interpreted to mean that a service pressure of 12-15 m at the stoptap must be maintained at all times (Hegarty 1994).

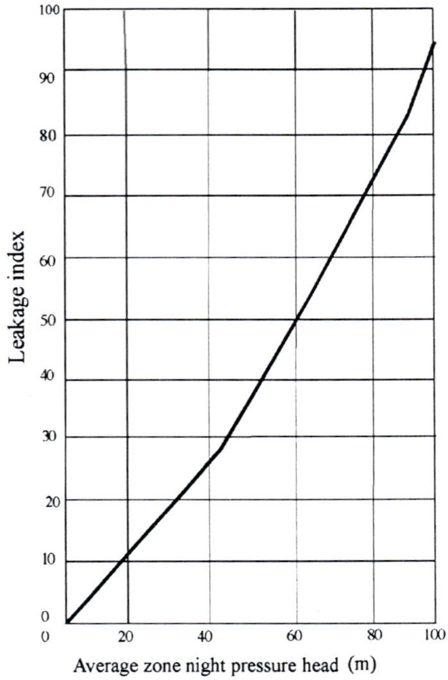

Figure 5. Relationship between pressure head and leakage

The relationship between leakage rates and pressure head is show in Figure 5., taken from the latest work by the NLI; it is derived from work carried out on selected areas where large leaks were repaired prior to the tests. Average Zone Night Pressure (AZNP) is defined as the mean of the lowest and highest pressures measured in the zone, generally weighted to take account of topography and layout of domestic properties. Leakage in each zone was measured in the approved way using net night flows before and after pressure control was applied. The relationship is shown in Figure 5. and indicates, for example, that if AZNP can be reduced from 60m to 30m, then the Leakage Index reduces from 45 to 20 or more than half, and this will be the approximate effect on the actual leakage rate in the zone. Thus, provided that the zone is suitable, quite large and cost-effective savings can be made; unfortunately, in real world conditions these savings often do not materialise because, for example, a customer on high ground within a zone or with a long service pipe requires that pressure across the whole zone has to be kept much higher than necessary for reduced leakage. Flow modulation is a relatively new feature which has been added to the more typical pressure reducing valves. It adjusts the action of the valve to produce a more stable pressure profile at a target point in the distribution system- where system headloss is significant, flow modulation can improve leakage savings by more than 20% over the non-modulated type of valve in such types of system.

The technology in pressure control has improved enormously in recent years and a wide variety of valves is now available for pressure reduction, pressure sustaining, with or without flow modulation, profiling and so on. In one example where flow modulation was employed average daily metered throughput to the district decreased by 20% and pressures at the target point were improved at peak demand - the results of this type of installation are reduced leakage, reduced incidence of burst mains and reduced customer complaints. Installations such as this also represent rapid pay-back and there have been improvements in the design of PRV's to improve flexibility and reliability and to reduce maintenance. A typical graph showing the relationship between flow and pressure head in a water supply network is shown in Figure 6.

More sophisticated control can be applied using electronic devices such as the Autowatt logger/controller now used extensively in Wessex Water. These may utilise remote pressure readings telemetered from a distant control point or may follow a local control setting stored in the controller which may be battery powered. Time switching and/or flow modulation can be used with this device, giving a high degree of flexibility and consequently higher savings.

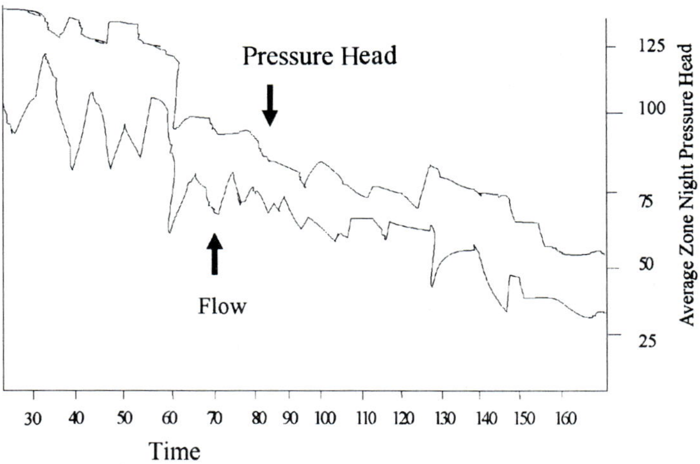

Figure 6. Relationship between flow and pressure

2.9. UPGRADE OF EXISTING SYSTEM

The previous sections of this paper have described some of the technologies now used in the diagnostics and rehabilitation of existing systems. We have seen how data available from telemetry and data logging systems is now collected routinely for monitoring and analysing pressures, flows and levels throughout the network. Wessex' **Aquaware** suite of programs has been developed specifically to automate the handling of this data as far as is possible. Some of the results of this work are described in Chapter II.

In addition, a number of techniques are used to assist in planning the upgrade of an existing water supply network, including modelling and data integration - if used properly these can lead to an optimised network where the capital investment yields the maximum return.

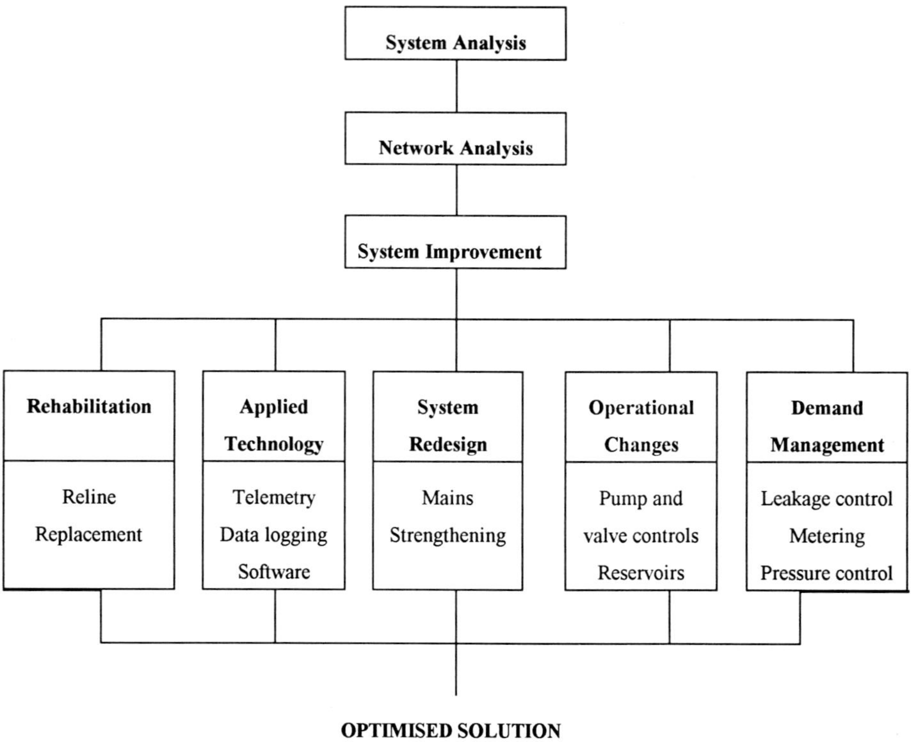

Figure 7. A methodology for upgrading existing water supply systems

For example, it may be that pressure control is more cost-effective than mains replacement for parts of a network, or that relining is better than replacement. Network modelling often reveals major bottlenecks or "black holes" in a network, where relatively small expenditure on system redesign is much more cost-effective than major replacement.

In these situations technologies for data capture, analysis, modelling, leak detection and location, logging, pressure reduction, metering and so on often means massive savings in capital expenditure on major schemes. Similarly, simple changes to controls on the network may lead to economies in capital expenditure. This can only be achieved by approaching the problem in an integrated manner.

The major elements in such an integrated approach are shown in Figure 7., and the major elements in the process are described below.

2.10. SYSTEM ANALYSIS

The first step is to establish the basic facts about the network and produce a clear simplified schematic showing all of the main elements and their interrelationships; the schematic should include all main pumping units, reservoirs, water mains, major metered customers and so on. Boundary valves and main metering points, PRV's, non-return valves etc. should also be shown clearly. This process is a vital first step in the process of analysing basic water balances in the network and involves using every source of basic data which is available. This includes the "local knowledge" of the supply staff (much of which usually turns out to be wrong on investigation), schematics used for valve operation, GIS records on paper or computer file and so on. Once this basic information is cross-checked, verified and put onto a schematic, the basic system operation can be analysed by using as much dynamic data as is available concerning flows, pressures, levels, pump and valve controls etc. These can be manual records, data collected via portable data loggers, or ideally, data from a telemetry system. It is useful if all of the data collected during these stages can be stored, collated and integrated in one software package. This has been one of the objectives in the development of WESNET, Wessex' network management software package, which allows schematics, graphics, GIS, telemetry and logged data to be integrated and analysed to establish the basic operation of the network. This data also establishes a database for the construction of a network model, if this is deemed appropriate, at a later stage in the process.

Collating this basic information is crucial to the overall success of the project and it is essential that, once collected and verified, the resulting documentation is put under proper document control procedures so that it can be maintained over a long period of time - this can be achieved by 2 simple means - firstly the basic schematic should be put under central control via, say, the corporate GIS, and secondly authority for changes must reside with one individual only - i.e. responsibility for the basic information must lie with one individual in the organisation.

The schematic will reveal the overall structure of the network and suggest ways in which the overall water balance can be established via the major inputs and outputs.

Other data also gives an insight into the system; burst main records overlaid onto the GIS background are a good indicator of possible pressure problems, while customer records on dirty water complaints give an indication of internal condition, for example, tuberculation due to iron deposits.

This stage of the exercise - System Analysis - should be properly documented and naturally leads to the next phase of the exercise - Network Analysis.

2.11. NETWORK ANALYSIS

This phase involves the construction of a **computer model** of the network and is vital with some types of network where the system is hydraulically sensitive, where we wish to experiment with different types of control or where options for mains replacement or system redesign need testing. These decisions often involve major expenditure and it is vital that a number of options can be evaluated hydraulically before such expenditure is made. Modelling is not always necessary, however, nor desirable in some cases. Often much can be deduced simply by analysing the basic data available and concentrating on, for example, isolating areas of major losses which can be rectified most cost effectively.

Network modelling is a specialised task requiring judgement, experience and knowledge, there is no widely accepted methodology which covers all situations in all networks though attempts have been made to produce codes of practice for, say, calibration.

WESNET, the network modelling and management package used in Wessex, has been developed specifically to aid the modeller in the construction, and specifically the calibration, of the model by integrating GIS, telemetry, data logging, and other data sources into one package in a way which is unique in the water industry. Some systems have considerable telemetry coverage and this invaluable data can be used to establish good basic calibration of the model - often, a field test will be needed where key data is logged on the system, usually over a period of several days, to establish the calibration of the model.

Once confidence in the model has been established it can be used to test a wide variety of operational changes and system redesign options, as well as confirming or otherwise areas of high leakage, bottlenecks or other anomalies including meter errors and other faulty instrumentation. A well-calibrated model is, in fact, a way of verifying the current state of knowledge of the network, much of which is often found to be wrong.

This phase, network modelling, leads to a set of proposals for improving the system. These will generally comprise:

- **Rehabilitation** - It may be apparent that certain areas would benefit from rehabilitation - replacement or relining. Replacement may be the only viable option in high pressure zones with old mains and high leakage rates, for example, associated with particular pipe materials and/or jointing compounds. Relining, a much cheaper option, may be appropriate where the structural condition is sound but flows are restricted by internal encrustation and pressures are low.

- **Applied Technology** - It may be apparent that certain instrumentation is faulty or missing, that certain points should be telemetered or logged, or that software systems could be used to great effect for data handling and analysis. Wessex' Aquaware package has been developed to provide the complete integrated software system for water supply network management. Low-level PC-based telemetry systems are now available in addition to the more traditional SCADA systems and these offer cost-effective data capture for the key data from the system.

- **System Redesign** - The investigation may reveal that the network can be redesigned at lower overall cost than any of the alternative options. This often applies to old complex systems which have evolved in piecemeal fashion over a long period.

- **Operational Changes** - A wide variety of pump and valve controls are now available and major savings can be made by implementing these changes. Obvious areas of saving include pump control and tariff selection, valve controls for pressure sustaining, reduction and so on. It is often the case that the effective life of the system can be extended by relatively simple operational changes rather than major capital expenditure on pipeline replacement.

- **Demand Management** - As stated earlier in the paper, in the UK at least, the major emphasis in system management is now in Demand Management, which includes leak detection and location, pressure control, metering and mains rehabilitation. Most of these topics have been described earlier but they should be considered as part of the overall integrated approach .

The result of this process is an overall plan for the effective upgrade of the network at lowest cost and inevitably this means changes and improvements in the management of the network; in Wessex the following key areas have been identified:

- **Information** - Key information must be made available to the staff running the system at regular intervals and basic disciplines established in the use of that data. .

- **Training** - Network management is becoming increasingly technical, through the widespread use of computers and also through the kinds of technology described in this paper - data loggers, telemetry, valve and pump controls, system analysis, modelling, relining techniques and so on. Most companies in the UK are now investing in GIS' (Geographical Information Systems), which are invaluable in more detailed analysis of geographically-based data, in addition to providing an automated means of record-maintenance. All of these techniques require fewer people but higher levels of technical ability and training.

- **Operating Manuals** - Tighter control of the network can only be achieved through better management disciplines and in Wessex this has been approached by introducing operating manuals for each network - in this way basic disciplines are introduced for rezoning, valve operation, pump controls and so on. Controlling demand effectively can only be achieved in this way though it can be a major culture change and difficult to implement.

2.12. CONCLUSIONS

Over recent years the technology for diagnosing problems with existing water supply systems has improved enormously; in the UK, where domestic customers are still predominantly unmetered, there has been major investment in instumentation, data capture equipment and information systems for producing accurate water balance data at various levels in the network.

The technology for rehabilitating these networks has also advanced, providing the management of such systems with a number of options for improving performance and stretching the operational life of the network. The optimum solution generally now involves a combination of techniques and the paper suggests that a methodology is needed which enables the management to assess this best combination.

2.13. REFERENCES

Technical Working Group (1985) Report on Leakage Control Policy and Practice, No. 26., Water Authorities Association, WRc ISBN 0 904561 95 X 1985.

Water Metering Trials (1993) The National Metering Trials Working Group.final report. ISBN 0 947886 33 8.

IWSA (1994) Rehabilitation. Regional Conference. Zurich.

National Leakage Initiative (1994) Managing Leakage. Final Report, WSA.

IWEM (1994) Accounting for Losses - The Bursts and Background Concept. IWEM Journal, Vol.8 No.2 April.

Biwater Industries (1993) Water Industries Manual.

AquaWare Systems (1996) WESNET User Guide, AquaWare Systems. UK.

The Water Mains Rehabilitation Manual (1989) WAA/WRc. second edition.

Hegarty M. (1994) The Strategic Role of Information Technology in Leakage Control. BICS Conference.

Chapter VI

The Application of Modelling in the Management of Water Supply Systems

Chapter VI

The Application of Modelling in the Management of Water Suppply Systems

INTRODUCTORY STATEMENT

Computer modelling in water supply is an old technique, often referred to as network analysis. In recent years a number of advances have been made in the application of this technology to the management of water supply systems and this chapter looks at some of these new concepts and applications. The basic principles of modelling are reviewed, including the traditional applications in Planning and Design.

The Chapter then goes on to look at the way modelling is being introduced into operational management, using examples from Wessex Water (UK), and the WESNET software package, and the advances that are now being implemented in integrating complementary sources of data, including telemetry, GIS, data loggers and post code data. The paper also looks at the introduction of a simple expert system module into the WESNET network modelling software package to improve the accuracy, speed and consistency in the calibration process. The conclusion is that the future for network modelling lies in making models relevant to everyday operational management, and in having links to live data for ongoing calibration checks - this data can be derived from traditional SCADA systems, data loggers or from the increasingly common telephone dial-up type loggers which are now commonly used in the UK for monitoring leakage at DMA level.

The second Section in the Chapter looks at new techniques in the computer software used in some models and the impact of object oriented programming techniques, using the HYPRESS package; the paper also looks at the way in which GIS can be used for presentation of results and as a complementary system to hydraulic modelling.

1. Water Supply Network Modelling and its Application in Design, Planning and Operational Management

Dušan Obradović
WINS Ltd.
Tower House, Parkstone Road
Poole, Dorset BH15 2DH
John Snoxell
WESSEX WATER
WESSEX House, Passage str.
BS2 0JQ Bristol
UK

1.1. INTRODUCTION

Computer modelling in water supply is an old technique, often referred to as network analysis. The applications and objectives of such techniques are well documented and will be found in the references; this Section looks at new ways in which these techniques are being used for operational management and the way that complementary data sources, from telemetry, GIS, data loggers and Post Code data are being used to facilitate this process by reference to work in Wessex Water in the UK.

Water supply has become a limiting factor in development, along with employment, food and transportation problems (UNESCO Workshop, Zandvoort, 1977). The community requires better levels of service, expansion of facilities to meet rising demand, but also tighter control over waste water treatment and disposal.

In recent years there has been rapid progress in the implementation of new technology into water supply management. It is generally accepted that technology is the means by which increasing pressures for better service and higher efficiency can be achieved (Brammer and Schulte, 1993, Cosgriff, 1985, Cubillo and Mercier, 1993, Cullen, 1987, Haddon, 1994, Halpern and Pascal, 1987, Huntington, 1984, Nguyen, 1994, Snoxell, 1994, Takagi, 1983, also other references).

NATO ASI Series, Partnership Sub-Series, 2. Environment – Vol. 15
Water Supply Systems – New Technologies
Edited by Č. Maksimović, F. Calomino, and J. Snoxell

The modernization of a water company, however, should not be a revolution. The service must continue without serious interruptions, while important changes within the company take place. All actions must be carefully planned, having in view both the desired ends and available means, then executed as efficiently and economically as possible. Errors can be very expensive with, in some cases, very damaging consequences. It is to be expected that they could be avoided by studying the methods of the most advanced companies and learning from their experience. This text will try to summarize the achievements in information technology and mathematical modelling as applied in such water companies today.

1.2. FUNDAMENTALS OF MATHEMATICAL MODELLING

A mathematical model is an artificial entity which describes the behaviour of a real system in various situations. In the case of a water supply system this means that the model can simulate variations of flow and pressure in a distribution network, the operation of pumping stations, the functioning of all devices such as control, check and float valves, changes in water quality at service connections, compute costs, forecast demand patterns, optimize the use of pumps, etc. - and all this in normal, peak demand or emergency situations.

1.2.1. Classification of Models

Obviously, the building and testing of a model will take time and effort, besides significant costs. No one will do it without good reasons, least of all a water engineer pressed with other urgent tasks. But such reasons do exist; to name just a few:

- the model will collate all known facts and data for the real system, with all known relationships between its elements,
- the model will permit the analysis of all operational data in much greater detail, helping to understand the complex process within,
- all interested parties: operators, designers, engineers and managers can participate in the building of the model, make a contribution and later use it for various tasks,

- the current control policy can be tested on the model, verified or improved, without risky experiments with the real system,

- the model should be a vehicle for design and planning, greatly reducing the time and costs for such activities,

- the response of the system in various normal, emergency and catastrophic situations may be analyzed beforehand and adequate strategies developed and perfected,

- the model is a wonderful tool for education and training of new hands: from operators to managers.

These arguments, as well the availability of computer hardware and software, have caused a proliferation of models and modelling techniques in the water industry in the last few decades. One can argue that modelling was always an integral part of the design process; the novelty was its application on real water supply systems.

This complexity has forced most of investigators to concentrate just on one facet of the problem, say, hydraulics - the other option being an unacceptable oversimplification of the system. Models therefore can be sorted in three main groups according to their main purpose (Obradović, 1993):

I. **Hydraulics**: models which simulate changes of flow and pressure in water supply and distribution systems;

II. **Water Quality**: models which simulate changes in water quality in ambient waters and in the water supply system;

III. **Economics**: models which optimize functioning of water supply systems.

Hydraulic models may analyze either steady state (one "snapshot" or a series of steady states interlinked with an integration scheme) or transient regimes (rigid water column models, water hammer, hydraulic vibrations, two-phase transient regimes) - see for example Bhave (1991), Hoogsten and Zwan (1985), Rao and Don (1977), Walski (1984), Zwan (1988).

Water Quality models may analyze the situation either in ambient waters (surface or ground water or both) or in the water supply system itself (transport of raw water, water treatment plants or deterioration of WQ in the distribution network) - see Clark (1994), Cohen (1990), Filip and Obradović (1986), Males et al. (1985), Nguyen and Montiel (1994), among others.

Economic models analyze either Operational Efficiency (optimum pump scheduling, water sales, operation and maintenance costs, specific consumption of various resources) or Planning and Investments (water demand in future, capital investments, development of alternative solutions for new systems, optimization of various plans and designs) in order to find optimum solutions. A good example are models for optimum control of water distribution systems, see Chase and Ormsbee (1993), Chase and Jones (1993), Coulbeck and Orr (1988, 1989), Fallside (1977), Jowitt et al. (1989), Schulte and Malm (1993).

Beside these models, specially developed for the water industry, many other software tools and general purpose programmes are used in water companies: CAD, GIS, demographic studies and others.

The models can be classified alternatively by their Scale, Time Step, Depth & Sophistication or numerical method applied. The user might not be aware of these technicalities; a more important distinction is between:

- Simulation Models, which **describe** the behaviour of a real system as closely as possible, and

- Optimization Models, which **prescribe** what should be done in order to achieve a certain objective.

Hydraulics and Water Quality models belong to the first group.

All these models can not be covered in one book. After all, they do not have the same importance and frequency of use in practice. This text will concentrate only on hydraulic simulation models, for steady-state analysis.

1.2.2. Structure of a Hydraulic Model

The current view is that any water supply system may be quite adequately represented by a set of **nodes** and **links**, arranged in the proper order.

A **node** is a location within the system where either head or inflow/outflow is known. The first group includes:

- **fixed-head nodes**, where water level remains unchanged during the simulation because the storage is infinitely larger than all withdrawals and/or inputs (examples: aquifers, lakes, large pools, springs, wells, etc.),

- **variable-head nodes**, where water level rises if the inflow exceeds the outflow and vice versa (examples: reservoirs, water towers, contact and balancing tanks).

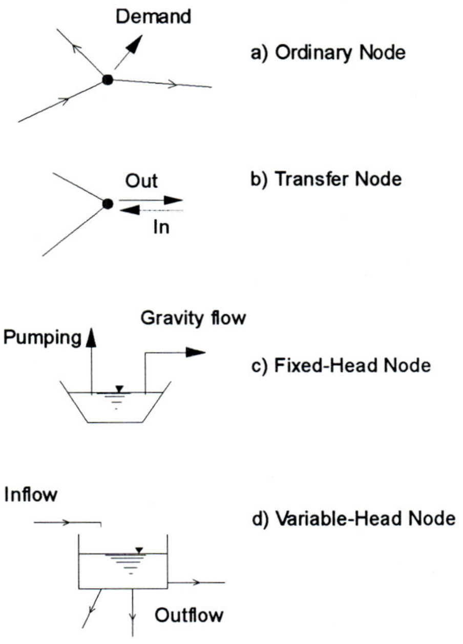

Figure 1. Nodes in a model

In both cases water level is a known quantity while outflows and inflows need to be determined through the computation.

The other group is called **ordinary nodes**. There the total sum of all inflows and outflows is known and the head is to be computed. Examples are:

- a transfer point, where a known quantity of water is either delivered to (or taken from) the system,
- a junction (branching point) of two or more pipes, with or without some local consumption,
- a service connection to an important user, where local consumption of water is known in advance,
- any other point within the system which is important due to any other reason.

Examples of nodes used in modelling are shown in Figure 1.

A **link** connects two nodes and can convey a certain amount of water. There are three types of links:

- a **pipe**, which simply transports water from one node to another,
- a **pumping station**, which adds energy to water,
- a **control valve** (any type), which regulates the flow of water by reducing its hydraulic energy ("throttling").

These names are mnemonic: they describe the behaviour of one element of the model by using the name of the similar physical device.

This does not mean that each valve in a water supply system will be represented by a "valve" in the model - this is seldom practicable! Quite the opposite, only a few important physical valves (control valves, for instance) will be shown as such in the model - and the bulk (zone valves) will be simply included as minor losses in their pipes.

Next, a "pipe" in the model represents any conduit, a tunnel, a water main, a large or a small pipe section - anything that is used to convey water from one place to another under pressure.

As an opposite example, a real pumping station which houses several pumps under the same roof may be represented in the model as several "pumping stations" (in modelling sense) if the suction or the delivery node is not the same one for all units.

Examples of links are shown in Figure 2. Note that hydraulic energy falls along the flow in the case of "pipe" and "valve", while it is boosted in a "pumping station".

It can be easily shown that all elements of any water supply system could be described by these items - either directly or by a suitable combination.

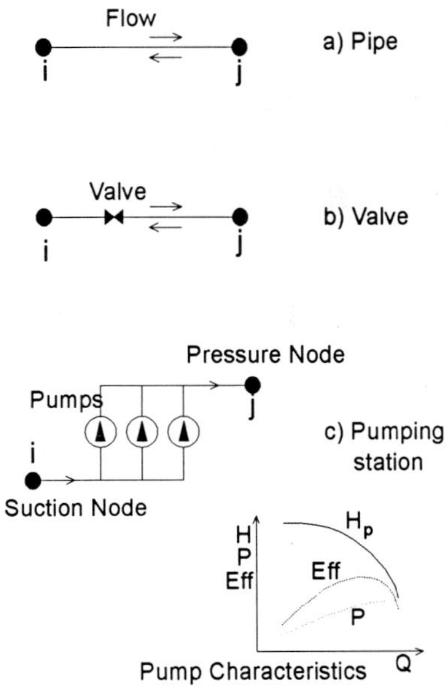

Figure 2. Links of a model

1.2.3. Modelling of Steady State or Gradually Varying Flow

Thus far it has not been defined whether the model will serve for analysis of steady or transient regimes. Further development will be limited to relatively slow changes - where inertial forces and the compressibility of water can be ignored. This assumption is valid for all normal and for most of the emergency situations in water supply systems, especially if longer time periods (up to several days) are being analyzed. Of course, when analysing transient states - which last just a few minutes at most - inertial forces, the elasticity of pipe walls and compressibility of water must be taken into account, using more sophisticated analytical tools.

A further simplification of the real system is about water withdrawals. It is clearly impossible to represent each tap in the model so whole buildings and blocks of houses are represented as "sinks" - nodes with aggregated demand (total consumption plus some losses). At the very best, this can be only a crude image of the real process - but still useful! Even with all these assumptions the task can not be solved without a computer.

The resulting system of non-linear equations can be solved by various numerical methods, see, for instance, Bhave (1991), Obradović (1993) or Walski (1984).

The first task is reading all necessary data: from model files or from telemetry files. The data are then cross-checked to discover all errors and inconsistencies - before the simulation. The user is warned and the procedure halted to permit the corrective actions from user's side.

After this stage has been successfully completed, the program formulates the initial state of the system, makes necessary preparations (rearranges the data, closes input files, opens runtime files, sets pointers, etc). Then the "clock" is set "on" and the simulation cycle starts.

The program solves steady-states in this system for given boundary conditions, using one of a few available methods described in the next section. The results are then analyzed in order to determine eventual changes in boundary conditions:

- compute the current water demand (option: take into account service pressure as well),
- increase/decrease of water level in all reservoirs, water towers, contact and balancing tanks,
- change of status of any pump - due to changed water level in control reservoir or pressure in the control node,
- change of valve opening - for the same reasons, etc.

Relevant messages are displayed - so the user may intervene through the interface - for instance, to review current results. Then the procedure is resumed: the clock increases time for a certain increment (from 1 to 30 minutes, rarely more) and determines the water demand at this moment before repeating the same cycle again.

The simulation lasts until the set time is reached - or until the user has stopped it, for any reason. The main results are stored in data files for analysis. The user may get information on extreme values of flows, pressures or water levels in selected nodes, the operating range and time for pumping stations and individual pumps, max/min openings and flows through control valves, various diagrams and snapshots at certain moments, etc. The data could be shown on maps (provided by GIS) and schematics; or set against the current or historic operational data.

1.2.4. Model Application

The abundance of options available by modern software is such that the need for additional on-line help in the form of an EXPERT system is recognized by many users.

The main applications of modelling are in:

- **Operational Management** of existing water supply systems,
- **Design & Analysis** of existing and future water supply systems,
- **Education**, **Study** and **Research** in hydraulics and water supply.

It follows that the principal users are:

- engineers and operators in **Control & Information Rooms** of water companies,
- engineers and technicians in **Design Bureaus**,
- engineers and students in **Universities**, **High Schools**, and in **Research Departments** and **Institutes**.

Typical applications for models are:

- analysis of how the system really operates,
- verification of current control policy,
- evaluation of storage reserves in emergencies,
- optimum pump scheduling,
- discovering 'bottlenecks' within the system,
- selection of post-chlorination sites,
- decisions concerning which tariff should be used,
- completion of an existing telemetry system,
- design of new facilities,

- design of enlargements and reconstruction of the system,
- design of a new control/telemetry system,
- 'war games' for the staff,
- training of personnel at various levels, etc.

Modern software products are labelled "user-friendly" because they do not ask for any special knowledge or training prior to use (this is only partly true). The characteristic features of such a program are:

- is extremely easy to **learn** and **use**, because it is tailored to suit user's demands,
- is completely **menu-driven**, so the user does not often have to refer to the manual,
- has an comprehensive system for **capturing errors**, with an explanation in plain language, so the user is protected against common blunders and mistakes,
- **on-line help** is available at any stage,
- can model a wide range of possible **control policies** for pump stations and control valves, but keeps all technicalities invisible to the user,
- an ability to keep the user **informed** about the state of the system during a simulation run,
- always provides **an answer** to the user regardless the size or complexity of the case - and never loses control!
- has good **computer graphics** available to the user both for input and output jobs,
- can **compare** operational data with simulation results on the same graph,
- permits the user to create **the output** according to his (her) particular wishes in a very versatile way,
- is a very efficient tool!

As for the user himself, it is assumed that:

- the user is **well informed** about the water supply system he/she is modelling, but does not have to be an expert in mathematical modelling or computers,

- the user should describe **the real system** as it really is, without "inventions" - and the programme must be versatile enough to cope with real-life complexities.

1.3. SIMULATION MODELS AS TOOLS FOR DESIGN

The design process has its creative phase, when alternative solutions for water supply systems are being formulated and its routine phase, when accepted solutions are elaborated in greater detail. These two phases cannot be completely separated, because every proposal has to be evaluated from various points of view: functionality, controllability, capital expenditure, operational & maintenance costs, feasibility in given circumstances, etc. This evaluation is based partly upon the

Designer's experience and judgement, and partly on computation. Here the models come into play, permitting the Designer to examine quickly and efficiently different alternatives with sufficient precision and detail. The idea is as old as designing itself, but Computer Science has opened new possibilities which did not exist before, as amply demonstrated in references - see, for example, Brammer (1993), Coulbeck and Orr (1988), Hoogsten and Zwan (1985), Obradović (1986, 1993), Rao and Don (1977), Walski (1984) or Zwan (1988).

1.3.1 A Simple Case

To illustrate this point a simple case is included here, see Figure 3. The new system has to supply four different communities (denoted as "Demand A" to "Demand D") from a single source. One possible solution is to build a main pumping station, four large reservoirs (I to IV), two booster stations (II and III) and control the pressure at two places (A and D), as shown below.

Obviously, this is not the only solution. Maybe Reservoirs II and III could be merged into one and then "Demand C" could be fed by gravity flow from either Reservoir II or III - or both.

Clearly many similar alternatives can be offered, each one with its advantages and shortcomings. Without modelling, one could only make arbitrary decisions (not necessarily the wrong ones!) which cannot be demonstrated to other interested parties, nor could the consensus be reached on the basis of facts. With models the situation is far better: every proposal can be examined on the model of the water supply system in relatively short time. All concerned can see for themselves why one solution is better than another. Expensive mistakes could be then avoided and a sound consensus reached at the earliest possible stage of the project development. This consensus cannot be taken for granted; the owner of the future system must be consulted almost from the beginning of the design process. Another simple case is included to illustrate the importance of good cooperation.

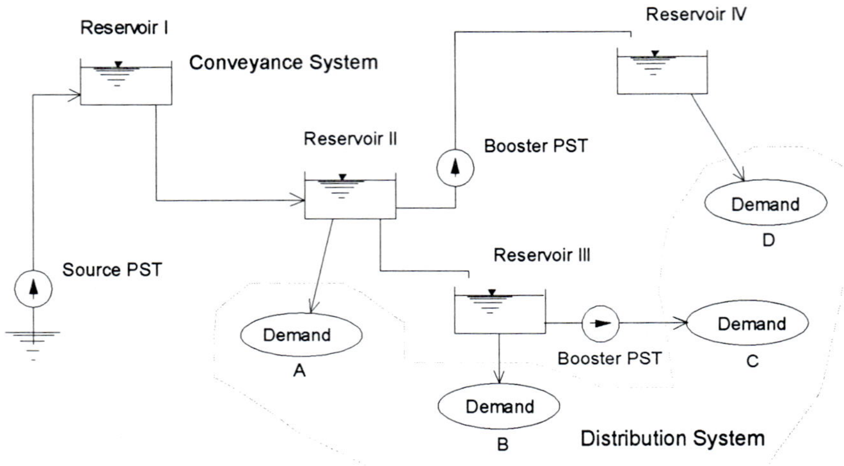

Figure 3. A water supply system

Figure 4. shows the simplest possible problem: satisfy a given Demand from a Source. Suppose that water must be pumped into the demand area; but there are still (at least) three possible solutions:

I: pump all water through demand area to a balancing reservoir at the opposite side;

II: pump all water first in a high reservoir from which it can flow to the demand area - the pressure can be controlled by a PRV;

III: pump all water directly to demand area by variable-speed pumps, without any storage space.

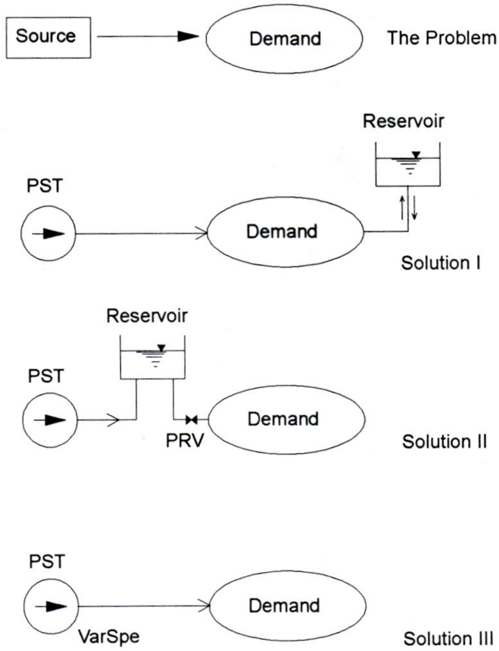

Figure 4. Alternative solutions

There are pros and cons for each solution (without even mentioning widely different capital costs); for instance, Solution I is simple, can be controlled easily, offers some security in emergencies - but the network will be submitted to relatively high and variable pressures. Solution III can be the cheapest to build, but operating costs could be high, and the maintenance of variable-speed pumps might be difficult. Moreover, there would be no reserve capacity if the power system fails (is this probable? How often?). Next, pressure management would be almost impossible in Scheme I and quite simple in the other two, etc. Obviously, the water company staff should be consulted and their preferences taken into account from the very beginning of the design process. The model offers excellent possibilities for quick and systematic evaluation of various alternatives so many costly mistakes could be avoided in good time.

1.3.2. The Design Procedure

The Designer is well advised to start the job by carefully examining the existing water supply system - the cases where there is none are extremely rare. The benefits are many: one can gather valuable information about local circumstances(example: specific consumption and seasonal effects), then make contacts with the local staff and mobilize their support for the project. Local experience is usually irreplaceable, so the additional effort is more than justified.

This exercise could be performed best on a mathematical model; in some cases this model will already be available in the water company, while in others it has to be made by the Designer. In any case modelling of the existing system provides plenty of opportunities for the development of cooperation and for collecting of information vital for the design. The list of tasks include:

- Analysis of demand in time & space: current level, trends, specific consumption for various uses, patterns of use, local peculiarities;

- Unaccounted-for-water levels: true losses, pressure management, billing system, non-billed use;

- Existing Control system: equipment, instrumentation, level of technology and expertise within the water company;

- Data Management: organisation, sensors, communications, available hardware and software;

- Control policy in normal and emergency situations: remote and local facilities, reaction time, spare capacity.

The good and bad sides of the existing system will surface in this process, enabling the Designer to make sound solutions with the maximum benefit/cost ratio. Even more important, the stages between the existing system and the future one will be clear to all concerned, not only to the Designer.

The steps in design procedure therefore should be:

I Make or update the mathematical model of the existing water supply system;

II. Calibrate the model in cooperation with the local staff using fresh operational data;

III. Analyze the existing system in great detail and find its good and bad features;

IV. Define several feasible alternative solutions for the future system;

V. Optimise acceptable alternatives by using multiple criteria like Internal Rate of Return, average and initial price of water, etc.;

VI. Elaborate stages of development from the present system to each alternative solution;

VII. Analyze various **Normal** and **Emergency** states on the model and discuss the relative merits with the staff;

VIII. **Decision** - select the most promising alternative, possibly with one or two major subvariants.

The model for design must include all sources, reservoirs, pumping stations and control valves, while the distribution network can be reduced to major pipes only.

The model must be verified before the beginning of any serious analysis. This is done best by simulating the performance of the real system during one or several days, and then comparing the results of simulation with observed values.

A computerised database with complete operational data (flows, water levels, pressures) will greatly facilitate this task; if such facilities are not readily available the Designer should organise data capture effort in cooperation with the local staff.

1.3.3. Calibration

One area where there has been major progress in recent years has been in the use of operational data for calibration.The results of simulation will not agree well with observation, certainly not at the first attempt.

The reasons are many, from imperfections of the model to false data. A careful analysis of differences can be very rewarding, showing which instruments are faulty, what data are wrong or which hypothesis was bad.

The model is then adjusted accordingly. This process of "fine tuning" is called **calibration** and must be repeated at regular intervals in order to keep the model well trimmed. Data from the real system can come from many sources and traditionally this may have been from manual records- over the last decade a wide variety of portable, inexpensive data logging instruments have become available which provide an easily accessible source of real data for calibrating models. These data loggers can now be interrogated remotely via telephone lines and this approach is now common in the UK for continuously monitoring leakage at DMA level (a DMA, or District Metered Area, is the basic zonal unit of around 5000 properties at which flows are measured continuously for analysis of Minimum Night Flows).

Most water supply networks are now monitored by telemetry and if this data source can be opened up for modelling then the process of calibration can be made even more efficient and cost effective. The natural consequence of this is also that modelling can realistically become part of operational management because all staff can see that the model is producing similar results to the real system.

Examples of results taken from a typical model using telemetry data for calibration in the Wessex region are shown in Figures. 5 to 7.

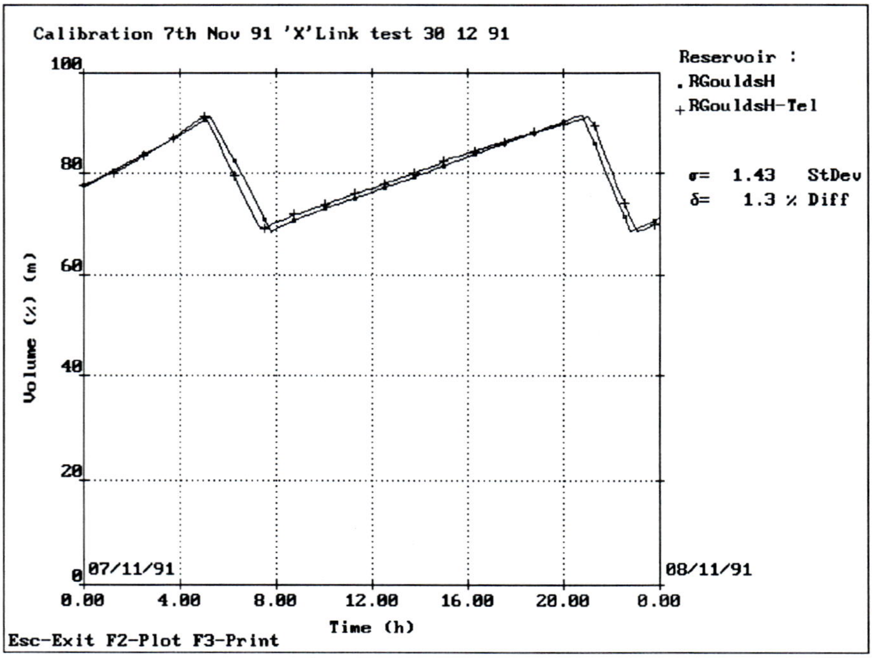

Figure 5. Variations of water level in "Gouldshill" reservoir

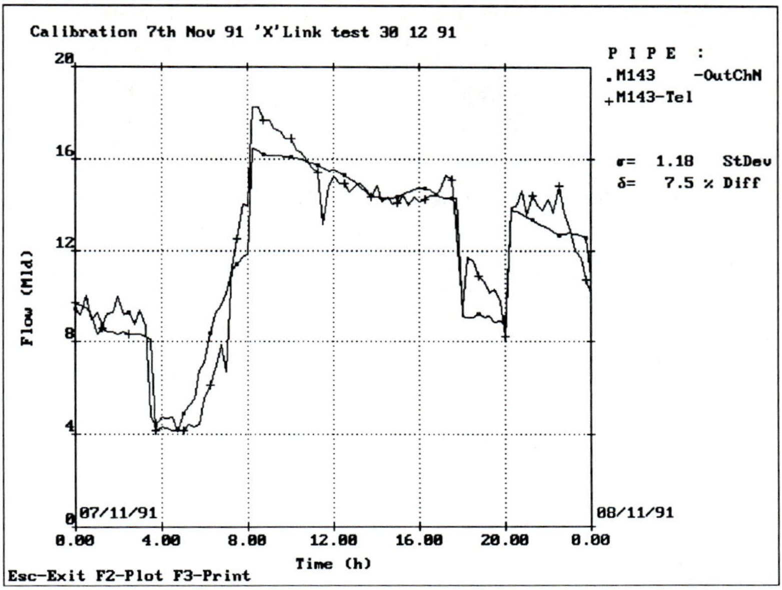

Figure 6. Outflow from Chabury Reservoir, Weymouth

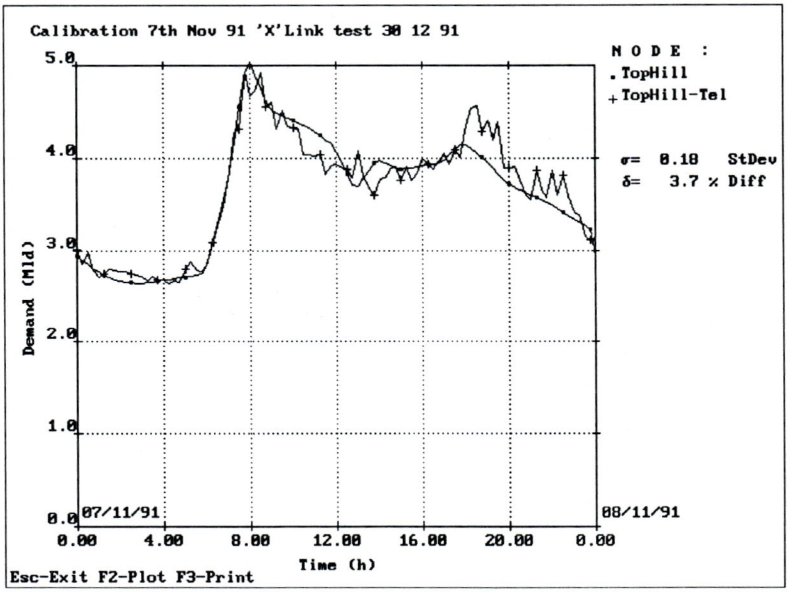

Figure 7. Water demand of Isle of Portland, Weymouth

Figure 5. compares computed variations of water level in a main reservoir with the data collected by the telemetry system for the same day. Figure 6. shows the computed and observed values of outflow from another reservoir. Figure 7. contains similar data for water demand in a zone within the model. Note good agreement between computed and observed values in all three examples - they are close to each other, but are not identical; the "noise" is peculiar to the observed data, quite understandably. By studying these examples one can get the idea about the range of differences between the real system and its model. Further calibration will not improve substantially our knowledge about the real system.

1.3.3. Using GIS Data in Modelling

Over the last decade many modern water companies have invested heavily in computerising their asset record data with mapping systems, now generally referred to as "GIS"- Geographical Information Systems. The link between such systems and network modelling has long been recognised but rarely achieved in any significant way (excluding, that is, research or demo projects). Wessex has made much progress in this area in the last 3 years in converting all of its major models into geographically correct format by using actual GIS data from its corporate system as a background to the model. The model can be constructed by tracing over the background and results can be displayed against the background, greatly enhancing overall understanding.

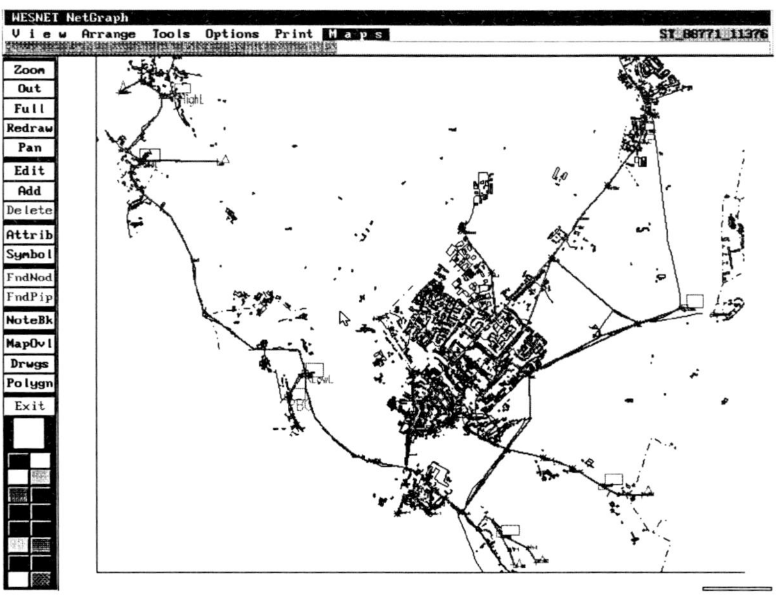

Figure 8. Using GIS in modelling

In addition, a wide range of other data can be integrated into the modelling environment to complement the results of modelling work, including , for example, burst main locations, customer complaint data, post code information and so on.

The example (Figure 8.) shows an example of a network model superimposed upon the GIS background (water main features and houses only have been included in this example).

The natural evolution in this process is to use the actual network data held in the GIS as the network geometry for the model and various demo projects are to be found in the literature-however, few, if any can be sustained in the real world and there are serious problems in doing this in a manner which is cost effective and manageable in real-world conditions. These problems are beyond the scope of this paper but it is likely that in the next 5-10 years this will indeed become properly understood, technically feasible and manageable.

1.4. NETWORK MODELLING IN OPERATIONAL MANAGEMENT

In reality Operations managers rely these days on telemetry and on judgement, experience and knowledge. Because of this the methodology for creating, maintaining and managing computer models for Operational management does not exist and many questions remain about the basic viability of such a management philosophy. It is worth considering this in 4 main areas:

- What are the benefits to Operational management of creating and maintaining computer models of water supply networks?
- What are the problems (including cost) in implementing this philosophy; why has it not been achieved?
- How can such a philosophy be implemented and what are the managerial and technical problems and solutions.
- What have been the actual experiences in Wessex Water over the period of implementation?

This Section describes how, through the development of WESNET, Wessex Water has introduced computer modelling into Operations management and the results and experiences of the exercise.

1.4.1. The Potential Benefits

These are generally recognised as follows:

Accounting for Water: Any major production operation in any branch of industry must know how much product is being produced, and where it is being consumed. In the case of water the model provides a factual base for this accounting process, including an assessment of how much water is being lost in each of the demand areas. To be able to do this regularly over a long period of time in a consistent manner is a very powerful tool in basic management.

Cost Control: A major cost in running the network is the pumping costs and these can be reduced significantly through a combination of source management (eg "cheapest first"), pump control (eg level, profile, etc) and tariff selection (eg "Demand", "Day/Night", STOD - Season Time of Day). The permutations are very complex, however, and a modelling aid is now essential to answer these questions.

Standards of Service: The Director General of Water has recognised formally that network modelling is a valid method of demonstrating compliance with pressure standards. An operational model would be an ideal vehicle for this.

Resources (Bottlenecks): It is often the case that resources cannot be made available at the point of consumption at peak demands because of bottlenecks in the system. Identifying and removing these is essential to the operational manager.

Emergencies - Burst/Contaminants: The ability to simulate bursts (ie exceptional demands), or pollutants introduced to the network has long been recognised as of major potential value to the operational manager in dealing with emergencies.

Capital Expenditure: An operational model could potentially identify situations where minor changes to the network could avoid major capital expenditure.

Duplication of Modelling Effort: One common feature of current practice is the wasted effort in duplicating models over a period of years, mainly because of a lack of ongoing maintenance. Operational models, kept live through telemetry, would provide a central library for others to use as a basis for other work, thus removing much of this duplication.

1.4.2. Problems in Implementing Computer Modelling in Operations Management

In the previous section we outlined the major areas where, potentially, computer modelling could have a major benefit in Operations management. The major question which arises from this is: Why has so little been achieved?

The experiences in Wessex Water through the development of WESNET have focused on some of the difficulties and also the solutions to these problems.

1.4.3. Software

Firstly, the software for creating and maintaining such models has been very limited. The available packages are not suited to modelling the actual dynamic behaviour of the network, including the wide variety of pump and valve controls now commonly in use on real networks. In practice it is common to find network modellers (who are usually engineers) using a wide variety of tricks to simulate the behaviour of the network; indeed it is often a source of pride and discussion how to "fiddle" the network geometry to reproduce a comparatively simple feature. Often people have become used to "hacking" data files to achieve their ends rather than use the software through the User Interface, partly because it is not very User friendly or because this "expertise" has become second nature to the individual.

This approach is simply not viable in the Operations environment where day-to-day changes have to be made quickly and reliably to keep up with the dynamics of the actual system. The software must reflect this environment and WESNET has been specifically developed to answer this need. If the User has to "fiddle" or "hack" then something is wrong - the software must satisfy the User's requirements, not require the User to be a computer expert.

Current software packages are very weak in simulating pumping equipment accurately, yet it is this which is often the most expensive single cost centre in the running of the network, and therefore of prime concern to the Operations manager. This is a reflection of the fact that hitherto modelling has been the exclusive province of the Planner or Designer who is concerned primarily with the hydraulic limitations of the distribution network, for example what are maximum and minimum flows and pressures under certain conditions of demand. It is quite common for such models to exclude most or all of the significant pumping equipment altogether; often models start at the service reservoir and go into immense detail in the distribution network, much of which adds nothing to the overall understanding of the network but which adds a considerable maintenance overhead.

It is a common experience to find that such models are, in practice, useless as a basis for an operational model and it is a mistake to try to convert them for operations' use.

1.4.4. Calibration

A major obstacle to the ongoing use of network models in operational management is the problem of calibration; there are two aspects to this: firstly the costs and feasibility of maintaining the model over a period of time, and secondly the actual process of calibration, which to date has been developed solely on a "snapshot" basis, a method which is inadequate for operational use.

The development of WESNET has been largely driven by the realisation that the first problem can largely be solved by the integration of telemetry and data-logger data into the modelling package, thus providing a constant, free source of data for the fine-tuning of the model over a long period.

The latest version of the software (V 6.0) now provides extensive capabilities for integrating logger data from DMA monitoring, thus providing an additional source of on-line calibration data from the low level telemetry which is now employed on many networks for leakage monitoring. The experience has also proved that the existing calibration techniques, as expounded in the WAA Code of Practice are not appropriate, or adequate, for operational modelling.

As a result Wessex has included in WESNET a new set of criteria for judging the adequacy (or otherwise) of the dynamic calibration of a model. This provides a much better guide to how well the model is actually simulating the behaviour of the real network, and speeds up the ongoing calibration process. WESNET produces a "verdict" in terms of "good", "fair" or "poor" for each specified calibration point on the network, based on statistical criteria over the whole period of simulation.

This approach formalises the rather subjective judgements which would be made about the agreement between the model and reality over a period of time, and also points up how important is the dynamic calibration, as opposed to the "snapshot" approach. It is clear that the industry should now start to encompass the dynamic approach in a national Code of Practice as a complement to the existing Code, perhaps based on the ideas in WESNET - the refinement of these techniques will only come through experience and actual implementation.

1.4.5. Mythology

A further problem in implementing modelling in operations management is that the subject has been surrounded by considerable mystique and mythology, partly through a lack of real data and experience. One common fallacy concerns the inherent complexity of demand modelling; it is argued that demand is made up of many components, each of which **has** to be modelled separately for the model to balance properly. If true, this would make the ongoing maintenance of the model quite difficult. Experience in operational modelling over the last three years has shown quite clearly that this is not the case, however, and demand patterns have not proved to be a major problem in the ongoing maintenance of the Wessex models. In practice, demand patterns have proved to be stable, repeatable, and easily manageable.

Another common view is that the system changes so frequently due to rezoning, mains reinforcement, pump changes, new sources etc, that it is impossible to keep the models up-to-date. In practice, with the link to telemetry this has not proved to be the case; the operations models act as a focus for different departments once they understand what the objectives are and communications can be set up to account for all such operational changes.

1.4.6. Operational Reliability

One sometimes hears telemetry data being derided, especially flow data, the implication being that telemetry per se is inherently unreliable and unusable for modelling purposes. This stems partly from the early days of water industry instrumentation, when certain devices did have design faults which led to "drifting" or other problems. Nowadays, however, most of these problems have been resolved and telemetry devices generally are reliable, provided they are installed and maintained correctly. Experience in Wessex is that telemetry is an invaluable and reliable source of raw data for the calibration of operations models; in practice the models sometimes show up anomalies resulting from transducer problems in a way that would be difficult to detect by other means and in this way act as a perfect complement to a telemetry system.

These are some of the problems that have prevented the perceived benefits of modelling being implemented in Operations management; it is a daunting list and it is no surprise that little real progress has been made. A corollary of this is that there is virtually no methodology for people to follow to implement this in reality. One assumption is that "network modelling" is just one technique - in practice there are now at least two classes of network models: "classical" as developed traditionally by Planners and Designers, and "operational" as developed and implemented through WESNET. They are quite distinct in character, implementation and use and the following section describes in detail the methodology developed in Wessex Water over the last three years to implement and realise this philosophy for operational modelling.

1.5. EXPERIENCE AND BENEFITS OF OPERATIONAL MODELLING IN WESSEX WATER

1.5.1. Model Construction

The initial phase when the model is constructed is a period when all current knowledge about the network is collated and verified. This invaluable information typically becomes history very quickly with conventional modelling, but with operational modelling, where the model will continue to be "live" for an indefinite period, this presents an opportunity to create a database which can be used as a source of basic information on a day-to-day basis. WESNET has been developed to exploit these possibilities in the following way:

- GIS data is available as background and is user selectable such that any key features can be included.

- Photographic data of key features is available.

- Schematics can be overlaid and organised in one place.

- Basic pump curve information can be organised and catalogued.

- User defined or imported polygon data can be used for a variety of spatial analysis tasks

Overlays of point data (eg customer data, burst or leak locations etc) can be incorporated into the model to provide greater insights into modelling results.

The major benefits are, therefore, in verifying current knowledge about the network and in organising and managing key information about the network.

This is a hitherto unexploited feature of network modelling which can only be used with confidence if the model is constantly updated, which, of course, is the key feature of the operational model.

1.5.2. Initial Calibration

This phase usually throws up numerous anomalies concerning current "knowledge"; typically these might be methods of pump control, zoning, closed valves, et cetera. The achievement of the overall balance will usually indicate areas of possible leakage - the Wessex experience is that the overall balance nearly always indicates such areas and that quite significant leakage can be found on pumping mains and trunk mains upstream of main service reservoirs, areas which are sometimes ignored with the traditional UK method of leakage control . The key factor in model construction is to ensure that the complete network is modelled, from source pumping to demand area.

1.5.3. Pumping Costs

A key cost centre in any operations budget is power; energy costs for water production in Wessex are £4-5 millions per annum. Containing and reducing those costs is one of the key objectives in maintaining an operational model and WESNET has been developed to reflect that need.

Pump controls and tariffs are User selectable so that any combinations can be tested and evaluated. Wessex estimates are that 2% reductions in electricity costs should be made annually (roughly equivalent to £100K per annum) through a combination of pump control, tariff selection and waste control; the savings may be far greater than this - Wessex has concentrated on energy cost reductions for some ten years and thus the annual savings still possible are around 2%.

1.5.4. Telemetry

Anomalies between telemetry and model results may be caused by transducer error, and WESNET provides a useful first-line check on flow-meter calibration. Flow meters may be configured in groups to balance against one another, thus highlighting possible errors. The Wessex experience is that telemetry data has proved generally reliable but that there will be occasional errors. Since 90%+ of flow data is now collected through telemetry (at a major saving in manual data collection) it is essential to have low-cost routine flow data verification - WESNET provides this facility.

In addition, WESNET provides an excellent means of analysing telemetry data off-line in its own right; that is without any reference to the model. Most telemetry systems are very limited in their ability to analyse historic data in a flexible way; the operational model opens up this information to users across a wide area at very low cost, for example using existing PC networks, and in Wessex is used almost exclusively to analyse historic data.

1.5.5. Standards of Service

A primary concern to the Operations manager is to achieve the set Standards of Service at least cost; but how does he demonstrate that this is being achieved? In the UK, network analysis has been accepted as a primary means of demonstrating this to the Director General of Water, and WESNET has been developed such that an operational model can do this on a regular basis. Each demand area has one critical node modelled in it specifically to represent the "worst case" - typically this may be a block of flats in a remote part of a zone, or properties on high ground. WESNET includes a module which enables the duration of a pressure below a user-selectable standard (usually 15m) to be recorded and viewed. This feature is invaluable in highlighting potential areas of low pressure since it can be used as a filter.

1.5.6. Emergency Management

The Operations manager has to live constantly with the risk of an emergency on the network. The most common are mains burst, power disruptions, and contamination of the network (pollution). It has long been recognised that a reliable operational model of the network should be a useful decision support tool in the management of the emergency. Experiences in Wessex have confirmed this view though the reality is that, in practice, real emergencies are not managed by running simulations of various scenarios at the time of the emergency. The model does aid the manager by enabling the major scenarios to be tested in advance rather than flying blind at the time; in particular the model confirms the limitations or bottlenecks in the system, enabling difficult strategies to be clearly worked through in advance.

1.5.7. Duplication of Modelling Effort

Finally, an unexpected benefit of operational modelling has been a reduction in duplication of modelling effort throughout the Company, because it is now possible for all departments to make use of an up-to-date, fine-tuned model of the major urban supply systems, rather than repeat the effort at regular intervals because confidence in previous versions has diminished. PC-networking now means that Users across the Company have access to a central library of up-to-day models on line, together with all of the relevant management information, including graphical, telemetry, data loggers etc. This is invaluable and greatly improves the efficiency of other departments.

In summary, therefore, the experiences in Wessex have demonstrated that the concept of using operational models in water supply management can be translated into reality; that the benefits are many and fully justify the small cost incurred in construction and ongoing maintenance - the benefits increase with time. Many other benefits have been realised which were not foreseen at the outset, and the narrow concept of operational modelling has now evolved to a much broader one of NETWORK MANAGEMENT; ie the model becomes an essential element in a much wider concept, that the supply network is actually effectively managed through the system, which becomes the managers "toolkit" for this task. This concept is now overlapping with that of Quality Assurance on the management of the network; the system confirms that the supply network is operating to targets of cost and standards and **demonstrates** that this is the case.

Perhaps the most telling evidence that these techniques can have a major impact on operational management is that Wessex has consistently avoided any restrictions on water supply to its customers during a period of consistently low rainfall from 1989-1996 and has been consistently regarded most highly among the UK plcs for its standards of service, in contrast to many of its neighbours.

1.6. REFERENCES

Beal D. (1988) Integrated Distribution Management Systems. Aqua, Great Britain, No. 5, pp. 267-276.

Bessey S.G. (1985) Progress in Pressure Control. Aqua, Great Britain, No. 6, pp. 325-330.

Bhave P.R. (1991) Analysis of Flow in Water Distribution Networks. Technomic Publishing Company Inc., Lancaster, USA.

Bland P., Townend J.C. (1987) An Integrated Approach to Telemetry - North West Water Authority - A Regional Telemetry Scheme. Symposium Water/Eau 2000, Nice, France, pp. 9-2-1 to 9-2-14.

Brammer L.F., Schulte A.M. (1993) A Joint US and UK Approach to Water Supply and Distribution Modelling. Journal of the Institution of Water and Environmental Management", Great Britain, Vol. 7, No. 5, pp. 471-480.

Chase D.V., Ormsbee L.E. (1993) Computer-Generated Pumping Schedules for Satisfying Operational Objectives. Jour. of American Water Works Assoc., Vol. 85, No. 7, pp. 54-61.

Chase D.V., Jones G.L. (1993) Lessons Learned from Application of Optimal Control Technology. Integrated Applications in Water Supply, ed. B. Coulbeck, Vol. 2, JOHN WILEY & SONS, pp. 377-394.

Clark R.M. (1994) Modelling Water Quality Changes and Contaminant Propagation in Drinking Water Distribution Systems: A US Perspective. Aqua, Vol. 43, Nr. 3, pp. 133-143.

Cohen J. (1990) The Development of A Dynamic Calculation Model for Drinking-Water Networks. Aqua, Vol. 39, No. 3, pp. 172-187.

Cohen J., de Visser P.M. (1992) The Development and Application of Monitoring Systems for the Distribution of Drinking Water. Aqua, Vol. 41, No. 6, pp. 352-359.

Cosgriff G.O. et al. (1985) Interactive Computer Modelling:Monitoring and Control of Melbourne's Water Supply System. Water Resources Research, Vol. 21, No. 2, pp. 123-129.

Coulbeck B., Orr C.H. (1988) Computer Applications in Water Supply", Volume 2: "Systems Optimization and Control. Reseach Studies Press, Ltd., Herts, Great Britain.

Coulbeck B., Orr C.H. (1989) Computer Programs for Real-Time Optimized Control of Water Distribution Systems. HYDROCOMP '89: Computational Modelling and Experimental Methods in Hydraulics, Elsevier Applied Science, London 1989, pp. 289-300.

Cubillo F., Mercier M. (1993) An Approach to Water Quality Modelling in the Paris and Madrid Distribution Networks. Water Supply, Vol. 11, Nos. 3/4, pp. 377-385.

Cullen N. (1987) Computing - Not Yet Half-Way There. Symposium Water/Eau 2000, Nice, France, pp. 3-3-1 to 3-3-12.

Fallside F. (1977) On-line Control of a Water Supply System. International Workshop on Instrumentation and Control for Water and Wastewater Treatment and Transport Systems, Paper No. 40, London-Stockholm, pp. 6-40-1/10.

Filip A., Obradović D. (1986) An Advanced Method for Prognosis and Control of Water Quality in Complex Water Distribution Systems. Proceedings of XVI Congress of the International Water Supply Association, Spec. Subject No.13, Rome, Italy.

Fontaine J. (1983) Towards Integral Automation in Water Distribution (in French). Aqua, Great Britain, No. 6, pp. 302-312.

Glasbrook D. (1993) The Practical Application of GIS, and its use for AMP2. IWO Conference 1993, Cardiff, Great Britain.

Grombach P. (1986) Centralized Monitoring and Control of Water Supply System of the City of Zurich: An Example of the Systematic Optimization of Operation by Computer. Colloque Eau et Informatique, Paris, France, pp. 216 - 225.

Haddon M. (1994) Ringing the changes for Meter Reading. Water Bulletin, Great Britain, No. 605, pp. 9-11.

Halpern O., Pascal O. (1987) Water Distribution System for the 3rd Millenium" (in French). Symposium Water/Eau 2000, Nice, France, pp. 1-1-1/14.

Hoogsten K.J., van der Zwan J.T. (1985) Distribution System Modelling. Aqua, Great Britain, No. 5, pp. 247-251.

Huntington R. (1984) Improvement of Cost and Performance of Distribution Systems. Proceedings of XV Congress of International Water Supply Association, Spec. Subject No.9, Monastir-Tunisia.

Huntington R. (1990) WESSEX WATER's Integrated Water Distribution Management System. Proceedings of the 5th IAWPRC Workshop, Japan.

Huntington R.(1993) Updating and Development of A Regional Telemetry Scheme. 6th IAWQ Workshop on Instrumentation, Control and Automation of Water & Wastewater Treatment and Transport System, Canada.

Jeffery J., Taylor C.R. (1993) Development in Leakage Control at North Surrey Water Ltd - A Case Study. Aqua, Vol. 42, No. 4, pp. 223-232.

Jowitt P.W., Garrett R.T., Cook S.C., Germanopoulos G. (1989) Real-Time Forecasting and Control for Water Distribution. Coulbeck B. and C. H. Orr: "Computer Applications in Water Supply", Vol. 2: "Systems Optimization and Control",Research Studies Press, Ltd., Herts, Great Britain.

Kado M., Itoh H. (1987)A Satisfactory Water Service Expert. Symposium Water/Eau 2000, Nice, France, pp. 2-1-1/20.

Lonsdale P.B. (1984) Automatic Control of Pressure in Distribution Systems. Wessex Water Co., Bristol, Great Britain.

Males R.M., Clark R.M., Wehrman P.J., Gates W.E. (1985) Algorithm for Mixing Problems in Water Systems. Transactions of the ASCE, Jour. of the Hyd. Div., Vol. 111, No. 2, USA.

Morita S., Arakawa T. (1989) Water Distribution Control System. WATER NAGOYA '89, ASPAC IWSA (1S-22-B22), pp. 626-635.

Nguyen B. (1994) Automatic Operation of the Water Distribution of the City of Paris. IWSA Regional Conference, Zurich, Switzerland, pp. 311-317.

Nguyen B., Montiel A. (1994) On-line Quality Control in Distribution Networks. IWSA Regional Conference, Zurich, Switzerland, pp. 299-309.

Obradović D., Kordić M. (1986) Studying a Disastrous Situation Before It Actually Happens. Proceedings of XVI Congress of the International Water Supply Association, Spec. Subject No. 4, Rome, Italy.

Obradović D., Filip A. (1986) The Value of Data Banks: The Case of Belgrade Water Supply System. Colloque Eau et Informatique, Paris, France, pp. 111-126.

Obradović D. (1989) Modernisation of Urban Water Supply Systems. Proceedings of the NATO Urban Water Resources Advanced Research Workshop, Isle of Man, UK, pp. 150-165.

Obradović D. (1993) Mathematical Modelling of Water Supply Systems. IHE - International Institute for Infrastructural, Hydraulic and Environmental Engineering, Delft, The Netherlands.

Rao H.S., Don W.B. Jr. (1977) ExtendedPeriod Simu-lation of Water Systems. Transactions of the ASCE, Journal of the Hydraulics Div., - part A, Vol. 103, HY-2, pp. 97-108, - part B, Vol. 103, HY-3, pp. 281-294, USA.

Robertson J. (1993) Optimum Network Management through Integration of Operations Processes. Integrated Applications in Water Supply, ed. B. Coulbeck, Vol. 2, JOHN WILEY & SONS, pp. 317-329.

Santoni A. et al. (1987) EXPERT Systems - New Tools for Water Treatment. Symposium Water/Eau 2000, Nice, France, pp. 6-1-1/8.

Schulte A. M., Malm A.P. (1993) Integrating Hydraulic Modelling and SCADA Systems for System Planning and Control. Journal of American Water Works Association, Vol. 85, No. 7, pp. 62-66.

Snoxell J.D., Obradović D., Čavor R. (1989) The Use of Real-Time Telemetry and Modelling in Conjunction for the Operational Management Water Supply Systems. Computational Modelling and Experimental Methods in Hydraulics - HYDROCOMP '89, Dubrovnik, Elsevier Applied Science, London 1989, pp. 301-310.

Snoxell J.D. (1991) New Technology in the Water Industry - Making It Stick in the Real World. Proceedings of the Pipeline Industry Guild, England.

Snoxell J.D. (1993) Network Management in Water Supply. International Conference "Computer Applications in Water Supply and Distribution", De Monfort University, Leicester, England, pp. 331-345.

Snoxell J.D. (1993) The Use and Value of GIS in the UK later Industry. Experience in WESSEX WATER Services Ltd.

Snoxell J.D. (1994) Management of a Widespread Multi- Plant Water Supply System.

Takagi T. et al. (1983) Distribution Network Control for Water Supply System. Hitachi Review, Japan, Vol. 32, No. 5, pp. 259-264.

Walski T.M. (1984) Analysis of Water Distribution Systems. Van Nostrand-Reinhold, New York, USA.

Zwan J.T. van der (1988) Computer Utilization in Distribution. Aqua, Great Britain, No. 1, pp. 9-13.

2. Object-Oriented Hydroinformatics Tools for Water Distribution Analysis

Petr Ingeduld
Vladimír Havlík
Stanislav Vanecek
Czech Technical University of Prague
Thakurova 7, 16629 Prague 6
Pavel Jiroušek
Charles University
Prague
CZECH REPUBLIC

2.1. INTRODUCTION

A notion proposed by Abbott (1991) concerning primarily the software and hardware development of simulation tools seems to have come true. New achievements in hardware and software, such as Dual-processors, high density media opens a space for new operation systems, Windows NT for example. The gap between compilers that can be used to design different modules of software, such as model management, computer graphics, database connections and the numerical algorithm itself, is rapidly deepening. The outlook of present word processors, spreadsheets, high standard level of pre-and-post processing techniques makes new demands on developers of mathematical models to change their programming paradigm. Knowledge encapsulation and integration of several tools into a numerical model may easily navigate a user through a modular part of a software. The authors would like to use the HYPRESS package as an example of possible way satisfying some of the required features.

NATO ASI Series, Partnership Sub-Series, 2. Environment – Vol. 15
Water Supply Systems – New Technologies
Edited by Č. Maksimović, F. Calomino, and J. Snoxell

2.2. WATER DISTRIBUTION

The problem of the analysis of water distribution networks consists in determining both flows in every link, such as pipe, pump, valve etc. and the piezometric head at every node in the system. Regardless of the fact that most of the recent published studies solving the water distribution problems are based on standard methods, such as Gradient Methods, Newton-Raphson method, The Linear Theory or The Optimization techniques, there exists an unique approach published by Verwey (1994) based on a Finite Difference Method. Since the idea seems to be applicable the authors expand on this technique in this Section.

2.2.1. Algorithm Based on Global Gradient Method

The steady flow in water distribution networks is based on the simultaneous fulfilment of a mass conservation and an energy conservation law. The flows in the links (branches) and piezometric heads at the nodes are interrelated via the head loss formula and the water distribution problem is restricted to determining either link flows or nodal piezometric heads.

An algorithm is based on one of the most advanced and robust technique, the Global Gradient Method with an incomplete Choleski conjugate gradient algorithm, as the base matrix is always symmetric, positive definite, highly sparse, large and Stieltjes type. This algorithm has been tested under several modes including Windows 3.1 enhanced mode, DOS 32-bit protected mode, in order to provide as short computation times as possible.

Pipe network consists of branches (pipes), nodes with unknown heads and nodes with fixed heads. It may contain hydraulic components located along pipes such as pumps and valves, but their influence is included in equations describing head loss in pipes. Nodal hydraulic components such as simple tanks, surge chambers, feeder tanks and air vessels are schematised by means of nodes with fixed or unknown heads. Steady flow in the network is achieved when both the heads in nodes and the flows in branches don't change in time.

2.2.2. Branches

There are two parameters characterising each branch, namely α and β. For a given flow rate Q be the flow in a particular branch, relationship between flow and head loss (or head difference) may be written as:

$$h_L = \alpha \cdot Q^2 + \beta \tag{1}$$

where α contains the summary information about roughness of the pipe, its length and local losses such as losses in fittings and valves, while β is an additional parameter which is non zero only if branch contains a pump.

If a branch connects two nodes with heads H_1 and H_2 and is characterised with parameters α and β, the flow Q in this branch (if the "positive" direction is from node with head H_1) must meet this condition:

$$H_1 = H_2 + h_L \tag{2}$$

where h_L is the head loss. This equation can be then expressed as:

$$H_1 = H_2 + \alpha \cdot Q^2 + \beta \tag{3}$$

2.2.3. Nodes with Unknown Heads

Each node is characterised by its consumption. Ground level does not have to be specified. Consumption can be either a positive or a negative number. Positive value means real consumption, negative one represent of a source with a fixed flow.

If there is n branches connected to a node and the the flows Q_1,...Q_n in these branches are positive or negative depending whether they are coming in or out, the mass conservation law for the nodal demand q must be satisfied:

$$q = Q_1 + \ldots + Q_n \tag{4}$$

2.2.4. Nodes with Fixed Heads

In order to solve the system of linear equations, at least one node with fixed head is required. These reference nodes (usually represented by reservoirs) can provide arbitrary flow (even keeping their water level constant). Consequently they do not have to satisfy the Equation (4) in the present form.

2.2.5. Algorithm

The recursive algorithm is used to generate an initial flow solution satisfying the nodal balance in the first iteration. The initial solution is generated by taking the identity matrix I (n=1). The linear system of simultaneous equations in the flows is coupled to the nodal balance linear equations and solved iteratively to produce consecutive flows. The explicit inverse of the matrix on the right-hand side of the equations is not needed, an iterative method based on a preconditioned conjugate gradient method can be effectively used for the solution of linear system of equations.

2.2.6. The Incomplete Choleski Conjugate Gradient Algorithm

As the coefficient matrix A is always symmetric, positive definite, highly sparse, large and it is of Stieltjes type, it seems to be the best way to solve the set of simultaneous equations using the incomplete Choleski conjugate gradient algorithm. Working with sparse matrix, only non-zero elements are stored in the computer memory to reduce the storage and computation time requirements.

2.2.7. Finite Difference Method

The domain of the finite difference method consists of branches connected one to another by means of nodes. Minimal number of grid points is generated along branches and they represent in fact the place, where governing equations are computed. Various hydraulic structures (such as simple tanks, surge tanks, feeder tanks, air vessels, valves, pumps, etc.) can be included any time at selected places of the domain (Figure 1).

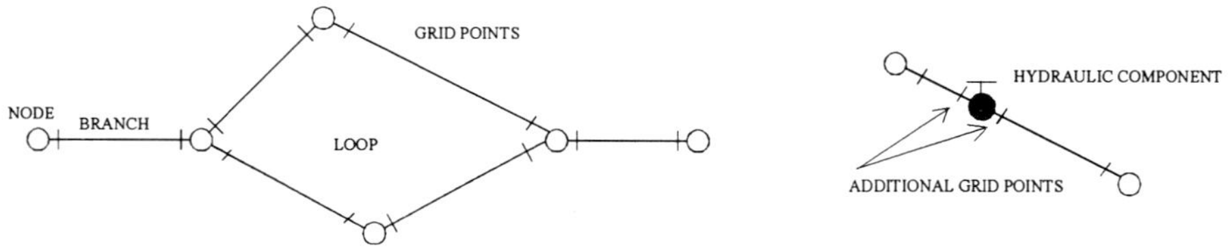

Figure 1. Discretization of network

Branches represent pipes of constant properties and may include various controls e.g. valves, pumps, etc. Nodes represent free branch ends, branch connections or a specific storage. Boundary conditions are defined at the nodes by setting up values of hydraulic gradient line or discharge as a constant value or as a function of time. For a branching junction the continuity equation is used and a common head is assumed since the minor losses are neglected. Simple tanks, surge chambers, feeder tanks and air vessels are located in the nodes. The system requires a computational grid to define small sub-sections on a branch over which the governing equations are numerically solved.

The governing equations describing one-dimensional unsteady pressurised flow in a pipe are the continuity and momentum equations:

$$\frac{\partial Q}{\partial x}+\frac{gA}{a^2}\frac{\partial H}{\partial t}=0 \tag{5}$$

and the momentum equation has been modified as:

$$\alpha_m\frac{\partial Q}{\partial t}+gA\frac{\partial H}{\partial x}+\frac{f}{2DA}Q|Q|=0 \tag{6}$$

where: Q discharge [m^3/s], H- piezometric head [m],

f Darcy-Weisbach friction factor [-],

D diameter of the pipe [m],

A area of the pipe [m^2],

x, t space and time co-ordinates [m], [s],

α_m momentum diffusion coefficient.

Governing Equations (5, 6) are solved by Preissmann implicit finite difference scheme. The space centred scheme is stabilised by the weight parameter θ equal to unity and high value of α_m - momentum diffusion coefficient. The value of α_m has its signifficant effect especially during the first several steps of fiding the equilibrium. The initial conditions are usually set as follows: zero discharges and constant pressure head corresponding to the highest pressure head in the system (including pump characteristics).

During the first ten time steps the boundary conditions achieve the desired value of discharges and pressure heads at the system boundaries.

The loop algorithm is based on the fact, that a loop network contains elements known as nodes which represent the confluence of several flow paths, some of which originate from other nodes, some from boundary points. A system of simultaneous linear equations is developed where the piezometric head H changes at each node serve as the only unknowns. The solution of this system by any matrix elimination technique yields the H values at each node. The H and Q values at all intermediate grid points between the nodes are expressed in terms of H variation at the nodes.

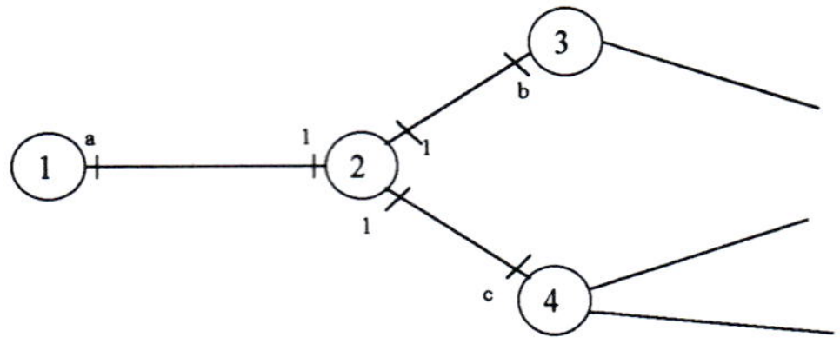

Figure 2. Scheme of computational grid

Supposing, that there are three links, 21, 23 and 24, along branch 23, *b* grid points and *c* grid points along a link 24 (Figure 2.). For any computational grid point the Equations (5, 6) may be written as:

$$H_i = L1_i H_1 + M1_i H_{jj} + N1_i \quad (7)$$

$$Q_i = L2_i H_1 + M2_i H_{jj} + N2_i \quad (8)$$

where L, M, N are functions of coefficients $(A, B, C, D, E)_{1,2}$ found through a double sweep elimination.

These Equations express the partial dependence of the unknown variables Q and H at any grid point on a branch on the value of H in the two adjacent nodes.

At internal nodes a compatibility condition must be satisfied. The most simple one is that of continuity and common hydraulic piezometric heads:

$$\sum_{k=i}^{m} Q_{ik}^{n+1} = 0 \quad (9)$$

$$h_{i1}^{n+1} = h_{i2}^{n+1} = \ldots = h_{ik}^{n+1} = \ldots = h_{im}^{n+1} \tag{10}$$

where n+1 indicates the (n+1)Δt time level in the solution, *k* is index of the links which emanate from the node *i*, and *m* is number of links. These relations can be written for each node of M nodes and leads to a system of M linear equations having as unknowns piezometric heads at each node.

$$[S]\,\{h\} = \{T(_L)\} \tag{11}$$

where [S] is a coefficients matrix, M x M elements, {h} is a vector of unknowns, M elements; {T($_L$)} is a vector of the free terms, M elements.

This system of linear equations may be solved by any matrix inversion techniques. Once the increments of piezometric heads H are known at the nodes, it is possible to recompute Q(i) and H(i) values for all intermediate grid points by means of Equations (7, 8).

The loop algorithm may be expressed by the following steps:

- the coefficients of the implicit Preissmann scheme discretize the governing equations between two successive grid points on a branch;
- local elimination method is used to express Q and H grid point values on each branch in terms of H at the branch ends (nodes);
- one equation for each node leads to the system of linear equations that is solved by a matrix elimination method;
- substitutions inside the branches yield the values of Q(i) and H(i) values for all intermediate grid points from the known values of H at the branch ends.

The solution of this system by fast topology-based searching technique (Verwey, 1994) yields the H values at each node. The required H and Q values at all intermediate grid points between the nodes are expressed in terms of H variation at the nodes.

2.3. DESIGN OF A NUMERICAL SIMULATOR

The modern software package architecture should consists of several independent modules that are coupled together by means of database and internal data files. The data pre and post-processing is based on a storage data in a standard database that gives QBE (query by example) and/or SQL (structured query language) options to classify groups of pipe diameters, roughness, wave celerities. The same facility may be used for the visualisation of computational results, such as pipe discharges, velocities etc. The graphical user interface may be designed on the basis of multi document interface. This implies, for example, that during editing of data the changes are visualised immediately in graphical windows. All stored data in the data base can be imported or exported in ASCII format. This can always be neatly organised for standard data, such as values associated with the description of pipes and nodes. The output processing part covers the most important features as time series of longitudinal profiles of pressures, discharges and pipe velocities with optional envelope drawing. Colour animation; time series plots of the same variables with interactive graphical selection of point location; colour animation of time series in the horizontal plan; zooming of any graphical screen and co-ordinate reading on any function or network plot by means of the cursor.

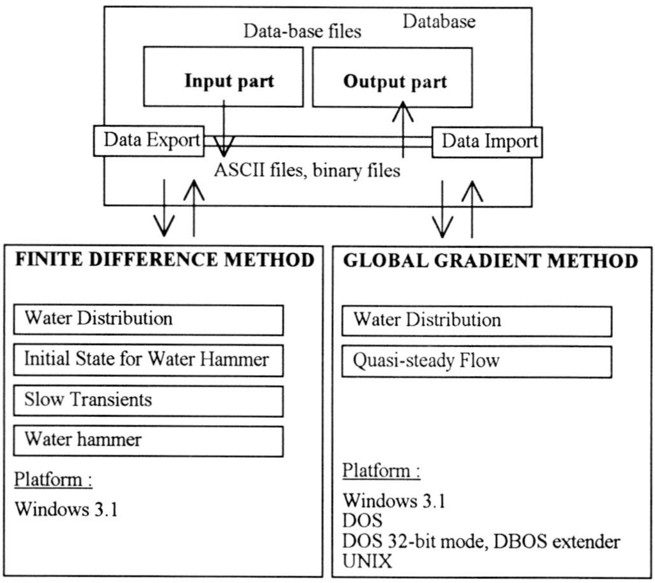

Figure 3. Software architecture

Computation models can perform water distribution calculations, slow transients and water hammer simulations. The models are independent programs that read data from internal data files prepared by data preprocessing part. The results are stored on hard disk in ASCII and binary files. These files are transferred into the output module. Software architecture with respect to computational models is shown in the Figure 3. describing an example the HYPRESS architeture.

This implies, for example, that during editing of data the changes are visualised immediately in graphical windows. The data post-processing part covers the most important features, such as:

- time series of longitudinal profiles of pressures, excessive-pressures, discharges and pipe velocities with envelope drawing and their colour animation;

- time series plots of the same variables with interactive graphical selection of point location
- colour animation of time series in the horizontal plan;

- zooming of any graphical screen and co-ordinate reading on any function or network plot by means of the cursor.

2.4. THE HYPRESS EXAMPLES

2.4.1. Loop Network

The following example shows the layout of a proposed pipe loop network for a fire fighting system (Figure 4.). Under conditions with zero hydrant discharge, a pressure of approximately 10 bar is maintained in the system by small pumps. In the event of a fire (one fire at a time is assumed) an electric duty pump supplies the demand flow to any of the hydrant points.

The steady-state head-discharge characteristics of a centrifugal pump has been approximated as $H = 101.9 - 875\ Q^2$. The rated speed is 1000 r.p.m. The pump sump is connected to one of the nodes. The results of computation for a steady state discharge $Q = 0{,}035\ m^3/s$ with flow distribution within the pipe network is indicated by arrows in Figure 4. Pump operating point is $Q=0.034\ m^3/s$ with the corresponding head $H=108$ m. The steady state calculation gives the initial conditions for a transient flow analysis which has to solve the pump trip, opening and closing of the hydrants, proper behaviour of a check valve and the air vessel performance.

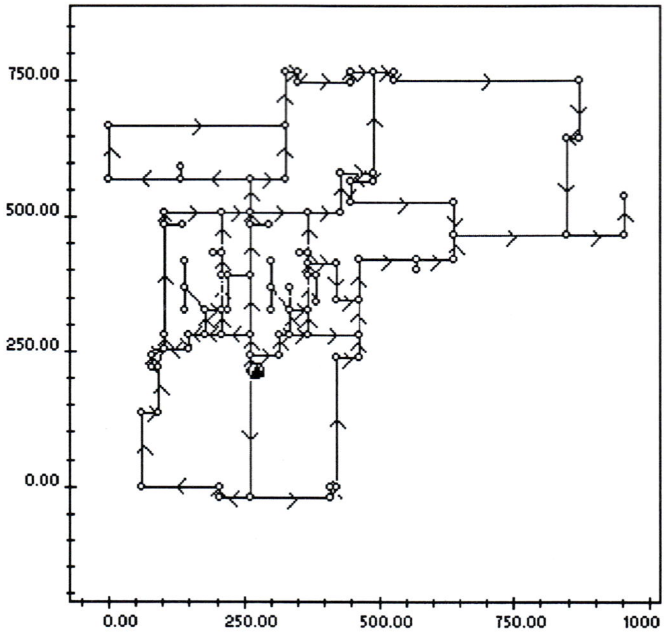

Figure 4. The layout of a proposed loop network

2.4.2. Large Networks

Algorithm was successfully tested on large pipe networks with several thousands of nodes and pipes (Figure 5.). The computational requirements were significantly lower using the Global Gradient Method than in case of numerical process based on the unsteady flow algorithm. Since both these algorithms have been implemented under different programming languages it is very difficult to get comparable results.

The Global gradient method with the Incomplete Choleski Conjugate Gradient Algorithm seems to be more robust and has a faster convergence. Basic tests using the Water Distribution Models have been caried out on the existing network topology of 2304 branches and 2255 nodes.

Several test covering different boundary and initial condition have been computed to demonstrate HYPRESS ability to simulate the staedy flow. The total length of the pipe branches was 74938 meters, the shortest pipe was 1.00 m long and the longest one was 236 m long; diameter varied between 0.10m to 1.00m. There were 574 boundary nodes with a given pressure head and/or discharge type of the boundary condition.

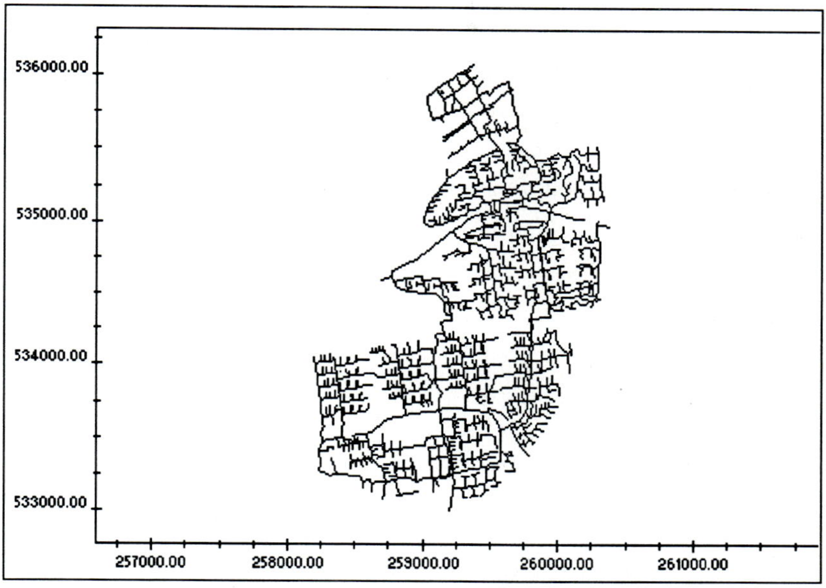

Figure 5. Large network with several thousands of nodes and pipes

2.5. COUPLING GIS WITH A NUMERICAL SIMULATOR

Numerical models for solving water distribution analysis are based on a certain topological schematization. The solution domain consists of branches connected one to another by means of nodes and of various hydraulic structures (such as simple tanks, surge tanks, feeder tanks, air vessels, valves, pumps, etc.) that can be included at selected branches or nodes. The models usually have their internal database in which the data are stored together with results of hydraulic calculations. Most of the models support import and/or export of ASCII data. Hence the data can be presented in specialized graphical packages or can be imported into geographic information systems. GIS model can be efficiently used for data preparation and verification and for data post-processing and visualization. Input part for the model may cover creating of digital model of terrain, map plans, digitizing, scanning and vectoring of data, creating cross-references for the numerical model etc. Output part may use GIS as a graphical data processor for displaying results from numerical simulator in the form of graphs, maps, thematic maps, 2D, pseudo 3D and 3D images, browsing results, using of grid-analyst tools to work on the computed data and for geocoding of data.

GIS commands for selecting objects with the help of querying, selecting and SQL Select enable the preparation of input data for the model and the analysis of the results. Using SQL Select it is possible to create query tables containing information which was only implicit in the data base tables. SQL utilisation in combination with other basic units greatly improves the efficiency of the application of simulation packages.

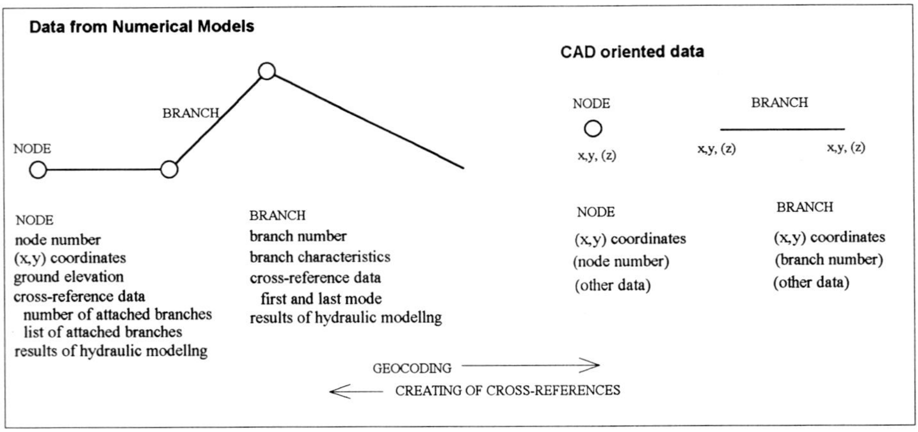

Figure 6. Two different types of the input spatial data describing a model topology

There are two different types of the input spatial data describing a model topology, we can distinguished between CAD oriented data models and data from numerical models (Figure 6.). CAD models usually work with entities such as points, lines, polygons and surfaces. Each object has its own spatial co-ordinates and the objects are stored and visualised independently. Descriptive data attached to these objects may contain attributes such as colour, shape of the objects or a data base attributes such as number of en element, values from catalogues etc. On the other hand, data from numerical models do not necessarily have spatial co-ordinates (x,y co-ordinates are usually used for pipe network visualisation but not really for the simulation itself), but ground elevation data may be included. There are several types of cross-reference data describing the pipe network topology (start and end node attached to a branch). Data exchange between these two models requires creating of additional operations such as geocoding, creating cross-references and data pre-processing. As a result the data can be either used in CAD systems and visualised with respect to their spatial co-ordinates or the same data can be used as an input into a numerical model.

Raster data are mostly used for creating an intelligent background of the map layers. The map layers contain additional information that neither is stored in a database nor available in a vector form. Raster data can describe descriptive data into more detailed level, image of a certain place of the pipe-network, schematic pictures, etc.

The choice of GIS modules should respect the specific requirements of the end-user, the type of data that will be maintained and the size of the projects. It is possible to distinguished between 'small' and 'large' GIS because the hardware and software requirements (the choice of GIS modules, database, hardware platform) limit the project.

The coupling of GIS with numerical models can be one of the following three types.

- GIS does not exists: In case that there is no GIS in the beginning of the project, it is possible to develop own GIS and use it both for data preparation, verification, analysis and for presenting of the hydraulic modelling results.

- GIS exists, but does not include layers with data we need for hydraulic modelling. In this case the existing GIS is a source of general information. We need to develop additional layers and data attributes and this layer is later on included into GIS.

- GIS exists and contains a layer with data we need. Then GIS is a source of data, there is a structure already created to contain results from the hydraulic model and the structure may be only slightly modified if necessary.

The Figure 7. shows a map consisting of different layers; pipe networks (drawn by lines between two points with known x,y co-ordinates), manholes (a symbol on x,y position), several layers with map information such as block of flats, green areas, river (drawn by the means of lines, polylines and areas). Despite of their common cross-references, the individual objects on the map are drawn independently. Since the descriptive data attached to graphical objects, the database information can be displayed interactively on the screen or stored in a browse table. Each graphical objects can be shaded by a value corresponding to the value or function of its descriptive or graphical attribute, for example the object length as shown on the picture. Descriptive data without any graphical representation can be geocoded, i.e using a common attribute they can be assigned to an existing geographical data and can be visualised on that data. It is also possible to create join tables or to aggregate data.

A powerful SQL language gives the possibility of selecting objects satisfying certain conditions. It is possible to display pipes within a given length range or with a diameter higher than a given value etc. Moreover, geographical operators "within", "intersects", "contains", "contain entire" can query data accordingly to their geographical location. Hence it is possible to display intersecting pipes, manholes that are located within certain street, etc. The results of SQL queries are stored in separate tables and they can be later used to perform desired operations, or for creating a new mapper. Graphs and simple statistical functions are also usually a part of GIS modules. Data is mostly stored in an internal or external database and can be imported or exported in several data types. ASCII and xBASE files describing both geographical and descriptive data are almost always supported, most systems usually recognise DXF files for graphical data.

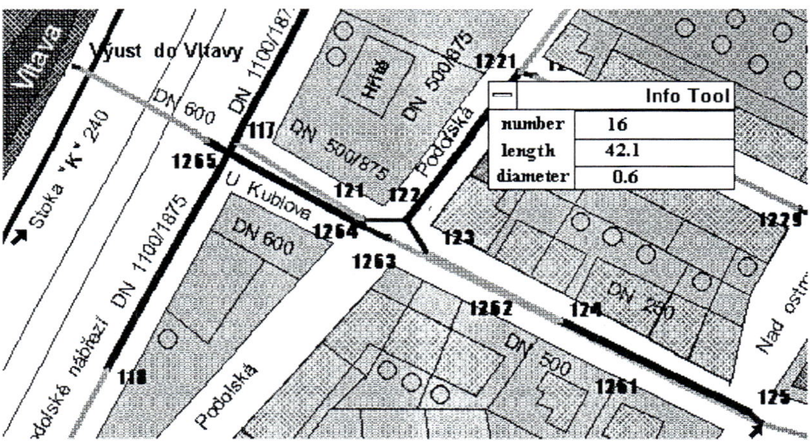

Figure 7. Map with different layers

The Figure 8. shows a connection between Digital Elevation Model and results of water distribution analysis. A part of pipe network consists of several pipe branches, nodes and a control valve. The results of water distribution model are pressure heads in nodes and flow rates in links. Digital Elevation Model (DEM) can create a terrain contour lines or can be used as well to create pressure isolines from the known pressure heads in the nodes. This additional information is drawn together with pipe network on a map and graphical objects-polylines may be linked to descriptive data of corresponding pressure head. It is possible to use SQL tools to query pipes and nodes according to the pressure isoline using standard GIS SQL tools. Coupling of DMT with a numerical simulator can bring a different view on computed results and can help user to carry out water distribution analysis. Different GIS modules as statistical tools, grid-analysts, databases and data visualization packages have very powerful general tools that can extend the use of numerical models and support analysis of the results.

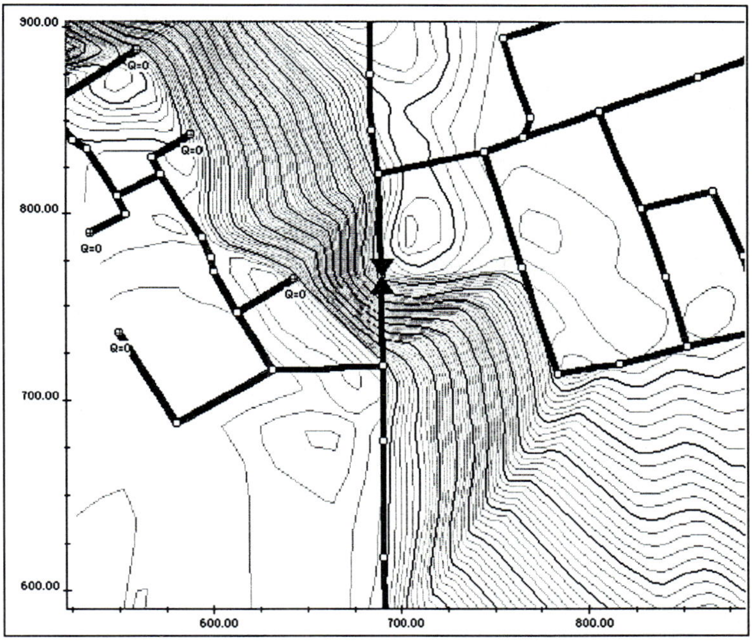

Figure 8. Connection between DEM and result of water distribution analysis

2.6. CONCLUSIONS

A modern software development is nowadays based on the technology of object linking and embedding, dynamic exchange of data between different applications, unifying database platform and user-friendly Application Programming Interface. The present technology of geographic information systems based on the visualisation of information gives new possibilities to handle an increasing amount of data. GIS as a part of hydroinformatics systems have the potential to perform a complex data analysis. Data visualization techniques and analytical tools can improve the everyday work of engineers, scientists and researchers. The link between these models and numerical simulators supports integration of modelling systems, knowledge-based tools and information systems.

A complex package for water distribution analysis has to cover hydraulic design and operation of the pipe networks. The software has been designed on the basis of interactive graphics support to data editing through a multi document interface. During editing, all logical and topological data types are processed immediately and, wherever appropriate, visualised. The information system on a pipe network is based on the storage of data in a standard database.

The numerical models can solve on a pipe network both water distribution and hydraulic transients including real time control emulation on a variety of hydraulic components such as pipes, pumps, valves, reservoirs, surge chambers, feeder tanks and air vessels. Two different numerical algorithms have been used for the steady-state simulation, a numerical solution based on a finite difference scheme with minimal number of grid points and an alternative algorithm based on The Global gradient method with an Incomplete Choleski conjugate gradient method, with examples demonstrated on the HYPRESS package.

2.7. REFERENCES

Abbott M.B. (1991) Hydroinformatics: Information technology and the aquatic environment. The Avebury technical, Aldershot, UK and Brookfield, USA.

Ajiz M.A., Jennigs A. (1984) A robust incomplete Choleski conjugate gradient algorithm. International Journal for numerical methods in Engineering, Vol. 20

Deakin R. (1994) Data to information : GIS and decision support for coastal engineering. Hydroinformatics '94 Delft, 1994, Balkema, Rotterdam.

Fabri A.G. (1992) Spatial Data Analysis in Raster-based GIS: An Introduction to geometric Characterization. In Introduction to GIS edited by C.R.Valenzuela, ITC, 1992.

Kershaw D. (1987) "The incomplete Choleski-conjugate gradient method for the iterative solution of systems of linear Equations", Journal of Computational Physics, 1987

Maksimović Č., Prodanović D., Elgy J., Fuch L. (1994) GIS (or GIM) in water projects - Tools or Toys. Hydroinformatics '94 Delft, 1994, Balkema, Rotterdam.

Salgado R., Todini E., O'Connell P.E. (1987) Comparison of the Gradient Method with some Traditional methods for the Analysis of Water Supply Distribution Networks. Leicester.

Tullis J.P (1989) Hydraulics of Pipelines. John Wiley & Sons, 1989.

Valenzuela C.R. (1992) Basic Principles of Geaographic Information Systems. ITC.
Verwey A., Yu J.H. (1993) A Space-Compact High-Order Implicit Scheme for Water Hammer Simulations. Proceedings of XXVth IAHR Congress, Tokyo, 1993
Verwey A. (1994) Linkage of Physical and Numerical Apects of Models Applied in Environmental Studies. Proceedings of the Conference on Hydraulics in Civil Engineering, Brisbane.
Tomplin C.D. (1990) Geographic Information Systems and Cartographic Modeling. Prentice Hall.

Chapter VII

Transient Flows

Chapter VII

Transient Flows

INTRODUCTORY STATEMENT

The transport of water in pipes normally involves some sort of control and, the imposition of certain constraints over the water flow. As the size of the water supply system grows, the close control of flow becomes more necessary. The system is designed to meet required demand profile, while on the other hand, water treatment and pumping could be carried out according to some other (different) schedule. Adjustment of system parameters by valve control and pump switching is required all the time.

In standard design procedures it is normally assumed that the system operates under steady state conditions and that transient flow (unsteady flow) occurs when the system is in transition between two steady state flows. Every change of flow rate in a pipe is associated with a pressure change, the intensity of which is related to the rate of flow change. Slow transients usually cause no problems to system operation, but fast transients, known as water hammer, may cause serious problems.

The design process of a water supply system normally includes an appropriate hydraulic analysis of all possible transient flow events and risk assessment of equipment damages and system malfunctioning. Pumping stations have the main role in transient flow analysis. Transients caused by them are usually the most severe, as boundary conditions in mathematical modelling they are the most complex, and the equipment in pumping stations is that which is hit initially by transients. The great variety of system layouts and pump arrangements in pumping stations (single, parallel, series) do not allow general predictions to be made about system behaviour in transients caused by pumps.

There are three classes of transient flow events that are important for a designer:

- **Normal events**, i.e., events under normal operational conditions such as: correction of flow or pressure by control valves, adjustment to changes in consumption, deliberate switching on/off of a pump, etc.

- **Emergency events**, i.e. pump power failures, malfunction of control devices, pipe bursts etc.

- **Catastrophic events**, i.e. earthquakes, strong explosions etc.

Normal events are not expected to cause significant pressure changes and to disturb the normal operation of the system. They can be planned and expected. However, their significance is usually overlooked especially in complex systems with several pumping stations.

The analysis of transient flow events in complex water supply systems with pumps, valves, air chambers, sophisticated control devices, etc., can be carried out efficiently only by computer simulations. Along with well established methods of transient flow analysis based on the method of characteristics, there are some attempts to use methods developed in other areas of Hydraulics and using new informatics tools. A method based on a high order compact implicit scheme and object-oriented programming has been explained in one of the Sections of this Chapter.

1. Pumping Stations and Transient Flow Analysis

Marko Ivetić
IRTCUW, Insitute of Hydraulic Engineering
Faculty of Civil Engineering
University of Belgrade
P.O. Box 895, 11 000 Belgrade
YUGOSLAVIA

1.1. INTRODUCTION

The determination of the capacity, location and configuration of a pumping station has to be based on a detailed analysis of the proposed system. Factors like forecasted demand, available supply, function of the pumping station in the water supply system etc., have to be thoroughly considered. Due to these requirements pumping stations in water supply systems (WSS) almost exclusively have more than one pump, and lots of other devices and equipment required for correct operation. In this analysis, however, only the schematic representation of pumping stations will be used (for example, Figure 1.) and equipment relevant for mathematical modelling of transient flow will be indicated.

There is a lot of possible arrangements of pump units within a pumping station designed to meet specific system requirements (parallel or series operation, variable speed units etc.). The main task of a pumping station is to increase fluid pressure downstream, but also it should be done at minimum cost and with an accepted level of reliability.

Regarding the suction side there are two main types of pumping stations. Pumps may take water from an open surface reservoir, or from a relatively long suction line. In the first case suction pressure varies in a narrow range regardless of the operating regime of the station. If the pumping station and intake structure are located within a surface reservoir or in a well, vertical turbine pumps with columns extending down into a suction reservoir are usually used. Classical solutions are pumps that are taking water from a nearby reservoir through a relatively short suction pipe. In this case, horizontal centrifugal pumps are usually used.

The other group of pumping stations is typical for use within distribution systems. They are so-called booster pumping stations. Suction pressure may vary across a much wider

NATO ASI Series, Partnership Sub-Series, 2. Environment – Vol. 15
Water Supply Systems – New Technologies
Edited by Č. Maksimović, F. Calomino, and J. Snoxell

range, which is dictated by the part of the system upstream of the pumping station. Consequently, the question of efficient and reliable control becomes more difficult and important than in other cases.

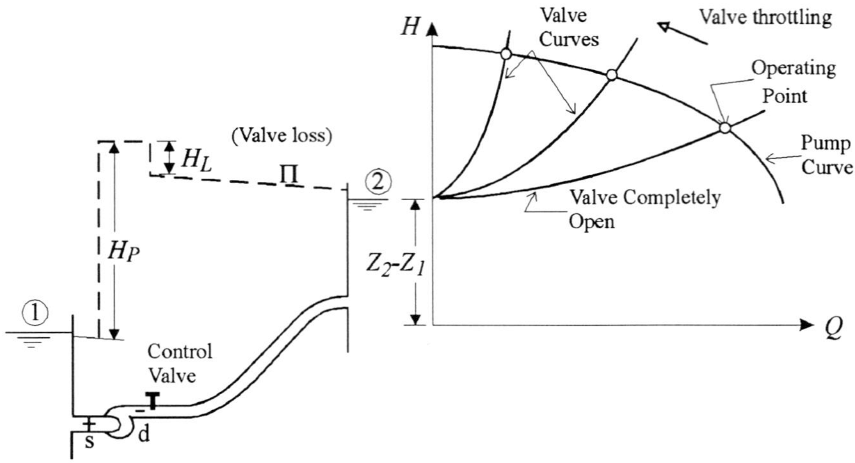

Figure 1. Schematic of a hypothetical pumping station, pump curve and flow control by throttling

Water supply systems are continuously growing in size and in complexity. New sources of drinking water are more and more distant from water consumption area. Requirements for efficient control and reliable operation become more restrictive all the time.

Pumping stations serve in water supply systems which are always in transition from one steady state to another. Some transient flow situations (actually, the most dangerous ones) are caused by pumps and it is necessary to be fully aware of their behaviour in some irregular flow regimes and possible consequences.

1.2. PUMPS

The main elements in any pumping station are pumps and this is by itself, a wide topic (the reader may refer to several textbooks, like Stepanoff, 1957, Karassik et al., 1976, Cherkassky, 1990, Dickenson, 1988, or those provided by big pump manufacturers).

The pump motor, through the torque applied on the pump shaft, delivers energy to the impeller to cause flow through the pump and to increase pressure from the suction flange (s) to the discharge flange (d) (Figure 1.).

In normal operation, the motor torque is balanced by reactive hydraulic torque of the fluid pumped. The total dynamic head, H_p, i.e., energy increase per unit weight of fluid is equal to

$$H_p = \Pi_d - \Pi_s + \frac{V_d^2}{2g} - \frac{V_s^2}{2g} \tag{1}$$

where, Π_d and Π_s, are piezometric heads or elevations of hydraulic grade line on each side of the pump. For a system with equal pipe area at (d) and (s), H_p can be measured as a pressure head difference by a differential manometer, provided that the suction and discharge flange are at the same elevation. A typical head-discharge relation, or pump curve, is given in Figure 1.

For the reservoir-pump-valve-pipeline system in Figure 1. energy equation from reservoir 1 to reservoir 2, yields the total pump head, H_p in terms of pipe and valve characteristics

$$H_p = Z_2 - Z_1 + \left(\Sigma\xi + \xi_L + \lambda \frac{L}{D} \right) \frac{V^2}{2g} \tag{2}$$

The lift, or static head, is a difference in elevation between supply level and the discharge level, $(Z_2 - Z_1)$, and the rest is due to the system resistance. λ is the Darcy-Weissbach friction coefficient, and ξ_L is the valve head loss coefficient. For a constant value of, λ , Equation 2. can be expressed as a parabolic function of Q

$$H_p = Z_2 - Z_1 + RQ^2 \tag{3}$$

This is a traditional way to represent graphically the dependance of required pumping head on system losses and static head, which make total system head.

Operating point of a pump or, of a pumping station, is at the intersection of pump characteristic and system head curve.

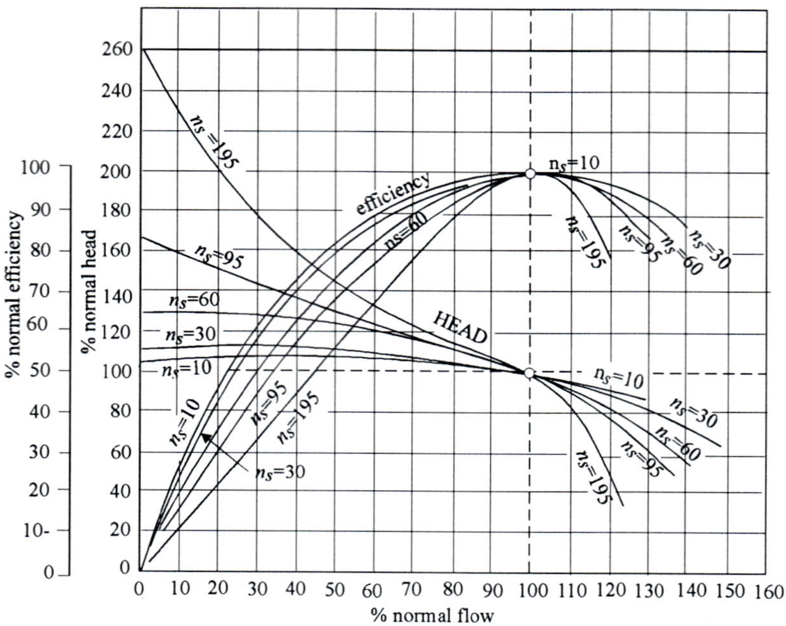

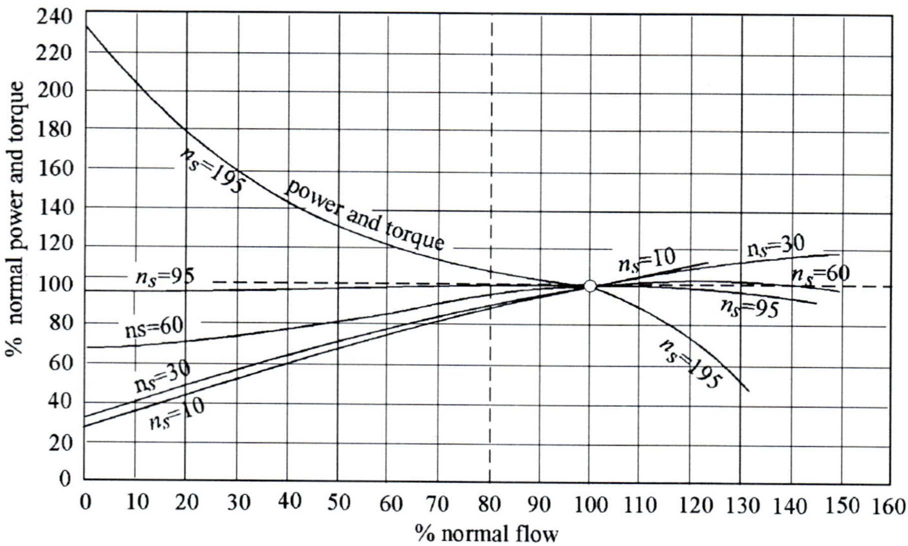

Figure 2. Standard pump characteristics

In addition to H_p and Q_p there are other quantities and performance characteristics related to a pump: namely, shaft power P, rotational speed n, efficiency η, and some other, non-dimensional parameters. Pump efficiency can be expressed as follows

$$\eta = \frac{\rho g Q H_P}{P} = \frac{\rho g Q H_P}{T\omega} \tag{4}$$

where T is the shaft torque, and ω is the angular velocity of the shaft (in rad/s). Instead of ω, rotational speed, N (in rpm), is also used.

Various types of pumps are classified by their specific speed, n_s,

$$n_s = \frac{N\sqrt{Q}}{H_P{}^{0.75}} \left(= \sigma \frac{30(2g)^{0.75}}{\pi^{0.5}} = 157.8\sigma \right) \qquad \text{in SI units} \tag{5}$$

where, N is in (rpm), Q, pump discharge in $(m3/s)$ and, H_P, pump head in (m) n_s is an old, dimensional definition, still widely used, while σ is non-dimensional.

Several researchers have provided standard Q-H-P diagrams for pumps depending on their specific speed. Those diagrams result from the statistical analysis of several centrifugal pumps. They cannot replace true pump characteristics, because they are not based on dynamic similitude (so-called homologous relations, which are discussed in the following sections). However, they do provide the basics for understanding pump behaviour in steady state analysis. For example, radial-flow pumps (specific speeds lower than 50) develop relatively small shut-off head and they are usually started with the downstream valve closed, while axial-flow pumps (specific speeds, n_s above 120) develop relatively high shut-off head and cannot be started against a closed valve.

The most reliable information can be obtained by measurements in situ, but unfortunately, this is not always possible.

1.3. MATCHING PUMPS TO SYSTEMS

Under constant flow conditions and where pressure requirements do not change, it is easy to select appropriate pump(s) because the pump needs to match the system at a single duty point. In complex systems, including all relevant regimes, water level fluctuations in reservoirs, forecasted system development, roughness changes due to aging of pipes, a pump is required to match the system at a number of different operating points. There is no unique system head curve and no single duty point. It is *the operating range* of a pump, which has to be defined.

Furthermore, there is a tendency towards local control of pumping stations (Record, 1988), which is based on the information available in the vicinity of the pumping station site. Instead of the system head curve in the sense of Figure 1. we may speak about pressure head required by the user at the control (monitoring, or, critical) point of the network (corresponds to the level Z_2) and about the discharge pressure head required from the pump due to system losses between the pump and the monitoring point.

These analyses have to be done by computer simulation, though for basic considerations the concept of system head curve will suffice.

NPSH (Net Positive Suction Head) requirements.

Also, it is necessary to check conditions on the suction side and to determine the net positive suction head (NPSH) available. The required NPSH of the pump has to be smaller then NPSH available, otherwise, it will result in very low pumping efficiency and in cavitation of the pump.

1.3.1. Operating Efficiency

The efficiency of the pump operating under normal conditions depends on how close the system requirements are to the pump characteristic curve. It is important to select a pump that will have its best efficiency within the operating range of the system and at the condition the pump will operate most often.

In Figure 1. three parabolas are constructed, each corresponding to one valve position. The vertical axis, zero flow condition, corresponds to the closed valve ($\xi_L = \infty$). The intersection of the parabolas with the pump curve ($H_P - Q$) represents the operating point for each valve position. This is a common method of flow control, and probably the least economical.

Typically for water supply systems (WSS), reduced flow is associated with reduced pressure requirements. It is not acceptable to leave the consumers to control pressure in the distribution network (high demand $\rightarrow$ low pressure, low demand $\rightarrow$ high pressure), because excess pressures will generate increased leakage.

Pumps operating at lower flow rates (high pressure) consume more energy (lower efficiencies) and their lifetime is shorter. Sometimes, a solution (not the cheap one) can be a pump equipped with control valve and bypass line leading to the suction reservoir. There are ways to overcome these difficulties and there is a price to be paid.

1.3.2. Reservoirs Serving as Buffers

With storage available in a downstream reservoir, pumps can be switched on and off so that pumps always operate at the same fixed operating point. This is not always acceptable because it requires larger reservoirs and may generate severe, and frequent, transients. Instead, multi-pump schemes and/or variable speed units are used.

1.3.3. Parallel and Series Operation

The installation of multiple pumping units operating in parallel or in series, enables more efficient use of pumps and allows finer, yet discrete, transitions in flow, whenever required. Parallel installations are more common in water supply systems (Figure 3.).

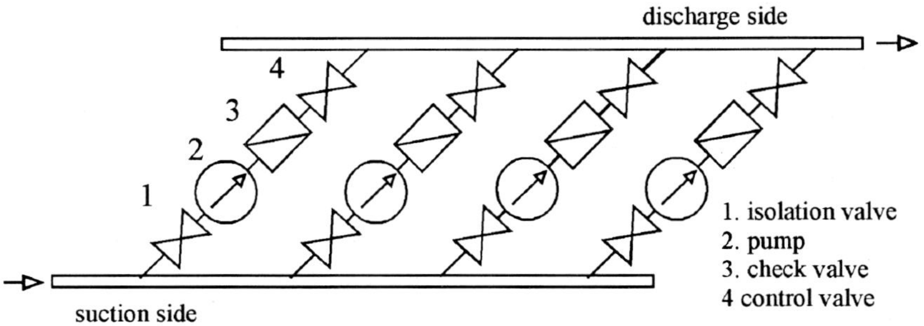

Figure 3 . Four pumps in parallel operation

Pumping station characteristics can be determined by adding the individual pump discharges associated with any particular head. Obviously, every additional pump in operation increases the discharge of pumping station. However, the operation of two identical pumps in parallel will not result in a discharge which is equal to doubled capacity of one pump, because the system head curve will increase. In Figure 4. the operating points of one pump in a 4-unit pumping station are indicated. Interval (a) is the operating range of the pump indicated as No. 1, while (d) is the operating point of a pump, when the fourth pump is switched on. Note that the pump No. 1, when operating alone, uses maximum power (this is true for centrifugal and mixed flow pumps, which are more common in WSS) - and may operate in the zone of dangerous cavitation (high NPSH required by pump).

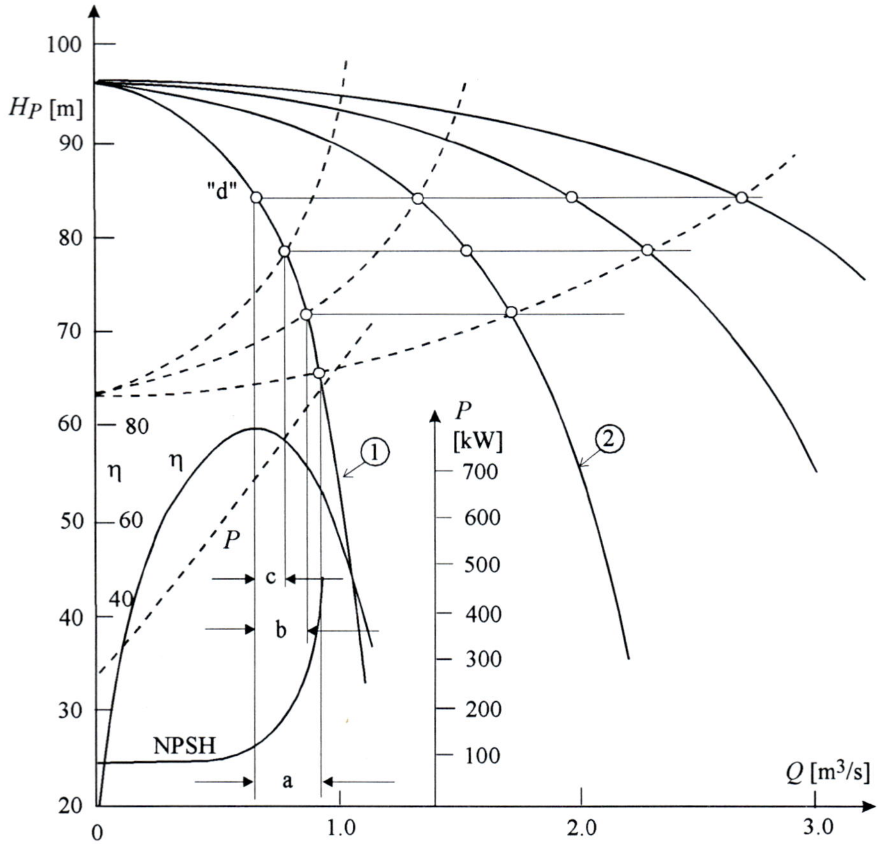

Figure 4. Operating range of unit no. 1 in a 4-unit pumping station

In determining pumping capacities of pumps in series operation (Figure 5.), pumping heads are added (Figure 6.). They are used in systems with steep system head curves, and also in booster pumping stations with a wide range of pumping heads. Two identical pumps with capacity Q_1 at H_1 of system head, will, if working in series, discharge Q_1 at $2 \times H_1$ of system head.

1.3.4. Variable Speed Applications

The flexibility and efficiency of a pumping station is significantly increased by adding a variable speed pump. The major disadvantage is the high capital cost of the variable speed device. The correct values of flow, head, power and efficiency could be determined only by tests. However, this is an area where model similarity is used with the highest confidence. The affinity laws give an acceptable answer to the question, through well known relations:

- flow: $Q / Q_0 = N / N_0$,
- head: $H_P / H_{P0} = (N / N_0)^2$,
- power: $P / P_0 = (N / N_0)^3$.

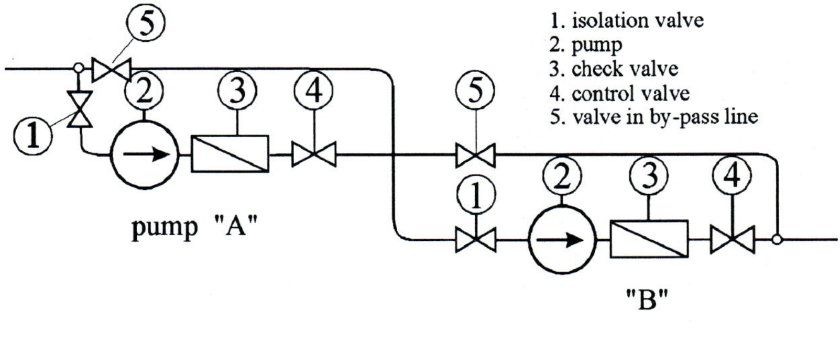

Figure 5. Two pumps operating in a series

The efficiency remains much the same, regardless of the rotational speed N (provided that it remains between 0.7 and 1.0 of the rated speed N_0).

The efficiency of this kind of flow regulation can be seen in Figure 7. where flow reduction from 55 l/s at operating point A, to 25 l/s is required. Pump characteristics are shown for three speeds, 1200, 1500 and 1800 rpm. Dashed lines are affinity lines, and lines with constant efficiency. v is non-dimensional flow rate (Q / Q_0), and, $\alpha = N / N_0$, non-dimensional rotational speed. The pump performance is compared with a fixed speed unit of 1500 rpm (although, the advantage over a higher speed unit is much more pronounced).

By throttling the system head curve can be changed, and the operating point would be C. However, with a speed change to 1200 rpm, the new operating point would be B. The difference between C and B is the energy loss caused by throttling and probably improved pump efficiency.

Also, this pump can cover higher flow rates without any problems (there will be some shift towards higher flow rates and lower efficiencies). This is a question of economy and requires careful consideration, because variable speed units are more expensive than fixed speed ones.

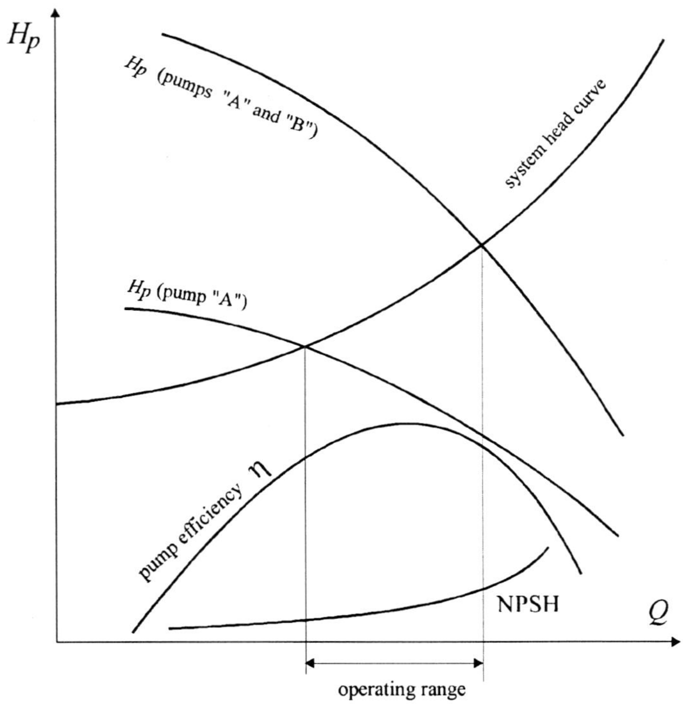

Figure 6. Operating characteristics of two pumps in series application

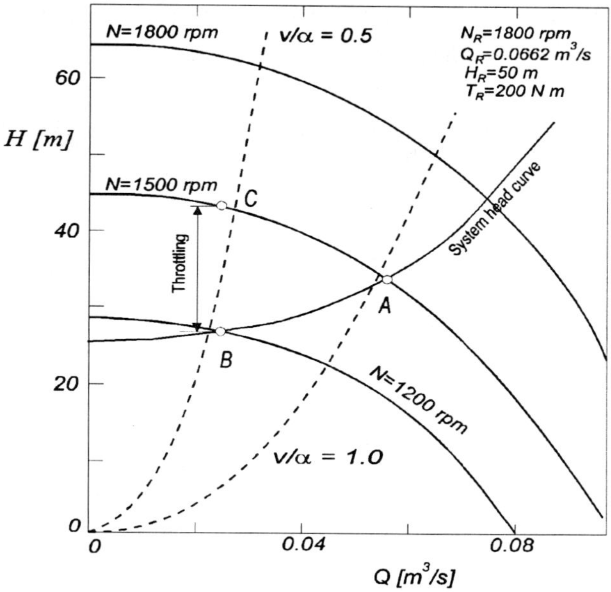

Figure 7. Operating characteristics of a pump for three rotational speeds

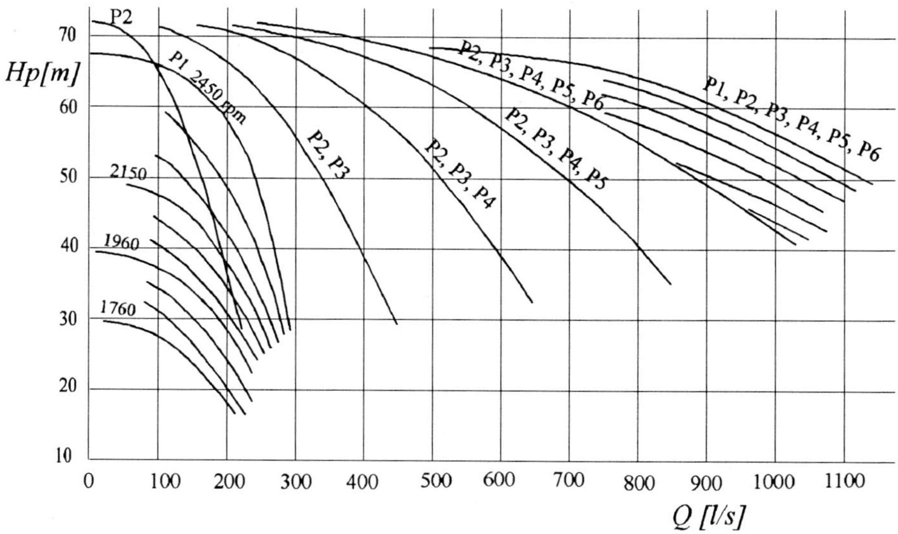

Figure 8. Operation of one variable speed pump in parallel with several fixed-speed units - Pumping station "Schwechat", Austria

The best way to control the flow in a multi unit pumping station is to combine one variable speed pump with several fixed speed ones, as in the case in Figure 8, where pump P1 is a variable speed unit.

1.4. TRANSIENT REGIMES

The analysis of transient regimes in systems with pumping stations is very important because of the great destructive forces associated with these phenomena, known as water hammer. This can not be achieved without reliable data on pumps in all possible operating regimes. In transient regimes pumps interact with flow in a special way and also, they may operate under very unfavorable conditions, which can hardly be identified in steady state analysis.

According to the frequency of occurrence, transient events can be classified into three groups:

1. *normal,* which occur very often, like pump start-up, flow adjustments etc.

2. *irregular*, which are expected, but uncontrolled, like pump power failure etc., and

3. *catastrophic*, with very low probability of occurrence.

The analysts usually overlook the risk associated with the first group of events and concentrate on the second group of events.

1.4.1. Methods of Analysis

Basic equations for transient flow can be expressed in the following form (Wylie & Streeter, 1978)

$$\frac{1}{\rho}\frac{\partial p}{\partial x}+\frac{DV}{Dt}+g\sin\alpha+\lambda\frac{V|V|}{2D}=0 \tag{6}$$

$$\frac{1}{\rho}\frac{Dp}{Dt}+a^2\frac{\partial V}{\partial x}=0 \tag{7}$$

Equation 6. is the equation of motion, and Equation 7. is the continuity equation. The derivative $\frac{\mathrm{D}}{\mathrm{Dt}}=\frac{\partial}{\partial \mathrm{t}}+\mathrm{V}\frac{\partial}{\partial \mathrm{x}}$ is for the slice of fluid having velocity V, p is centerline pressure, and a, elastic wave speed in the pipe.

The bulk of the methods used in transient flow analysis are based on the method of characteristics, although there are successful attempts at solving basic equations by finite element methods and finite differences methods (Verway & Yu, 1993, Ingeduld et al., 1994) as well.

After some simplifications related to the introduction of piezometric head, Π, instead of pressure, p, and, flow rate, Q, instead of velocity, V, through simple transformations, basic equations for the method of characteristics can be written in the following form

$$\frac{d}{dt}\left(\Pi \pm \frac{a}{g}V\right)=\mp\lambda\frac{aV|V|}{D2g} \tag{8}$$

which are valid along lines

$$\frac{dx}{dt}=V \pm a \tag{9}$$

in the simulation domain (x,t).

These Equations are approximated by the finite difference method

$$\left[\Pi \pm \frac{a}{g}V\right]_{A/B}^{P}=\mp\lambda\frac{\Delta L}{D}\frac{V|V|}{2g}\bigg|_{A/B} \tag{10}$$

The expression in square brackets represents the difference between values of Riemann quasi-invariants at point (P) and point (A), for positive characteristics, and point (B), for negative characteristics. The right hand side is taken from the point A, for positive characteristics, and from point B, for negative characteristics.

In Figure 9. two variants for the method of characteristics with interpolation are given. The method with time-line interpolations has certain advantages over the other, although, it requires more data to be stored.

If the right hand side of Equations 8. and 10. is dropped out, the well known Joukovsky expression is obtained.

However, the most important in the simulation of transient flow is the way in which boundary conditions, especially pumps, are presented.

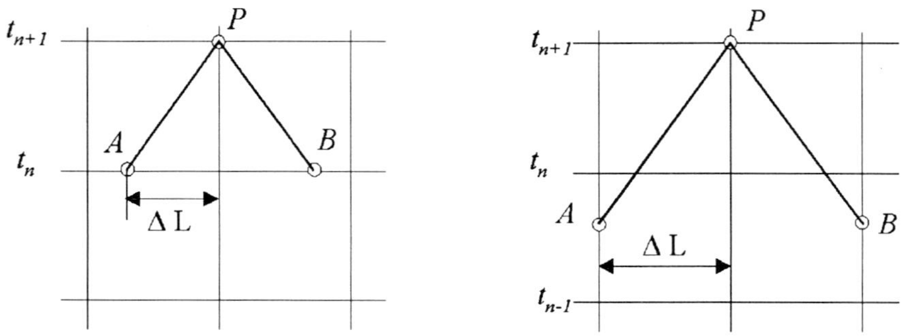

Figure 9. Spatial and time-line interpolation method in the method of characteristics

Simplified analysis. As an example of the potential of simplified methods, the trajectory of the operating point of two pumps caused by switching off the third unit, is given in Figure 10. Before reaching the new working point C, the flow rates of other two units are temporarily increased, moving down, almost vertically, to point B. Due to inertia of the water column, pressure increases in the suction line and decreases in the discharge line. Due to friction losses and after several wave reflections, the pumps will get to point C. The adverse effects of this maneuver are more pronounced after switching off the second unit, which, according to the NPSH curve, may cause transient cavitation and temporary motor overloading.

The most interesting point is that this event belongs to so-called normal operating conditions with very high frequency of occurrence. Detailed analysis using complete pump characteristics can provide further information on the possible consequences.

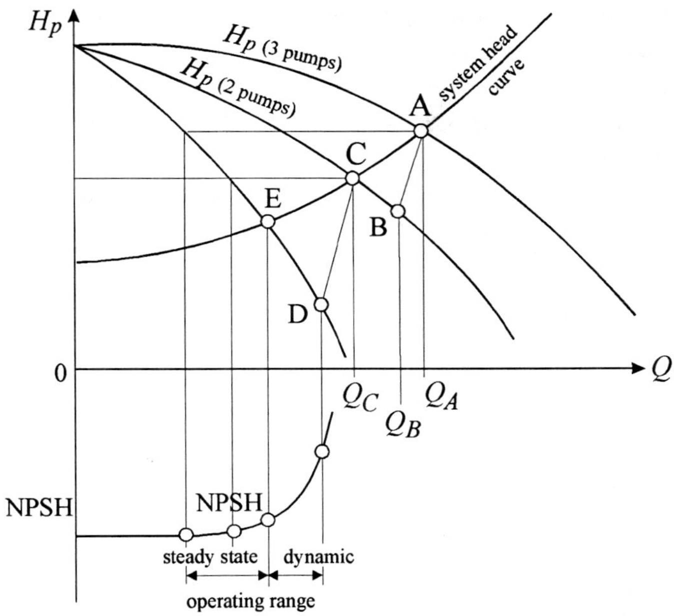

Figure 10. Steady and dynamic operating range during pump switch-off

Almost generally, water hammer is more dangerous for systems with low head pumping stations than for systems with high head pumping stations. Steady state velocities are of the same order, as well as the changes of velocity at boundaries, which are directly related to the pressure changes. Therefore, pressure changes of the same magnitude are not the same in relative units scaled with pumping head or with steady state working pressure. Furthermore, low-head high-capacity pumping systems are much more vulnerable to low pressures and possible column separation, caused by pump power failure, than by high pressures.

The sequence of events following pump power failure is shown in Figure 11. The system consists of 6 pumping stations connected to a common discharge pipeline with diameter equal to 500 mm. The distance between PS6 and PS1 is 2000 m. Pipe diameter is increased to 600 mm downstream of PS1. The initial discharge is equal to 650 l/s. The most upstream pumping station, PS6, is switched off at discharge equal to 118 l/s, causing a negative wave to propagate downstream. Instantaneous elevations of Hydraulic Grade Lines (HGL) are given at 1 s interval, the first one related to 0.5 s after initiation of the pressure wave. According to simplified analysis, one can conclude that the initial wave, equal to $\Delta\Pi = -60m$ (in downstream pipe $\Delta\Pi = -42m$), will cause column separation in the whole of the downstream pipeline.

However, the envelopes of HGL in Figure 11., diamonds indicate maximum and squares indicate minimum, show that nothing serious would happen. The initial wave is reduced becuase of pump inertia and the effects explained in Figure 10., i.e. temporary discharge increases in neighboring pumping stations. Therefore, in order to assess the risk associated with flow transients, a more detailed analysis of pump behaviour in transient regimes is required.

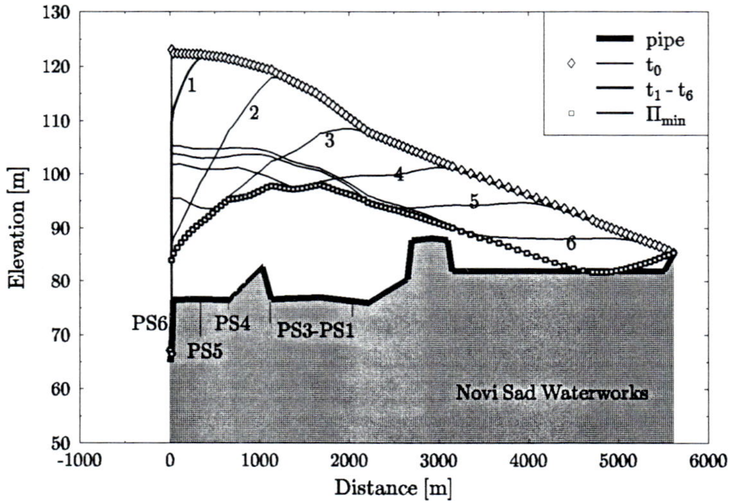

Figure 11. Pressure wave propagation caused by pump power failure

1.4.2. Pump Characteristics in Four Quadrants

Simplified analysis may indicate problems in the operation of the pumps, but the problems indicated will not be solved. More detailed data on pump behaviour in *abnormal* situations are needed. Namely, after pump power failure, the pump may experience either reversal in flow, rotational speed, or both. It is also possible that the head and torque may reverse.

These data are usually arranged in dimensionless diagrams, as given in Figures 12., 13. and 14. Basic variables are dimensionless discharge, $v = Q/Q_0$, and rotational speed, $\alpha = N/N_0$, while dependent variables are, pump head, $h = H_P/H_0$, and hydraulic torque, $\beta = T/\ T_0$, where subscript (0) indicates the values at the point of best efficiency. Similarly, relative efficiency is obtained as $\eta/\eta_0 = vh/\alpha\beta$.

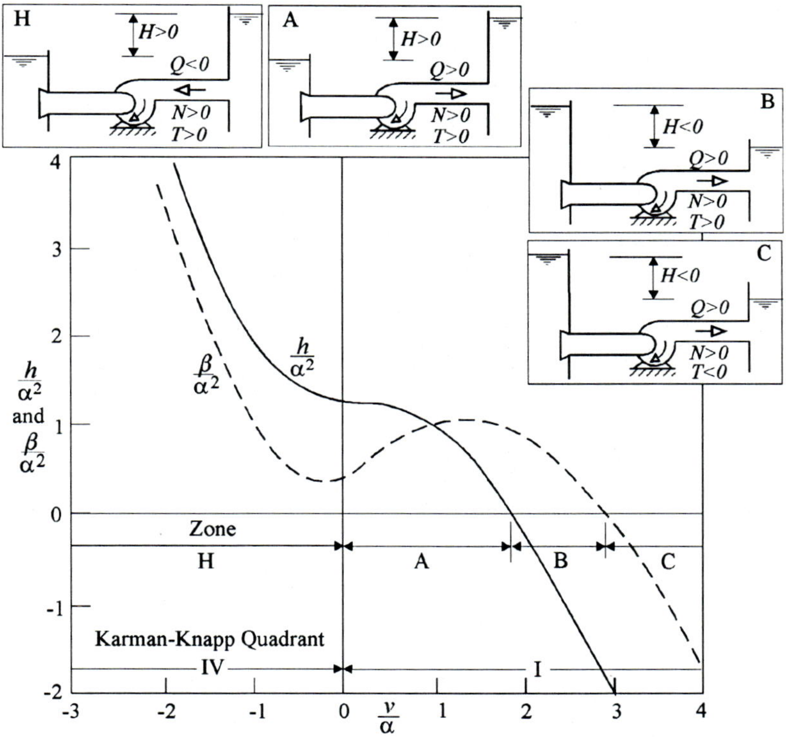

Figure 12. Homologous head and torque characteristics for radial flow pump (n_s = 24.6, SI) for positive rotation ($\alpha > 0$)

These are usually called "four quadrant diagrams" because they can be represented in the plane (v, α), with positive and negative values of v and α. Following the work of Knapp (Knapp, 1937), which is widely accepted (Martin, 1982, Stepanoff, 1957), eight possible zones of operation, denoted by letters from A to H, will be analysed. Four of them belong to the normal pump operation, when α>0 (Figure 12.), and four to the negative rotational speed (Figure 13.).

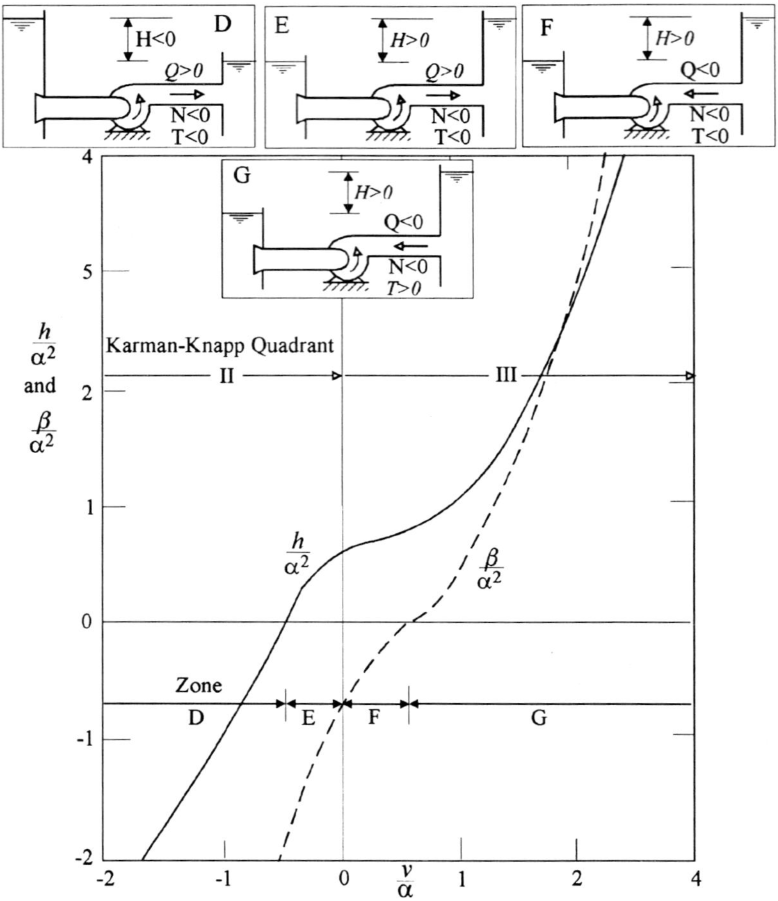

Figure 13. Homologous head and torque characteristics for radial flow pump (n_s = 24.6, SI) for positive rotation (α< 0)

Zone A (normal pumping) depicts a pump under normal operation for which all four quantities, v, α, h, and β are positive. Also, η is positive.

Zone B (energy dissipation) is a condition of positive flow, positive rotation, and positive torque, but negative head, giving $\eta<0$. A machine could operate in zone B by a sudden drop in head during a transient caused by power failure, or being overpowered by another pump or by a reservoir.

In zone C (reverse turbine), it is possible for a pump to generate power. Flow and rotation are positive, but head and torque are negative. The maximum efficiency in utilizing this as a turbine would be relatively low because of improper entrance conditions.

Zone H, called energy dissipation zone, is often encountered after pump power failure. In this zone, the combined inertia of the rotating parts of pump, motor and entrained water, still maintain positive rotation, while flow has been reversed by positive head.

During power failure the pump will enter zone G (normal turbining) provided that reverse rotation is not prevented by some mechanical devices. A pump rotating freely in a negative direction is not generating power, but it is exactly the mode of operation of a hydraulic turbine. Head and torque are positive as for normal pump operation while flow and speed are negative.

Zone F (energy dissipation) can be encountered after continual operation of a machine earlier failed as a pump. The difference between zones F i G is that the torque has changed sign for zone F, which causes a reduction in speed. In fact the real runaway condition is attained at the boundary of the two zones, where torque, $T = \beta = 0$.

The two remaining zones, D and E are very unusual in practical application (they can be encountered in the test loop). They are possible in case of improper wiring of electric motor.

Figure 14. integrates two diagrams from Figures 12. and 13. for the same pump. Curves of h/α^2 for $h=1$ and $h=-1$, and straight lines for $h=0$, are represented by continuous lines. Lines for $h=0$ and $\beta=0$ are used for separation of zones of operation.

Figures 12. and 13., with variables, h/α^2 and β/α^2, as a function of v/α, are very useful for physical explanation of pump performance, but inappropriate for computation because (v/α) becomes infinite as the pump passes through, or remains at, zero speed $(\alpha = 0)$.

Marchal, Flesh and Suter solved that problem by introducing a new variable as abscissa, $\theta = \pi = \arctan(v/\alpha)$, instead of, (v/α), and also, $h/(v^2+\alpha^2)$ and $\beta/(v^2+\alpha^2)$, instead of, h/α^2 and β/α^2. In this way complete pump characteristics are represented by two single curves, as given for the same pump in Figure 15.

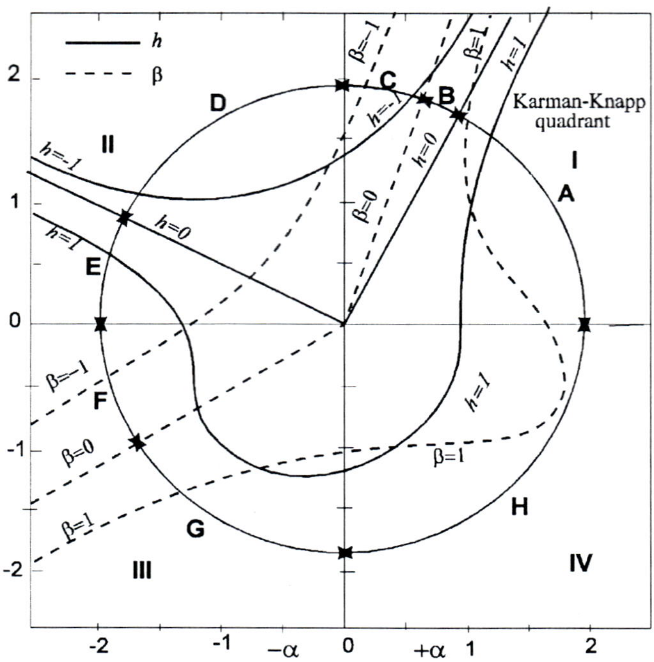

Figure 14. Karman-Knapp circle diagram (n_S = 24.6, SI)

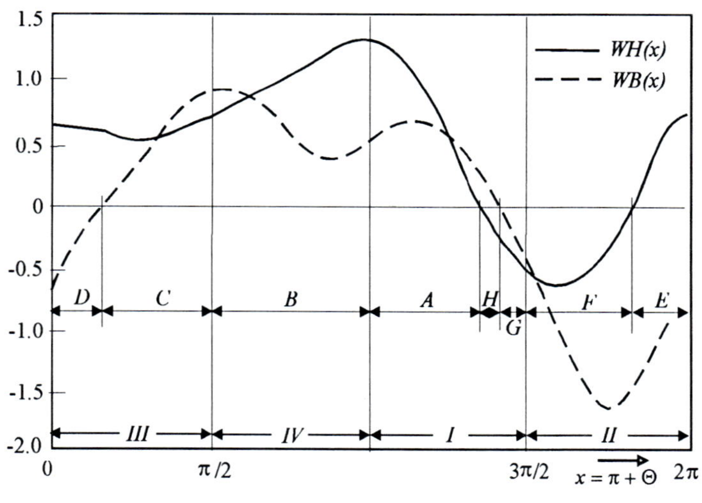

Figure 15. Suter representation of head and torque characteristics for radial flow pump (n_S = 24.6, SI) in four quadrant

1.4.3. Pump Power Failure

This is one of so-called irregular, or uncontrolled, events because no adequate action can be undertaken to prevent it. It is usually caused by pump power failure, but also by automatic operation when necessary actions, like the closure of a discharge valve, are not completed. All transient flow analyses include this event and with reliable data on the dynamic characteristics of pumps and valves, and their condition, the results of simulation are acceptable.

This problem will be illustrated by a case study, BVK Makiš- Mladenovac, a booster pumping station with a long suction line, the results of which are presented on separate pages. In order to secure trouble-free operation of the system, several methods can be recommended depending on the objectives of the analysis.

Characteristic flow rate is 1200 l/s, and pipe diameter 1200 mm. In the beginning of the analysis glass reinforced plastic (GRP) pipes of the nominasl pressure (NP) 16 bars were envisaged, and the only problems expected were low pressures at high points of the pipeline. Acceptable protection could be achieved by inexpensive surge relief valves. The results of the analysis obtained by software package HYTRA (Ivetić, 1990) are given in Figure 16. The first diagram represents the longitudinal profile of the pipeline and envelopes of extreme piezometric heads. The numbers in small boxes indicate the positions of sections where flow rates and piezometric heads are monitored and given in the two remaining diagrams.

However, it was later decided to use reinforced concrete pipes of NP 12 bars, which are resistant to low pressures. Instead of a surge relief valve, an inclined stand pipe (more expensive, and more reliable) had to be used (Figure 17.). The problem of pump power failure may be further complicated by taking into consideration the dynamic behaviour of check valves, which is analysed in the next subsection.

Check Valve Dynamics

This problem has been analysed on a high-head pipeline (length ≈14 km, diameter 330 mm, flow rate 80 l/s) for raw water (Figure 18.). The designer of the system was mainly concerned with almost certain column separation which was about to happen shortly after pump power failure about 9 km from the pumping station. An air chamber was proposed, and the analysis had proved that the pipeline was protected from column separation. What is common for these analyses is the rather unrealistic assumption that check valves close automatically as the flow changes sign (used for results in Figure 18., as well). However, that is a characteristic of a small number of check valves, let say, of nozzle check valves in this analysis. In this case, there was a classical swing check valve, which is much slower in response, though admitting significant backward flow. It turned out that an air chamber would worsen the situation and the air chamber was dropped off the project. In this case column separation was not a problem. There were no serious over-pressures following vapour cavity collapse. Pipe wall thickness was determined to sustain normal operating pressures.

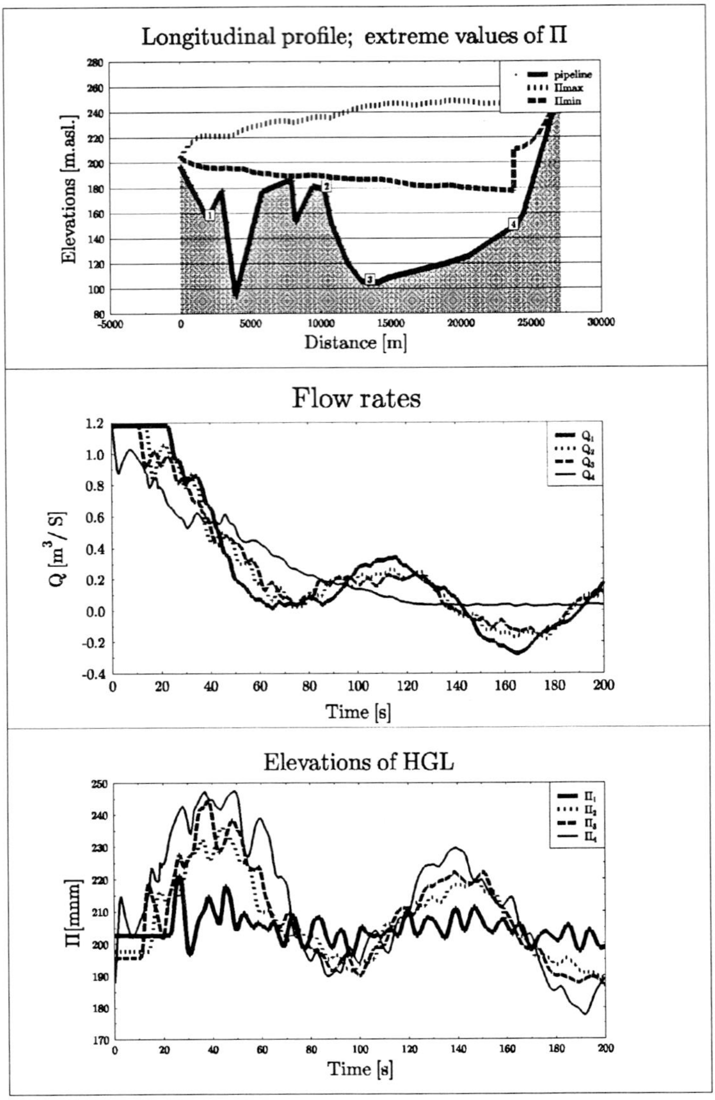

Figure 16. Results of the simulation of pump power failure: BVK Makiš-Mladenovac; surge relief valve upstream of the booster pumping station

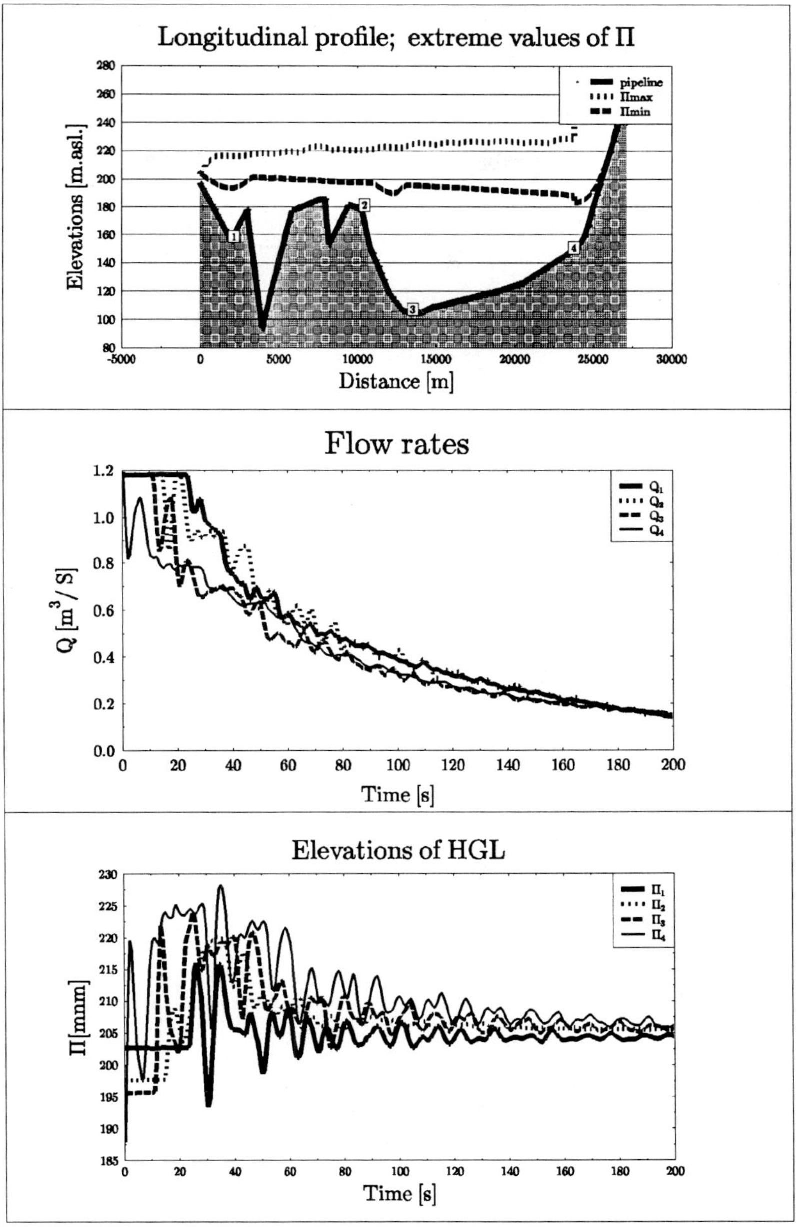

Figure 17. Results of the simulation of pump power failure: BVK Makiš -Mladenovac; stand pipe (or exhaust pipe) upstream of the booster pumping station

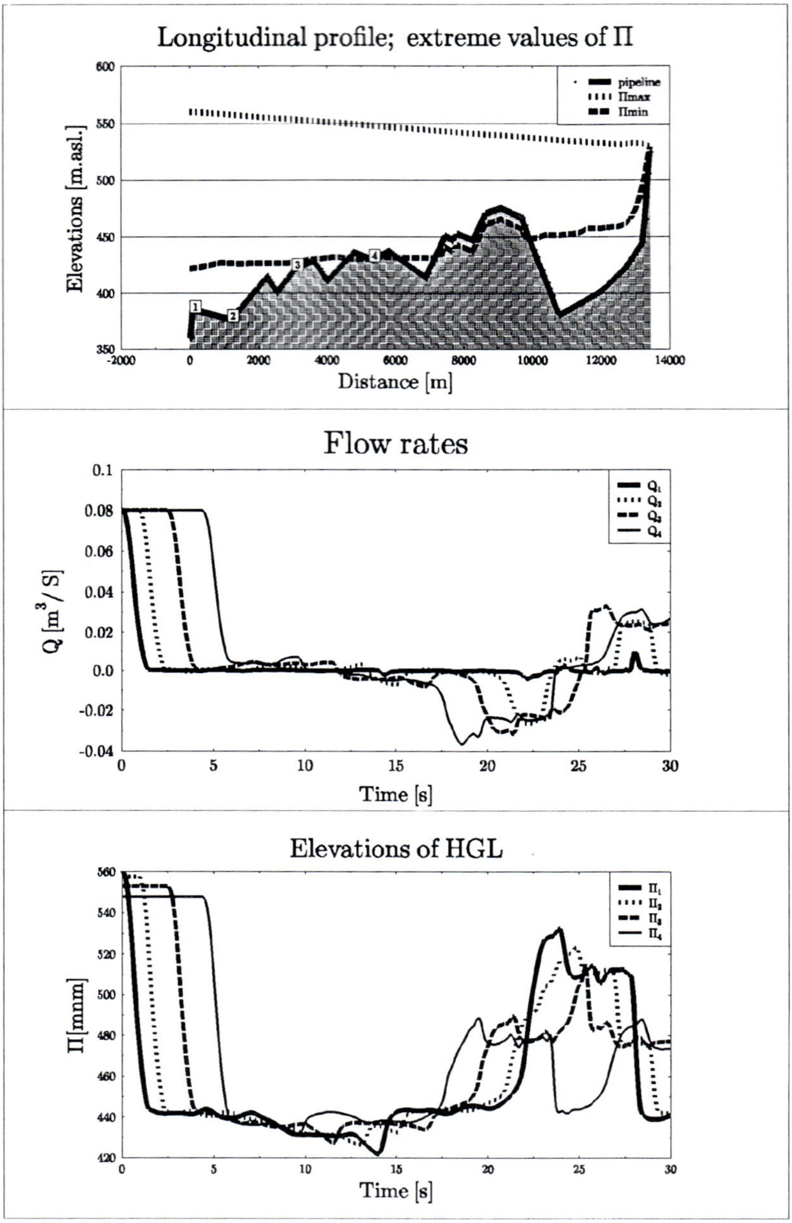

Figure 18. Results of the simulation of pump power failure: Bor; usual assumption of instantaneous check valve closure

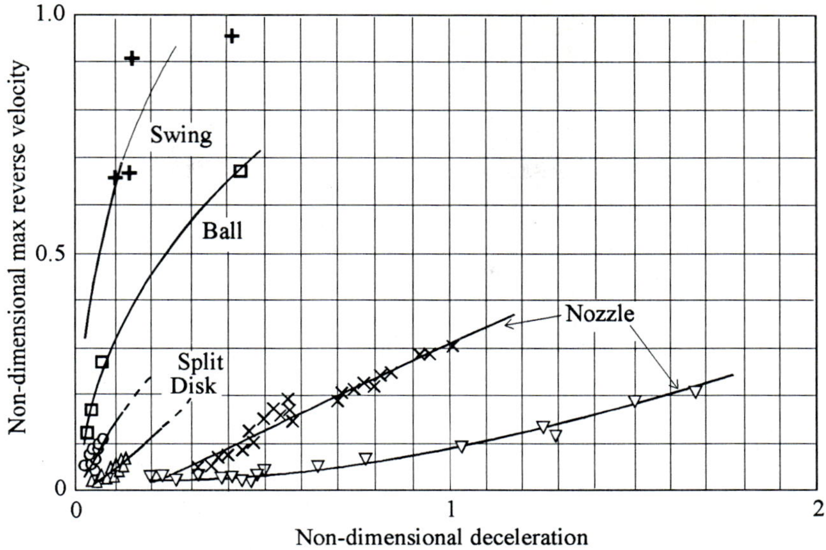

Figure 19. Dynamic characteristics of check valves of different types and sizes (Thorley, 1991)

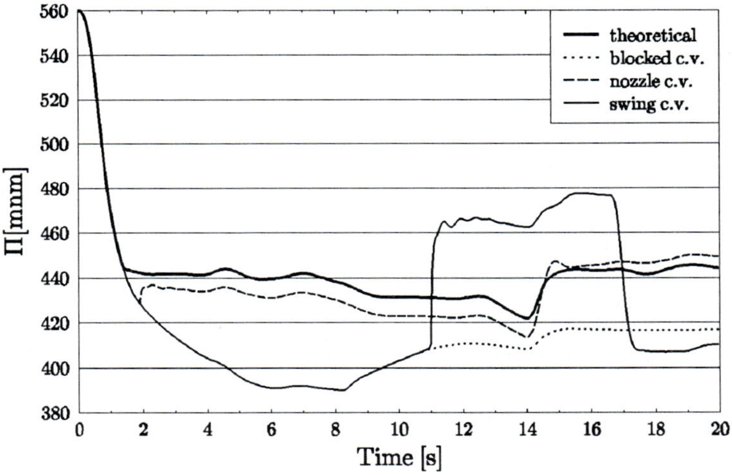

Figure 20. The influence of different presentation of a check valve behaviour

In Figure 20. several ways of presentation are compared concerning check valve behaviour in the pumping station, through the changes of piezometric heads downstream of the pumping station.

One of them, indicated as theoretical, is the same as in Figure 18.

The dotted line represents the case when the check valve remains open, with negative rotation of pumps prevented. The dashed and thin lines represent nozzle and swing check valve, respectively. Dynamic characteristics of check valves are taken form Figure 19. (Thorley, 1991).

1.4.4. Pump Start-Up

This is a regular event, neglected by many analysts. Usually, no problems are expected, but, unfortunately, they do happen.

Starting Against a Closed Valve

This is a recommended operation for radial flow pumps, but not for axial flow pumps.

Starting Against a Check Valve

This is a usual mode of operation. Problems may arise due to the malfunction of the check valve in high head systems.

Starting a Pump Running in Reverse

This is an unwanted but possible situation at unmanned pumping stations with automatic control. Overloading and overheating of the motor are possible.

1.4.5. Available Surge Control Devices

It should be noted that there is no typical solution for surge protection. Every system is unique and requires a unique approach in analysis. Some of the most common devices used in WSS are shown in Figure 21.

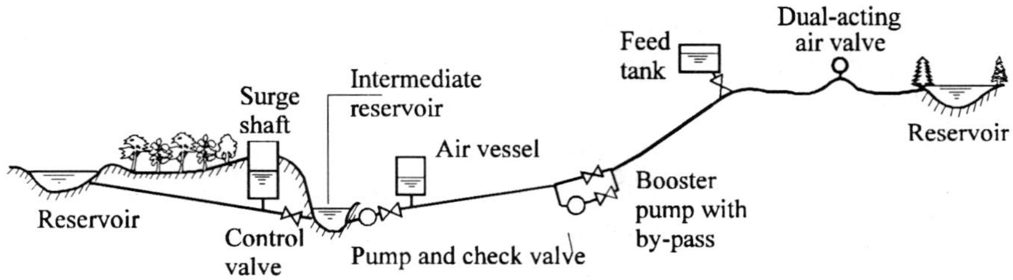

Figure 21. Protection devices against pressure surge commonly used in water supply systems

1.4.6. Flow Control and Energy Recovery by Pumps

The problem of flow control in long gravity pipelines is closely related to energy dissipation by control valves. It is accompanied by cavitation, vibrations, sometimes severe transients etc.

An inexpensive solution represents a pump operating in turbine regime, more exactly in zone F (Figure 14.). With small modifications to the motor it can work as a generator. Because of inertia of the electrical system frequency control is usually not required.

One of the situations where the flow control in long gravity pipelines is required is pumping over levees in flood risk zones with small or negative static heads. Relatively high head is required during the priming of the system, while later, pumps are usually throttled to prevent drifting of the operating point to the right end of pump curve. Compared to standard micro-hydropower schemes this one is much cheaper (Apfelbacher and Etzold, 1989). According to some analyses, recovered energy may exceed 50 % of the energy used for pumping.

1.5. CONCLUSIONS

The problem of transient flow analysis related to the operation of pumping stations will remain one of the most important in the future.

The emphasis will probably shift from irregular events (like pump power failure) towards transients under normal operation conditions due to the ever-increasing complexity and size of water supply systems. Apart from the standard schemes of pumping stations with nearby suction reservoir there will be more booster (single-unit) stations, locally controlled.

In order to be able to predict the behaviour of real systems, it is necessary to acquire full knowledge of equipment characteristics (control valves, pumps, check-valves etc.) preferably under dynamic operating conditions.

New pipe materials and wider usage of some existing materials (PVC, PE, GRP etc.), which have much lower elastic wave velocity, instead of traditional ones (AC, concrete, even steel, etc.), will decrease the risk associated with transient phenomena.

As a solution for resolving severe operating conditions (cavitation, vibrations etc.) of control valves in long gravity pipelines and to recover significant amount of energy dissipated for the purposes of control, pumps operating in turbine mode can be used.

1.6. REFERENCES

Apfelbacher R., Etzold F. (1989) Energy-saving, shock-free throttling with the aid of a reverse running centrifugal pump. KSB Technische Berichte 24e, Frankenthal.

Cherkassky V. (1980) Pumps, Fans, Compressors. Mir Publishers Moscow.

Dickenson C. (1988) Pumping Manual. The Trade and Technical Press Ltd., Morden, Surrey, UK.

Ivetić M. (1990) HYTRA - Hydraulic Transient Analysis. IRTCUD Software.

Karassik I.J., Krutzsch W.C., Fraser W.H., Messina J.P. (1976) Pump Handbook. McGraw-Hill Book Company.

Knapp R.T. (1937) Complete Characteristics of Centrifugal Pumps and Their Use in the Prediction of Transient Behavior. Transactions ASME, Vol. 59, New York.

Martin C.S. (1982) Representation of Characteristics of Hydraulic Machinery. Intercambio International sobre Transientes Hidraulicos & Cavitacao, Sao Paulo.

Record R. (1988) Local Intelligent Pumping - A Standard Approach, Computer Applications in Water Supply. Volume 2, eds. B.

Stepanoff A.J. (1957) Centrifugal and Axial Flow Pumps. 2nd Edition, John Wiley & Sons.

Thorley A.R.D. (1991) Fluid Transients in Pipeline Systems. D. L. George Ltd.

Verway A., Yu J.H. (1993) A Space-Compact High-Order Implicit Scheme for Water Hammer Simulations. Proc. XXVth IAHR Congress, Tokyo.

Wylie E.B, Streeter V.L. (1978) Fluid Transients. McGraw-Hill Inc.

2. Object-Oriented Tools for Transient Flow Analysis

Petr Ingeduld
Evžen Zeman
Czech Technical University of Prague
Thakurova 7, 16629 Prague 6
CZECH REPUBLIC
Adrian Verwey
IHE, Delft
P.O. Box 3015, 2601 DA Delft
THE NETHERLANDS

2.1. INTRODUCTION

Today there is a need for an engineering software package for the simulation of steady or unsteady-flow in pipe networks, providing all information about pressure distribution, solving slow transients and water hammer (Abbott, 1991). An object oriented package can provide a flexible description of a variety of pipes and hydraulic structures such as pumps, valves, reservoirs, surge chambers, feeder tanks and air vessels. The modular structure of such a tool may be based on graphical facilities for the definition and checking of input data, including accurate and fast numerical solvers and an engineering approach to presenting output data. A package structure may benefit from a database platform enabling query-by-example and other modern data handling facilities.

The numerical modules may even cover equilibrium problems such as water distribution in pipe networks and slow and fast transients including real time control simulation. The complex and robust parts of the numerical code allow the user to describe a wide variety of hydraulic components and their interconnections within any kind of network. The software package may benefit from an object-oriented design of the code. The recently completed HYPRESS package based on the above mentioned approach has been used to demonstrate some of the characteristic features of new technologies.

NATO ASI Series, Partnership Sub-Series, 2. Environment – Vol. 15
Water Supply Systems – New Technologies
Edited by Č. Maksimović, F. Calomino, and J. Snoxell

2.2. THEORY

The solution domain of pipe network usually consists of branches connected one to another by means of nodes. The authors take as an example the algorithm based on a finite difference method. Grid points are generated along branches where they represent the place, where the governing equations are computed (Figure 1.). Various hydraulic structures (such as simple tanks, surge tanks, feeder tanks, air vessels, valves, pumps, etc.) can be included at any selected place of the domain.

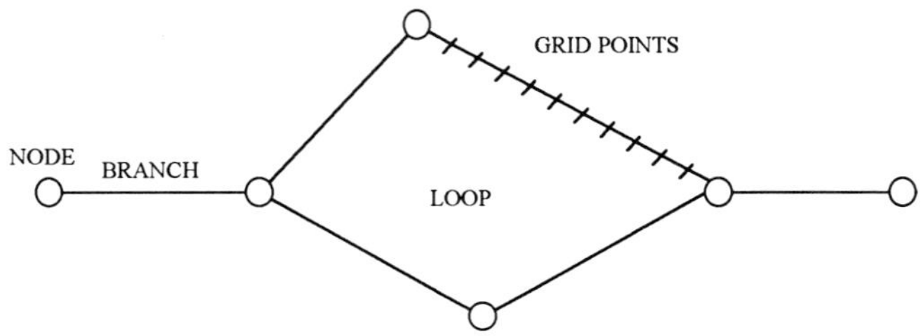

Figure 1. Discretization of network

- Branches represent pipes of constant properties. In the pipe network, branches may include such elements as hydraulic controls, e.g. valves, pumps,

- Nodes represent free branch ends, branch connections or specific storage. Boundary conditions are defined at the nodes by setting up values of water pressure or discharge as a constant value or as a function of time. At nodes connecting several branches together, pressure compatibility is assumed. Simple tanks, surge chambers, feeder tanks and air vessels are located at the nodes,

- Grid points are generated along the branches and they represent the computational grid where the values of water pressure and discharge are solved, or specify points where data input or output is required. The system requires a computational grid to define small sub-sections in a branch over which the governing equations are solved numerically.

The governing equations describing one-dimensional unsteady pressurised flow in a pipe can be derived, for example, in the form of the continuity equation (Streeter and Wylie, 1983; Chaudhry, 1987, Tullis, 1989):

$$\frac{\partial Q}{\partial x}+\frac{gA}{a^2}\frac{\partial H}{\partial t}=0 \qquad (1)$$

and the momentum equation:

$$\frac{\partial Q}{\partial t}+gA\frac{\partial H}{\partial x}+\frac{f}{2DA}Q|Q|=0 \qquad (2)$$

where:

where:	Q	discharge [m^3/s],
	H	piezometric head [m],
	f	Darcy-Weisbach friction factor [-],
	D	diameter of the pipe [m],
	A	area of the pipe [m^2],
	x, t	space and time co-ordinates [m], [s].

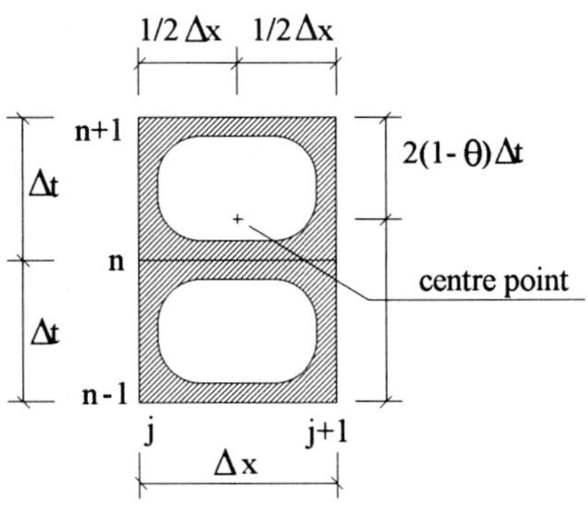

Figure 2. Operator for space-compact implicit scheme

The governing Equations (1) and (2) are solved by the advanced fourth-order-accurate, space-compact implicit finite difference scheme (Figure 2.),Verwey and Yu, 1993. The scheme was developed by adding two Preissmann type operators (Cunge et al., 1980) over two successive time levels and by subtracting all third-order derivative terms of the truncation error. This was achieved by converting all third order derivative terms to a form $\partial^3 t/\partial x \partial t^2$, which can be defined in finite difference form on a domain between two successive sections on a branch and three time levels.

The space-compactness of the scheme assures the same formal accuracy of computations all along the pipes, including the sections near the boundaries. The completely centred scheme is stable for all Courant numbers less than or equal to unity and the scheme can be stabilised for higher Courant numbers by increasing the weight parameter θ.

The general implicit finite difference form of Equations (1) and (2) is:

$$A1_j H_{j-1}^{n+1} + B1_j Q_{j-1}^{n+1} + C1_j H_j^{n+1} + D1_j Q_j^{n+1} = E1_j \tag{3}$$

$$A2_j H_{j-1}^{n+1} + B2_j Q_{j-1}^{n+1} + C2_j H_j^{n+1} + D2_j Q_j^{n+1} = E2_j \tag{4}$$

where coefficients A1,B1,C1,D1,E1 for the continuity equation and A2,B2,C2,D2, E2 for the momentum equation are derived from the implicit high-order scheme.

The looped algorithm for the finite difference scheme is based on the fact that a looped network contains elements known as nodes which represent the confluence of several flow paths, some of which originate from other nodes, some from boundary points. A system of simultaneous linear equations is developed where the piezometric head H at each node serve as the only unknowns.

The solution of this system by a fast topology-based searching technique (Verwey, 1994) yields the H values at each node. Required H and Q values at all intermediate grid points between the nodes are expressed in terms of variation in H at the nodes through a transformation of the set of Equations (3) and (4).

For water distribution under steady state conditions and/or slow transient states, a similar algorithm should be applied (Salgado et al., 1987). In this case, however, there is no need for a refined grid or for a high-accuracy scheme.

These computations are based on a Preissmann-type scheme with grid points defined only at the pipe ends, or at additional points around structures, wherever these are present along a pipe.

2.3. OBJECT-ORIENTED DESIGN OF THE ALGORITHM

An object-oriented design brings the advantages to the developer to create more efficient code that can be more easily shared by a team of programmers (Booch, 1991, Meyer, 1988). Combination of the traditional and the object-oriented way of programming can be optionally used for the computational modules programming as discussed on the basis of a following example (Ingeduld, 1993).

If a hydraulic structure is present in the domain of the solution, Equations (1) and (2) applied by two succesive grid points must be replaced by other equations defining such a hydraulic structure. Different hydraulic structures can be coupled together. The closing of one valve can determine the operation of another valve, for example. In the case of the existence of a link between such hydraulic structures, communication must be maintained and controlled by the main algorithm. This message has to be attached to the object in such a way that it resembles the reality (Baugh and Rehak, 1992). The overall numerical algorithm, described briefly in Section 2., can be schematised into the following steps:

Coefficients Routine

The coefficients of the implicit high-order scheme discretize the governing equations between two successive grid points on a branch.

An object-oriented design is used for the derivation of the ABCDE representing hydraulics components located on a branch.

Local Elimination Routine

This is used to express Q and H grid point values on each branch in terms of H at the branch ends (nodes).

Node Routine

Applying the boundary conditions along the nodes lead to one equation for each node.This system of linear equations is solved by the matrix elimination method;

For hydraulics components located in nodes, again an object-oriented approach is used.

Recalculate Routine

Substitutions inside the branches yield the values of Q(i) and H(i) for all intermediate grid points from the known values of H at the branch ends.

An object-oriented design can be effectively used for the **real-time connection of branch and node components.**

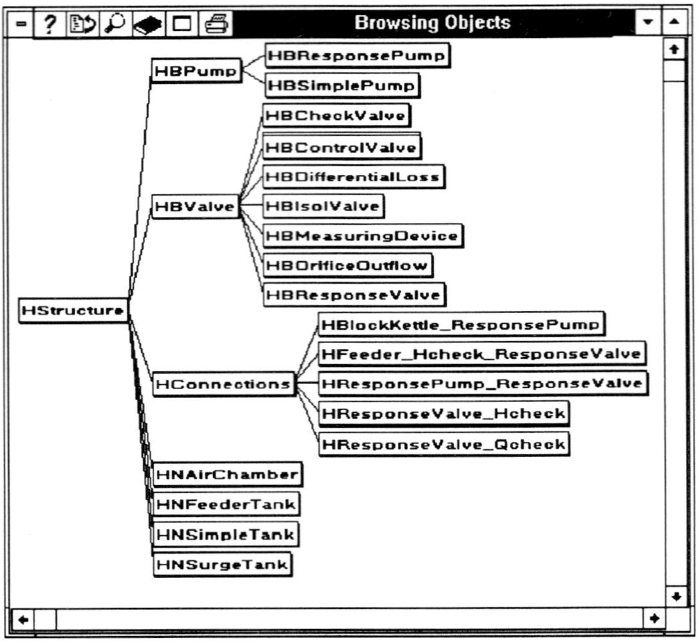

Figure 3. HStructure hierarchy

An example of HStructure hierarchy consisting of descendants of hydraulic controls is presented in the following Figure 3. The hierarchy covers different classes that have different function in the numerical algorithm.

HBValve: branch located hydraulic components. Replaces A,B,C,D,E coefficients of high-order implicit scheme in Coefficient routine.

HBPump: branch located hydraulic components. Replaces A,B,C,D,E coefficients of the high-order implicit scheme in Coefficient routine.

HNAirChamber, HNFeederTank, HNSimpleTank, HNSurgeTank: node located hydraulic components. They modify parameters describing a type of boundary condition in a node. HNAirChamber, for example, solves 6x6 equations each time it is invoked and these equations operate on its private data. The complexity is hidden inside the object definition and methods are encapulated with their data. A misuse is nearly impossible.

Hconnections: send messages to certain descendents of the classes mentioned above. These messages, based on simple rules, change private data of corresponding objects.

The object-oriented features (inheritance, encapsulation, polymorphism) can be shown on an example of a class HBValve. There are 3 main objects collections, **BranchStructures, NodeStructures, Connections,** that contain descendants of corresponding classes. Using ForEach@ iterators it is possible to evoke the virtual method such as for GetABCDE coefficients and this message is passed along a hierarchy path as the proper descendant is found. The same principle is applied to GetErrorMessage, GetReport functions common to all HStructures or GetBoundary which are only valid for HNAirChamber, HNFeederTank, HNSimpleTank and HNSurgeTank.

GetABCDE method in fact invokes GetA,GetB,GetC,GetD,GetE methods and returns a pointer to a record containing A,B,C,D,E coefficients; this record is used in Coefficient routine in the main numerical algorithm. Particular GetA,GetB,GetC,GetD,GetE methods are virtual method.

Inheritance: a relationship among classes, wherein one class shares the structure or behaviour in one (single inheritance) or more (multiple inheritance) other classes. Inheritance defines a "kind of" hierarchy among classes in which a subclass inherits from one or more superclasses. The following example is valid for all descandants of HBValve.

GetE is inherited from HBValve and is unchanged, GetE=0 for each.
GetB is inherited from HBValve and is unchanged, GetB=0 for each.
GetD is inherited from HBValve and is unchanged, GetD=1 for each.
GetA and GetC are inherited from HBValve and are overwriten.

For example, GetC=0 for a closed IsolationValve.
As another example GetC=$C_d\sqrt{1/\Delta H}$ for Orifice Outflow, (Cd is a flow factor of a valve and ΔH the pressure difference).

Moreover, many report and error functions, functions allocating the valve on a pipe, functions used for communication among objects are declared for HBValve and inherited by its descendents.

Polymorphism: A concept in type theory, according to which a name (such as a variable declaration) may denote objects of many different classes that are related by some common superclass; thus any object denoted by this name is able to respond to some common set of operations in different ways.
GetA method is overwritten in each descendent Report functions also have a typical polymorphical behaviour. Different hydraulic objects, such as valves, pumps, measuring devices create their own report during the computation.

Encapsulation: the process of hiding all of the details of an object that do not contribute to its essential characteristics; typically, the structure of an object is hidden, as well as the implementation of its methods.
Each function GetA for example when activated is applied to private data of a particular object. Functions used for communication among objects send messages along the hierarchy path. This message is recognized by corresponding descandants and sets the private data of this object. Then, when the object is asked to GetABCDE coefficients it acts accordigly to these changed data. Object-oriented programming helps a programmer to create a more safe interface to the numerical algorithm, which low level operations that remain the same, are hidden inside objects. This model brings the advantages of a modularity and hierarchy relations into a numerical model. As a result, the model is more secure and can be easily changed in the future.
On the other hand the main risk of the object-oriented approach in the form of start-up costs might be easily declared now. If the particular object-based or object-oriented programming language is used for the first time, the start-up costs associated with this object-oriented design may prove to be a real barrier to adopting the method (Dubois-Pelerin et al., 1991a&b). The creation of such a model, at least its object-oriented layer behind which the algorithm works, takes time and requires extensive process of verification of the model, even if such a model already exists expressed in a conventional programming language.

2.4. HYPRESS EXAMPLES

2.4.1. Single Pipe

Basic tests of water hammer simulation were made on a single pipe of lenght L=1524 meter, diameter D= 0.76m with water hammer celerity c=1239m/s and carying a steady state discharge Q= 0.695m^3/s. The pipe connects two adjacent reservoirs with water levels H1=202.80 and H2=198.20m. The valve is located at the end of the pipe and is closed linearly in time over a period of 1.0 second.

The initial head loss in the valve is neglected. The results show water hammer during the first 20 seconds in the form of water pressure history in front of the valve (Figure 4.).

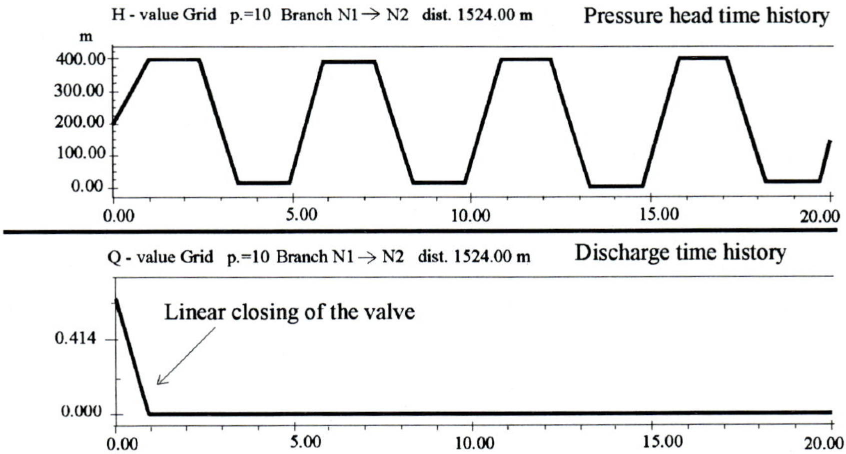

Figure 4. Pressure changes caused by linear valve closure

2.4.2. Surge Tank

This study shows the functioning of a surge tank as a water hammer protective device, located on a pipe line between two reservoirs (Figure 5.). The network topology consists of two pipe branches, L1=100 and L2=200 meters long, each with diameter D=1.25 m, the wave celerity in these pipes is 1000 m/s. Water levels in two adjacent reservoirs are H1=960 and H2=958.70 m. The initial discharge is Q=1.997m^3/s. The surge tank is a rectangular basin with dimensions 3x4m. The initial water level of this tank is 959.13 m. The water hammer phenomena is caused by the instantaneous decrease of the discharge at the lower outlet.

The following figures describe the network topology, the graph of pressure changes along the pipe branches during the first minute and the corresponding water level changes inside the surge tank for the same time period.

The initial state was computed using the HYPRESS Initial Module. The subsequent water hammer simulations were made with the Water Hammer Module with the a step Δt=0.05 seconds.

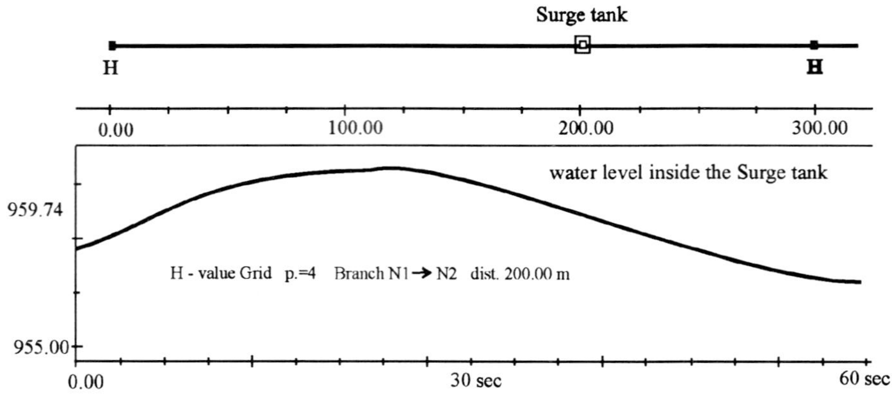

Figure 5. Network topology and water level change in surge tank

2.4.3. In-Line Control Valve

In-line control cone valve is located in this example at the end of a pipe network. The network consists of 10 pipe branches connecting two reservoirs (Figure 6).

The total length of the system L=6096m, the diameter D=1.0m and the water hammer celerity C=974.90m/s. The last pipe is 5457m long and the control valve is situated at its end. The valve characteristics are described by the polynomial function, the valve is initially open at 13% giving a steady state discharge Q=0.2055 m^3/s. The control valve is linearly closed over a period of 14 seconds.

The water hammer phenomena is documented in the following figures. The layout of the system is shown in the first picture. The next two figures describe the water pressure and discharge time history in front of the control valve. The pressure profile in the last pipe branch is shown on the last figure.

The initial state was computed by the HYPRESS Initial Module and the subsequent water hammer by the Water Hammer Module.

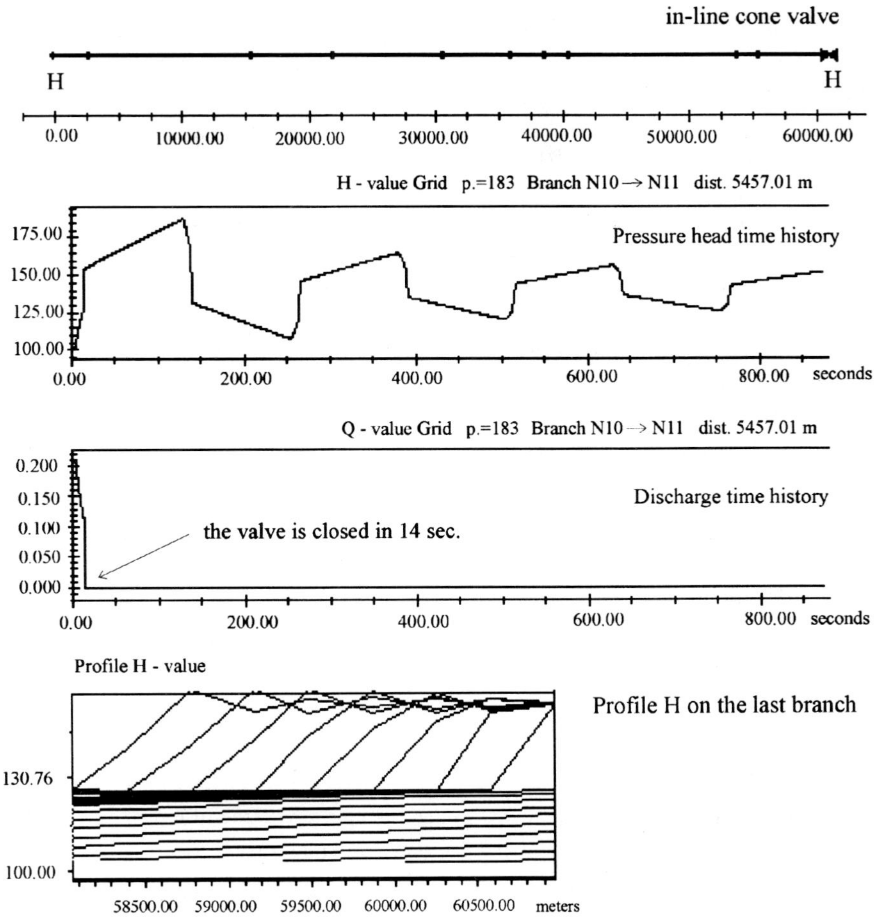

Figure 6. Network topology and results of valve closure simulation

2.5. CONCLUSIONS

This software package has been designed on the basis of an interactive graphical support to data editing through a multi document interface, during editing. All logical and topological data types are processed immediately and, wherever appropriate, visualised. The information system on pipe networks is based on storage data in standard database.

The numerical modules solve pressure and flow distributions in pipe networks as steady state, or slow and fast transients including real time control emulation on a variety of hydraulic components such as pipes, pumps, valves, reservoirs, surge chambers, feeder tanks and air vessels.

2.6. REFERENCES

Abbott M.B. (1991) Hydroinformatics: Information Technology and the Aquatic Environment. The Avebury technical, Alderschot, UK and Brookfield, USA.

Baugh J.W.Jr, Rehak D.R. (1992) Data Abstraction in Engineering Software Development. Journal of Computing in Civil Engineering, Vol.6, No.3, July, 1992, 82-301ASCEE.

Booch G. (1991) Object Oriented Design with Applications. The Benjamin/Cummings Publishing Company, Inc.

Cunge J.A., Holly F.M.Jr, Verwey A. (1980) Practical Aspects of Computational River Hydraulics. Pitman Advanced Publishing Programm.

Dubois-Pelerin Y., Zimmermann T., Bomme P. (1991a) Object-Oriented Finite Element Programming: I. Governing Principles, Computer Methods. In Apllied Mechanics and Engineering, 98 291-303.

Dubois-Pelerin Y., Zimmermann T., Bomme P. (1991b) Object-Oriented Finite Element Programming: II. A Prototype Program in Smaltalk, Computer Methods in Apllied Mechanics and Engineering, 98 (3) 361-397.

Chaudhry M.H. (1987) Applied Hydraulic Transients. Van Nostrad Reinhold, 1987.

Ingeduld P. (1993) Elements of Object-Oriented Simulators. M.Sc Thesis H.H 155, IHE, Delft, The Netherlands.

Meyer B. (1988) Object-Oriented Software Construction. Prentice Hall.

Salgado R., Todini E., O'Connell P.E. (1987) Comparison of the Gradient Method With Some Traditional Methods for the Analysis of Water Supply Distribution Networks. Leicester.

Streeter W.L., Wylie E.B. (1983) Fluid Transients. McGraw Hill, NewYork, 1983, 1978.

Tullis J.P. (1989) Hydraulics of Pipelines. John Wiley & Sons.

Verwey A., Yu J.H. (1993) A Space-Compact High-Order Implicit Scheme for Water Hammer Simulations. Proceedings of XXVth IAHR Congress, Tokyo.

Verwey A. (1994) Linkage of Physical and Numerical Apects of Models Applied in Environmental Studies. Proceedings of the Conference on Hydraulics in Civil Engineering, Brisbane.

Chapter VIII

Use of GIS and Remote Sensing in Water Supply

Chapter VIII

Use of GIS and Remote Sensing in Water Supply

INTRODUCTORY STATEMENT

Remote sensing and Geographical Information Systems (GIS) are two complementary technologies that can be applied to the problems of water supply.

The technologies of GIS and Remote Sensing that are relevant to water supply, water body monitoring and water quality will be described in the first Section of this Chapter.

The second Section deals with applications and case studies using remote sensing and the spatial modelling capabilities of GIS for applications as diverse as determining urban irrigation demands and detecting leaks from buried pipelines.

The third Section describes a specific example of how the digital elevation model and other geographic data can be used in design and operational management of water distribution systems.

Finally, in the fourth Section the application of GIS and digital mapping to a major water supply company covering a population of 1 million people will be discussed.

1. Relevant GIS and Remote Sensing Technology: Theoretical Considerations

John Elgy
Thomas William Charnock
Peter David Hedges
Department of Civil Engineering
Aston University
Aston Triangle, Birmingham B4 7ET
UK

1.1. INTRODUCTION

The Geographical Information System is well suited as a tool for assisting with water supply problems. It not only carries out the inventory functions, as to what lies where, but assists with the manipulation of the data and the extraction of information for modelling and decision making. This section will describe the fundamental features of GIS directly relevant to water supply projects, and the ways that data are stored in a GIS. The Digital Elevation Model (DEM), which is of great importance to hydrology and water supply, will be considered in some detail. In terms of cost for implementing a GIS solution: manpower, data acquisition, software and hardware rank in about that order. Manpower is a very difficult aspect to specify for a project, so it will only be discussed briefly. Manpower depends as much on existing skills, training requirements and workloads for existing staff as on the GIS. The other three aspects are discussed with respect to the current state of the art; however technology is likely to change in the future.

1.2. GEOGRAPHICAL INFORMATION SYSTEMS

Geographical Information Systems (GIS) can be used to store, retrieve, manipulate and display information having attributes that can be related to a position in a co-ordinate

NATO ASI Series, Partnership Sub-Series, 2. Environment – Vol. 15
Water Supply Systems – New Technologies
Edited by Č. Maksimović, F. Calomino, and J. Snoxell

system, specifically information relating to the Earth's surface. In this respect it is identical to a Computer Aided Design (CAD) system. It is the nature of the manipulation that distinguishes GIS from CAD. A definition of GIS is;

'an integrated system to capture, store, manage, analyse and display information relative to concerns of a geographic nature' Carter (1989).

With such a definition it can be argued that many organisations that systematically deal with maps and mapping can be said to have a GIS, whether the process involves the use of computers or not. It is the availability of inexpensive, powerful computers that have made the full power of a GIS readily available. It is now financially possible to carry out detailed studies that were impossible a few years ago.

GIS makes it possible for the engineer or resource planner to produce rapidly a combination of maps and tables that show together "where" (location of resources), "what" (kind of resources), and "how much" (quantity/limitation of resources). Resources may be taken here to include: reservoirs, water supply network, land cover, percentage impermeable area, contributing areas, etc.

The GIS is an institutional entity, reflecting an organisational structure that integrates technology with a database, expertise and continuing financial support over time. Thinking of the GIS in this way takes it beyond being just a package of technology - hardware and software capable of performing GIS types of function.

There are few books and periodicals on GIS available at the moment and the reader is recommended to consult the excellent book by Burrough (1986) on the principles of GIS in land resource, which covers the theory of GIS very well. This chapter will only consider theory where directly relevant to water supply applications.

Computing technology, hardware and software is changing rapidly (this is particularly true for GIS software), new products are announced almost every year, thus making it very difficult to make any recommendations as to software purchase, or indeed what is possible. The reader must keep abreast of current developments by studying the specialist press, for example the *International Journal of Geographic Information Systems* or *Mapping Awareness*.

The function of any information system, including a GIS, is to improve a users' ability to make decisions in research, planning and management. An information system involves a chain of steps from the observation and collection of data, through analysis, to the use in some decision making process. The distinction between data and information is important; that which reduces uncertainty, and enables decision making, is information; data is merely its transmission medium.

1.2.1. GIS Concepts

There are two distinct classes of GIS application and two differing technologies for those two classes. The different technologies are explained in the next section on data structures. The two main areas of application are as follows. Facilities Management systems, which are aimed primarily at managing infrastructure such as roads, sewers, customer information and other objects in the built environment. Another term for this type of GIS is Automated Mapping/Facilities Management (AM/FM). The vector based systems as opposed to the pure raster based systems are generally preferred for this kind of GIS. Land Information Systems (LIS), which are aimed primarily at natural resources applications frequently use image based data and are good at manipulating thematic map data. The raster based GIS are generally preferred for these systems, (the terms raster and vector are discussed in the next section and illustrated in Figure 2).

The two different branches of GIS are not mutually exclusive; one GIS package may be suitable for both types of application. There are also other branches of GIS concerned with weather forecasting, transportation routing, navigation and modelling systems.

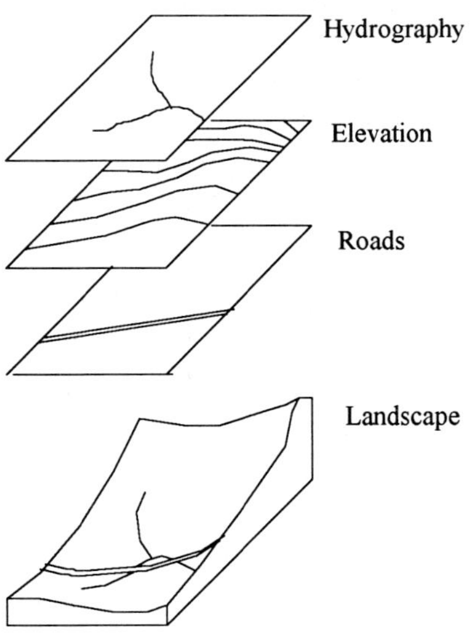

Figure 1. The overlay concept, the landscape is simplified into layers, each one representing an attribute of interest

Though digital mapping is an important concept in GIS, the principle that makes GIS useful and exciting is that different themes can be extracted and stored as separate map overlays, see Figure 1. The themes can be combined logically and mathematically to form new themes of interest by:

(i) an AND combination of two other themes, for example, a well drained soil and small surface slope,
(ii) an OR combination of two other themes, for example, a well drained soil or a sandy soil,
(iii) generation of corridor information, for example, areas within 10m of a domestic dwelling may be liable to urban irrigation,
(iv) use of mathematical models linking existing themes, such as erosion potential modelling,
(v) calculation of areas, lengths, etc.

1.2.2. Data Structures in GIS

There are five kinds of data to be represented in a GIS.

(i) Point features, e.g. location of soil samples, boreholes, manholes, rain gauges, burst water mains, pumping stations, trees, buildings.
(ii) Linear features, e.g. roads (on small scale maps), rivers, pipe lines, power lines, elevation contours.
(iii) Areas (polygons) with common properties, e.g. pressure zones, catchments, contributing areas, soil association mapping units, climate zones, administrative district areas, buildings and other land cover.
(iv) Actual or potential surfaces, e.g. ground elevation, variation of mean annual temperature, spatial distributions of rainfall, population densities.
(v) Temporal elements, e.g. changes in land use over time, changes to a pipe network.

There are two common generic data structures that are used by GIS to store these data; the vector and the raster data structures, see Figure 2. Any given GIS may use one or both of these data structures. Generally the vector structure is used when most of the data can be structured as points, lines or polygons, and the raster structure is used if most of the data represents areas of land, or if the data origin was an image. A third structure, the tessellated structure, is used for representing surfaces in a vector system; in particular the Triangulated Irregular Network (TIN) structure is favoured for elevation or surface data.

a

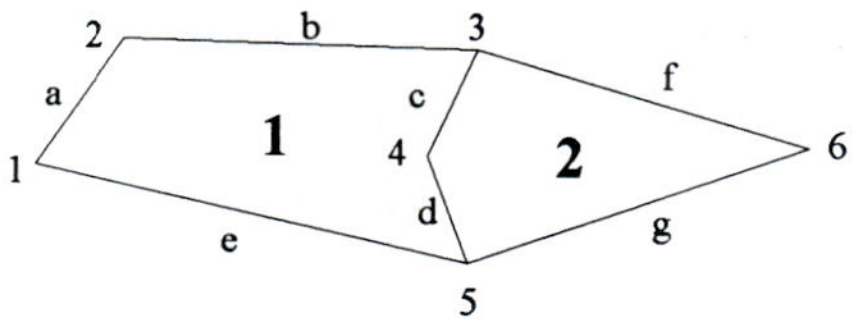

Link	Right Pol.	Left Pol.	Node 1	Node 2
a	**1**	**0**	1	2
b	**1**	**0**	2	3
c	**1**	**2**	3	4
d	**1**	**2**	4	5
e	**0**	**1**	1	5
f	**2**	**0**	3	6
g	**2**	**0**	6	5

node	x	y
1	23	5
2	25	7
3	34	7
4	33	6
5	34	4
6	41	5

Polygon	Owner	Landuse
1	Forestry Commission	Coniferous Forest
2	Forestry Commission	Mixed

b

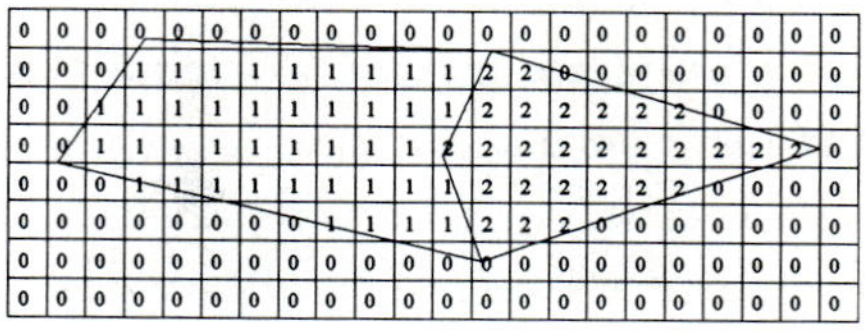

0 No data
1 Coniferous Forest
2 Mixed

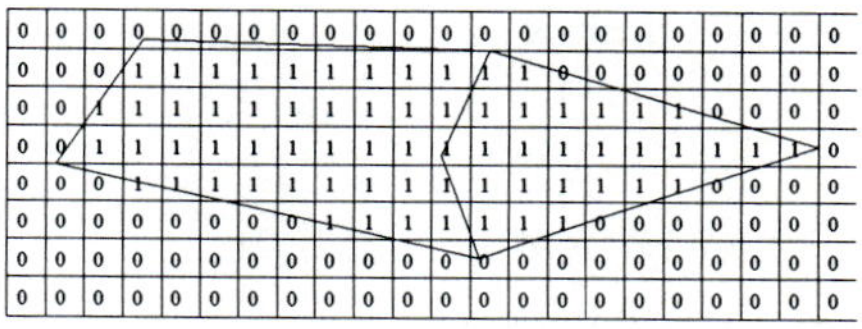

0 No data
1 Forestry Commission

Figure 2. (a) Simple vector representation, using the topologic model presented by dangermond (1982), more complex structures are used to improve access times,
(b) Raster representation, a raster layer is required for each attribute to be represented

1.3. DIGITAL ELEVATION MODEL

The common type of map used in GIS for inventory storage and land evaluation is the choropleth map; within each region of the map the value of the attribute represented is considered constant. However, in many situations we wish to use other data that varies continuously over the land's surface. This representation of a surface is often called $2^1/2$ dimension representation, or, since it is often used to represent elevation data, a Digital Elevation Model (DEM). Sometimes the term 'digital terrain model' (DTM) is used to describe this surface. Since the term 'terrain' is also used to describe the land cover as well as the elevations, we will consistently use the term digital elevation model to describe the surface alone and reserve 'terrain' for the combination of elevation and land cover.

DEM can be used to represent other surfaces than topography; peizometric surfaces, cost surfaces (the expense incurred when moving across the landscape) or the surfaces of geological units are just three examples.

In water supply and hydrology DEM's can be used; to calculate slopes and aspect, which can in turn be used to calculate contributing areas; to calculate pressure zones; to determine drainage networks and stream lines; to calculate erosion potential and to aid the design of networks.

There are many ways of representing surface data, but only the two described below are in common usage in GIS now. These are both local patch methods: one based on a regular grid of height values and the other based on an irregular triangular tessellation. Both methods have their own advantages and disadvantages.

1.3.1. Grid DEM

Perhaps the simplest way to represent the elevation is by way of a regular rectangular patch called an altitude matrix. In this method the area under study, the coverage, is divided into a number of rectangular cells like a raster, and each cell contains an elevation. Some of the features inherent in altitude matrices can be seen in Table 1.

Table 1. A DEM altitude matrix

7.8	7.8	8.0	8.2	8.3	9.4	10.5	12.0	9.1	6.0
7.8	7.8	8.1	8.3	9.3	9.5	11.1	13.2	12.8	9.1
7.9	7.8	8.0	8.2	8.3	9.8	12.2	14.5	13.0	12.1
7.8	7.8	8.1	8.5	10.4	13.1	16.2	13.3	12.7	8.1
7.8	7.8	8.0	8.2	10.8	14.1	17.5	15.2	12.1	6.0
7.8	7.8	8.1	8.1	10.0	12.2	22.4	14.8	9.2	5.8
7.9	7.8	8.2	8.2	9.7	11.2	18.6	12.6	15.4	7.1
7.8	7.8	8.1	8.8	9.1	14.3	14.9	16.0	17.4	8.8
7.7	7.8	8.1	8.3	9.2	14.2	15.6	17.2	17.2	7.9
7.9	7.8	8.0	8.7	8.8	13.9	16.2	18.1	19.4	5.1

The grid size is largely fixed by the most complex varying part of the model. In Table 1. the grid spacing has been chosen in order to represent the rapidly varying right hand side of the coverage, this means the left hand side has adjacent cells that only vary by a decimal point or so. To represent the subtlety of the surface required for urban catchments the grid size needs to be small and real numbers must be used to differentiate height from cell to cell. Grid DEM data is difficult to compress using techniques such as run length encoding, without losing information. To store the data required for a typical water supply catchment of 10km by 20km at a grid spacing of 20m will require (500x1000x4bytes) 2 Mbytes of storage.

Just what height should be allocated to a cell? Is it the mean altitude of the cell, the height of the centre point of the cell or the highest or lowest point within the cell? The height selected has important consequences for determining drainage basin boundaries in almost flat terrain, especially with large (50m plus) cell sizes.

Data derived from different sources will be of different qualities. The elevation data can be readily produced using modern photogrammetric techniques. However, if the data is extracted from existing maps, either as contour lines or as point elevations then spatial interpolation techniques need to be used. False pits and hollows often arise from this interpolation process.

Ideally, for hydrological studies, the height of each cell should be the lowest point in the cell, but since much of the data is currently derived from interpolation of contour data (isolines) the error of this interpolation exceeds the within cell error. It should also be remembered that in many situations the contour data itself is an interpolation of spot levels recorded by a surveyor.

Two of the most important products that can be derived from a DEM are slope and aspect, Equations (1) and (2). Like the DEM these can be manipulated just as any other layer in the raster GIS.

$$\text{slope} = \sqrt{\left(\frac{\partial z}{\partial x}\right)^2 \Big/ \left(\frac{\partial z}{\partial y}\right)^2} \tag{1}$$

$$aspect = \frac{\partial z}{\partial x} \Big/ \frac{\partial z}{\partial y} \tag{2}$$

1.3.2. Triangular Irregular Network (TIN)

A common alternative to the altitude matrix is the trianglular irregular network (TIN). Elevation is represented by a series of irregular triangular patches, Figure 3a. A set of points can be triangulated in any number of ways, but most TINs are based on the Delaunay Tessellation of irregularly spaced nodes (see Sibson, 1978, for a mathematical treatment of the properties of the Delaunay Tessellation). Values at points interior to a triangle are found by interpolation of the vertices, and the Delaunay Tessellation ensures that these values use the most local vertices possible.

The Delaunay Tessellation requires that a circle passing through the vertices of any triangle must not contain any other vertices, Figures 3b. and 3c. It can be seen from this definition that the Delaunay tessellation is the graphical dual of the Thiessen Polygon with the intersections of the bisecting lines between points being the centres of the circles, Figure 3d.

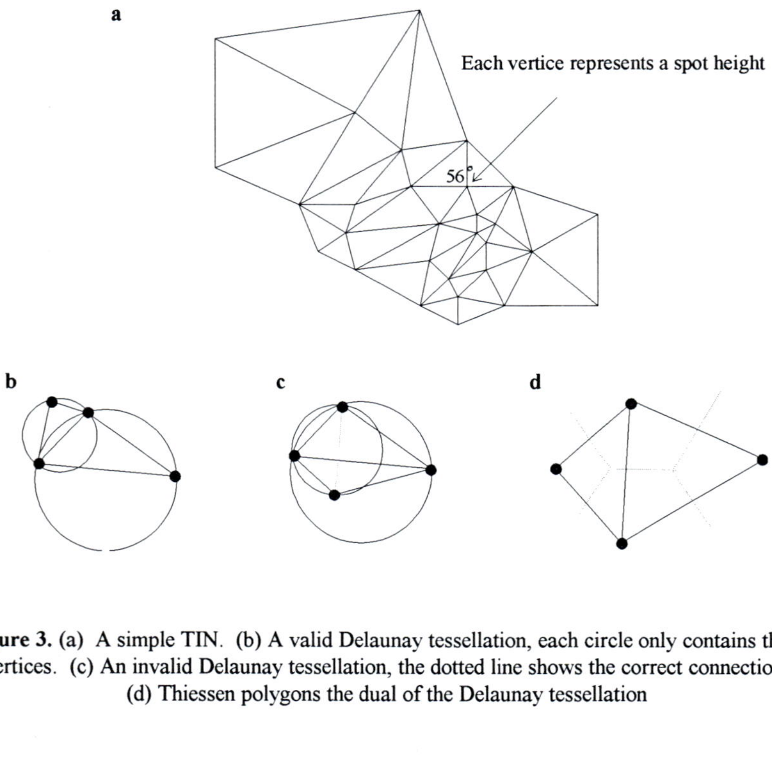

Figure 3. (a) A simple TIN. (b) A valid Delaunay tessellation, each circle only contains three vertices. (c) An invalid Delaunay tessellation, the dotted line shows the correct connection, (d) Thiessen polygons the dual of the Delaunay tessellation

In areas of gently varying topography the data points can be widely spaced, whereas in rapidly varying, or areas of great interest, the nodes can more densely placed. This means TIN's do not suffer from the data redundancy of the grid DEM. They can be stored as vectors, but, because they are much more constrained than the general vector representation, simpler structures can be used. A TIN list records the nodes in each triangle; then a further list records the *x, y, z* co-ordinates of each node.

In some ways the TIN makes computation easier. Using linear interpolation; contours consisting of straight lines can be readily drawn. Also as each triangle has a single slope and aspect these can be readily calculated. The false pits and hollows that often arise in interpolating the altitude matrix do not occur. However, like other vector structures, there is considerable computational effort involved in linking the TIN and its derivatives with other land cover information.

1.3.3. Applications of the DEM

Many useful themes can be derived from the DEM; drainage basin boundaries; potential stream lines within the catchment; input parameters for hydrologic models; erosion potential and possible sediment yields. Though these can all be calculated from the TIN, only the methods used for the altitude matrices will be described here.

To determine the catchment area draining to a particular point consider Figure 4a. which shows the direction of flow within each cell (the aspect). It can be readily seen that water from the north and north-west cells will flow into the target cell and that water from all the southern cells will flow away from it. But what about the north-east cell and the cell to the West? The flow directions are determined by whether or not the slope direction falls within the contributing angle θ (Figure 4b.).

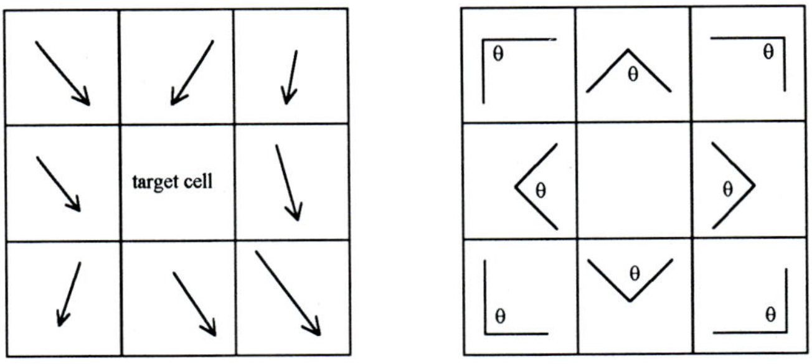

Figure 4. (a) Flows into the target cell (b) Contributing angles

Once a cell has been allocated to the catchment it becomes a target cell. This process continues until no more cells can be added to the catchment. The contributing angle is typically set at 90 degrees but may be increased to make more cells belong to the catchment. If a cell has zero slope then it may be considered to flow in all directions, or contribute to all cells surrounding it. In practice it is found that allowing flow in any direction from any cell with a slope of less than 2 degrees alleviates some problems with pits and hollows, see Elgy et al (1993).

Determination of potential stream networks can, conceptually, be carried out simply by calculating the area of the drainage basin upstream from every cell and excluding all areas less than a certain threshold from this accumulation map. The cells that define the resulting stream lines show the area draining into them and hence indicate the relative size of the resulting rivers. Problems arise from this simplistic approach:

- computational effort is extreme (unless more sophisticated algorithms are used),

- in valley bottoms that are wide, the slope perpendicular to the stream is zero and parallel rivers are often calculated, which must be removed by suitable heuristics.

By inverting the rules for flow into a target cell to flows from a target cell, irrigation command areas can be evaluated.

If the universal soil loss equation is employed, by considering slope landcover and rainfall, the potential for erosion of each cell can be calculated. The sum of all these potentials for a particular catchment gives the expected sediment yield for a river (see Bradbury et al, 1993).

1.3.4. Hardware and Software of GIS Systems

Figure 5. shows the typical layout of a single user GIS system. In most installations there will be more than one user and, in such cases, duplication of parts of the system will occur, for example a number of workstations may be networked and the console and CPU elements duplicated.

If the migration of source data is tracked through the GIS to the usable outputs, the hardware requirements can be considered in turn.

Data arrives at the GIS in one of five forms: paper maps, digital maps, remotely sensed images (either satellite or aerial photographs, digital or analogue), tabular data, and from existing data bases.

Digital maps are becoming more readily available in the industrialised world but are still not very common. The data when available is often very expensive. Transferring such data into the GIS requires a suitable device to read the supplied medium, and software to structure it into the database format. Paper maps are by far the most common source of data and need to be digitised to be incorporated into the GIS. There is a proliferation of defined and undefined standards for the transfer of digital data.

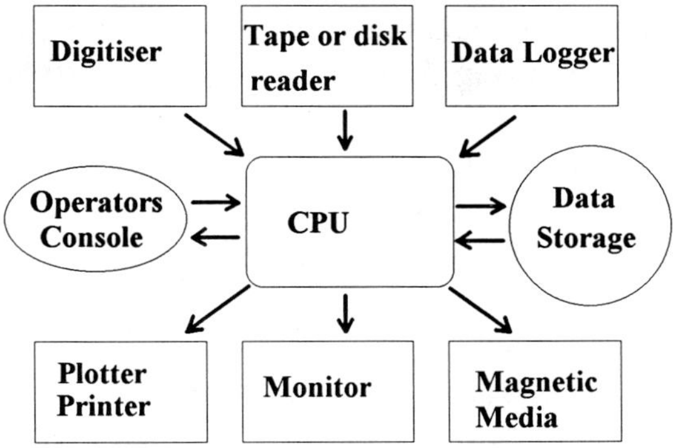

Figure 5. Hardware components of a typical GIS

To input remote sensed data requires either a magnetic tape drive or some other medium, such as an optical disk. Imagery in the form of photographs, or negatives should be scanned, or video digitised. Airborne videography is becoming more common for gathering certain types of imagery, and is especially useful for urban studies.

It is vital to record the source of the data as it is incorporated into the GIS, as well as other information such as revision history, accuracy of source data, map projections, scale of original maps, date of original survey and of digitisation, etc. Data, once incorporated within a GIS, tends to stay there for very long periods and is often passed from user to user with little quality control along the way.

Geographic Information Systems themselves run on almost all types of computer and the selection of hardware rests more with the software selected than a simple choice of the best computer hardware. Some GIS software requires the use of proprietary hardware; these should be avoided if possible. Generally the vector based GIS systems run on mini-computers or workstations, partly because of their memory and processor power requirements and partly because they tend to be the older GIS systems developed before micro-computers had sufficient power or networking facilities to share data. By running on multi-user/multi-tasking systems one user can be inputting data whilst another user manipulates other data in the system, and a third undertakes plotting.

If a micro-computer system based GIS is chosen then separate processor and memory boards may have to be inserted to perform image processing. The special chips on these boards allow for large amounts of information to be displayed upon the screen and to be processed very quickly. A mouse or tracker ball pointing device is an essential data input facility for the operator. The software supplier will give a specification as to the requirements of the host computer.

If, as recommended, a team of people is working on the GIS project it makes considerable sense to link the machines together with a Local Area Network (LAN) so that expensive peripherals and data can be shared.

Thought should be given to how the large amounts of data, which GIS generate, are to be stored. As an example, for a raster based GIS without data compression and assuming a 20m grid size, 5 themes, (land cover 1 byte, pipe network 1 byte, elevations, slopes and aspect 12 bytes in total) for an area of 20km by 20km will require approximately 14 Mbytes of hard disk storage. However, experience has shown that for urban drainage studies about 3 times this amount is required as a working store. This pushes the limits of micro-computer based GIS, even though with data compression techniques reductions to 10% or 20% can be achieved.

The hard disk storage of the computer provides the working store. Archival storage should be provided from either: magnetic tape storage, streaming tape, WORM or similar optical disk drive system

Output from the GIS takes a number of forms, not all of which are essential for the system, but there must be a way to visualise at the computer monitor the results of actions being performed and some way to produce hard copy. A colour display monitor for real time manipulation and analysis of the data is essential and this can be part of the control system of the computer or a separate monitor. Colour printers and plotters are useful for producing hard copy output, though a standard 35mm camera pointed at the monitor screen can in certain circumstances produce excellent (though not dimensionally correct) illustrations for presentations, reports, etc. There is still, however, considerable difficulty in printing colour reports and many publishers cannot handle raster images.

GIS software comes in a variety of packages. The two main types, as already described, are the vector based system and the raster based system. More modern systems permit the total integration of raster and vector data, allowing the advantages of both methods to be enjoyed, with few of the disadvantages.

Vector systems are often supported by traditional DataBase Management Systems (DBMS). The most common conform to the relational model, see Avison (1992). Arc-Info, the most widely used vector GIS package, follows this approach, Info being a relational DBMS in its own right The relational model is the basis of most DBMS used in organisations and businesses. This underlies the vector model's principle use as an asset or resource inventory system. A DBMS should allow access to appropriate parts of the database to different types of user, and prevent unauthorised viewing or changing. It should also maintain data concurrency, provide archive facilities and present a simple interface to the user for manipulating the data

Raster systems generally do not employ such strict data management. They have developed from image processing systems and are often used by a single user. Clearly these are generalisations, and many packages will embody aspects of both systems.

The most up-to-date systems are described as 'object oriented'. The distinction of object oriented systems is that all data items are described as being of one or more object type; e.g. a linear feature, a point, a vector polygon, a regular raster, a raster cell, a TIN, a DEM, etc. In addition to storing the description of the object, the methods of displaying, plotting and general manipulation are also carried with the object type, this is known as encapsulation.

Objects are hierarchical; rivers, roads and pipes will be objects that are descended from the linear object, each will, therefore, have the properties, behaviour and methods inherited from the linear feature, such as length. However they will each have behaviour and properties that are distinct; roads will have classes (i.e. 'A' roads and motorways); pipes and roads will not be able to connect to form a network.

The object oriented paradigm is currently of great interest to the computer science community. Object oriented programming languages, databases and, of course, GIS are under development, (Worboys et al, 1990).

There are several advantages that are stressed by advocates of the object oriented approach:

(i) it is intuitive as people naturally think in terms of objects,

(ii) by specifying behaviour, inconsistencies in the database can be reduced, for example sewers and water mains objects exhibit different behaviour and should not be part of the same network,

(iii) developing applications is easy; by having a hierarchical structure new objects are easily created.

1.4. REMOTE SENSING

Remote sensing in the broadest sense is defined as:

' the science and art of obtaining information about an object, area or phenomenon through the analysis of data acquired by a device that is not in contact with the object, area or phenomena under investigation'. (Lillesand and Kiefer, 1987)

This section will concentrate on remote sensing systems that use electromagnetic radiation for studying the Earth's surface.

1.4.1. Background

Remote sensing systems using electromagnetic radiation, have four basic components (Figure 6.): an electromagnetic source, an interaction with the earth's surface, an interaction with the atmosphere, and a sensor. The source of the electromagnetic radiation may have one of three origins; radiation generated by the sun, thermal radiation emitted from the earth and man-made radiation such as RADAR.

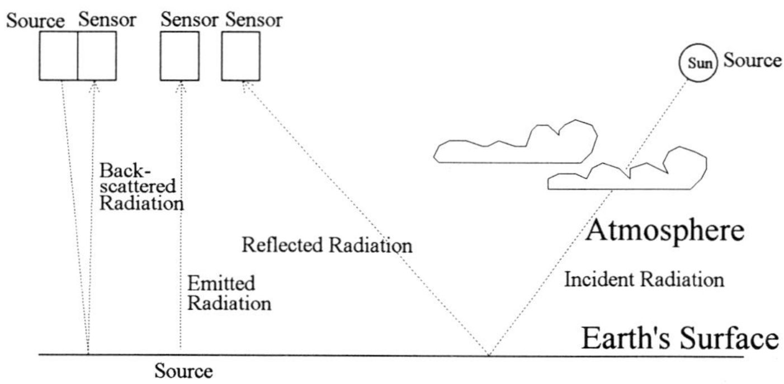

Figure 6. Relationships between atmosphere, earth, sources and sensors

1.4.2. Electro-Magnetic Radiation

Visible light is only one form of electromagnetic energy; the full spectrum includes radio waves, heat, ultraviolet rays and X-rays. Properties of electromagnetic waves are linked by the equation:

$$c = \lambda f \tag{3}$$

where *c* is the speed of light, λ is the wavelength and *f* is the frequency.

In remote sensing it is customary to describe electromagnetic waves by their wavelength. emote sensing is primarily concerned with the following bands:

the visible spectrum		0.4 to 0.7μm
with subdivisions into		
	blue	0.4 - 0.5 μm
	green	0.5 - 0.6 μm
	red	0.6 - 0.7 μm

Adjoining the visible red is infrared (IR), which is divided into three categories:

near - IR	0.7 - 1.3 μm
mid - IR	1.3 - 3 μm
thermal - IR	beyond 3 mm.

At much longer wavelengths are microwaves, 1mm to 1m.

The amount of energy an black body can radiate is a function of the object's surface temperature, and is expressed by the Stefan-Boltzman Law:

$$M = \sigma T^4 \quad (4)$$

where M is the total radiant excitance from the surface of the material (in Watts/m^2), T is the absolute temperature of the emitting material in ^oK and σ is the Stefan-Boltzman constant (5.669×10^{-8} W m^{-2} $^oK^{-4}$). The sun emits radiation similar to a black body of 6000 ^oK, by comparison the earth emits radiation similar to a black body at 300^oK. Hence it is harder to detect radiation emitted by the earth, and the thermal sensors which detect this energy tend to have low spatial resolution.

1.4.3. Atmospheric Interactions

All radiation detected by remote sensors passes for some distance or "path length" through the atmosphere. The path length will vary widely depending upon the sensor and platform combination used.

As the radiation passes through the atmosphere specific bands of wavelength are absorbed by particular gases. As a consequence there are certain wavelengths whose intensity is significantly reduced or even missing when the solar radiation reaches the earth's surface having passed through the atmosphere.

This has important implications for remote sensing; only those regions of the spectrum where the radiation is not blocked out by absorption can be used. These regions of the spectrum are called *atmospheric windows*, and sensors are designed specifically to 'look through' these windows.

1.4.4. Interactions at the Earth's Surface

When electromagnetic radiation interacts with the Earth's surface it can be reflected, absorbed or transmitted through the surface it strikes. Different wavelengths will interact with the same surface in different ways and thus the proportion of each wavelength that is reflected back will vary according to the surface. These changes yield specific spectral patterns, which are related to the surface with which the radiation has interacted.

1.4.5. Spectral Reflectance

The reflectance characteristics of earth surface features can be quantified in terms of the relationship between the incident radiation and the reflected radiation.

$$\text{Spectral Reflectance} = \frac{E_{R(\lambda)}}{E_{I(\lambda)}} \times 100 = \frac{\text{energy of wavelength } \lambda \text{ reflected from object}}{\text{energy of wavelength } \lambda \text{ incident on object}} \times 100 \qquad (5)$$

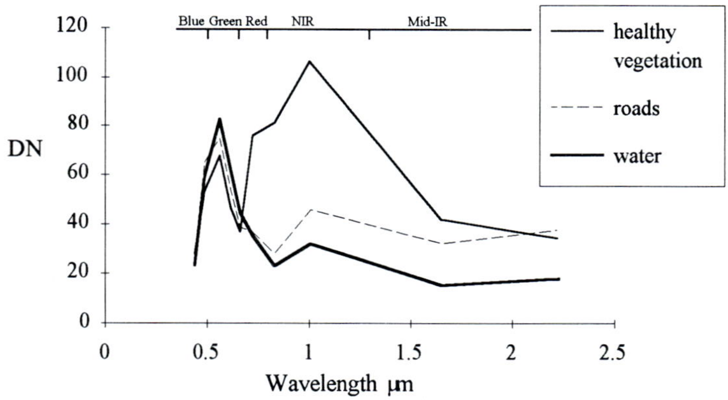

Figure 7. typical reflectance curves for three different surfaces, taken from an air-borne scanner thematic scanner (ATM) image of an urban area

The spectral reflectance is expressed as a percentage and is determined for specific wavelengths. Experience has shown that many of the earth's surface features can be identified and studied on the basis of their spectral reflectance in different bands. Figure 7. shows some typical reflectance curves for three features in different bands.

1.4.6. Thermal Infrared Radiation

All bodies above 0°K (absolute zero) emit radiation (Equation 4). Devices sensitive to the appropriate part of the electromagnetic spectrum (3 to 50 μm) can record some of the energy emitted from the Earth's surface - the thermal infrared.

There are four factors that influence the amount of thermal infrared energy emitted and hence received by a sensor:

Emissivity
All features differ in the amount of incident radiation that they re-emit. Thus objects may be distinguished by their emissivity.

Kinetic Temperature (T)
The temperature as recorded by a thermometer or a direct thermal sensing device.

Thermal Properties of the body
The thermal capacity, conductivity, diffusivity and inertia together indicate the ability of a body to retain, conduct or diffuse heat.

Rate of Heating
This may depend on aspect, as the rate of heating is a function of both the rate of absorption of solar energy, and the amount falling upon a body.

1.4.7. Spectral Signatures

The characteristic spectral response patterns of various bodies and features to incident or emitted radiation are often referred to as *Spectral Signatures.*

Although it is true that many features exhibit very distinctive spectral signatures, they must be regarded with caution and only used as guides in image interpretation, since both temporal and spatial effects will influence the spectral response of individual features.

The spectral reflectance for healthy green vegetation usually shows the peak and valley characteristics of Figure 7. These are caused by the different absorption and reflectance characteristics of the different components of leaves.

In the visible spectrum the valleys are dictated by the pigments in the leaves absorbing energy. Chlorophyll strongly absorbs energy around 0.45μm and 0.67μm - for this reason vegetation is seen as green.

At about 0.7μm the reflectance of the near-infrared (NIR) increases dramatically, with the leaf typically reflecting 40 to 50% of the incident energy between 0.7μm and 1.3μm - most of the remaining energy is transmitted with only about 5% being absorbed. This high reflectance results primarily from the internal structure of the plant leaves. Since structures vary between plants, the reflectance varies; this enables differences between plants to be distinguished.

It is often possible to tell if vegetation is stressed. Stress may be caused by water shortage, water-borne pollution, air-borne pollution, restricted root growth, root damage, or seasonal effects. This may result in the production of chlorophyll being decreased or stopping entirely, or damage to the internal structure of the leaf, either shifting the reflectance towards the red, or causing a change in near infrared reflectance. This is illustrated in Figure 8.

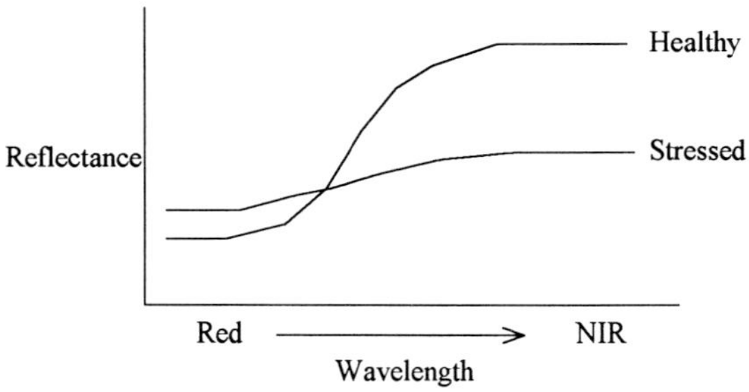

Figure 8 The difference of reflectance between red and near infrared (NIR) of healthy and stressed vegetation

The difference between healthy and stressed vegetation, and between vegetated and non-vegetative surfaces can be accentuated by using a ratio of the red and NIR bands, to produce a new image, the normalised difference vegetation index (NDVI):

$$\text{NDVI} = \frac{R - IR}{R + IR} \tag{6}$$

The NDVI is generally scaled to fit a suitable range of values, i.e. 0 to 255.

1.4.8. Photography

The technology of photography is well known. It should be noted that, as well as the usual monochrome and colour film, the near infra-red part of the spectrum can be captured to give a false colour image.

Video cameras can be used in much the same way as traditional colour films. The images are, however, easier to capture in digital form for image processing.

1.4.9. Linescanners

A linescanner measures the radiance of the earth's surface along a line perpendicular to the path of the platform - an aircraft or satellite. As the platform moves forward successive lines are scanned enabling a picture to be built up. There are essentially two types of linescanner; the optical-mechanical or 'whiskbroom' scanner, Figure 9b., and the 'pushbroom scanner', Figure 9a.

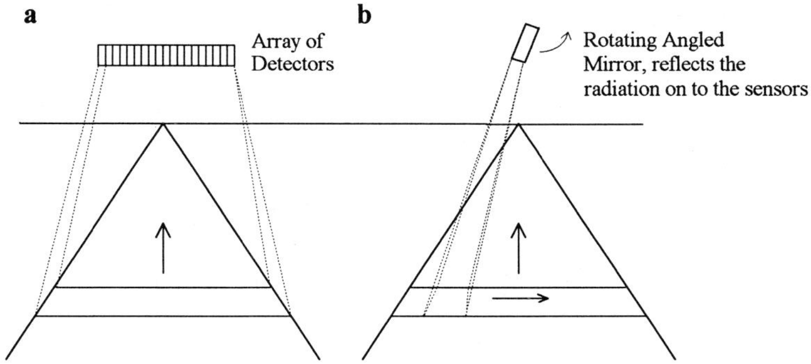

Figure 9. (a) Pushbroom scanner (b) Whiskbroom scanner

Linescanners were developed to escape from the spectral limitations of photographic and vidicon methods, essentially 0.3 to 0.9 μm, since they also allow data to be collected for the mid and thermal infrared parts of the spectrum.

The multispectral Scanner (MSS) is an example of a whisk broom scanner (Figure 9b.). It contains mechanical components that are subject to failure. Whiskbroom scanners have other limitations too (eg poor signal to noise ratio) and they cannot have both high spatial and high spectral resolution since the time spent receiving radiation from any particular point during a sweep of the mirror is so short.

The 'push broom' linescanner gets round these problems. It uses tiny radiation sensitive cells - called charge-coupled devices (CCDs). Currently these can operate up to 2.4μm, so the full visible and Near IR range can be monitored - however Thermal IR is not yet accessible. The CCDs are arranged in thousands along a linear array. The radiation is directed on to the CCDs by an optical system, so no moving parts are required to build up an image (Figure 9a.). Each CCD of an array measures directly the radiation from an individual segment of the ground swath as the platform moves over it.

1.4.10. Microwave Imaging Systems

The wavelengths used in microwave imaging systems are in the range 1mm to 1m; some thousand times longer then thermal infrared. There are two distinctive features that characterise microwave energy from a remote sensing standpoint;

- microwaves can penetrate the atmosphere under almost all meteorological conditions; haze, light rain and snow, clouds and smoke,

- microwave reflections from the earth's surface have different characteristics from their counterparts in the visible or thermal portion of the electromagnetic spectrum; as a result of the longer wavelengths involved they see roughness in a different way.

There are two forms of microwave imagery systems; passive microwave systems and active microwave systems or radar.

Passive microwave sensors employ an antenna to receive the natural radiation emitted by all earth's bodies. This is much weaker than within the infrared region but is still a function of temperature.

A passive microwave system comprises an antenna to pick up the radiation, an amplifier and a detector. The ground resolution (the detail that can be seen) depends on the antenna size - the larger the antenna the better the resolution. This is, of course, a problem when trying to carry the instruments aloft in an aircraft.

Radar is an acronym for Radio Detection And Ranging and was developed during World War II for military purposes. Just as with passive microwave sensing, the spatial resolution of a radar system is determined by the size of the antenna. To at least partially get over the problem of having large antenna, most airborne radar remote sensing systems use an antenna fixed below the aircraft and pointed to the side - it is therefore known as side looking airborne radar or SLAR.

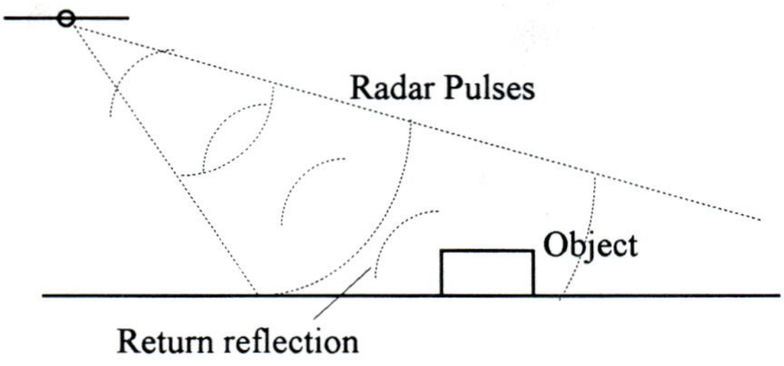

Figure 10. The principles of SLAR

The system operates by emitting discrete pulses that are microseconds long, which travel to an object and are reflected back. Figure 10. illustrates the principles involved.

To maximise the resolution of an image, we need to minimise the radar beam width so that only a narrow strip is illuminated. Beam width (β) is directly proportional to the wavelength (λ) and inversely proportional to the antenna length *(L)*.

$$\beta \propto \frac{\lambda}{L} \tag{7}$$

It can be seen that to minimise the beam width the antenna length must be maximised. Systems that achieve this with real antenna are referred to as real-aperture radar. However, the same effect can be achieved by electronically simulating long antenna. Such systems are called synthetic-aperture radar (SAR) systems and are employed in high altitude aeroplanes or in satellites.

1.4.11. Platforms

Platforms vary from ground level hand held instruments to geostationary satellites. As the sensor moves further from the object increased spatial coverage is achieved at the expense of spatial resolution.

Ground level observations enable the most detailed information about an object or area to be obtained; ground truth data for calibrating, classifying and confirming airborne or satellite imagery observations; detailed visual record at a specific time and place (i.e. a photograph), or local spectral data using a radiometer.

By far the most common platform used in aerial photography is the Aeroplane. Generally there is a deliberate overlap in images of about 60% along the flight path. This allows two adjacent photos to be viewed through a pair of binoculars (stereoscope) to obtain a 3D view which facilitates intepretation and allows ground height information to be extracted.

To many people the term remote sensing is synonymous with satellite remote sensing. Satellites are classified according the altitude of their orbits. If a satellite is positioned at an altitude of 35,900 km above the earth's surface it will have an orbital speed that matches the earth's rotation - termed geosynchronous. A satellite placed in such an orbit will appear fixed above the same point on the earth - i.e. geostationary . Geostationary satellites include METEOSAT (Europe), GOES (USA) and GMS (Japan). These weather satellites provide linescanned images that have considerable application in the provision of real time meteorological information and in estimating rainfall volumes over large areas and time spans - catchments of hundreds of km^2, and time periods of months rather than days (see Siyyid, 1993).

A polar orbit takes a satellite over or near the North and South poles at an altitude above the Earth between 500 and 1000 km. The satellite's orbit moves progressively around the globe and, over a period of time, coverage of most of the earth's surface is obtained. The repeat cycle is the time taken for the satellite to return to the same position over the Earth's surface - expressed in days. Polar orbit satellites have spatial resolutions within the range $10m^2$ to $100m^2$. Polar orbit satellites include LANDSAT 1 - 5 and the SPOT series, that supply linescanned data. Data from high resolution military satellite with resolutions down to 2-5m is slowly coming onto the market, though data about sources, etc. is difficult to come by.

1.5. REFERENCES

Avison D.E. (1992) Information Systems Development: A Database Approach. 2nd Edition, Blackwell Scientific Publications.

Bradbury P.A., Lea N.J., Bolton P. (1993) Estimating Catchment Yield: Development of the GIS-based Calsite Model. Report OD125, April 1993, HR Wallingford.

Burrough P.A. (1986) Principle of Geographical Information Systems for Land Resources Assessment. Clarendon Press Oxford.

Carter (1989) On Defining the Geographic Information Systems. Fundamentals of Geographic Information Systems: A compendium, edited by Ripple WJ, pp3-6.

Dangermond J. (1982) A Classification of Software Components Used in Geographic Information Systems. Proc. US - Australia Workshop on the Design and Implementation of Computer Based Geographic Information Systems, Honolulu Hawaii, pp70-91.

Elgy J., Maksimović Č., Prodanović D. (1993) Using Geographical Information Systems for Urban Hydrology. International Conference on Application of Geographical Information Systems in Hydrology and Water Resources, Vienna, Austria.

Lillesand T.M., Kiefer R.W. (1987) Remote Sensing and Image Interpretation. 2nd Edition, Wiley.

Sibson R. (1978) Locally Equiangular Triangulation. The Computer Journal, v21 n3, pp243-245.

Siyyid A.N. (1993) The use of METEOSAT data for rainfall/runoff modelling. PhD. Thesis, Aston University, May, 1993.

Worboys M.F., Hearnshaw H.M., Maguire D.J. (1990) Object-Oriented Data Modelling for Spatial Databases. Intention Journal of Geographical Information Systems, v4 n4, pp369-383.

2. Use of GIS and Remote Sensing in Water Supply: Applications

John Elgy
Thomas William Charnock
Peter David Hedges
Department of Civil Engineering
Aston University
Aston Triangle, Birmingham B4 7ET
UK

2.1. INTRODUCTION

This Section applies the theory of the preceding Section to the problems of water supply systems. The first section looks at the use of GIS with environmental process models. The second section covers the use of GIS and remote sensing techniques in demand evaluation and forecasting. The third section looks at two current applications of remote sensing for water supplies; the detection of buried objects and leaks, and the assessment of water quality. The last section considers the issues of using GIS and remote sensing in a utilities environment.

2.2. ENVIRONMENTAL PROCESS MODELLING AND GIS

Environmental process models are now common in research environments, these techniques have been applied to modelling erosion, fluvial systems, groundwater and ecological systems. This section looks at the use of GIS with environmental process models for water supply systems in general, it then examines two case studies in detail; the use of GIS with urban drainage models and the linkage of a particular groundwater model with a GIS. These serve as examples of the considerations involved in linking any model to a GIS.

NATO ASI Series, Partnership Sub-Series, 2. Environment – Vol. 15
Water Supply Systems – New Technologies
Edited by Č. Maksimović, F. Calomino, and J. Snoxell

If a model is to be used within an operational environment for decision support then the administrative effort must be minimised. Linking a model or several models to a GIS can provide this, particularly if a corporate GIS already exists within the organisation, (see the section on GIS Within a Utilities Environment). The advantages of linking a GIS to environmental process models are:

- data management functions are subsumed by the GIS,
- several models can be used together in an *ad hoc* manner, without reconstructing the data,
- different models draw on the same version of the data,
- datasets can be easily constructed,
- consistent user interface across models,
- output can be visualised in combination with other data within the GIS.

For a water company the operations that such a system might aid are:

- assessment of environmental impacts of water resources or supply system development or operation,
- gaining information about the processes affecting the water resources upon which the system draws and which might affect supplies, such as siltation of reservoirs, saline intrusion into aquifers, etc.,
- modelling in-system processes such as flow processes, algal growth etc.

There has been an explosion in the use of environmental process models with GIS, though these are mainly being developed and used in research environments. Chairat and Delleur (1993), linked the hydrological model TOPMODEL with the GIS GRASS, to model hillslope runoff production and gain an understanding of the effects of soil variability. On a large scale De Lange and Van der Meij (1993) describe a national groundwater model for the Netherlands, based on the custom GIS REGIS that manages a huge hydrogeological dataset of Holland. Kern and Stednock (1993) use hydrochemical models to identify surface waters that exceed standards of heavy metal concentrations.

One of the strengths of this approaches is that models can be linked in an *ad hoc* manner as they share similar input and output formats. Klaghofer et al. (1993) linked the GIS Idrisi to the soil erosion model EPIC and to the Agricultural Non Point Source Pollution Model (AGNPS), developed by the US Agricultural Research Service and the University of Minnesota, to estimate sediment and nutrient produced by an alpine catchment.

Ross and Tara (1992) describe a system designed to be used in an operational environment from regulatory authorities to small consultant companies. It involves the linkage of several hydrological models; a groundwater model, surface water models and specially written code with the commercial GIS SPANS. They praise the GIS for performing the time consuming and tedious data management tasks such as georeferencing, and claim considerable savings in time and money:

"what would have taken many months of effort in terms of model parameter definition previously can now be accomplished using digital data and the GIS".

Other processes are also amenable to modelling. McDonnell and Macmillan (1993) use a set of different models not only to study the environmental aspects of large dam development but also social and economic aspects of the system.

2.3. PRACTICAL IMPLEMENTATION OF GIS TO URBAN DRAINAGE MODELS

The development of general purpose GIS systems has urged and encouraged their application in various fields of engineering, particularly in the area of complex urban infrastructures. This application is spreading in two major directions:

i. generation of maps and inventories,

ii. links with simulation models and the graphical presentation of results.

Although it is difficult to draw the dividing line between them it could be generally stated that the final product of the first type is a digital map and it's accompanying database. The second group is more complex and it requires specific target oriented information processing (modelling). It is the second application that is of interest here.

Current urban drainage models are well established and generally adequate for the simulation and design of urban drainage networks. There is, however, a problem with the acquisition, manipulation and storage of data for these models. The combination of the technologies of remote sensing and geographical information systems allows a solution to these problems. This section describes a project in which GIS are being used to provide the input data for two drainage areas: one a pilot study of the Miljakovac catchment in Belgrade is described in detail, and one a full scale study of Dresden is referred to. Smith (1992) presents a different way of linking a GIS and an urban drainage model where he employs the drainage model as an intrinsic part of a GIS. This presents new difficulties in the calibration of finely grained distributed models that are beyond the scope of this text. Further descriptions of these methods can be found in Elgy et al. (1993), Elgy (1992) and Prodanović (1992).

Many details of urban drainage networks are already stored in GIS systems of type (i) above, but the most effective use of that data relies on type (ii) applications. This section examines how four commercial GIS packages were used to create input suitable for two standard urban drainage models. The packages concerned were: the raster-based GIS Idrisi, and the vector based; ARC-INFO, INTEGRAPH, and the CAD package Autocad which, with suitable macro programs, offers a sort of GIS capability. The outputs from each of these GIS were then to be passed to two urban drainage models: WALRUS, from Hydraulics Research UK, and BEMUS, from IRTCUD, the University of Belgrade. The University of Belgrade has a well-instrumented and documented catchment at Miljakovac in Belgrade and this was used as the study area to investigate the approach. The detailed objectives of the project were:

- to form the land cover and DEM for a catchment,
- from the DEM identify the contributing areas to each inlet to the sewer system, and determine how these can be modelled in each GIS,
- investigate the effect of flow diversion structures, for example, walls, road chamber etc.,
- calculate the statistics of each subcatchment required for WALRUS (now replaced by SPIDA) and BEMUS,
- compare the statistics from each method with traditional hand methods,
- prepare input files for the models from the GIS,

- investigate the different performance of the model for each method of deriving the input files,

- consider the sensitivity of output to each method of generating input files,

- assess the resources required for each approach in terms of manpower, training, computer power and software costs.

However, due to time limitations many of these objectives have yet to be met. This section describes the use of the Idrisi GIS system (Eastman, 1992), and how the data stored on this system was used to create input files for the BEMUS and WALRUS urban drainage models. Only the BEMUS model has produced output for analysis so far. The work continues.

The vector-based GIS's proved difficult and slow for data acquisition and modelling. Their complex data structures meant that it was also difficult to write suitable programs to extract the required data for the urban drainage models. Almost no progress was made using ARC-INFO, and AUTOCAD proved useful only as a tool for digitisation and to move data in the DXF format to Idrisi. Digitising the maps for INTEGRAPH proved very slow and though the accuracy of the maps produced was very high, failure to correctly brief the operators of the system meant that many of the features required for urban drainage modelling, such as topologically consistent polygons, were not included in the database. Correct briefing meant that the correct maps were produced for the Dresden project (Maksimović et al., 1993)

2.3.1. Choice of GIS

Idrisi is a low cost raster GIS available from Clarke University in the USA (Eastman, 1992). It is very popular as a teaching system, and its simple file structure means that new modules can be readily written for the system. Two new modules were written for this project. The first was to determine the contributing area to each inlet in the sewer system - similar modules could be written to determine irrigation command areas and areas that have potential to be fed by gravity from reservoir, (Chapter VIII.1.) The second was to take these contributing areas and write the output files suitable for the urban drainage models, with features such as percentage impermeable area, area of contributing area, length/width ratio, average slope of catchment etc - again similar modules could be written to output data for pipe network analysis.

In addition, the relevant Idrisi module, which forms a digital elevation model from point data, was rewritten to make use of a transputer parallel processing board added to the micro computer system for faster execution. This improved the execution time from around 12 hours to 20 minutes.

The only other modules from the original Idrisi package used for this project were: the viewing module, COLOR; the slope and aspect module, SURFACE; the reclassifying module, RECLASS; the area calculating command, AREA; and the geometric correction module, RESAMPLE. The cross tabulation command CROSSTAB was used to verify the output files calculated from the post-processing program, generating direct inputs for the urban drainage models. Other modules can be used for traditional GIS applications.

2.3.2. Results and Recommendations

Much work still needs to be undertaken to complete the objectives mention above and only one successful run of a simulation program has actually taken place. This run (on the BEMUS storm water drainage package) indicated that after tuning the percentage roofs connected to the sewer system, the model accurately matched the measured outflow hydrograph - as did the traditional hand method. Provided that most of the catchment had been allocated to an inlet the exact boundaries seemed to matter very little. There were very large discrepancies between the contributing areas to particular inlets derived from the hand and from the automatic methods.

The lessons learnt from the GIS approach can be summarised as:

- initially the GIS approach is much more time consuming than drawing boundaries roughly by eye and measuring areas by planimeter, but this may be largely due to the steep learning curve of developing and using new technologies,

- it took 2 days to digitise the 4 map sheets covering the Miljakovac catchment, though if this had only been one sheet half a day would have been sufficient - having gained experience and a larger digitiser, it was faster for the Dresden catchment,

- using aerial photography took only a few hours to create the land cover map - digitisation took 2 days, aerial remote sensing offers the advantages that land-cover data can be obtained, that is up to date and easily ported into a GIS,

- formation of the DEM is easier from stereo pair aerial photography than by any other method,

- the GIS approach is more objective than traditional approaches,
- the availability of digital maps makes the GIS approach faster;
- existing urban drainage models do not need rewriting to take account of GIS technology, though direct interaction would be beneficial it would limit the choice of GIS,
- grid based GIS are easier to deal with than vector GIS for this purpose,
- once a GIS study has been carried out, changes to the catchment or to the sewer system can be readily investigated,
- more work is still needed on subcatchment boundary delineation, particularly with regard to flow diverting structures,
- levels of accuracy of the digitised maps and of the resulting output still need the judgement of an engineer; completely automatic techniques still do not exist,
- if the source of the information is an existing paper map then this is still the weakest part of the system,
- field visits are essential, if only to discover which roofs are directly connected to the sewer system; this data could be directly entered into a portable computer,
- more work needs to be done on examining other GIS approaches and urban drainage models,
- very flat catchments still prove a problem when determining contributing areas, see the Section 1. in Chapter VIII.

In this example Idrisi proved a very flexible and useful tool for developing an objective approach to obtaining the input files for the traditional urban drainage model, BEMUS. The programs developed for calculating the contributing areas from the DEM and to create the input files for BEMUS and WALRUS are very useful tools. With further development the GIS approach could provide the standard method for gathering information for urban drainage models.

2.4. GROUNDWATER MODELLING WITH A GIS

In the Seventies several large groundwater developments were initiated around the UK. However, the effect of the drawdown on soil moisture and the resulting impacts on crops were given little attention. Hedges (1989) investigated this issue with reference to the Tern River Catchment within the Shropshire Groundwater Scheme.

Hedges found that in areas where the soil is in hydraulic continuity with the aquifer and where the root zone is within a critical height from the water table, crops could experience a loss in available water due to the drawdown of the water table. GIS technology gives a method for defining areas that are vulnerable to this effect. Hedges produced maps of vulnerable areas, this was achieved by combining map layers representing agriculture, critical height, solid and drift geology, depth to water table and drawdown.

The current research effort is to produce a system that will dynamically model, groundwater, soil moisture and crop yield. This section describes the linkage of a GIS (GRASS) with a groundwater model (Modflow).

The Geographic Resources Analysis Support System (GRASS) was developed at the US Army Construction Engineering Research Laboratory (USACERL), (Shapiro et al., 1993). The principle data structure used by GRASS is the raster, though it can handle vector data. It was chosen for this project for several reasons:

- GRASS is public domain software, it is freely available, along with extensive documentation and, significantly, the source code is readily available,

- the standard C programming language is used as its implementation language, and GRASS comes with large libraries of functions for developing new applications, which means that any shortcomings can be programmed around and new modules written,

- a raster data structure is used, that is more suitable than the vector structure for problems that involve the handling of features whose attributes can vary continuously over space, such as the transmissivity in an aquifer,

- GRASS shares features with many other GIS; it is atemporal and attempts to handle vector data as well as raster,

- there is a large co-operative user community, and easy contact with users and developers is possible via active internet discussion groups.

Modflow was written by the US Geological Survey (McDonald and Harbaugh, 1988), and models groundwater flow in three dimensions using a finite difference numerical method. It is stable, well developed, has a large user community and is arguably the standard groundwater model used throughout the world. The source code, in FORTRAN, is public domain, well written and easy to understand and modify.

The issues that must be addressed, concern both the conceptual and practical features of Modflow and GRASS. This section concentrates on the conceptual issues that are likely to be common to most GIS, in particular reconciling the differences in spatial and temporal data representation between the process model and the GIS.

2.4.1. Spatial Representation

The GIS and the model represent two-dimensional space in slightly, but significantly, different ways. Both use a grid structure, but GRASS uses a raster structure that imposes more constraints on the data. In GRASS the grid lines are normally oriented north-south east-west, there are regular intervals between grid lines, but these intervals may differ between north-south and east-west (Figure 1a.). Modflow has fewer constraints; like GRASS it uses a rectilinear grid but this can be oriented in any direction and grid intervals may vary along either direction (Figure 1b.).

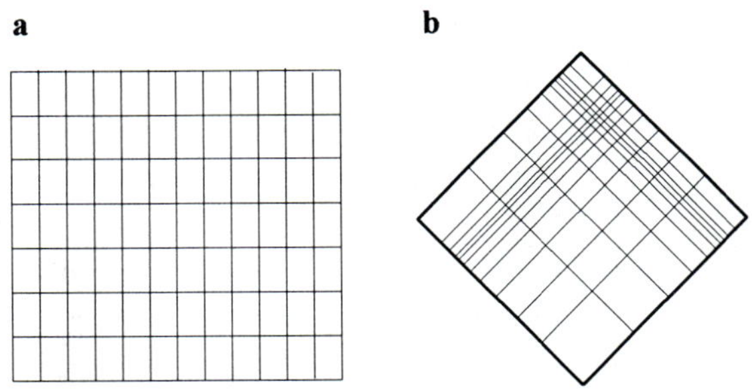

Figure 1. (a) A raster grid (b) A typical modflow grid

Modflow allows variable grid interval and orientation for important reasons. Within the study area there will be regions where the gradient of hydraulic head changes rapidly within a few metres, such as around an operational borehole, conversely there are other areas where the gradient is nearly constant through space over tens of kilometres. In areas where there is a rapid change in gradient, error, between the model and reality, can be reduced by minimising the size of grid elements, see Rushton and Turner (1977). However, this is unnecessary in those regions with a small change in gradient. A large grid interval can save considerable computation time by reducing the number of calculations required. Simple functions were therefore developed to sample GRASS data to the more irregular Modflow style grid, and to resample the output back.

The resampled grid is still oriented north-south, east-west. Where flow is through an anisotropic medium this may not give an adequate representation of reality. Several authors have stressed the importance of having the directional axis of the equations upon which Modflow is based, parallel to the principle axis of permeability (Rushton and Redshaw, 1979). However, this problem will not be faced in the current project as the model will be applied where isotropic conditions can be assumed in the x,y plane.

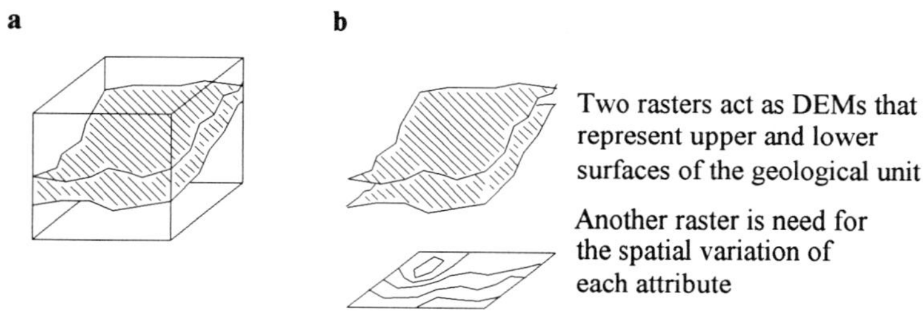

Figure 2. (a) Simply geological situation. (b) Upper and lower surfaces of each unit are represented by a DEM, another raster is needed to carry the attribute information

For this project GRASS is required to manage data that can vary through three-dimensional space, in particular it must cope with geological data, Figure 2a. The raster structure is fundamentally 2 dimensional, though it can be used to represent three dimensions as an altitude matrix (see the Section 1. of the Chapter VIII., section on DEM). The method chosen represents the surfaces of geological units with a raster DEM; each pixel holds the value of the height of the surface at that point. Another raster is required to hold the value of each attribute as it changes across the geological unit Figure 2b. The onus is on the user to use such data correctly, as GRASS has no structure to logically link rasters used in this manner. This method has several advantages. (i) It is easy to produce; as most of the raw data is in the form of borehole logs, interpolation functions can be easily applied to estimate the surfaces of the geological units. (ii) The use of functions such as kriging or splines to interpolate parameters is valid across a geological unit but these functions do not necessarily hold across geological boundaries. By explicitly separating the attributes of different units into different rasters these functions can be applied more easily. (iii) Most importantly, this approach closely resembles the way Modflow handles three-dimensional data. In Modflow the third dimension is represented as layers, and for each layer attributes such as transmissivity and specific storage are represented as Modflow grids. Layers are not necessarily flat but can be adjusted to follow geological boundaries.

This kind of modelling, which is known as $2^1/_2$ dimensional surface modelling, has the draw back that complex geology, such as extreme folding, cannot be easily represented.

2.4.2. Temporal Data Representation

A significant issue is how to deal with temporal data - particularly that which varies at different time scales. In this study geology can be considered static. Agricultural land-use can be considered static over a growing season. Weather can change from day to day. Groundwater levels can vary within a few hours next to an operational well but remain nearly constant over a season elsewhere.

Modflow is able to cope with transient flow situations; groundwater flow that varies with time. Time is divided into stress-periods that are further divided into timesteps. However, as yet GRASS has no explicit model of time (this is in common with most traditional database models, spatial or otherwise, see Kowalczyk and Kemp, 1993). This makes the analysis of temporal phenomenon cumbersome. The usual solution is to treat raster overlays as snap shots in time, and give them appropriate names; e.g. *watertable.june.84*

As with three-dimensional data, the onus is on the user to manage these data correctly and analyse them in meaningful ways. For example, if a parameter is required between available snap shots, the user must devise a method of calculation using the raster manipulation facilities available.

The research into temporal GIS is still largely at a theoretical stage, see for example; Kowalczyk and Kemp (1993) or Wachowicz and Healey (1993). Research has generally been aimed at solving the traditional database problems of inventory and asset management systems, see the final section. Al-Taha et al. (1993), give a useful bibliography of current research. Only a few of the authors have touched on the issues of linking any kind of environmental model to a temporal GIS, Hazelton (1991) is an example.

2.4.3. Remotely Sensed Data

Remotely sensed data on its own is of little value to a groundwater model. It requires the functionality of a GIS to make use of this data source. For example; Modflow models recharge to the aquifer simply as a grid of values that is added to a given layer at the start of each stress period. It is up to the user to calculate these values. A GIS greatly facilitates this process.

Land-use influences recharge to the aquifer by modifying the infiltration rate and evapotranspiration. Remote sensed images, such as landsat TM, can be classified to provide the data source for land-use using the image processing modules of GRASS. This data set will be further manipulated and combined with other datasets, such as geology, drift, soil and rainfall, to produce maps of infiltration rate and ultimately aquifer recharge.

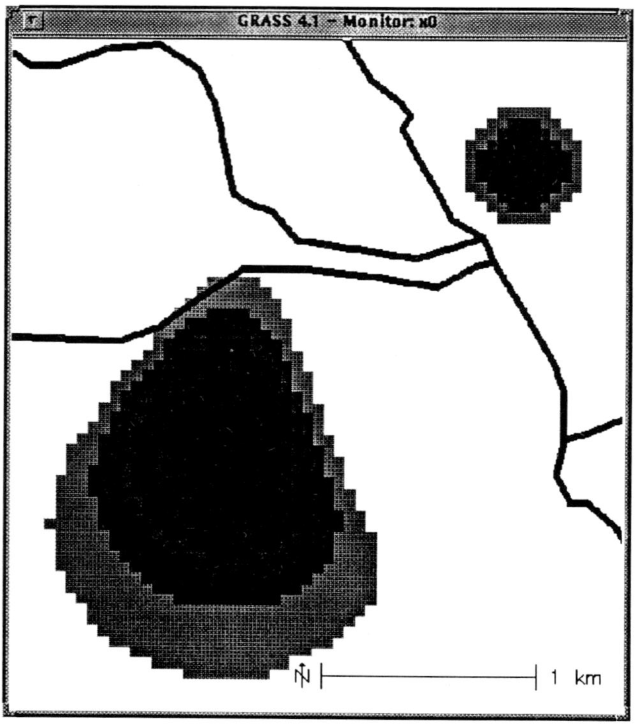

Figure 3. Typical output from the GRASS-modflow system
Here two wells have been modelled giving, after 5 days, two areas of drawdown in a raster layer. The river Tern has been overlaid. Where the river passes over the drawdown, some loss of surface water is expected (depending on the conductivity of the riverbed) and the efficiency of the augmentation scheme is impaired

2.4.4. Discussion

Linking environmental process models with GIS can provide a powerful tool for environment managers. Ross et al. (1991) found that considerable savings in time and money could be made when they linked a groundwater model with a GIS to investigate pollution from a mine. However, the task is not trivial, particularly as the current generation of GIS is only beginning to deal with the dimensions of height and time. These difficulties are likely to become less as GIS become more sophisticated, Hazelton (1991) suggested that as a GIS becomes more fully functional in 4 dimensions it "...subsumes dynamic modelling almost by default". Figures 3. and 4. demonstrate some of the output from the GRASS-Modflow system.

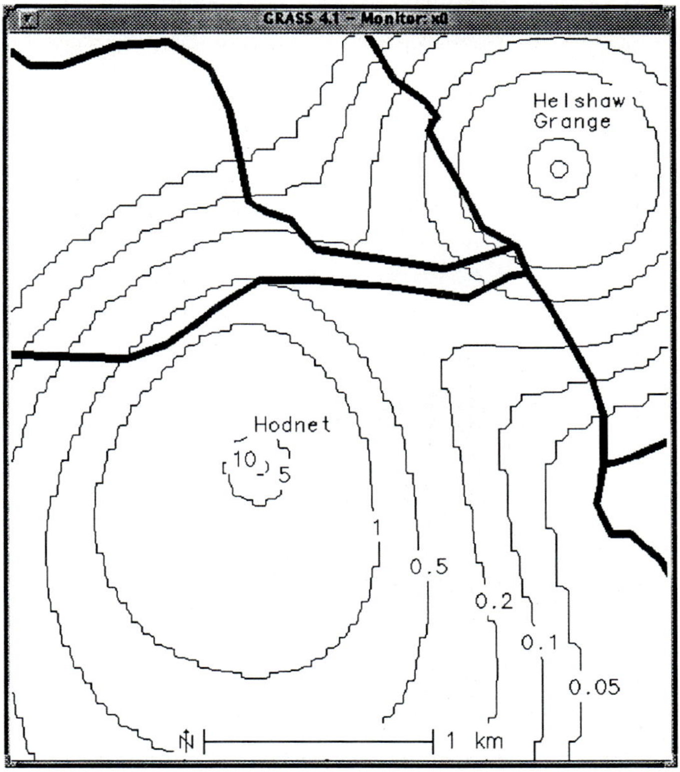

Figure 4. Typical output from the GRASS-modflow system. Vector contours have been derived from a raster layer of draw down after 22 days pumping, again the river Tern is overlaid, as well as name labels for the wells. After 22 days of pumping the areas of drawdown are considerably expanded, and the efficiency will be further impaired.

However, users should not be insulated from the considerations of the model construction, otherwise there is the risk of using the model in inappropriate ways. For example, it would be inappropriate to use the GRASS-Modflow link in its current state in areas where there is considerable anisotropy in the x,y plane. Reddi (1990) discussed some of the pitfalls to be avoided when using groundwater models and lists the following issues;

- using an over complex model, "over-killing", that may not be warranted by the problem it is being applied to or the quality of the data available,
- misconceptualizing the system,
- using inadequate parameters,
- misintepreting the results.

Groudin and Lite (1990) discuss the problem of scale with regard to groundwater models; regional groundwater models can miss important local effects and local models can miss regional trends. Groudin et al. (1990) stress the importance of understanding the principles and the limitations of modelling techniques, a necessity which may be lessened in future with the application of expert systems that can aid the modeller construct the model and interpret the results. Grayson et al. (1993) warn that a user may have an "..unrealistic sense of reality of model accuracy" induced by the sophisticated nature of the user interface and data handling and visualisation functions. Clearly these are considerations that can be applied to all dynamic models, and should be kept in mind when linking these models to GIS.

2.5. GIS AS A TOOL TO PREDICT WATER DEMAND

In order to improve the design of water supply systems it is necessary to obtain an accurate forecast of the demand the system would face. This prediction can be based either upon extrapolation of the existing demand without reference to the components of that demand, or upon demographic studies and economic models of water consumption. Demographic methods of forecasting future water requirements have ranged from simple projected population multiplied by projected per capita usage, to multiple regression analysis of different types of user and their demands for water. The crudity of the first is obvious, but the second suffers from the same problem in that without knowing how these users are distributed it is little more than a better projection of the aggregate consumption of water. The detailed spatial modelling of GIS allows for each small area to have an accurate demand profile associated with it, and this demand can then be incorporated into the design of the water distribution system.

For the best predictions of water demand we need to know three things:

i) the location of users,
ii) who the users are,
iii) what is the average water consumption of each type of user.

GIS can assist in determining i) and ii), and remote sensing can also assist in acquiring much of the raw data. As for iii), the modelling of demand functions is beyond the scope of this chapter, but is discussed in a paper by Hughes et al. (1994). Hughes' approach is to derive demand functions in linear, semi-log etc. form, and to determine the parameters of these equations by linear regression using as input parameters such items as housing density, lot size, family size, etc. For example, Hughes presents the following equation for summer demand:

$$Q_s = b_0 b_1^{PPH} b_2^{LOT} b_3^{AV} b_4^{WHC} \tag{1}$$

where Q_S is the unit capita demand, PPH is people per household, LOT is lot size, WHC is soil type and AV is the assessed value of the property. Interestingly they do not include household income within their prediction model. Bishop and Hughes (1993), offer a more comprehensive paper describing their GIS Water Planning Model (GIS-WPM).

In the UK small area statistics are available from the national census, and this data includes such items as age, martial status, occupation, housing and car ownership. The data can be spatially referenced in accordance with the national mapping agency (the Ordinance Survey, in the UK) system to give the data for each enumeration district (typically a few tens of households). Extracting information from this data is a complex process but with suitable regression models an estimate of each enumeration district's water demand can be obtained. The use of such data is, of course, widespread in marketing (particularly direct postal mailshots) and in planning shopping and recreational facilities.

A number of commercial companies have come into existence to take census data and other data, such as credit and debt data, share ownership data etc, to arrive at measures of affluence and demand for a particular type of product, for each enumeration district (Slieght and Leventall, 1994). These companies could just as easily provide likely demands for water as for soap powder. Similar geodemographics studies are available for the rest of western Europe (Watts, 1994).

As an example the process of calculating demands for imaginary pressure zones in Birmingham using a GIS will be illustrated. For illustrative purposes a much simplified equation will be used. Assuming per capita consumption within a given region is $2m^3$ and each household contributes an additional $4m^3$ them demand per month is:

$$\text{demand per month} = (2 \times \text{population}) + (4 \times \text{households}) \qquad (2)$$

Figure 5. shows raw 1991 census data for Birmingham: (a) is the population and (b) the number of households, by ward. Average population and households per cell is found by dividing each ward by the number of cells in that ward, and to find the demand per cell Equation (2) is applied. Figure 6a. shows imaginary pressure zones, these were created by digitising around areas, on a DEM of Birmingham, where height is approximately equal, (Figure 6a.). The domestic demand within a pressure zone is the sum of the demands of the cells contained within it. Figure 6b. shows the results of this calculation.

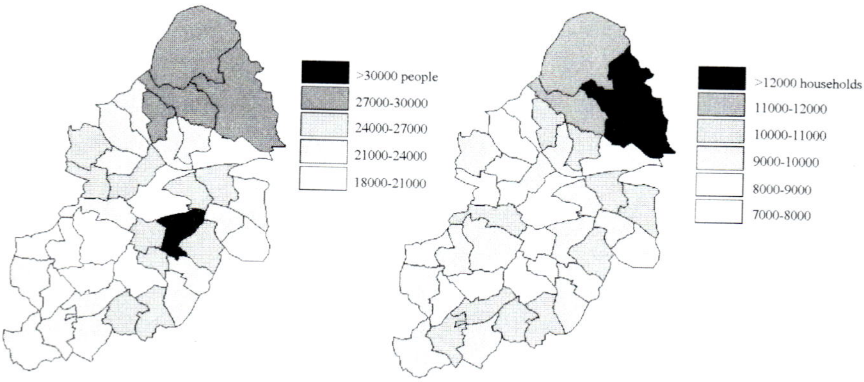

Figure 5. (a) Population by ward (b) Households by ward in Birmingham

This method depends on the assumption that population and household density are constant across a ward, which is seldom true. A more reliable result may be obtained by using smaller units; in the UK census data is available at the enumeration district level, which may contain as few as ten houses.

Once we have the domestic demand layer created it is necessary to create the industrial layer. As with domestic consumption we take the available data and use that. Such data takes the form of either existing consumption data or projections based upon the development plans for the region and the location of the industrial consumer.

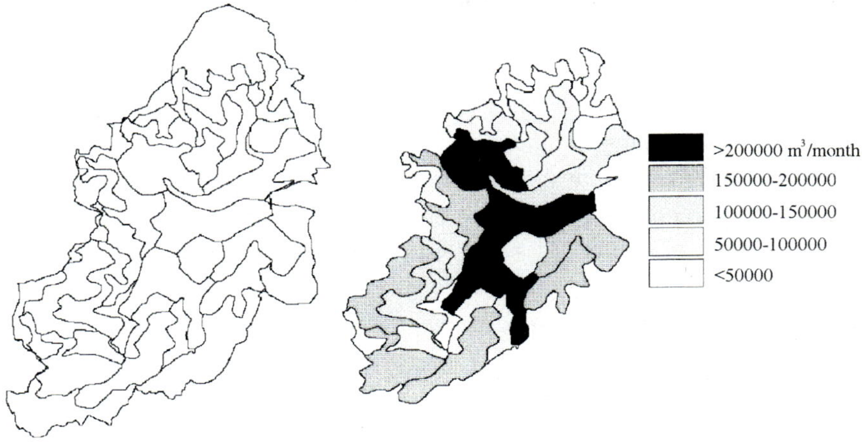

Figure 6. (a) Imaginary pressure zones within Birmingham (b) Domestic water demand by pressure zone

Hughes et al. (1994) identify a third layer of water demand: urban irrigation. In many parts of the world the watering of gardens and parks is the significant part of the demand for water. In Utah it is common for this irrigation water to be provided, untreated, in a dual system by specialist irrigation companies. This contribution needs to be taken into account when forecasting water demand.

In some areas, plot size (included in the regression model) gives a measure of the demand for irrigation water, but a better measure can be obtained by remote sensing. Satellite remote sensing (pixel resolution about 30m by 30m) has too coarse a resolution for accurate measurement but use of a vegetation index (as described in Chapter VIII.1.) will give an indication of the proportion of each pixel covered by vegetation. Airborne remote sensing has a better resolution (about 1x1m) than satellite imagery, and this can be used to map urban vegetation, see Blagejević et al. (1994). A vegetation index image is shown in Figure 7a: note that the vegetated areas are considerably lighter in shade than the non-vegetated area (roads, houses and factories etc). Figure 7b. shows the same image with a simple reclassification into two classes; areas with and without vegetation. Knowing the average irrigation rates of the principal vegetation, we can then calculate the water demand per pixel or polygon - though there is evidence that smaller plots may be irrigated at higher rates than large expanses of lawn.

Urban irrigation is associated with domestic houses, and if information regarding the spatial spread of houses is available, then an estimate of demand can be made. This information can be extracted from maps, which is particularly easy if the map is in digital form. Alternatively remote sensing provides a more timely source of information. Traditionally remote sensing classification is based on spectral signatures, however increasingly other information such as context and texture available from the image is being used to give a better classification - this will be illustrated in the following example.

Figure 8a. is a thermal image of an urban area from an air-borne sensor which includes residential and urban areas. In Figure 8b. the image has been classified into two classes; pixels with a radiance less than 70DN (relatively cold pixels) are given a value of 1 and pixels with a radiance greater than 70DN (hot pixels) are given the value 0. As a result, as can be seen, roads and buildings stand out clearly.

It is possible to further enhance Figure 8. by reclassifying it on the basis of size or shape rather than on spectral properties. Figure 9a. shows the same image further enhanced, small specks and very large areas (which cannot be houses) have been removed. Another reclassification and size filter was subsequently applied to produce a layer of roads, which was used as a mask to remove the slivers of road that had been mis-classified as houses. In most cities the difference between residential and commercial areas is sharp, cities are generally zoned. Using this information or visual interpretation of the image, it is possible to cut out the industrial areas which will not contribute to domestic urban irrigation, but which may irrigate their own green space extensively.

A further rule can be applied to give a first estimate of irrigable land, if we assume that all land within 20m of a house is irrigated we get the raster layer in Figure 9b. and an estimate of 0.26km^2 irrigable land within this 0.75km by 0.75km sample area. Clearly this estimate could be improved; for example the intersection of this image and a classified image of vegetation could be used, or roads could be excluded.

The above approach relies on the distinctly different spectral characteristics of vegetation and man made structures, and on the different shape and size characteristics between residential buildings roads and other buildings. It works because each house in this suburban zone is surrounded by vegetation and this allows each house to be automatically recognised and treated as an object distinct from the road network.

In the absence of accurate census data, remote sensing may also be used to determine the type of housing (or other land-use), and the housing densities within the distribution area. This process is, however, far from automatic and will need further research.

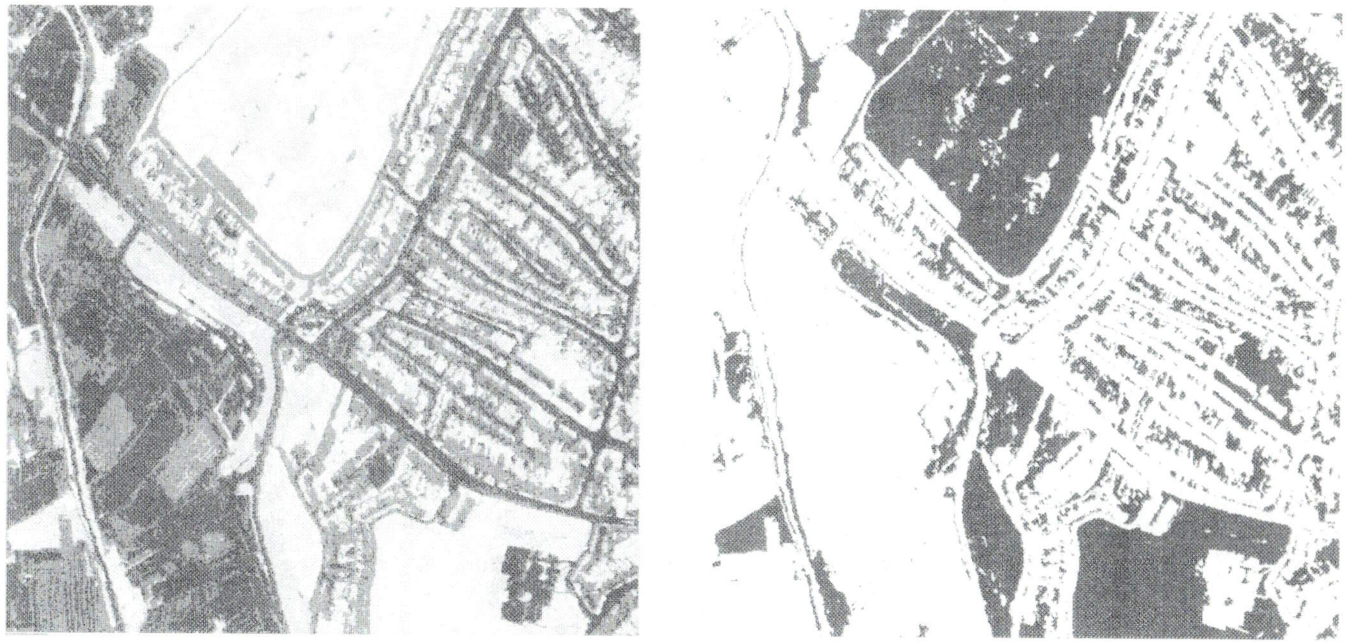

Figure 7. (a) Part of an image of the Normalised Difference Vegetation Index (NDVI) band, derived from air-borne thematic mapper red and infrared bands, of an urban area. The printing process has degraded the image, in the original each pixel can potential take a value of 0 to 255
(b) A simple reclassification of the NDVI to highlight areas of vegetation, black areas show vegetation white areas little or no vegetation

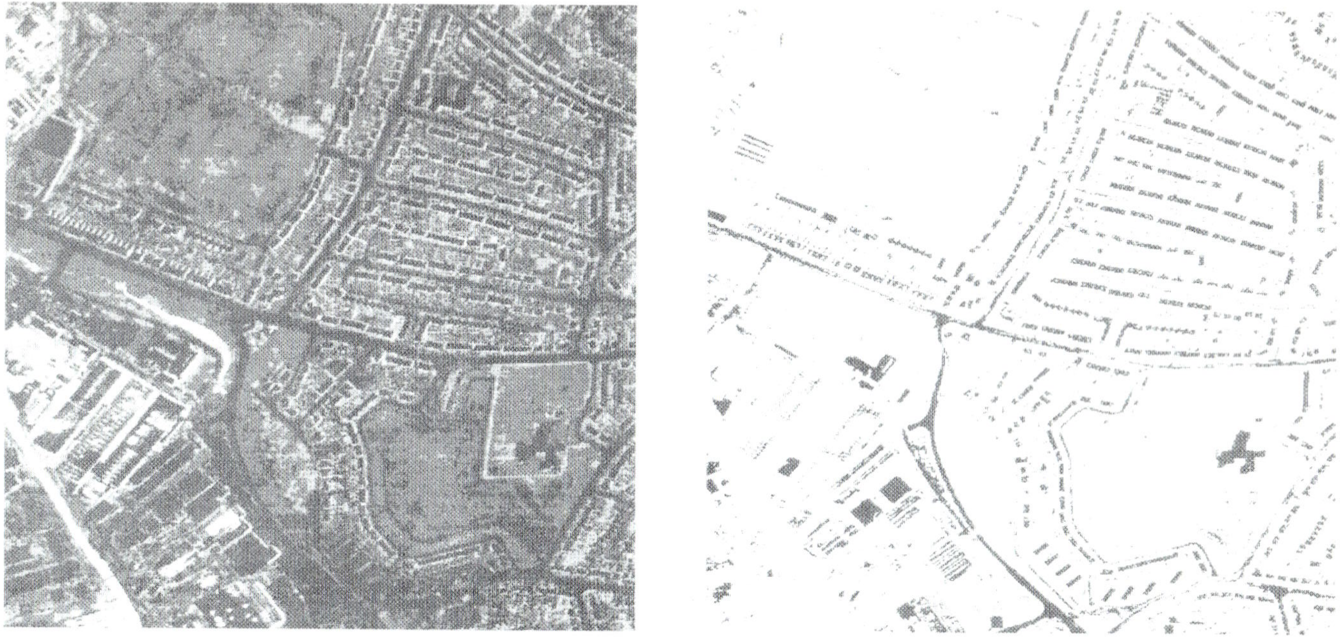

Figure 8. (a) A thermal image of Dudley, an Urban Area in the Midlands of UK. Again the printing process inevitably leads to a loss of information, in the original image each pixel may take a value of 0 to 255 depending on the surface emittance of thermal radiation
(b) A simple reclassification of the same image, black pixels are these with a radiance of less than 70DN and white pixels have radiance greater than 70DN.

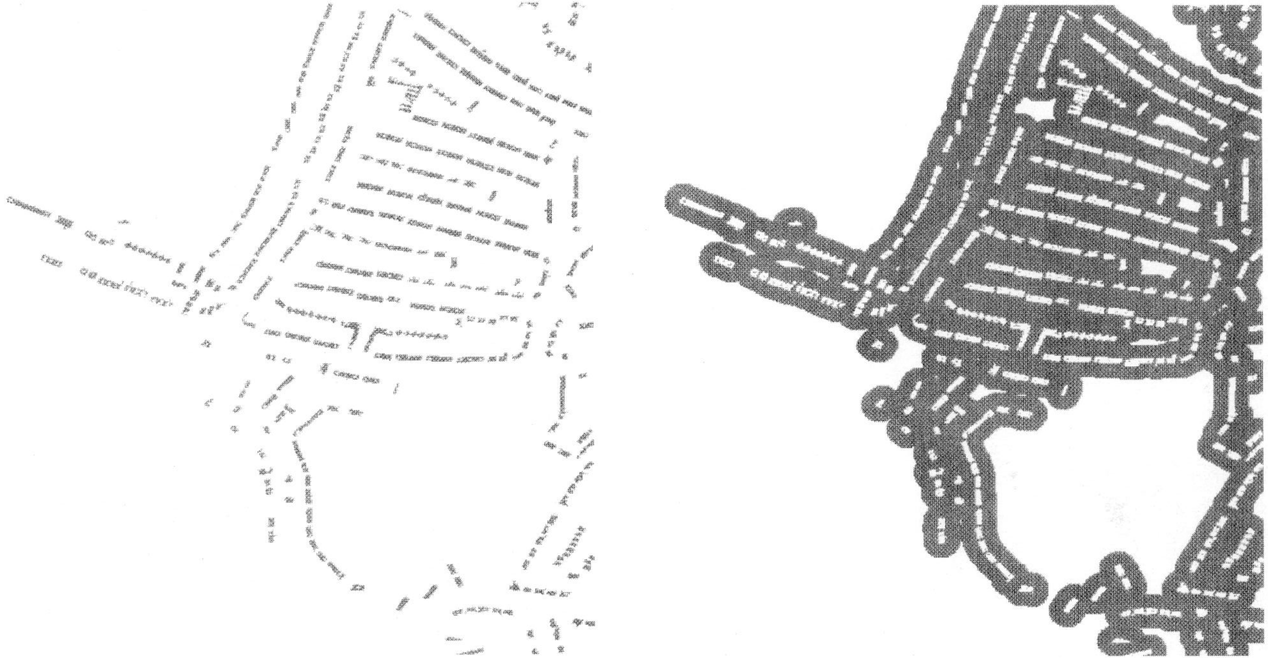

Figure 9. (a) Cleaned-up layer of houses, small areas and very large areas have been removed with a size filter, the industrial zone has been remove by cutting it out, roads have been removed by using a layer of roads as a mask.
(b) Areas within 20m of a house, a rough estimate of irrigable land of 2.6km^2

The combination of the different layers of demand is accomplished in a raster GIS. In the resultant image each pixel will carry an estimate of the value of total water demand for it's area. We have seen how pressure zones may be overlaid onto that total demand layer, to give total water demand for each zone. This method is not presented as an alternative to demand projections based purely on historical use, rather as a complimentary technique to reduce uncertainty. It can be argued that the more information and knowledge you have the more accurately you can predict the future or understand the present. Furthermore this approach will allow areas where there is a discrepancy between the expected demand (given by the profile of different users) and the actual demand (a large proportion of which may be due to unidentified leaks or some other undesirable process) to be identified as areas which need further investigation.

2.6. REMOTE SENSING APPLICATIONS IN LARGE WATER SUPPLY SYSTEMS

The discipline of remote sensing is in a period of rapid development. Sensors are improving, and new problem domains are being explored. This section will cover the current use of these techniques in applications that have relevance to water supply systems; the detection of buried objects and the assessment of water quality. The user is referred to such publications such as *Remote Sensing of the Environment* or *Photogrammetric Engineering and Remote Sensing* (PE&RS) to keep abreast with with future developments.

2.6.1. The Use of Remote Sensing to Detect Buried Pipelines and Leaks

The advantages of locating objects and leaks before excavation are clear - there is a financial saving because of the increased speed and mininal effort of the operations. Remote sensing systems can aid in: the location of pipes and other buried objects for maintenance, the identification of possible leaks, the detection of free regions underground into which a new pipe might be placed, the detection of voids that might affect construction or imply a cave-in, and the mapping of networks.

As systems age, as the competition from different utilities for underground space increases, particularly in urban areas, and as comprehensive GIS based inventories become more common the use of remote sensing techniques are likely to become more important. Current techniques use ground penetrating RADAR and thermal infrared sensors. Vehicle or handheld platforms are generally used, the large spatial coverage provided by aircraft and satellite platforms has yet to be regularly exploited because of the restrictions on spatial resolution.

2.6.2. Thermal Infrared Techniques

Thermal infrared techniques rely on temperature differences at the ground surface as an indication of what may lie underneath. For a pipe to be detectable it would have to have a significant temperature difference to its surrounding material. Weil and Graf (1991) report on the use of thermal imaging to detect the presence of a hot water pipe and to detect voids. It is in the detection of leaks that this technique would be most useful. Infrared techniques have been used very successfully in the past to detect the discharge of springs into water bodies such as streams or lakes (Brereton and Downing , 1975; Brereton and Hall, 1983). Brereton and Downing (1975) listed two factors which determined whether a discharge can be identified:

- the temperature difference between the groundwater and the surface water,
- the dilution ratio.

Their work suggested a temperature difference of at least 2°C and a dilution ratio of 1:2 groundwater to surface water was required - indicating that a temperature difference is 1°C can be detected.

The temperature difference between the saturated soil (caused by a leak) and the normal ground surface temperature can also be used to identify either water leaking from pipelines or natural springs. Weil (1993), reported on detecting leaks in buried pipelines using an infra-red imaging camera - a plume of water flowing from a pressurised pipeline causes a reduction in ground temperature.

Both Weil (1993) and Weil and Graf (1991) rely on either handheld or vehicle mounted cameras and work at very high spatial resolution. A more useful system would involve an aircraft and have a resolution of about 0.5m.

Deeply buried pipes with slow leaks where the water will not migrate to near the surface are not likely to affect surface temperatures. The sensitivity of imaging thermographic cameras is around 0.1^{o}C, so these methods are only suitable for large leaks. Their use is likely to be in the detection of unreported large leaks as part of a leak reduction strategy, rather than the pin-point locating of leaks that have been reported by other means. Tracy and Walton (1989) report on the application of airborne thermal infrared imagery to monitor about 233 km of a New York state canal and about 691 km of a major covered channel in California. The method was to detect moisture along the channel edges, it was found to have particular advantages in reaching isolated stretches.

Surrogate data may be used as an alternative to direct moisture monitoring, in particular vegetation data. Within an area of water-stressed vegetation, a patch of lush vegetation might indicate an abnormal source of soil water possibly a leak. The remote sensing of vegetation was discussed in Chapter VIII.1. and using a normalised difference vegatation index (NDVI) the difference between stressed and non-stressed vegetation is clear. This method requires both a knowledge of the position of the pipe and for that pipeline to be traversing an area of homogenous stressed vegetation. Clearly it will take time for an effect to be discernable so again this is a method that might be used as part of a leak minimisation strategy to identify pipes that might warrant further investigation.

2.6.3. RADAR Systems

A different approach to detecting buried phenomena is the use of ground penetrating RADAR (GPR) systems. Unlike thermal techniques these are active systems, which emit microwaves that are reflected by the ground surface and by other surface such as a pipe, a soil layer or a water surface beneath the ground. The return signal must be processed to remove noise and to construct an image that may be intepreted and used.

As with all radar systems resolution is effected by antenna length, this must be balanced by the need for portabilty. The image quality is also affected by soil properties, particularly the soil's conductivity and it's dielectric properties. These will vary in three-dimensions within the soil, meaning that considerable processing may be required to obtain a usable image. Several authors report strategies for dealing with this: see for example, Osumi and Ueno (1988) and Ho et al. (1993). Kikuta and Tanaka (1989) give figures for the performance of their system; they successfully detected 59 out of 64 gas pipes and state that "a 200 mm diameter pipe is clearly detected at a depth of 2.3 m".

The different networks beneath the ground coexist in three dimensional space, there is a need for the information to be presented to the user so that the spatial relationships between objects, including depth, are explicit. Caldecott et al. (1988) describe a system which will give vertical slices through the ground, as well as plan views.

Currently handheld or vehicle mounted platforms are used. The increasing performance and decreasing size of computers allows them to be portable and enables processing to be performed on the spot, giving real-time information. As yet airborne platforms are not regularly used, however the potential for subsurface objects to be accurately detected from an aeroplane has been demonstrated by Clark et al. (1993) who use RADAR to detect buried landmines, both metal and plastic.

Weil et al. (1993) use a combined method; thermal infrared images with a large spatial coverage are examined to locate and roughly gauge the position of subsurface objects and regions of interest in contaminated waste sites. Ground penetrating RADAR was employed to give the fine detail that is need for excavation, this proved to be an efficient strategy. Whilst searching for suspected petrol tanks beneath a Petrol Station, they also found a void which led to the discovery of the cracked sewer main which had caused it.

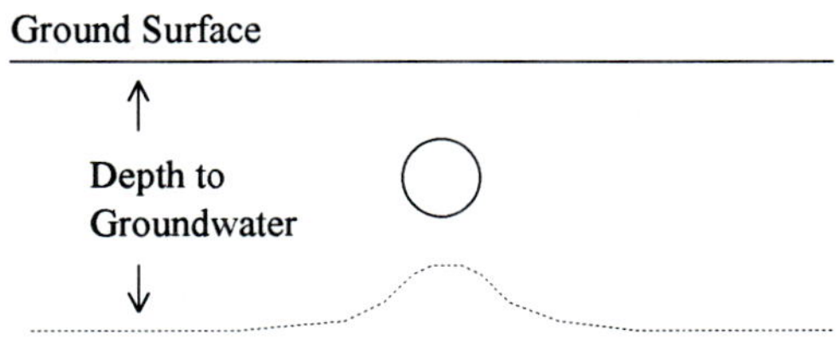

Figure 10. The distorted ground water table beneath a leaking pipe

For leak detection there are several approaches that may be employed. One method is to monitor the depth to saturated surface water beneath the Earth's surface. The principle is to look for anomalies in the vicinity of known buried pipelines - i.e. a raised phreatic surface, Figure 10. A second method is to search for changes in soil properties such as conductivity, which are detectable by RADAR and are influenced by the presence of water within the unsaturated zone.

Again handheld or ground vehicle mounted sensors are the norm however there are examples of air-borne techniques in the literature. In hydrological studies, air-borne sideways looking RADAR (see previous Section of this Chapter) is frequently used for measuring the soil moisture content in the top few centimetres of soil - anomolous soil moisture is another indication of possible leaks.

Sutinen et al. (1994) were able to delineate a plume of water from a landfill site, using low frequency radio waves mounted on an aeroplane flown at 40m. Delineation was possible due to the presence of pollutants which caused a detectable change in conductivity between the plume and the surrounding medium.

2.7. REMOTE SENSING AND SURFACE WATER QUALITY

Phenomena such as turbidity and algal blooms, which influence the quality of surface source water, and the processes sedimentation can be monitored using remote sensing. This has the advantage that isolated and large bodies of water can be easily and frequently assessed. This approach provides a time-series of data which can aid the understanding of underlying processes.

The approach of Dekker et al. (1992) was to use specially selected bands for monitoring water quality. They used different bands for the sensing of turbid or algal waters than for clear waters, since the particulate matter causes a high degree of scattering and the organic matter, particularly the algae, causes high absorption. Nine bands were selected, each was 10-20μm wide within the spectral wavelength range of 500 to 800μm. This waveband was identified as the range of "prime interest" when investigating turbid waters. A purpose built sensor was developed and flown successfully in the Netherlands and it was found that optical water parameters could be determined accurately.

Many investigators have attempted to use images from standard sources for water quality monitoring. Lathrop (1992) reports on a study to relate standard Landsat thematic mapper (TM) images with water quality, particularly turbidity. Ground truth, in the form of total suspended solids and Secchi disk depth, was collected almost simultaneously with the landsat images. Using regression techniques they where able to derive a model which proved valuable in estimating total suspended solids and water transparency. However, the algorithms derived could not be applied to other study areas.

In a study of the wetlands of Norfolk, UK, Baban (1994) used two TM bands (TM1 the blue band and TM6 a thermal band) to assess the influence of circulation on the distribution of temperature and turbidity in large slow moving water bodies. Areas with significant currents could be distinguished from those without flow from the distribution of these parameters.

Marcia et al. (1991) also collected ground truth simultaneously with a landsat pass. Like Lathrop (1992) regression techniques were used, to relate temperature, Secchi depth, total suspended solids, chlorophyll concentration and nutrient concentration to the landsat TM bands 1, 2, 3 and 4. Significantly they showed a relationship between total phosphate, total nitrogen, suspended solids and chlorophyll to the optical water quality parameter Secchi depth. Sedimentation is a process closely linked with turbidity. Wagle (1985) was able to map turbid water zones in coastal waters, using landsat. These zones correlate closely with the distribution of seabed sediments.

Though these techniques are still largely under research, as they improve they will no doubt enter the operational environment. The advantages are: the spatial coverage, the ability to build up time series of data on important processes, the identification of trouble spots and, the ability to collect data quickly in response to developing situation.

2.8. GIS IN A WATER UTILITIES ENVIRONMENT

Currently in the UK the widest application of GIS in water companies is as an inventory system for asset management; sophisticated spatial modelling is usually a secondary consideration. On first appraisal the problems of acquiring and maintaining an inventory of the components of a water supply system may seem trivial. However, the shear amount of data means this is not the case, Moutal and Bowen (1991) report that they had to digitise 6000 miles of water mains in New York city. For water companies the task is essential and many are turning away from paper systems and putting computer based GIS at the centre of their strategies.

A water company must aim to provide water 24 hours a day, the whole year round They must maintain the current system, integrate new components into the existing system, plan for disasters and satisfy various statutory requirements. This clearly implies an extensive and easily assessable database, describing not only location of pipes, valves, pumping stations and reservoirs, but other facts such as depth, size, pressure and engineering specifications. For an inventory, complex spatial analysis functions are not required, rather an effective database that provides all those facilities that any database would provide along with the spatial dimensions. Clearly the vector model, which allows each spatial object to be linked to a database record, is appropriate, though in future object oriented GIS may be used, see previous Section of this Chapter. The issues involved in inventory and asset management can be split into technical and organisational considerations and the economics of the system. Though operations management is the principle aim of GIS development at the moment it is possible that decision support will become more important than at present.

2.8.1. Technical Considerations

Technical considerations include: allowing multiple users to access the data whilst maintaining security; linking to other databases; coping with a very large dataset, and protecting the data physically. The system must provide for these considerations whilst giving an adequate system performance, which means careful system analysis and design including the selection of hardware and software. The system design must recognise the problem of uncertainty and incomplete data sets.

The inventory database will be used by a considerable number of users; some will require only specific bits of the data, others will need to make updates, and managers may make *ad hoc* queries. For some field operations it may be useful to have the data in the form of a map. At the same time data that is not of concern to the user should be hidden from them, not only does this present a simpler view of the data to the user, but it protects data that may be sensitive. It is the DataBase Management System (DBMS) that must provide all this.

Athough water supply networks cover large regions, users may require the data at street scale. This will clearly require considerable storage. If the data is required on-line or access is to be distributed around a number of sites, or even to mobile terminals, then the demands are further increased. All these considerations are similar to any corporate database; there is a whole body of literature concerning the engineering of databases and information systems, see for example, Avison (1992).

Uncertainty may be a considerable problem. Where the responsibility for water supply has changed hands, as in the UK, some information about the system may be lost, paper maps may be damaged or may be compiled to differing standards and to different scales. Jacobs et al. (1993) describe the construction of an inventory. The main source of pipe specifications came from observations made when the pipes were repaired, this process began in 1975 so any pipes not repaired since then will not have their specifications recorded in the database.

The temporal dimension is one of the current areas of GIS research. A water supply system will change over time, customers may change, pipes are replaced or re-routed. Computer scientists have developed a concept of time that differentiates between world time and database time: world time concerns the actual time of an event; database time is the time that the event was recorded in the database. The recording of database time allows the user to find out when the change on some object was known - this may be as important as the time of the change itself. The 'snap-shot' model of time exploited in the section on Environmental Process Modelling and GIS is not adequate to represent this multi-dimensional view of time.

Another associated problem is dealing with future time; different designs and associated data of future components must be held, but there must be no confusion, these are objects that do not and may not exist. For further information about temporal databases and GIS see Wachowicz and Healey (1993).

2.8.2. Organisational Considerations

Organisational considerations include: planning for the transitional period; providing training; imposing the system on an existing system and on existing personnel, and bringing in new personnel; establishing new routines and procedures. It is a common experience in information system development, that failure in these considerations can cause the system to fail, corporate GIS is no exception.

A database is more than a computer package and a set of data - it's use in an organisation requires changes to organisational systems. It is clearly advantageous if each department and individual have access to the same version of the data, and that the data is the most up-to-date available. This implies consistent systems of updating data across the organisation and the sharing of data across departments. This may go against corporate culture, where different departments have different procedures and where data is viewed as the property of the department, or even the individual.

Several authors have stressed the importance of the users in both system development and in operation, see Avison (1992) or Senn (1992). If the system has been badly designed then in operation users may adopt a behaviour of avoidance; the department or individuals find ways of performing their functions without using the information system, so the new system becomes an expensive waste of money and envisioned benefits do not happen. This can occur for various reasons; in particular poor usebility or learnability of the system, and inadequate training.

Davies and Medyckyl-Scott (1994) conducted a survey that uncovered some interesting patterns specific to GIS use within various organisations. Vector systems, which are the most suitable for the inventory tasks, were found by users to be the least usable; single-user PC based raster systems scored the highest usability. Raster based systems were "significantly more likely than vector based systems to be used for complex analysis and modelling tasks". Davies and Medyckyl-Scott found evidence of "gross underuse" as well as low usebility of large corporate vector systems. This may reflect the users of each type of system; in the corporate system the user is more likely to be performing a job that existed before the GIS arrived, whereas the single user may have adopted GIS to perform a new operation.

To an extent corporate GIS have to be complex and so inevitably have a low usebility - they must, after all, serve the needs of a large population of users with vastly varying needs. Althought these corporate systems have the greatest potential for customisation, Davies and Medyckyl-Scott found that customisation in general "seems to have not radically improved usability". Despite these findings, Davies and Medyckyl-Scott, also found that utilities reported a higher usability than any other type of organisation. Several conclusions can be drawn.

- Corporate GIS have been oversold on the basis of sophisticated functions and analysis, whereas much of this functionality is never used,

- The addition of more and more functions, whether through customisation or upgrade, does not necessarily improve productivity. A more realistic approach might be to customise to a crucial subset of functions, an approach adopted by certain software houses. For example, Arc-view, which uses only a small subset of Arc-Info's functionality, has been developed to allow users to explore data,

- Clearly the needs of users are a central consideration in systems analysis and design. Ideally the user should be included in the development process. The reader is referred to the standard information systems texts such as Avison (1992), Senn (1992) or Hawryszkiewycz (1991). Papazian and Reddingius (1992) describe a process of human factors analysis to produce user friendly GIS.

2.8.3. Economics of Corporate GIS

Any development must be economically justifiable. Traditionally this justification has taken the form of a cost-benefit analysis and if the benefits outweigh the costs then the development can go ahead. The risk must also be assessed. How likely is it that the proposed system will fail? If the risk is high, then despite a positive cost-benefit analysis, it many not be advisable to pursue the project. There are many standard texts that discuss these issues in depth, see Hawryszkiewycz (1991) or Senn (1992) for an introduction with respect to information systems.

With regard to a GIS in an operational environment several issues impinge on these considerations.

In terms of costs, the cost of data is far greater than that of software and hardware. In the UK to purchase a single tile of vector data for an urban area covering 500m by 500m costs about £134 single payment, plus £10 maintenance per year. To acquire national coverage of Great Britain with 1:50,000 raster map tiles costs £39,500, copyright fees are extra[1]. To get a single 1:10,000 aerial photo covering $4km^2$, from an archive can cost about £15; a flight for airborne thematic mapper can cost around £20,000 and a single current LANDSAT 5 scene with 7 bands covering 185 by 170km area can cost about US$4000.

---

[1]Prices taken from 1995 Ordnance Survey Price list

If the data do not exist then they need to be gathered by some form of survey. In order to keep the database current this will need to be an ongoing process. There is also a considerable administrative cost in handling large dataset and converting data from one form to another. It is thus important to note that data is an ongoing cost.

The costs of personnel are hard to quantify, but are again likely to be considerably more than hardware or software. A large corporate GIS will require a full time manager and staff to develop and maintain it. The costs of personnel compound the costs of data. For example to digitise a theme from a single map sheet requires approximately 1 man/day. Remote sensing data must be processed; it takes 2-3 man/hr to extract a theme from a single air photo or stereo-pair.

The benefits of moving to GIS are likely to be less tangible than the costs, this is a feature of most information systems, as benefits generally derive from indirect improvements in efficiency and service. Having a good street level GIS that enables buried pipes and valves to be located more quickly has several advantages; a tangible but probably negligible saving in waste water; a saving in man hours and energy in searching for the pipe or valve and, an intangible perception of a better (or less bad) service in the minds of the customers affected by the maintenance work.

Pay back times, the time taken for benefits to outweigh the costs, can be much longer than is usually acceptable in business environments - and it may not even be possible to define a payback time because of the intangible nature of benefits.

2.8.4. Decision Support

Currently what most organisations want from GIS is an asset management or inventory system to assist with operations management. Decision support is a lesser priority. However, as the issues of inventory are resolved the advantages of using GIS for decision support may be realised, particularly the possibilities of system modelling applications such as network analysis.

An important part of the water supply system will be the network of pipes. Network analysis is a specialised field of GIS. Although pipe network analysis is considerably different from the usual traffic network analysis, there are some examples in the literature, such as Djokić and Maidment (1993). Network analysis uses the topology of the system rather than spatial relationships; distances are functions of the resistance to flow in the pipes as well as the length; resistance will depend on, amongst other parameters; diameter, roughness and slope. Clearly effective network analysis can only be accomplished with a complete database and will require the vector model or, in future, the object oriented model.

Rather than incorporating pipe network analysis into the GIS, an alternative approach for existing GIS, is to design the GIS interface to output files for inclusion with existing standard pipe network analysis programs (see the section on Environmental Process Modelling with GIS in this Section).

2.8.5. Discussion

GIS provides solutions to the existing tasks of water-supply inventory that companies have previously undertaken using paper systems. Silbovitz and Jolls (1989) found that the availability of GIS led to an increase in the quality of service, better decisions by management and a lowering of the cost.

GIS also affords a whole range of new operations that can increase operational effectiveness throughout the company. French (1993) used the data from GIS inventory to estimate damage to a network in an earthquake scenario. Wang et al. (1991) demonstrated that a GIS inventory can be used in earthquake contingency planning. It is not only in disaster analysis that GIS can allow companies to take on a more pro-active stance, a temporal GIS should allow managers to identify parts of the system that are getting old or that may be under stress from damaging activities or increasing demand.

As with any corporate database, changes will need to be made throughout the corporation to take full advantage of the GIS. The costs are high and there are risks, but the GIS should not be allowed to become an expensive map drawing package. The issues are as much managerial as technical.

2.9. REFERENCES

Al-Taha K.K., Snodgrass R.T., Soo M.D. (1993) Bibliography on Spatiotemporal Databases. SIGMOD ACM, v22 n1, pp59-67.

Avison D.E. (1992) Information Systems Development: A Database Approach. 2nd Edition, Blackwell Scientific Publications.

Baban S.M.J. (1994) Mapping Turbidity Surface Temperature and Water Circulation Patterns with the Aid of Satellite Imagery. Journal of the Institution of Water and Environment Management, v 8 n 2 pp 197-204.

Bishop A.B., Hughes T. (1993) GIS Water Planning Model for Urban and Agricultural Areas. Management of Irrigation and Drainage System: Integrated Perspectives 1993, ASCE New York.

Blogojević B, Elgy J., Chen Z., Maksimović Č. (1994) Airborne Videography as Data Source for an Unban Hydrological Model. Remote Sensing and GIS in Urban Waters, UDT 94.

Brereton N.R., Downing R.A. (1975) Some Applications of Thermal Infrared Linescan in Water Resources Studies. Water Services v79, pp91-96.

Brereton N.R., Hall D.J. (1983) Groundwater Discharge Mapping at Altnabreac by Thermal Infrared Linescan Surveying. IGS Report FLPU 83-7, British Geological Survey, UK.

Caldecott R., Poirer M., Scofea D., Svoboda D.E., Terzuoli A. (1988) Underground Mapping of Utility Lines Using Impulse RADAR. IEEE Proceedings Part F, Communications, Radar and Signal Processing, v135-F n4 pp343-353.

Chairat S., Delleur J.W. (1993) Integrating a Physically Based Hydrological Model With GRASS. Application of Geographic Information Systems in Hydrology and Water Resources Management, Eds Kovar K. and Nachtebel H.P., Proceedings of the International Conference HydroGIS 93 Vienna 19-22 April 1993, pp143-150.

Clark G.A., Sengupta S.K., Sailes K., Sherwood R.J., Hernandez J.E., Buhl M.R., Schnaich P.C., Kane R.J., Barth M.J., Del Grande N.K. (1993) Sensor Feature Fusion for detecting Buried Objects. Proceedings of SPIE - International Society for Optical Engineering v1942, pp178-188.

Davies C., Medyckyl-Scott D. (1994) GIS Usability; Recommendations based on the Users View. International Journal of Geographic Information Systems v8, n 2 pp175-189.

Dekker A.G., Malthus T.J., Wijnen M.M. (1992) Spectral Band Location for Remote Sensing of Trubid and/or Eutrophic Waters. Proceedings of the 1st Thematic Conference on Remote Sensing for Marine and Coastal Environments, New Orleans, USA.

De Lange W.J., Van der Meij (1993) Application of Geographic Information Systems in Hydrology and Water Resources Management. Eds Kovar K and Nachtebel HP, Proceedings of the International Conference HydroGIS 93 Vienna 19-22 April 1993, pp333-344.

Djokić D., Maidment D.R. (1993) Application of GIS Network Routines for Water Flow and Transport. v119 n2, pp229-245.

Eastman J.R. (1992) IDRISI User Manual. Clark University, Graduate School of Geography.

Elgy J. (1992) TEMPUS Project, JEP 2424-91/1, Final Report on 1991/92 Activities. Department of Civil Engineering, Aston University.

Elgy J., Maksimović Č., Prodanović D. (1993) Using Geographical Information Systems for Urban Hydrology. Application of Geographic Information Systems in Hydrology and Water Resources Management, Eds Kovar K and Nachtebel HP, Proceedings of the International Conference HydroGIS 93 Vienna 19-22 April 1993, poster session.

French S.P. (1993) Infrastructure Modelling with GIS, Geographic Information Systems and Their Application in Geotechnical Earthquake Engineering 1993, ASCE, pp45-48.

Grayson R.B., Bloschl R.D., Barling R.D., Moore I.D. (1993) Process Scale and Constraints to Hydrological Modelling in GIS. Application of Geographic Information Systems in Hydrology and Water Resources Management, Eds Kovar K and Nachtebel HP, Proceedings of the International Conference HydroGIS 93 Vienna 19-22 April 1993, pp83-92.

Groudin G.H., Lite Jr K.E. (1990) Numerical Model Uses and Limitations for Ground Water Management. Transferring Models to Users, American Water Resources Association 1990, pp121-129.

Groudin G.H., Gannett M., van der Heijde P.K.M., Patt R.O. (1990) Critical Errors that Hydrogeologic Professionals can Make with Computer Programs, Transferring Models to Users. American Water Resources Association 1990, pp149-157.

Hawryszkiewycz I.T. (1991) Introduction to Systems Analysis and Design. Prentice Hall Australia.

Hazelton N.M.J. (1991) Extending GIS to Include Dynamic Modelling. Proceedings of the 3rd Colloquium of the Spatial Research Centre.

Hedges P.J. (1989) The Impact on Agriculture of Drawdown of Shallow WaterTables. PhD Thesis, Aston University UK.

Ho G., Kawanaka A., Takagi M. (1993) Image Quality Improvement for Underground Radar by Block Migration. Proceedings of SPIE - The International Society for Optical Engineering v1942 p105-116.

Hughes T.C., Bishop A.B., Fullerton H.H. (1994) Application of GIS and Remote Sensing to Projecting Urban Water Demand. UDT-94 Moscow, Russia, Sept 12-16, 1994.

Jacobs P., Goulter I.C., Davidson J. (1993) Water Distribution GIS from Fragmented and Incomplete Information. Journal of Computing in Civil Engineering, v7 n3, pp372-386.

Kern T.J., Stednick J.D. (1993) Identification of Heavy Metal Concentrations in Surface Waters through Coupling of GIS and Hydrochemical Models. Application of Geographic Information Systems in Hydrology and Water Resources Management, Eds Kovar K. and Nachtebel H.P., Proceedings of the International Conference HydroGIS 93 Vienna 19-22 April 1993, pp559-567.

Kikuta T., Tanaka H. (1989) Ground Probing Radar System. IEEE Aerospace and Electronic Systems Magazine v5 n6 pp14-16.

Klaghofer E., Birnbaum W., Summer W. (1993) Linking Sediment and Nutrient Export Models with a Geographic information System. Application of Geographic Information Systems in Hydrology and Water Resources Management, Eds Kovar K and Nachtebel HP, Proceedings of the International Conference HydroGIS 93 Vienna 19-22 April 1993, pp501-506.

Kowalczyk A. and Kemp Z. (1993) Incorporating the Temporal Dimension into a GIS Application, GIS Research UK Proceedings 1993.

Lathrop R.G. Jr. (1992) Landsat Thematic Mapper Monitoring of Turbid Inland Water Quality. Photogrammetric Engineering and Remote Sensing, v 58 n 4 pp465-470.

Maksimović Č., Fuchs L., Prodanović D., Elgy J. (1995) Full Scale Application of Standard GIS Packages with Urban Storm Drainage Simulation Software, Kurier abwasserung.

Marcia E., de Moraes Novo L., Zuccar C., Braga F. (1991) Use of TM/Landsat Data to Estimate ChloroPhyll Concentration and Turbidity in the Barra Bonita Reservoir. Proceedings 24th INternation Symposium Remote Sensinf of the Environemnt, Rio De Janeiro, Brazil, v 1 pp389-401.

McDonnell R.A., Macmillan W.D. (1993) A GIS -based Hierarchical Simulation Model for Assessing the impacts of Large Dam Projects, Application of Geographic Information Systems in Hydrology and Water Resources Management. Eds Kovar K and Nachtebel HP, Proceedings of the International Conference HydroGIS 93 Vienna 19-22 April 1993, pp409-418.

Moutal H.P., Bowen D.R. (1991) Updating New York's Sewer and Main Distribution Systems: Practical Applications of GIS. Second National Speciality Conference on Civil Engineering Applications of Remote Sensing and Geographic Information Systems, 1991, ASCE, p155-164.

Osumi N., Ueno K. (1988) Detection of Buried Plant. IEEE Proceedings, Part F: Communications, Radar and Signal Processing v135-F n4 p330-342.

Papazian B., Reddingius N.H. (1992) Human Factors Analyses of GIS use Patterns. Geographic Information Systems: Proceedings of the Seventh Annual GRASS Users Conference, March 16-19 Lakewood, Colorado, pp195-202.

Prodanović D. (1992) Automatic Delineation of Subcatchments. TEMPUS JEP 2424/91-1, Institute of Hydraulic Engineering, Faculty of Civil Engineering, University of Belgrade, 1992.

Reddi L.N. (1990) Potential Pitfalls in Using Groundwater models, Transferring Models to Users. American Water Resources Association 1990, pp131-140.

Ross M.A., Schwartz M.A. Tara P.D. (1991) Groundwater Modelling with GIS: An Example. Proceedings of the International Symposium on Ground Water in Practice. Nashville Tennessee, pp244-249.

Ross M.A., Tara P.D. (1993) Integrated Hydrologic Modeling with Geographic Information Systems. Journal of Water Resources Planning and Management, v119 n2 pp129-140.

Rushton K.R., Turner A. (1977) Permissible Mesh Spacing in Aquifer Problems Solved by Finite Differences. Journal of Hydrology, v34, pp63-76.

Rushton and Redshaw (1977) Seepage and Groundwater Flow. Numerical Analysis by Analog and Digital Methods, John Wiley and Sons.

Senn J.A. (1992) Analysis and Design of Information Systems. 4nd Ed, McGraw-Hill Book Co-Singapore.

Shapiro M., Westervelt J., Gerdes D., Larson M., Brownfield K.R. (1993) GRASS 4.1 Programmers Manual. US Army Construction Engineering Research Laboratory.

Silbovitz A.M., Jolls J.D. (1989) Computerised Infrastructure Mapping and Management Systems at Low Cost. Journal of the New England Water Works Association, v103 n3, pp151-155.

Slieght P., Levenehall B. (1994) Meeting the Neighbours: a Review of Britain's New Classifications Systems. GIS Europe v3 n4 May 1994.

Smith M. (1992) A Distributed Parameter Hydraulic Model for Urban Storm Water Protection. Dr. Sc. dissertation, FAGG Ljubljana.

Sutinen R., Hannula P., Hyvonen E., Kontio M., Lanne E., Turunen P. (1994) Airborne and Terrestrial EM Detection of Landfill Plume in Suburban Rovaniemi, Finland. Proceedings of the International Airborne Remote Sensing Conference and Exhibition, 12-15 Sept 1994, Strasbourg, Vol 3 pp215-222.

Tracy J.P., Walton P.A. (1989) Hydrologic Investigations in Canal and Aqueduct Systems Using Airborne Thermal Infrared Linescanning. Digest - International Geoscience and Remote Sensing Symposium (IGARSS) v5 pp2815-2819.

Wachowicz M., Healey R.G. (1993) Towards Temporality in GIS. GIS Research UK proceedings 1993.

Wagle B.G. (1985) Distribution and Dispersion of Turbid Waters on the Western Continental Shelf of India - An Intepretation of LANDSAT images. International Journal or Remote Sensing, v 6, n 11, pp1681-1684.

Wang L.R.L., Ishibashi I., Wang J.C.C. (1991) Inventory and Seismic Loss Estimation of Portland Water/Sewer System. GIS Application to Buried Pipelines. Proceedings of the 3rd US Conference on Lifeline Earthquake Engineering 1991, pp490-499.

Watts P. (1994) European Geodemographics on the up. GIS Europe, v3 n4, May 1994.

Weil G.J. (1993) Non-contact, Remote Sensing of Buried Water Pipeline Leaks Using Infrared Thermography. Proc. 20th Conf. on Water Management in the 90's, May 1-5 1993, Seattle USA.

Weil G.J., Graf R.J. (1991) Infrared Red Thermography based Pipeline Leak Detection. Proc. 2nd National Speciality Conf. on Civil Engineering Applications of Remote Sensing and Geographic Information Systems. Washington May 1991.

Weil G.J., Graf R.J., Forister L.M. (1993) Non-Destrucitve Remote Sensing of Hazadous Waste Sites. Proceedings of the Conference on Digital Image Processing: Techniques and Applications in Civil Engineering, pp279-284.

3. Application of GIS in Analysis and Design of Water Supply Systems

Čedo Maksimović *
Dušan Prodanović *
IRTCUW, Faculty of Civil Engineering
University of Belgrade
P.O. Box 895, 11 000 Belgrade
YUGOSLAVIA

3.1. INTRODUCTION

In the preceding Section of this Chapter the application of basic principles of GIS (Geographical Information Systems) have been presented along with the application of these techniques to the analysis of the effect of drawdown of groundwater, water demand consumption analysis and prediction of water demand. This Section deals with the particular application of GIS matched to hydraulic analysis of the water supply network and for interpretation of the results of hydrodynamic modelling. The examples used in this Section are based on the application of an inexpensive GIS package IDRISI (Eastman, 1992). For large scale systems and full scale projects more powerful GIS packages may be more appropriate.

3.2. SOURCES OF INFORMATION FOR APPLICATION OF GIS IN WATER SUPPLY SYSTEMS

The use of GIS in Water Supply Systems is linked to:

- access to general purpose data bases (sources of data) dealing with spatial features,

---

* computer support: D. Štrbac, RGZ, Belgrade, M. Drašković, University of Belgrade, YUGOSLAVIA

NATO ASI Series, Partnership Sub-Series, 2. Environment – Vol. 15
Water Supply Systems – New Technologies
Edited by Č. Maksimović, F. Calomino, and J. Snoxell

- assessment of natural water resources for design and management,
- assessment of water demand,
- assessment of pollution sources and consequences,
- creation of DEM for matching with simulation models (zonning, etc.),
- creation of land use maps for population density, height of buildings, urban irrigation, etc.,
- layers of inventory with infrastructure, network, structures, etc.,
- leakage detection and presentation of the results obtained by mains burst statistics,
- presentation and analysis of results of simulation,
- design (drafting, plotting),
- presentation of other results, publishing, etc.

The most widely distributed source of information that can be used for simulation models are traditional, maps and drawings on paper that have to be either digitized and/or scanned and processed.

However, there are other possible sources of information, such as:

- satellite images,
- images obtained by photogrammetry and videography,
- automatic (total stations) field surveying,
- data collected by surface born platform and sensors (for example thermal infrared for leakage detection).

To match these data to the commercially available GIS packages such as ArcINFO, Intergraph, SQL, GRASS, etc., a series of user friendly interfaces; gluing routines (Elgy et al.,

1993), have to be developed and customized. Some examples of such routines have been presented and discussed in the preceding Section.

3.3. INFORMATION NEEDED FOR MATCHING GIS WITH SIMULATION PACKAGES

For practical reasons we will concentrate only on applying GIS to simulation packages for hydraulic analysis of water supply networks.

One of the major advantages of matching simulation models with GIS is the ease in which the results can be presented in human readable form and used for analysis and further improvements of the system's design and operational management.

For this purpose three major groups of information are required:

- underground services (utilities) such as pipes, valves, meters, etc. and their position, status, characteristics, etc.,
- information on water consumption in all network nodes,
- land use, digital elevation model, topography, surface devices.

It is assumed here that the information on the underground utilities required for hydrodynamic modelling (pipe characteristics, valves, etc.) have to be obtained from the relevant data bases and that they are available. Information on water consumption in nodes is also assumed to be known. The means of obtaining and analysing the data are presented in the preceding Section.

Information on topography (elevation) is handled in the form of digital elevation models. In addition to the other applications they can be used for analysis of pressure changes during the period of simulation or during daily operational management.

With the above sets of data one can create appropriate input files for running simulation models. The results from simulation models are then superimposed over the digital terrain model for obtaining spatial distribution of piezometric levels, pressure fields, etc.

3.4. MATCHING GIS WITH RESULTS OF MODELLING

Standard GIS packages can be used either for presentation of results or (more preferably) for interaction with modelling. The authors favor the second approach (Maksimović et al. 1994).

In what follows use will be made of the figures obtained by the interaction of the results of hydrodynamic modelling with the IDRIDI GIS package. For this purpose a simple hypothetical Water Supply network has been created. The network is shown overlain over the contour map in Figure 1. and the DEM in Figure 2.

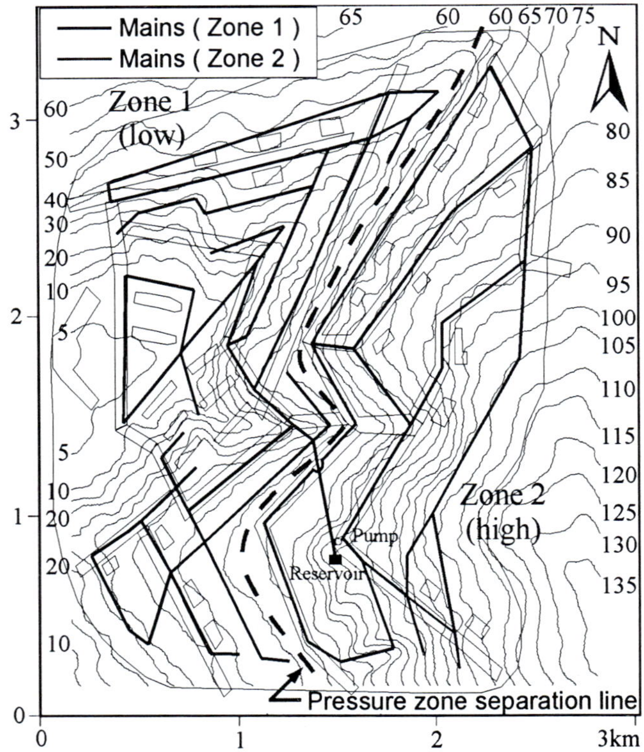

Figure 1. Digital Elevation Model presented in the form of contour lines with a hypothetical, two zones WS network layer superposed

A hypothetical network consisting of two pressure zones has been imposed on this DEM. It is assumed that water is supplied to the first (lower) zone which contains one reservoir. From the reservoir water is further pumped to the second zone by booster pumps. It is assumed that there is no reservoir in the second zone. This is often the case in smaller networks. In these cases pressure in the networks reacts more intensively as the consumption in the network changes.

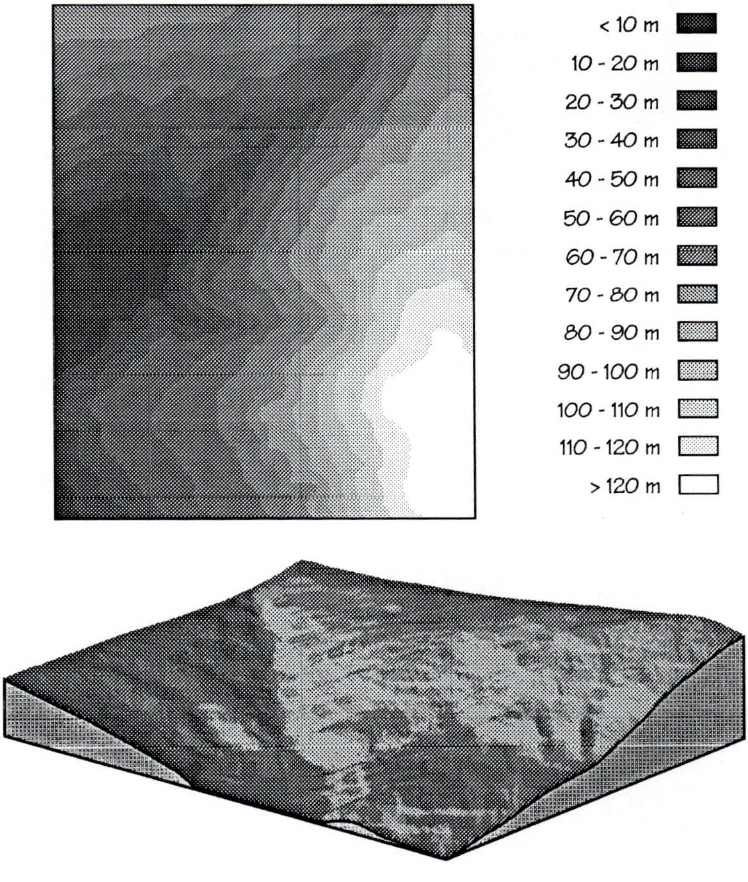

Figure 2. Plan and ortho view of the DEM

The network (containing two pressure zones) has been shown on the ortho view (Figure 3.) and plan view (Figure 4.).

With the above mentioned data prepared in digital form a hydrodynamic model for simulation of daily water consumption has been run.

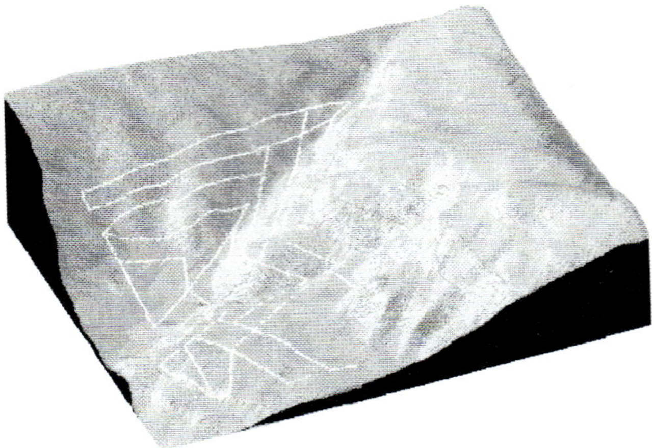

Figure 3. WS network systems for two pressure zones

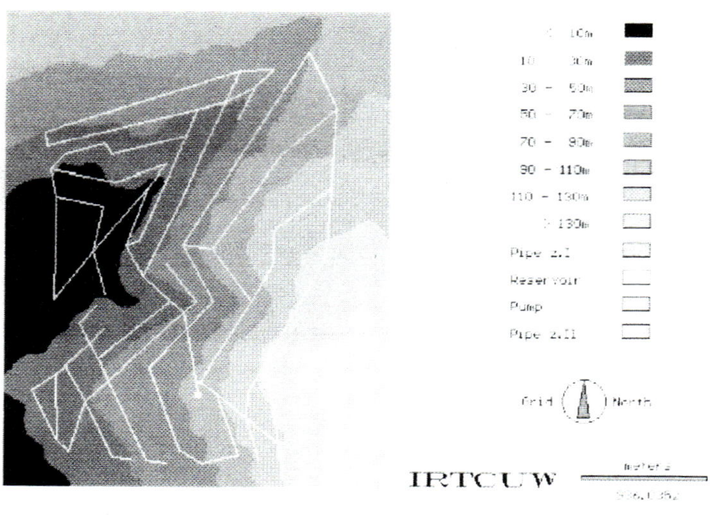

Figure 4. Digital Elevation Model with network system (plan view)

Some results of the hydrodynamic simulation (piezometric field) are shown in Figure 5. The first results obtained from the hydrodynamic analysis are the piezometric levels in nodes. They vary in both space, (from one node to another) and in time (as a result of variable consumption). Thus they are very suitable for presentation by GIS tools.

The piezometric fields shown in Figure 5. are two snapshots (maximum and minumum consumption).

By an overlay operation of GIS the piezometric field has been transformed into pressure field as follows:

$$\frac{p}{\rho g} = P - Z$$

where: p is pressure

ρ is water density

P is piezometric level

Z is elevation.

The result of this overlay operation is shown in Figure 6. for two snapshots (hours of maximum and minimum consumption). The image has been created by using only the nodal values of pressure heads.

For the sake of simplicity the nodal negative pressure heads have been allowed to assume physically unrealistic values (-22.90 - for example).

This means that the part of network with such values is filled by air and the network analysis should be altered accordingly.

The changes in the pressure field are illustrated in Figure 7. for four different snapshots (8 a.m., 2 p.m., 8.p.m. and 2. a.m.).In this Figure the whole range of pressure heads has been subdivided into only four categories as follows:

- dry areas (air in mains),
- areas with insufficient pressure ($p/\rho g < 20$ m),
- areas with acceptable and favourable pressure heads (20 m $< p/\rho g <$ 55 m),
- areas with high pressure ($p/\rho g > 55$ m).

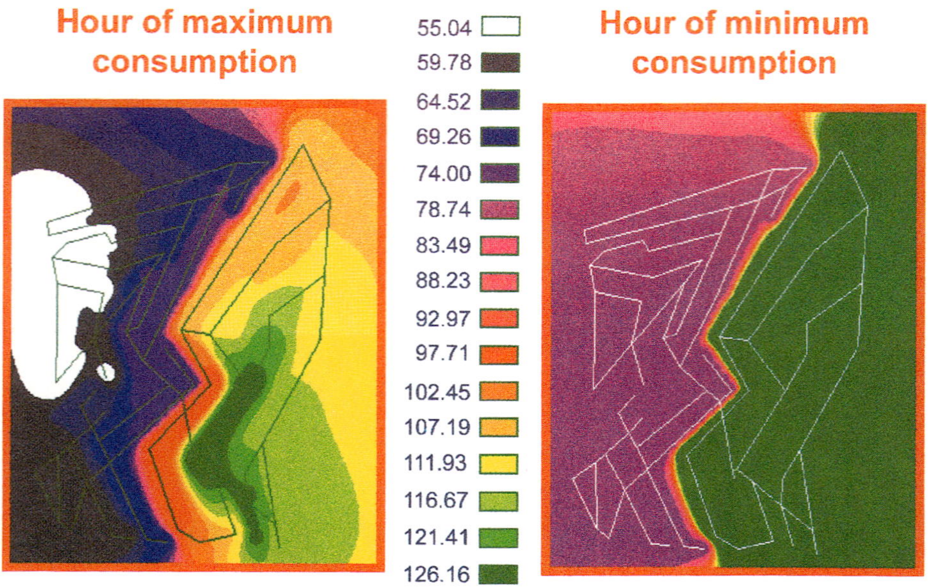

Figure 5. Piezometric fields

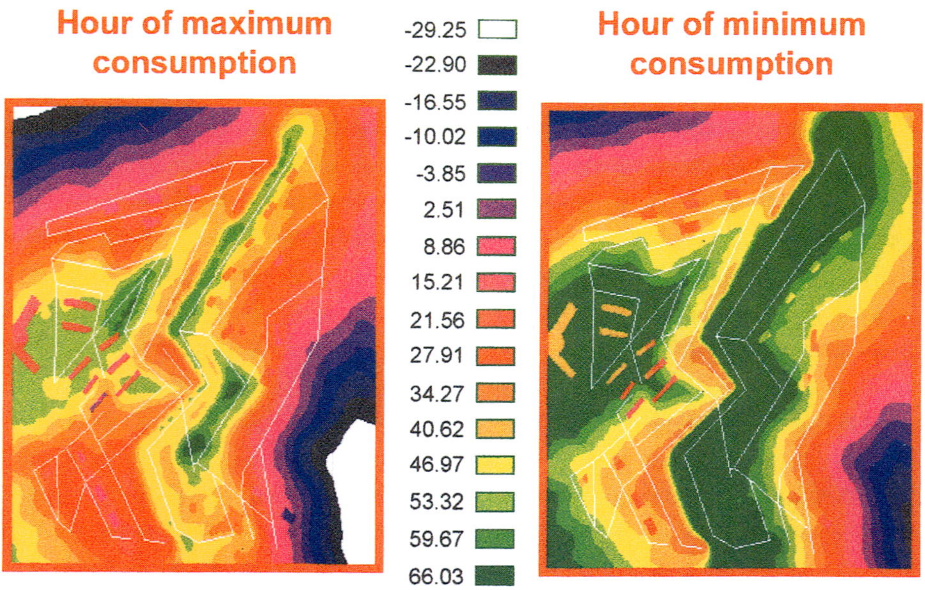

Figure 6. Hypothetical pressure head fields

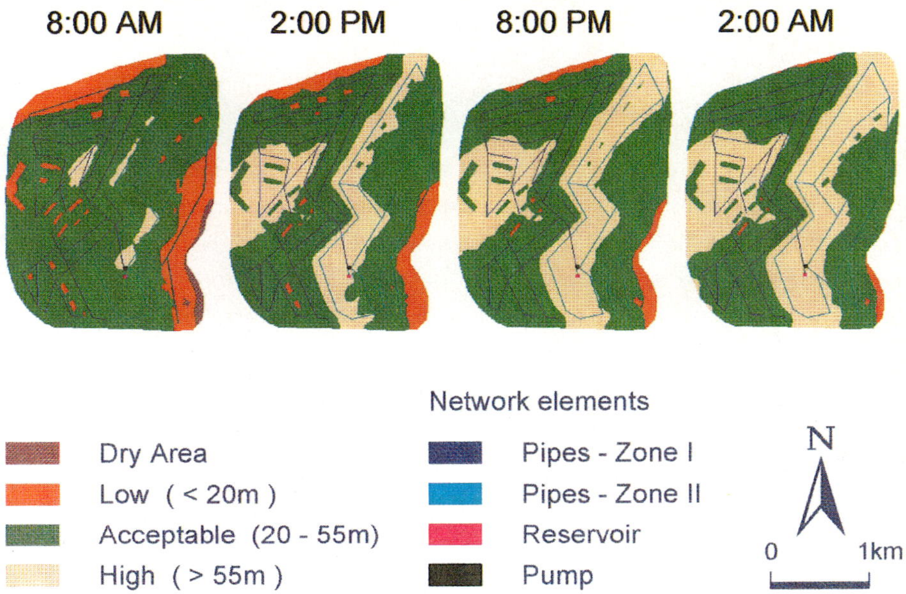

Figure 7. Daily dynamics of pressure head fields

This presentation of results enables an easy and reliable selection of reconstruction and rehabilitation options, because it is apparent that areas continously suffering fron either low or excessive pressure should undergo structural redesign, such as pipe replacement, distribution relocation, installation of pressure reducing or pressure sustaining valves or new booster stations, etc.

By making use of DISTANCE, a GIS operator, the shortest length from each pixel to the nearest pipe in the map can be determined. This information can be used for energy losses calculations in small diameter pipes, which are normally neglected during network analysis:

$$\Delta E = \lambda \frac{L}{D} \frac{v^2}{2g}$$

where: ΔE is energy losses,

L is length obtained (Figure 7.),

v	is velocity,
λ	is friction factor,
D	is diameter.

For this case all other values (λ, D, v) are kept constant. In normal operations a user can make connections between these values and cover information using other GIS tools. In this way the pressure field obtained is more realistic and it is much easier to delineate the zones subject to low and high pressure.

This information has been produced by ordinary GIS operators (Borroughs, 1986) and does not require any special programming.

The coupling of the GIS and hydrodynamic simulation, shown in the above figures, could be extended further with animation to describe the behavior of the system in an even more friendly manner.

3.5. REFERENCES

Borroughs P.A. (1986) Principles of Geographical Information Systems for Land Resources Assessment. Claredon Press, Oxford.

Eastman J.R. (1992) IDRISI User Manual. Clark University, Graduate School of Geography.

Elgy J, Maksimović Č., Prodanović D. (1993) Using Geographical Information Systems for Urban Hydrology. Proc. of HydroGIS 93, IAHS Publ. no 211.

Maksimović Č., Prodanović D., Elgy J., Fuchs L. (1994), GIS (or GIM) in Water Projects-Tools or Toys. Hydroinformatics '94. Proc. of the First Int. Conference on Hydroinformatics, Publ. by Balkema, Delft.

4. Geographical Infomation Systems and Digital Mapping for Water Supply Inventory

Thomas William Charnock
John Elgy
Department of Civil Engineering
Aston University
Aston Triangle, Birmingham B4 7ET
John Snoxell
WESSEX WATER
WESSEX House, Passage str.
BS2 0JQ Bristol
UK

4.1. INTRODUCTION

Water companies are complex organizations with many spatially distributed assets; pipes, plants, reservoirs and sewers. Furthermore they must communicate and interact effectively with a constantly changing base of hundreds of thousands of customers; industrial, agricultural and domestic. The statutory framework is continually evolving and companies must, more and more, consider such issues as the environment, liaise and report to regulatory bodies and operate in an increasingly competitive market. An efficient and secure information system is essential. More than most businesses, water companies have to manage uncertainty. In the UK in 1995, for example, a drought, followed by a cold snap caused many burst pipes and consequently increased leakage.

With the increase in power and flexibility of computer technology, water companies are turning to computer based information systems and because of the spatial nature of much of the necessary data, companies are either implicitly or explicitly building or buying Geographic Information Systems (GIS).

The activities that the information system must support can be divided between core operations at the heart of the companies business, secondary concerns, such as statutory duties and extraordinary operations caused, for example, by drought. Also the use of

NATO ASI Series, Partnership Sub-Series, 2. Environment – Vol. 15
Water Supply Systems – New Technologies
Edited by Č. Maksimović, F. Calomino, and J. Snoxell

computer technology makes possible many activities such as sophisticated data analysis and system modeling which were not possible or practical with a manual system. The benefits of these additional operations, in particular, are difficult to quantify and cost.

Water companies have the objectives of supplying water, handling sewage and charging the customer. Supplying water involves collecting and treating water, maintaining the pipe network, monitoring reserves and estimating future demand and supply. Handling sewage involves maintaining the sewer network and treatment plants and monitoring the receiving environment. Charging the customer involves estimating or monitoring customer usage and sending out bills. Furthermore there are support operations such as the payroll or maintaining the vehicle fleet. All these operations occur within a legislative framework and, increasingly, a competitive environment. In order to support these core operations some form of information system whether manual or computerized is required. Many of the core operations involve information that is fundamentally spatial in nature; the distribution of the pipe network, the customer base for example. Increasingly water companies in the UK are utilizing GIS to provide the basis for an information system that will effectively support these operations.

Having a single store of information which is then moved to individual computers ensures that every one in the organization is working to a consistent set of data. This prevents the issue of out of date maps or data where a modification has been carried out by one part of the organization but not everyone within the organization has the same information.

As well as the core operations, water companies have other activities that must be performed. In the UK water companies must liase closely with the National Rivers Authority, with the 'watchdog' bodies which regulates pricing, with local authorities that regulate access to facilities buried in the highway and with other utilities. Water companies in the UK have been privatized so there is greater need for business and investment planning based on sound information.

Water supply is a crucial part of the infrastructure and the economic viability of a region. Failure of supply for even a short period, such as a few hours can cause large loss of revenue due to lost production. Here the information system can help in three ways. Firstly by identifying vulnerable parts of the water supply system and allowing the company to be proactive in dealing with problems, this chapter will briefly discuss how a water company uses GIS to identify vulnerable parts of the network. Secondly by enabling the development of contingency plans and the performance of scenario analysis. Thirdly a flexible system including portable PC's can enable a very efficient reactive response to incidences.

In addition to the primary and secondary business needs of the organization which can, usually be quantified, GIS offers a number of tertiary unquantifiable advantages. The first of these is the idea of simply doing a job in the best possible manner. A leap of faith. It is always reassuring to know that the quality of data being used in any part of design or maintenance is as high as possible and a good GIS with documented and implemented quality control aspects gives that reassurance. A word of caution is required here; just because the data is stored in a GIS does not guarantee that the quality of the data is high. A method of quality control must be implemented at the setting up of the GIS.

Once the data is stored in a portable form (in this sense portable means the ability to move the data from one computer or application program to another) other methods of analysis become available that were not thought of before. For example studies can be carried out on the recorded incidence of disease, such as cancer, and the water treatment process used.

The introduction of any new information system, including GIS, to an organization involves costs and risks. The costs include: the hardware, the software, data acquisition, data translation, staff training and the costs associated with the change over. The risks are that the new system will not deliver the intended benefits, that the costs will be greater than expected (very common in any IT system) or that the system detrimentally affects the organization's operations. The establishment of a corporate GIS within a water company has very tangible, immediate and also ongoing costs; the benefits, however, are comparatively intangible involving projected increases in the effectiveness of core operations, improvements in secondary operations, possible commercial advantages and the benefits involved with more sophisticated data analysis and system modeling. These benefits occur away from the GIS in other parts of the organization and the GIS may be seen as a cost centre. It is becoming common for companies to establish internal markets to deal with this problem, whereby departments buy in the services of the information systems department and thus the benefits become apparent. This can be taken further and the information services department may sell services outside the business, indeed this chapter will describe the work of both Wessex Water and the Dwr Cymru's GIS departments.

Many authors have dealt with the problems, issues and strategies of introducing corporate information systems to organizations, see for example Senn (1992). But the issues concerning data should be emphasized. It can not be overemphasized that data capture is the over-riding cost of Utilities GIS, the cost include; manual digitizing paper maps, buying external sources of data in particular Ordnance Survey digital maps and the collection and inputting of attribute data.

This Section focuses principally on Wessex Water, a water company in South west UK. It also briefly describes the ways GIS is being utilized in Welsh Water (Dwr Cymru), a much larger organisation and in South Staffordshire Water which is much smaller.

4.2. WESSEX WATER

The UK Water Industry underwent a major reorganisation in 1974 which resulted in 10 large, catchment-based Regional Authorities responsible for complete water-cycle management (i.e. Water Supply, Sewage Treatment, Land Drainage). Wessex Water is one of the smaller plc's with a total workforce of around 1,200 supplying about 1 million water supply customers and 2 millions sewage treatment customers (2 independent water companies exist within the Wessex area, supplying around 1 million supply customers).

One advantage of the 1974 reorganisation was the possibility for large economies of scale, including computing facilities, and the 1970's saw major investments in mainframe provisions, telemetry systems and communications infrastructure. The legacy of record drawings from the predecessor organisations (e.g. Water Boards, Local Government Sewage Treatment agencies) was immense and it became apparent that some form of computerised solution was necessary to control, rationalise and utilise the enormous amount of data involved (10,000 km of water mains - 13000km sewers).

A further stimulus to the development of GIS systems was the perennial Water Industry problem, shared by the other major utilities (Gas, Electricity and Telephone), of damaging each others underground plant when digging holes in the highway. Computerised records were seen as offering a means of reducing such damage through shared data and better communications. A national working group, NJUG (National Joint Utilities Group) was formed in 1977 to promote such a solution.

Thus the 1980's saw widespread investment in GIS systems, generally called at the time "digital mapping" systems. Wessex Water acquired a Synercom system in 1985 following several years of research and data collection and the Company now has over 15 years experience in the use and value of such systems. The development of the system can be divided into several distinct phases and these are examined in turn, reflecting on what the original objectives were with actual experience in each area; the lessons learned are described.

Wessex decided that Mapping was an inevitable long-term investment for managing geographical information as early as 1978. The following potential benefits were identified:

- Rationalisation and verification of all existing records from the predecessor organisations.

- Proper centralised control of all such information (i.e. updating).

- Savings from improved handling and access of paper-based information for all departments; avoidance of staff costs associated with manual record updating.

- Savings in design, planning and new works departments resulting from improved information and rapid access.

- Greater understanding (and thus efficiency) of spatially-based information (frequency of burst mains, for example).

- Potential for reduced plant damage through shared information with other utilities.

- Greatly improved Asset Management information resulting in improved investment programmes.

- Wide variety of operational management improvements.

- Potential links to other systems (network analysis, customer databases, telemetry).

- Generally, progress has been made in all of these areas, but some of the perceived benefits have not materialised as first thought.

Wessex started the process of computisation of records by establishing the key workstations and methods of digitising records. Work started on water supply data in 1979 and, in 1991, around 95% of the distribution network has been digitised. 75% of the network has been physically inspected to verify the data on the ground and was completed, updated and verified by 1993. One early problem was that of background data. The UK National Mapping Agency, the Ordnance Survey, had a long-term programme of digitising all background data over the period 1980 - 2020, with the Wessex Water area scheduled for 2020. Wessex entered into an agreement with the Ordnance Survey to bring this forward for the Wessex area and currently 60% of background data is now digitised on the system. Progress in digitising sewers has been slower and currently 20% of the network (of 13,000 km of sewers) has been digitised with 4% verified on the ground. The situation is more complex in sewerage because Wessex, in common with all the UK companies, has Agency agreements with Local Government organisations who manage a number of the sewerage networks - this has undoubtedly slowed down the development of GIS in sewerage. The remaining part of the sewerage network is scheduled for completion by 1997.

The Company has 11 workstations around the Regional area, with 15 personnel working full-time on GIS, (9 on Water Supply and 5 on Sewerage), plus an estimated 2 personnel on hardware/software support.

Tables 1. and 2. give a breakdown of the time spent on the major activities currently ongoing in Water Supply and Sewerage.

Table 1. GIS activities - water supply

Water Supply GIS (9 Personnel)	
Activity	**Time (%)**
Routine updating of network	30
Ongoing amendments (new developments)	20
Map Production	12
Operational Support	4
Planning/New Works support	9
OS database maintenance/development	10
GIS development	15

Table 2. GIS activities - sewerage

Sewerage GIS (5 Personnel)	
Activity	**Time (%)**
Routine updating	5
Map Production	10
Development	50
Planning/New Works support	15
Contact supervision/admin	20

Total annual running costs for the Regional system including depreciation of all hardware and the acquisition of OS background is currently around £690K per year. This includes external sales of OS background.

The experience gained in the early 1980's led to a number of realisations:

- Updating procedures needed careful management.
- GIS systems are specialised and this has consequences for training, development and access.
- The original expectations of wide on-line access to geographical information did not materialise.
- The sharing of data with other utilities did not generally materialise.

The major benefits during the period are now seen to be:

- GIS has imposed basic standard disciplines in validating asset data which can now be used with confidence by all departments in the Company.
- The asset database has greatly improved the capacity for planning.
- The capability to produce bespoke maps of specific areas with selected information, presented in colour, has proved very useful, in almost all areas of the Company.

A number of other factors affected the perception of GIS in the 1980's. The UK Industry was privatised in 1989 and a key feature of this process was the price regulation mechanism, which involved a factor (K), above the Retail Price Index by which prices to the customer could be allowed to rise. The K factor depended crucially on an Asset Management Plan (AMP) and GIS systems played a key role in the formulation of this plan.

The GIS effort on this exercise is now estimated at 1 man-year for Water Supply alone; the importance of producing accurate information for the Asset Management Plan is reflected in the fact that expenditure on water mains and sewers in Wessex over the next 5 years is likely to approach £100 millions.

On a technical level the 1980's was the decade of the PC, and this explosion has created a generation of "PC literate" users familiar with a variety of database and graphics packages. Some of these are now overlapping with tasks originally envisaged for GIS systems and generally there has been a need to converge GIS, CAD and general graphics packages in a sensible way. There are now several PC-based mapping systems and GIS available.

In addition, other forms of "mapping" have appeared, notably video-disc based systems which offer rapid access and simplicity of use for scanned records. As experience has been gained, the potential applications of GIS systems have expanded.

It is now clear that many types of information are best understood when presented geographically. In the Water Industry this applies to such things as pressure, or quality variations across a network, burst mains, certain customer data (e.g. renal dialysis users) and many more; the list of potential applications is enormous, limited only by the imagination of the users. Unfortunately the basic cost-benefit equation for these applications is not always very favourable and it is cost-effectiveness which will determine the future of GIS.

For those now considering GIS systems for the first time the choice of hardware, software, and the multiplicity of potential applications is enormous, probably beyond the understanding of any one individual. At a conservative estimate there are now around 1000 companies offering GIS-related goods or services (Dangermond 1991). In this situation it is worth examining the current situation in the UK Water Industry, the areas where GIS systems have been successfully implemented and where they have not lived up to the original expectations.

The current state of GIS system implementation can be summarised thus:

- Most of the larger PLC's already have some years experience with established GIS systems, either on a pilot basis or full-scale implementation. A number of the smaller Water Companies (former Statutory Water Cos) likewise now have GIS systems, though some of the smaller companies have not embarked on such systems; others are evaluating the options and costs. Wessex Water has a well-developed system, with over 15 years experience in full-scale implementation.

- Of the original objectives identified for GIS systems, good progress has been made in basic record management (e.g. map production) and a sound foundation for future development has been provided. This has been particularly beneficial in Asset Management exercises.

- The original expectations concerning the "paperless office", shared data with other utilities, and widespread access for potential users have not been realised.

- The 1980's has seen an explosion of PC-based systems which overlap with conventional GIS systems and a new generation of computer users familiar with graphics and database concepts.

It is worth reflecting on why some of the expectations have not been realised. The concept of shared data with other utilities has made little progress. The major motivation behind the idea was to minimise the disruption caused by the major utilities damaging each other's underground plant and the duplication of excavation in the "common highway", a feature of urban life which attracts ridicule from all sides. In the UK there has long been a requirement for any utility wishing to dig a hole in the "common highway" to notify all other parties (highway authorities and utilities) by post of their intent. This has spawned a complex paperwork bureaucracy, known in the UK by the acronym PUSWA (Public Utilities Street Works Act) involving the sending of over 15 millions messages to cover the estimated 3 millions such holes dug every year. It is proposed that a Computerised Street Works Register (CSWR) will be created which will reference each hole at an accuracy of 20 millions initially, and 1million on reinstatement. The potential for GIS is clear. It is not clear currently how this will interact with typical Water Company GIS systems but it may provide a new stimulus and development path, though in recent years the development of trenchless technology and improved methods of excavation has to some extent reduced the problem. Privatisation has probably not aided the process of data sharing.

The reasons for the lack of progress in this area are probably twofold: managerial (typically National Working Groups achieve little), and technical in that the lack of standardisation among GIS manufacturers has made the process of data exchange much more complicated than it should be. Wessex Water is currently involved with one electricity supplier to instigate a simple exchange of GIS data as a trial using PC's networked over the public telephone network.

A disappointing feature of current systems is their lack of accessibility to non-specialist users. There are two still largely different uses and views of the GIS, one, that it is a modelling tool for design and analysis and secondly, that it is a sophisticated digital map available for ready everyday use. This second use of GIS, that it would be available for users to access quickly, zoom into a particular area and manipulate data for a variety of everyday purposes has not been realised. GIS is still a specialised area with access limited to relatively expensive workstations. Computer users who have learnt to use PC's with common software packages such as spreadsheets and wordprocessors are now used to a high standard of user interface which current GIS systems simply do not provide. Most users have discovered that GIS systems need considerable ongoing expenditure on development and data aquisition to open up their potential for even relatively simple tasks. Even then the system may perform the task once very efficiently but then lack the flexibility for further progress. It is quite common for the end result of a development to be simply a paper copy of a map which is then stored for everyday use in the field. The notion that geographical information would be stored, used and transported electronically in the field is still some way off. Speed of access can still be relatively slow (dependent on hardware and system loading) and in day-to-day use paper maps (albeit updated through the GIS system) are still almost universal; indeed, one major task carried out by Wessex' GIS department is that of producing updated paper maps for smaller depots and/or field use.

It is interesting to reflect on this situation in the UK with that within the USA where vast quantities of digital map data are readily accessible over the internet for anyone with a telephone, modem and computer.

The fundamental problem is related to the way in which GIS data is stored and manipulated. If the GIS is to be used only as a digital map then it is simply a picture which is used quickly by staff to locate something in the real world in relation to something else (with underground assets this is particularly important since much time and money can be wasted searching or excavating to find the particular item). At this level it is not necessary to access a database describing the detailed attributes of each item - doing so via a typical GIS system is very time-consuming and is far less efficient than using a detailed paper map. Similarly, unless the map is portable (in the sense that the data can be carried into the field) it is of little use at this basic level. Most GIS systems do not satisfy these requirements of speed, portability, and simplicity, and for this reason GIS information on computer has not, to date, found its way into day-to-day operations in the way that was first envisaged and this has undoubtedly created something of a negative image for GIS; the high running costs are not reflected by widespread evidence of GIS in everyday use. It is better to design a flexible powerful digital mapping system than carry the burden of the unused functions of the GIS. Wessex has approached some of these problems as described in the following section

It is clear that the conventional GIS system of the 1980's as installed in Wessex, has major strengths but a number of weaknesses. Its strengths are in the control and maintenance of the central database, controlled updating mechanisms and Regional Standardisation of records. The database provides very powerful tools for asset management. The major weaknesses of the system are in access, speed of response and the user interface (in comparison to PC software standards). The cost of improving some of these shortcomings using conventional GIS hardware is disproportionate to the benefits. Consequently, Wessex is addressing some of the shortcomings through the development of WESNET, a PC-based network modelling and management package which is making major advances in integrating a variety of key data sources, one of which is spatial data. The fundamental concept of WESNET is to integrate on a standard PC platform, the 3 key data sources for overall water supply management: Network data, telemetry data and mapping data. These 3 key sources can then be manipulated to provide true integration. The PC allows local processing, a high standard of simple user interface, portability and the ever expanding possibilities of networking using standard hardware which is continually reducing in price.

GIS information is made available through the production of CGM files (Computer Graphics Metafile), the international standard for graphics information as an output option from the Regional GIS system. Such files may be manipulated in WESNET (or such packages as Freelance +), opening up the information to a wide variety of users for local information. The recent arrival of high quality (VGA) colour graphics displays on portable PC's now means that for the first time field staff can carry complete necessary GIS information in a van on site to the location where there is a problem. Advances in networking mean that maps can be downloaded via the public telephone network if necessary. This is one further step towards making staff independent of reporting/administration centres thus reducing travelling and other lost time.

These improvements in access and speed of processing have opened up many new potential uses - it is possible to use GIS data in a fast moving Control Room environment as a support to remote field staff, and in parallel to video disc based systems now commonly used in Control Rooms for vehicle tracking.

The basic GIS remains as the means of updating and controlling records and access to the database. A library of CGM format maps is available regionally over a PC network for users to download as required. These maps can be used for a wide variety of purposes according to the user but the crucial point is that for the first time GIS information is available throughout the organisation via inexpensive PC hardware which is portable and easy to use.

As discussed in the previous section a number of trends have merged in the 1980's which are likely to influence development of GIS in the 1990's. Two important factors in particular are discussed in this section: systems integration and cost benefits.

System integration: This concept is now widely understood but has made little real progress in the 1980's. GIS systems have long been recognised as potentially much more than platforms for manipulating geographic information. It is obvious that a computerised map might also be used to access and display data from a wide variety of sources; a few examples are listed below:

- Telemetry data: Real time data on flow, level, pressure or quality could be displayed on a GIS background, theoretically making it easier to relate such data to the real world, rather than a simplified telemetry schematic.

- Modelling data: Results of network analysis simulations (both water supply and sewerage) could be displayed at appropriate points on a GIS background as an aid to understanding those results in relation to the real world.

- Customer related data: Customer contact data, on, for example, taste complaints, leaks or burst mains, often held on mainframe systems, could be displayed on a GIS background to aid understanding of these geographically related events. Some are already held as an overlay on GIS systems and the results certainly illustrate the value for such presentation methods. There are many other such customer-related examples, for example billing, metering and water quality sampling zones.

While all these potential applications are attractive and can all be achieved with current technology there are some obvious drawbacks with current hardware. Firstly, the required links have to be made to the appropriate systems (e.g. telemetry) but more important, considerable software development is required to make such integration possible in a practical way for the user. Secondly, even if such data links are made, for such a system to be used widely, access must be opened up to a much wider audience than at present and with current hardware this is prohibitively expensive. It is quite common to see ideas or demonstrations of applications for GIS which are very impressive technically but which miss the obvious point that for everyday use the system must be accessible to the potential user. It is for this reason that many of the potential applications are never developed for real-world use. In addition, the real benefits have to be identified quite objectively and not in terms of doing innovative things with new technology for its own sake. If we consider the potential for displaying telemetry data on a GIS background, for example, the cost of a workstation for GIS access is many times greater than that for telemetry access. The benefits of displaying such data in this way are not so clear - it is sometimes much better to use a simple schematic to display such data rapidly than to include much irrelevant map data; one skill which is at a premium in a modern water company is the ability to simplify and perceive important events in a mass of detail.

Wessex has approached the question of systems integration through the development of WESNET; some practical examples are illustrated below. The underlying principle of WESNET is to integrate water supply modelling, telemetry and mapping data on a PC platform in a manner in which these data sources are complementary; there is no point in integrating data unless the "whole" is greater than the sum of the parts. Data transfer is by simple standard data files - no attempt is made to link telemetry hardware to GIS hardware or modelling hardware of any particular type. The system is entirely PC-based, leading to a high level of interactivity, access, portability and low cost thus opening up many new areas for truly cost-effective operation.

Telemetry data is available for direct comparison with model simulation results, making on-line calibration of network models very cost-effective - this is a genuine example of integration actually leading to a new and more effective type of operational management style. All of the major urban water supply networks in Wessex, around 15, are now modelled with a complete integration of telemetry data such as water level in a reservoir, flow at a particular point in the network and pressure at a node in the network. These are compared with the results of a network analysis simulation.

With this constant supply of telemetry data the models can be updated and calibrated with minimal resources over long periods leading to greatly improved understanding of the network and cost savings which stem therefrom. WESNET can be used as a simple mapping system in its own right (as described earlier) and also in combination with CGM files produced by any other package. Within Wessex, Lotus Freelance (+) is used extensively for a wide variety of general purpose graphics applications, and through the CGM file format any such drawings can be integrated into WESNET. CGM files can be used to display results from a network simulation in a variety of ways - thus results can be displayed on schematic or a GIS background. For example, a CGM mapping file, with complete background, could be used to display pressure and flow data from network analysis simulation; one disadvantage of this approach is immediately apparent. The amount of detail in "complete" GIS sometimes makes understanding more difficult than with a simplified schematic.

Cost Benefits: One major obstacle to the realisation for some of the potential applications in GIS is that often cost-benefit equations are either difficult to compute, or very unfavourable. There are two major reasons for this; the nature of the benefits and the high costs of hardware and software with many current GIS systems.

It is generally true that the benefits of GIS are difficult to quantify. The economies of scale resulting from the Regionalisation of the UK Water Industry in 1974 enabled major computer developments to be implemented, including GIS, but to estimate the financial benefits requires information on the costs of record maintenance of the predecessor organisations and estimates of wasted resources resulting from inaccurate or inaccessible records. This type of information is notoriously difficult to track down. Often the benefits of GIS are in hidden efficiencies through reorganisation - Wessex, for example, has progressively rationalised and streamlined its management structure, yet has maintained access to the central database of asset records throughout, irrespective of the location, of say, Planning and New Works departments, and this has greatly improved the efficiency of those departments. Wessex recently carried out a desk study to estimate the costs of providing a manual record-maintenance system to replace GIS and estimated it to be around £0.5 million per annum; this was simply to provide record maintenance, without any of the database facilities or specialised map production capabilities now in use. The value of these features is very difficult to quantify but is highly significant. Prior to privatisation each Company had to produce an Asset Management Plan upon which future investment levels would be based. Wessex' GIS system not only enabled this to be carried out quickly, but also with a high degree of confidence, essential for future business planning.

Under any criteria, GIS is expensive, both in initial investment (the immediately apparent hardware and software costs are small compared to data, personel and training costs) and running costs. Wessex' current system costs around £0.7 million per annum in running costs. There is now a clear realisation that GIS is a long-term investment and that there must be a commitment to make GIS work.

A major hindrance to GIS development which is now clear is the complete lack of standardisation (other than items such as SDTS, NTF, DLG which are not universally adopted, or ARC/INFO ungenerate format, which is appearing as a defacto satndard) and among the major manufacturers; each pursues its own internal standards with a view to making customers captives in the future. It is to be hoped that the trend elsewhere in the Computer Industry, of Standards, and competition (and thereby reducing prices) will be mirrored in the GIS in the future.

The major applications and uses which have evolved in Wessex' experience are broadly as follows:

- Record updating, map production and asset records. These tasks are all carried out via the GIS system and will continue. The support for Planning, New Works and Operations for ad hoc applications will also be via this system.

- Access to GIS at depot level, portable digital maps in vans. This will be on a PC (networked) base through the production of simple graphics files (e.g. CGM), and simple user-friendly software; Wessex has developed WESNET to fulfil this task.

- Rapid access emergency site location GIS. This will be via PC-based optical disc systems (e.g. AIMS).

- Data integration using GIS. This will be achieved at two levels: Via the standard GIS database for customer-based information, and on PC for technical integration of modelling, telemetry and mapping data using simple file exchange mechanisms and fully utilising the portability, networking and graphics capabilities of the modern PC. WESNET has developed these capabilities to a high degree.

4.3. OTHER EXAMPLES

Wessex water is not the only water company in the UK to implement computerized assets management. Welsh Water (Dwr Cymru) and South Staffordshire Water Company have also introduced similar schemes.

Dwr Cymru, Welsh Water, owns some 24,400 km of water mains and 16,200 km of sewage pipes as well as 163 treatment works and 850 sewage treatment plants (Lindsay and James 1995). In 1990 a 3 year deadline was set for the development of an information system to manage all the data generated by these assets; the Asset Information System (AIS).

Lack of digitized Ordnance Survey (OS) maps presented a considerable problem, but in agreement with OS Dwr Cymru digitized paper maps to OS standards, in all some 21260 large scale maps. Lindsay and James report that "At the peak [of digitizing], around 110 staff worked in shifts 21 hours a day, to meet the two year deadline for data capture".

AIS supports Dwr Cymru's core operations. It provides maps for all water network maintenance operations. It is important to note that operation procedure allows data within AIS to be continually improved and updated by reentering amendments to the map made by field workers back into the system.

AIS has a role as an executive information system (EIS). To rapidly provide managers with the information necessary for business decisions. Lindsay and James (1995) report that information provided by the AIS " was used to develop a plan for refurbishing water mains and sewers throughout the companies operating area".

As well as supporting operations and business needs, the AIS enables useful analysis that would not be practical with a manual system. For example Dwr Cymru have developed a software package now used by other organizations, "Background Leakage and Burst Estimates (BABE)" that uses pressure zone boundaries, OS digital elevation data together with address data and zone pressure data from the AIS as part of a leakage control strategy. Lindsay and James state that "AIS is a key element in the company's [Dwr Cymru's] business strategy."

South Staffordshire Water are using a GIS in order to improve the effectiveness of their pipe maintenance and replacement programme (Rogers, 1995). A Sysdeco GIS is used to maintain an inventory of the pipe network and includes pipe specifications and installation details. Laboratory analysis of pipe samples taken whenever a pipe is uncovered provides an assessment of the risk of pipe failure for different pipe types, ages and under different conditions such as soil type. Using this laboratory analysis in combination with the spatial distribution of pipes; areas most prone to failure can be identified and treated before the problem occurs. In this way South Staffordshire Water can adopt a proactive approach to pipe network maintenance.

4.4. CONCLUSIONS

Because of the spatial nature of much of a water companies assets a GIS is an appropriate response to the company's information needs. A GIS provides opportunities.

- To enhance the effectiveness of core operation such as pipe maintenance, for example Dwr Cymru's generation of maps for each maintenance task or Wessex's links to telemetry and hydraulic modeling.
- To support secondary activities such as statutory report generation or as in the case of Dwr Cymru providing the information required for investment planning.
- To assist in extraordinary operations such as drought or contamination incidence management.

A GIS also enables operations that utilizing a manual system would be impractical, such as South Staffordshires approach to proactive maintenance of the mains network. However there are risks, just as there are when introducing any corporate information system to an organization, and the money involved is large. It appears that the large investment is in providing a high quality data inventory and for the production of digital maps. The use of such data in complex spatial analysis is yet to be fully realized.

4.5. REFERENCES

Dangermond (1991) The Commercial Setting of GIS Geographical Information Systems. Vol 1 - Longmans. PP 55-65

Lindsay P., James A. (1995) Running Water in Wales: GIS meets the Business Challenge. Mapping Awareness, Sept. 1995 pp28-30.

Senn J.A. (1992) Analysis and Design of Information Systems. 4th Ed. McGraw-Hill Book CO Singapore.

Rogers A. (1995) Mapping out the Battle Plan. Surveyor, Nov. pp15-19.

WESNET User Manual (1991) Wessex Water Services Ltd.

WESNET Technical Reference Guide (1991) Wessex Water Services Ltd.

Chapter IX

Institutional Social and Economic Aspects

Chapter IX

Institutional Social and Economic Aspects

INTRODUCTORY STATEMENT

Very different are the topics connected with water supply systems, and those concerning the institutional framework are normally neglected. Social and economic aspects are relevant at various levels, and decision making procedures as well as technology transfer and management methods should be considered. The success or failure of very big projects very often depends on non-technological elements, and the co-operation of many specialists is essental.

Even some fundamental concepts of water law principles as well as fundamental historical regulations have to be highlighted, because water administration is an interdisciplinary subject, which requires that engineers could usefully collaborate with lawyers, whose task is to provide institutional solutions without necessarily an understanding of the technical issues.

In both developed and developing countries water supply management is rapidly changing, so that governments have to solve the most crucial problem, which is to provide good services at reasonable prices. Some European countries' consumption data show how water prices are strictly linked to specific consumption.

In order to make the decisions of new countries easier it seems useful to look at the management options by briefly showing the solutions adopted in some countries, among which England, France, Germany and Italy may have relevant structures.

The goal of the next years becomes the setting up of modern systems in which water resources will be managed in terms of efficiency, environment and economy. This goal will be hindered by municipal inertia especially present in many Eastern European countries, where a special effort must be undertaken in order to find a balanced way to pass from uneconomic management to public or private or mixed economic enterprises.

Another important task is to unify drinking water quality standards as well as the quality of water services, both because of increasing water transfers among bordering countries and the growing water world market.

European Nations will have to face a very big financial effort: joint ventures should be set up and national laws should promote and encourage systems by which, on the one hand, public interest and safety are protected and, on the other, private enterprise interests are guaranteed as well.

Public or privately managed enterprises have to be encouraged in such a way as to master many services, in the sense both of a drinking water only service, in large districts, and of many water services, including more phases of the water cycle, starting from water supply to sewage disposal and effluent treatment.

The implementation of a modern legislative and administrative framework concerning water services is often complicated both by existing laws on drinking water and other water uses, and traditional, largely accepted practices. In other words, all water usages must be harmonized, by taking into account all the needs and differences of the involved subjects. Any changes will probably be resisted by those people, who have long benefited from customary water rights, and by municipalities whose will to be independent often contrasts with an efficient and economic management.

There is no pre-established form, since economic and social aims can be reached by either direct public or private service.

Several elements are fundamental to correctly build and manage drinking water services.

First of all, a wide ***water service organization*** has to be fostered, which comprises both large scale territorial associations and all civil water cycle phases. It will be so easy to plan how to exploit the resource and how to use it in different ways. Charging policy is crucial: external funds should be avoided and at the same time investments promoted in order to modernize the obsolete distribution network facilities, by replacing deteriorated pipes, installing water meters, setting up remote controlled networks, building dual networks in the largest cities. Without any specific evaluation, tariffs cannot be fixed, because many factors widely influence them. The tendency is to work out normalized tariffs, in which water charges depend on the following elements: water cost, if the resource is purchased from others, electricity cost, if pumping facilities are involved, water losses, water resource strategy such as spring, river or other, treatment methods, distribution layout, ratio between delivered volumes and consumption network length, trend of recent investments. Balanced prices could also be obtained by establishing tariff bands, with lower prices for the required domestic consumptions and higher ones for exceeding consumptions.

Secondly, ***planning and control of water service performances*** are crucial, because it is necessary to guarantee both the economic interests of managing companies in terms of fullfilment of investment plans and the consumer requests in terms of service quality. The quality of water services falls into the broad framework of *quality of life,* which is more and more attracting the interest of the citizens. The so called *levels of service* consist both in the management style of water companies and in the standard of the service provided to customers. The discussion on quality of management is now accelerating due to the ongoing water services privatisation. Particular attention has to be devoted towards efficient control mechanisms, which guarantee good competition by assessing and comparing levels of service. Water services should be processed as an enterprise, following the same principles adopted for other market products. In order to get objective criteria some means are needed: implementation of quality management instruments, such as regulation, standardisation, metrology and certification; building of rules and rights which are mandatory for both producers and consumers; creation of a quality system, which should cover all consumer expectations. Even if the discussion on how to assess the performances of water suppliers is very broad, some fundamental criteria can be assumed: it should be established that consumers can have access to supplier documents in order to examine records of water distribution and so on; companies should make the levels of service of their performance public, in order to guarantee a minimum level in terms of service reliability. Good indicators for levels of service are the following: quality of delivered water; water availabilty over a fixed minimum threshold; pressure of mains water over a fixed reference level; number of customers subjected to water restriction; number and duration of service interruptions due to failures; number of accidents due to back flow from sewer system; response to billing queries. In any case it is necessary that levels should be based on characteristics of globality, simplicity, independence of the type of managing institution and objectivity, so that each customer can assess indicators which are easily measured.

Finally, ***training of specialists*** has to be promoted at various levels in order to coordinate the different subjects involved in water supply matters. At a national level government and ministries often conflict with each other, causing contradictory regulations and a lack of information. At the educational level the complexity of water supply systems requires integration and coordination of traditional "curricula studiorum", which must comprise programs of resources management, economy, environmental protection, sanitary engineering, information processing, and so on. Urban water projects often fail because of the poor coordination of the many specialized engineers (hydraulics, chemical, structural, electrical, mechanical, automation) in the design phase and of the difficulty in understanding among the same engineers and the operational team in the management phase. The role played by professional associations should be strategic: they could promote thematic meetings, courses, seminars to encourage innovation of knowledge and transfer of new technologies. Consulting companies should adopt more reliable data and perform projects based on deeply studied existing systems. Water supply companies are in a transition situation, since the change of ownership has been experienced in many countries recently. Generally speaking a low awareness of the importance of collecting data is common, and it is crucial to remove the barrier placed by senior managers against the introduction of new technological tools.

1. Historical and Institutional Aspects

Paolo Veltri
Dipartimento di Difesa del Suolo
Università della Calabria
87040, Montalto Uffugo (Cs)
ITALY

1.1. HISTORICAL ASPECTS

"Water is indispensable to human life, because it satisfies both pleasures and needs of daily use" (Vitruvio, 1978).

The importance of water in all the human activities is unarguable. Starting from the earliest human communities the more was water available the faster was the progress towards civilisation. It is neither easy nor relevant to this Section to analyse the enormous historical material dealing with the legislative aspects of water: it is however necessary to start from socio-legal considerations in order to understand the complexity and unified nature of water problems and to handle the principles of water administration.

First of all water has to be studied from a physical point of view, in terms of the hydrological cycle, precipitation, surface and groundwater flow, flow control, collection and transfers. When passing from these last ones to the possible water utilisations it becomes difficult to distinguish between engineering, management and legal tasks: in other words water becomes a highly interdisciplinary subject. In this field it becomes important to give engineers a basic knowledge of the social and legal aspects, so that they could have a useful dialogue with lawyers, whose task is to provide institutional solutions but who may not have a full understanding of technical matters.

In order to link "natural context" and "human context" with each other water has to be regarded both in terms of a single resource and as part of all the natural resources; moreover from a political and geographical point of view it is very often not possible to separate the origin and exploitation of water from one single country to another.

If we consider only freshwater as a single resource, the following potential usages must be emphasized: water can be used for civil purposes, irrigation, hydroelectricity, industry,

NATO ASI Series, Partnership Sub-Series, 2. Environment – Vol. 15
Water Supply Systems – New Technologies
Edited by Č. Maksimović, F. Calomino, and J. Snoxell

transportation and recreational uses. Each of these usages might have been considered more relevant than other ones in a specific country so that the historical context and, consequently, laws and administrative regulations have reflected this situation.

On the other hand water uses could be in conflict with other natural resources, such as land, atmosphere, subsoil. It becomes clear that some lawful social and economic interests could favour or not some particular water usages and also other different resources. You may think, for instance, of water misuse, both in quantitative and qualitative senses, in exploiting caves and mines: the results may affect the downstream users, and so on.

Finally, national interests in the available resources might be in conflict with those of bordering countries.

Starting from the earliest human settlements it appears that towns developed around river valleys or in humid areas, so that it is common to classify those ancient civilisations as "fluvial or hydraulic civilisations": the history of Egyptians is the history of Nile river people, that of Assyros and Babyloneses is that of Tigris and Euphrates river people, and so on for Hindu, South Eastern Asian and Chinese people.

All these people had the need to regulate in some ways the use of water, both in regions in which this was abundant, in order to prevent damage and control floods for irrigation purposes, and in arid regions, in order to make better use of the precious resource.

The first water regulations reflected situations of single uses, that were implemented just to solve conflicts or to penalize users responsible for offences. During the centuries the argument has assumed more complex features, whereby each individual legal body rises from the preceding one modified by custom, religion, policy as well as modern events.

The early *Egyptian* water regulations originated in a context in which the ownership of water and land had a divine origin: so the Pharaoh or in turn the high social hierarchies had control of the whole resource: but, in any case, they provided for controversies regulation and collection of the revenues deriving from water taxes through a scattered system.

In *Mesopotamiam Civilisations* the dominion of water was not private and the use of water by private owners was restricted by early regulations. In the Hammurabi Code some principles were clearly stated: these concerned the rights to use water as well as corresponding sanctions against people who either did not maintain and repair the banks of irrigation channels or were responsible for breaches and failures.

Also ancient Talmudic laws stated for *Hebrew* people that rivers and springs belonged to every man.

Breaking off other equally important early regulations, it is necessary to focus our attention on *Roman* water law principles. Almost all the existing legislative systems around the world derive their bases from Roman law. Starting from the origins and passing through Republican and Principate periods until the fall of the Western Roman Empire an impressive and evolving legislative body looked at water use. The sources can be found (Caponera, 1992) in customs, statutes (leges), edicts (edicta), decisions of the Senate (Senatus Consulta), imperial constitutions (edicta, decreta, rescripta, mandata), opinions of the jurists (responsa prudentium). The earliest rules on drinking water supplies originated during the Roman Republic.

In that very impressive legal matter some important principles can be summarized. Regarding ownership, Roman law distinguished between common water, public water and private water. Those resources which did not have any owner, either public, private, individual or collective, belonged to the first category. Most water was common; only an eventual right of use was allowed. Public institution water, which was also subjected to the right of use, belonged to the second category. Finally private water, which usually was considered as a part of the land and was restricted to a few cases, belonged to the third category. In contrast with modern tendencies in the ecological field, it must be highlighted that Roman law did not consider the potential negative effects of water use: in fact people who had the ownership as well as the water rights on their land were free to use it without any consideration for neighbours.

Another legislative body which influenced hundreds of millions of inhabitants in all the *Moslem* countries is that deriving from Koran as well as Islamic tradition. The water is considered as a property of all Moslems and a right explains in a clear manner the philosophy of that law: the so-called *thirst right* statutes that everybody can take water to quench his thirst. A specific water legislative system based on the Prophet Mohammed teachings is the code Mejelle (1870), that was in force in all the countries of the Ottoman Empire. It is halfway between the Koran and modern laws in Moslem countries, but excluding some which have only recently re-established the original principles of Koran. For many aspects the Mejelle code emphasizes even more the natural rights of water, in fact it extends the thirst right to animals. According to water administration in Moslem countries, few cases of organised systems can be found: in fact the administration is nowadays based on local customs, which are derived from traditional rules as well as from colonial heritages, imposed by France, England, Italy etc.

The modern water legislative systems in Western industrialised countries result from the differences created by the *Napoleonic Code (1804)*. According to this Code we can in fact distinguish between *Civil Law Countries* (France, Italy, Spain, Netherlands, etc.) and *Common Law Countries* (England and former English colonies). The difference arises from water classification: in civil law countries there are two water categories, that is private or public waters; in common law countries the Roman principle of common water is maintained, constituting the basis of the so-called riparian doctrine. Both these legislative principles deeply influenced the incoming legislations of the colonies, even if the resulting codes took into account local customs as well as religious and traditional codes. Just to give a remarkable example, the water doctrine in the USA is not as severe as the original riparian one.

1.2. INSTITUTIONAL ASPECTS

According to the civil law the ownership of water can be either public or private, and in particular water necessary for domestic purposes is classified as public; as a consequence the rights to use public water have to be legally granted. These rights have a fixed duration and have a corresponding price. Some restrictions exist also for private water which, according to the recent trends in ecological matters, has not to be polluted or misused. As regards water administration it is not possible to classify the institutions which are in charge of the control and management of the resource, this task belonging both to many national boards and local institutions. Anyway, as evidenced again by Caponera (1992), the prevalent tendency is now to regulate by rules, permits and other legal tools both public and private water. Moreover the differences among procedures either in common law or in civil law countries are disappearing, especially when we are dealing with problems of managing and distributing drinking water for domestic purposes.

This is a relevant consequence of the always increasing importance of the ecological issues, whose political evidence can be highlighted even by the replacing of legal terms: the recent Italian "Legge Galli" on Water Resources states in fact that water is public, and also in the French law the term private water has been changed into non-domainal water. The differences between civil law and common law countries are less and less evident, and in particular the countries which belong to the EU share common regulations and directives. But the reasons of an overall water policy arise from more and more frequent water transfers, so that European countries will have to make legislative and financial efforts in order both to extend to countries out of the EU the standards, controls and regulations which are already mandatory for the member states, and to get water supply systems comparable in terms of the quality threshold of the delivered resource as well as efficient and economic distribution services. Due to the events which have occurred in Eastern Europe after 1988 and because of either lawful inertia and resistances or consolidated customs and rights, it will presumably not be an easy task to build in a short time a real water market economy. On the other hand, it is difficult to find the same features in all the EU countries and many different management systems may co-exist within a single member state.

In order to better understand the effort which the Eastern European countries will be dealing with, it must be considered that the water administration in those states were, and in part still are, derived from socialist principles. Water was considered as a whole with land and the State had on water either ownership or all the rights. If we look at the previous USSR system in particular, even administrative aspects were to be seen in a framework of national economy, with many agencies responsible for the service, both at national and regional levels. The result is that many problems can now give rise to a social and economical context in which the contemporary existence of central financial support and municipal monetary resources must not hinder the development of private companies, among which the international ones are in a profitable position.

1.3. REFERENCES

Caponera D.A. (1992) Principles of Water Law and Administration. Balkema, Rotterdam.
Vitruvio P. (1978) (ed.) Dell' architettura. Giardini Editori e Stampatori, Pisa.

2. Management Options

Paolo Veltri
Dipartimento di Difesa del Suolo
Università della Calabria
87040, Montalto Uffugo (Cs)
ITALY

2.1. INTRODUCTION

The features of drinking water management systems extend as a logical consequence from the general legislative principles. Into a general law framework many rules regulate the water supply service. Very often the success of an economic and socially acceptable water management system is based on the nature of the institutions which manage the resource and then ensure, both at national and local levels, control and investments.

The first thing we have to emphasize is that almost everywhere and regardless of the water legislation in force the uses of drinking water for domestic purposes have the top priority over all the others.

The subjects we are dealing with are obviously those related to public drinking water systems, which must be defined in terms of a minimum number of customers and which operate for a minimum of days on each average year.

From a strictly economical point of view it appears that in many countries all around the world the drinking water supply system has unbalanced prices, which only partially take into account market rules. Besides, referring to European countries, even in each state many managing configurations are present. In all the countries, in fact, except for England, many and different management systems co-exist.

The water prices are often affected by elements which are out of the market. We must in fact recognize that some elements make the water system original. The drinking water service can be seen as "a natural monopoly", in the sense that for an already managed area (municipality, district, basin or others) there are no competitive elements among different contractors, and moreover no comparisons are possible among distant water management systems. In other words the situation is contradictory: if a management company holds water

NATO ASI Series, Partnership Sub-Series, 2. Environment – Vol. 15
Water Supply Systems – New Technologies
Edited by Č. Maksimović, F. Calomino, and J. Snoxell

rights, then it needs to get them for long periods of time, in order to program investments and forecast credits; but in this case the plurality of many contractors is missing.

If, on the other hand, the rights last for a short period of time, then the contractor can neither program investments nor forecast anything.

It must be highlighted that each water supply system is a prototype and the exchanges between contiguous systems are at present not very common.

For these reasons many existing management systems do not have purely market-regulated water prices: other social, political and cultural factors play a role. Some economists talk of a "fictitious market", in order to justify the rules adopted to determine water charges.

By means of a fictitious market it is possible to put together rules and procedures which allow the comparison of the performances of different and distant water supply systems.

2.2. FEATURES OF DRINKING WATER MANAGEMENT SYSTEM

Referring to the European context and in particular to the EU member states (EUREAU, 1993, Rillaerts, 1994, Zabel and others, 1994), they show a great diversity of management systems, even within the same country. It is a very hard task to classify the different systems according to a few features.

In fact we should detail forms, procedures, structures, powers and models of both central and local governments, etc. As for the forms, either public or private or mixed ones exist; as regards procedures, they depend on adopted rules; structures can be created both to manage water supply only and to administer also all the phases of water civil uses, including the drinking water services and those concerning sewage disposal and wastewater treatment.

There are many cases in which also electricity and gas distribution services are run by the same company.

Anyway, even if differences can be found still in the terminology of operating entities, a broad classification of the possible management models can be established by reference to the most relevant features of the service.

a) Except for England, a public body, which can be either the State (this is the case of Northern Ireland) or a regional/district/municipal body, is responsible for the water provision service.

b) The management entity, which is responsible for the distribution service to the end-users, can be either the same or a distinct body from the responsible one.

c) The organization and the form of the distribution systems fall (EUREAU, 1993) into one of the following categories:

- public direct management,
- direct management by inter-municipal/regional structures,
- delegated public management,
- delegated management with mixed or private capital,
- private management.

In the case in which responsible and managing bodies are distinct, different circumstances can occur in terms both of ownership of facilities and equipment and investment financing. There may be, in fact, situations in which facilities are owned either by the responsible body or by the management body, and investment is either allocated by the management body itself or supported by the responsible body by means of subsidies as well as external financing. In the following Tables 1. and 2. (EUREAU, 1993) the major management types of drinking water services and statistics of the same types in the EU member states are reported. In order to better understand the meaning of the data summarized in Tables 1. and 2. more information about some existing water management systems in some selected European countries will be given in Section 4. of this Chapter.

d) The water prices, recognising that the drinking water supply is a public utility service, can be determined (Ieno, 1993) according to the following methods:

- method based on a cost recovery system, that is water prices must cover the total service costs,
- method based on water prices initially fixed at the risk of the management body (established in a contract and eventually subjected to an automatic revisional formula),
- method based on the control of the rate of return on invested capital.

Table 1. Major management types of drinking water services in EU member states (EUREAU, 1993)

Management Types	Belgique	Danmark	France	Deutschland	Ireland	Italia	Grand Duché de Luxembourg	Nederland	Portugal	Espana	United Kingdom England and Wales	United Kingdom Northern Ireland	United Kingdom Scotland
1. Direct Public Management													
Direct Management	x	x	x	x	x	x	x	x	x	x		x	
Autonomous Board	x		x	x		x			x	x			
2. Direct Supramunicipal Management													
Direct Management	x		x	x		x	x			x			
Autonomous Board	x					x				x			x
3. Delegated Public Management													
Cooperatives Companies		x											
Public trading Companies	x		x		x		x	x	x				
4. Delegated Private or Mixed Companies													
Private Companies	x		x	x		x			x				
Mixed Capital Companies	x		x	x					x				
5. Direct Private Management													
Private Companies											x		

Table 2. Statistics of management types of drinking water services in EU member states (EUREAU, 1993)

	Direct Public Management	Direct Supramunicipal Management	Delegated Public Management	Delegated Private or Mixed Management	Direct Private or Mixed Management
Belgique	5		90	5	
Danmark	67		33		
Deutschland*	35	20	30	15	
Espana	48	11	12	29	
France	23		2	75	
Grand Duché de Luxembourg	100				
Ireland	100				
Italia	72	23	1	4	
Nederland	15		85		
Portugal	92		8		
United Kingdom	3	9			88
Average Weighted by Population	37	11	15.5	20.5	16

United Kingdom:

England and Wales:	100% Direct private management
Scotland:	100% Regional direct management
Northern Ireland:	100% Direct public management

*Remark: The percentage distribution corresponds to the population of the former Federal Republic.

Morever, the prices can be either balanced, if they cover the total costs in the balance, or unbalanced, that is water charges do not cover the total cost of the service, the remaining cost being financed by either municipal budget or external subsidies.

The cost recovery system is adopted in Germany and Italy, while the well known "price-cap method", adopted in England and France with some variations of the original features, belongs to the method based on initially set prices. The third method is not very common in European water supply services, whereas it is sometimes adopted in the USA.

As for the *Cost Recovery method*, the managing entity depends in any case on Government, which usually fixes both the cost analysis methodology and the tariff structure. The costs include personnel, maitenance and operational costs, technical and financial depreciation, passive interest on investment, reward of the invested capital (if the management is carried out by private companies). The prices can be then determined by referring to either the annual balance or the many-years balance, and to either provisional or consolidated balances. The Cost Recovery method could fail if the Government control establishes restrictive clauses in determining charges, in which case the service quality and the water system improvements could suffer from negative effects.

As for the *Price-Cap method* it is largely used in England in all public utility services with a monopoly regime. The method is usually applied by a year by year up-dating of the initially set water price P_o by means of a simple formula:

$$P_{0+1} = P_o (1 + i + x)$$

in which: P_{0+1} is the up-dated price,

i is an inflation coefficient, being equal in England to the retail prices index,

x is the productivity coefficient, being positive if the service, as in the case of water supplies, is characterized by a growth of the service quality with productivity loss (Meucci, 1993).

The initial price P_o can be determined by taking separately into account operative expenditure, capital maintenance expenditure and return of capital.

Water companies in England assume $x = -1.0 \rightarrow 3.5$ %, with the compliance of the Office of Water Service (OFWAT), which has to set the water price limits. A contract between company and customers fixes service quality standards, expected goals and penalties in the case of non-compliance. The Price-Cap method gives many advantages:

the users understand very easily how much the service prices are and they are moreover guaranteed against any shortcomings in the management company, the public authority control and procedures are very simple, the surplus remains part of the enterprise economy.

Some improvements in the Price-Cap method have been introduced in France, by means of particular arrangements in the contracts between companies and public authority.

e) The management control policies usually concern price control and management, service quality and health standards conformity controls. The control entities can be different and belong to the central Governments as well as local authorities.

2.3. REFERENCES

EUREAU (1993) Management Systems of Drinking Water Production and Distribution Services in the EC Member States in 1992. EUREAU editor, Brussels.

Ieno E. (1993) Le tariffe qualità alla luce delle esperienze europee. Atti delle giornate di studio sul tema La misura nella gestione delle infrastrutture idrauliche. CUEN, Napoli.

Meucci F. (1993) Tariffe e price-cap. H_2Obiettivo '90, Federgasacqua - Cispel Emilia Romagna, Bologna, 29-31 marzo.

Rillaerts F. (1994) Structure and Management of the Drinking Water Service. Hydrotop 94, 12-15 avril, Marseille.

Zabel T.F., Rees Y., Correia F.N., Neves E.B. (1994) Institutional Arrangements for Water Management in Selected EC Countries. Hydrotop 94, 12-15 avril, Marseille.

3. Quality Management in Water Supply

Mario J. N. Neves
Universidade do Porto
Rua Jerónimo Mendonça 65
4200 Porto Codex
PORTUGAL

3.1. INTRODUCTION

In recent years, a growing concern about *Quality* has been evident at all levels. On one hand, this is connected to the quality culture that is being developed among the public, which determines a greater pressure with respect to suppliers. On the other hand the suppliers themselves are becoming aware that quality is a decisive factor for success in highly competitive societies, and that therefore they must develop their own mechanisms to assure the quality of their products or services.

Following concerns with the quality *of products*, preoccupations are now also being directed towards the *quality of life*. Environment is one of those manifestations, and the interest of the public in environmental impact assessments is clear. The quality of services is another topic that is attracting the interest of society, as is the case in health systems, education, transport, etc. In some cases, comprehensive evaluation systems are becoming common practice, for example for universities, quality assessment is now widely used.

Quality assurance in water supply systems, the discussion of which started in the late eighties, is described by Neves (1994). Manifestations of this concern can be seen in the management style of water companies, for example, and also in the quality standards of the service rendered to the customers, the so called *levels of service*.

Privatisation of water services is another factor that is accelerating the discussion of levels of service. Privatisation has been implemented in some countries, and it is being considered in many others.

On one hand, authorities that have signed or are planning to sign contracts with private companies, wish to be sure that the consumers are in "good hands". The enormous social importance of water supply justifies a particular attention towards an efficient control system.

NATO ASI Series, Partnership Sub-Series, 2. Environment – Vol. 15
Water Supply Systems – New Technologies
Edited by Č. Maksimović, F. Calomino, and J. Snoxell

On the other hand, competition between companies is stimulated by the levels of service evaluation. Besides the intrinsic reasons, the designation of the "best practice" company is, of course, very important in marketing terms.

Also financial reasons should be highlighted. Good performance can be seen as an argument for increasing water prices. On the contrary, low quality services may invoke penalties, as occurs in England, for instance.

The feeling that international legislation in this area will be prepared soon is also a reason to accelerate the discussion concerning levels of service. As in other cases, for example both drinking and bathing water directives, it is natural to expect that the European Union (EU) will introduce some rules concerning levels of service in water supply systems. Many people, at various levels, wish to be prepared for the construction, discussion and implementation of the legal framework. International authorities, national authorities, private companies, universities and other institutions, will certainly be involved in that process.

3.2. INSTRUMENTS OF QUALITY MANAGEMENT

The European Standard EN 29 000 (1987) defines quality management as an aspect of the general function of management which determines and implements the quality policy. That responsibility belongs to the managers of the organisation, but all its members must be involved in the process of quality attainment. Quality management includes strategic planning, resource allocation and other systematic activities related to quality, such as operational activities and assessments.

This methodology, firstly used in high technology domains such as nuclear and special engineering, was adopted by industry and is now being extended to many areas, including water services.

For instance, the Working Group on the Operation and Maintenance of Water Supply and Sewerage Systems created by the World Health Organisation, recommends (Hueb, 1993) that water services should be processed as an enterprise, following the same principles adopted for other products available in the market. The same author reminds us that the functions usually carried out in these institutions can be grouped in organisational systems, whose proposed configuration is shown in Table 1. He emphasizes that permanent assessment by means of management information systems is crucial for planning and correction of quality deviations and the use of indicators for operational and management performance is fundamental for that purpose.

Quality must be present in all phases and components of water supply systems, including planning, design, construction, rehabilitation and exploitation. Some means to achieve that objective include (Matos,1993):

(i) Implementation of **quality management instruments**, such as **regulation, standardisation, metrology and certification**.
Regulation is the production of documents with obligatory character, which set the essential requirements of products and services. In the European Union they can be **community directives**, which have to be transposed to the Member States law, and the "technical regulations", which are specific to each Member State.
Standardisation is the production of mandatory reference documents, which describe methodologies to render a service or the characteristics of a product, defining tests, control proceedings and criteria for acceptance or rejection.
Metrology is an activity concerning the quality assurance of apparatus and measuring equipment, involving the proceedings for their calibration and reliability.
Certification is the last stage of the quality cycle, involving inspection, quality control and auditing, activities which presuppose the existence of reference documents and the systematic use of measurements.

(ii) Definition of attributes and responsibilities of the principal **agents**, which are the **producers** (products and services enterprises), **the consumers** (citizens at individual or collective level, or their representatives) and the **State**, who must enforce the established rules and preserve the rights and obligations of the parts involved.

(iii) Creation of a **quality system**, understood as the set of the organisational structure, responsibilities, proceedings and resources to implement quality management (EN 29 000).

Since 1987, ISO, the *International Standard Organisation*, has published a set of standards related to quality management, the famous 9 000 series. These standards have been adopted by the European Committee for Standardisation under the form of European Standards, the EN 29 000 series.

EN 29 000 (1987) "Quality Management and Quality Assurance Standards" defines some concepts and indicates the guidelines for selection and use of International Standards. A distinction is made between standards for the external quality assurance (EN 29 001, EN 29 002, EN 29 003) and standards for the internal quality management (EN 29 004).

Table 1. Organizational systems (Hueb, 1993)

INFORMATION SYSTEM

HUMAN RESOURCES ADMINISTRATION AND DEVELOPMENT SYSTEM

COMMERCIAL SYSTEM	BILLING AND COLLECTING
	CONSUMPTION MEASUREMENTS
	CONSUMERS RECORD
	COMMERCIALISATION
OPERATIONAL SYSTEM	PROJECTS MANAGEMENT AND CONSTRUCTION
	OPERATION
	MAINTENANCE
PLANNING SYSTEM	PHYSICAL PLANNING
	ECONOMIC AND FINANCIAL PLANNING
	ORGANISATIONAL PLANNING
	PROGRAMMING
	CONTROL
LOGISTIC ADMINISTRATIVE SYSTEMS	SUPPLIES ADMINISTRATION
	PATRIMONY ADMINISTRATION
	TRANSPORTS ADMINISTRATION
	SOCIAL COMMUNICATION
FINANCIAL SYSTEM	FINANCIAL ADMINISTRATION
	ACCOUNTANCY

HUMAN RESOURCES ADMINISTRATION AND DEVELOPMENT SYSTEM

INFORMATION SYSTEM

Emphasis is given to the requirement that quality systems should be documented and demonstrable. Documentation may include manuals, proceedings, audit reports and other records. Demonstration means the capability of proving the existence of an adequate quality system, and the aptitude of obtaining products or services in conformity with the specified requirements.

The titles of another group of standards are self-explanatory:

- EN 29 001 (1987), "Quality Systems. Model for quality assurance in design/development, production, installation and servicing",

- EN 29 002 (1987), "Quality Systems. Model for quality assurance in production and installation",

- EN 29 003 (1987), "Quality Systems. Model for quality assurance in final inspection and test".

Finally, EN 29 004 (1987), entitled "Quality management and quality system elements", provides guidelines for the conception and implementation of quality management systems. The role of quality manuals and periodic assessments is strongly emphasised and attention is also devoted to economic aspects, marketing, products conception and specification, provisioning, production control, tests, corrective measures, handling and post-production activities, documentation and quality records, staff, security, responsibility and use of statistic methods.

Also relevant are EN 30 011-1, EN 30 011-2 and EN 30 011-3 (1993), transposed from ISO 10 011 (1990), concerning Quality Systems Auditing. Guidelines for the preparation of Quality Manuals can also be found in standardisation documents (e.g. NP-2732, 1986).

With respect to technical aspects related to water and sewerage services, the relevant EU Directives are the following, divided by thematic areas (Matos, 1993):

i) **Consumption water**

80/777/EEC	Exploitation of natural mineral waters
80/778/EEC	Quality of water for human consumption

ii) **Water for several uses**

75/440/EEC	Quality of surface waters for the production of drinking water
76/160/EEC	Quality of bathing waters
78/659/EEC	Quality of fresh water for fish life
79/869/EEC	Analysis methods of surface waters for the production of drinking water
79/923/EEC	Quality of conchiferous waters

iii) **Discharge of dangerous substances**

76/464/EEC Directive on dangerous substances discharged into the water and specific
Directives 82/86/EEC, 83/513/EEC, 84/156/EEC, 84/491/EEC, 86/280/EEC, 88/347/EEC

80/68/EEC Pollution protection of underground waters

91/676/EEC Protection of surface waters against nitrate pollution

iv) **Industrial control**

78/176/EEC Residues from the titanium dioxide industry, altered by 83/29/EEC

82/883/EEC Proceedings for verification of residues from the titanium dioxide industry

87/217/EEC Asbestos pollution prevention

v) **Products control**

73/404/EEC Detergents, altered by 82/242/EEC and 83/94/EEC

73/405/EEC Biodegradability of anionic surfactants

vi) **Waste water and sludge**

86/278/EEC Agricultural use of sludge from waste water treatment

91/271/EEC Municipal waste water treatment

vii) **Information exchange**

77/795/EEC Information exchange concerning fresh water quality, altered by 86/189/EEC

83/189/EEC Information concerning technical regulation

viii) **Public market of works and products**

71/305/EEC Public works contracts, altered by 89/440/EEC

77/62/EEC Contracts with suppliers, altered by 88/295/EEC

90/531/EEC Proceedings for products acquisition

ix) **Other Directives**

85/337/EEC	Environmental impact assessment of certain public and private projects
85/374/EEC	Responsibility for damages caused by defective products
88/320/EEC	Laboratory good practice
87/216/EEC	Serious natural accidents, altered by 82/501/EEC
89/106/EEC	Construction products

x) **Proposed Directives**

COM (84)303	Titanium dioxide pollution reduction
COM (85)373	Waste discharges into the sea
COM (86)733	Objectives of water quality concerning chromium
COM (89)162	Security of products.
COM (89)303	Environmental European Agency
COM (89)478	Alterations of 75/440/EEC, 76/160/EEC, 79/869/EEC, 80/778/EEC
COM (90)297	Application of EC rules to the proceedings for products acquisition
COM (91)145	Use of EC conformity mark
COM (91)347	Proceedings for services acquisition.

3.3. ASSESSMENT OF WATER SUPPLIERS PERFORMANCE

3.3.1. Existing Systems

The preceding paragraphs were devoted to some instruments for quality management, with particular emphasis on water services. Now, let us centre the discussion on the criteria for the evaluation of the quality of their product, that is, the quality of the water supply, considering the aspects usually seen as relevant by the customers.

The English system to evaluate the performance of water suppliers is very comprehensive and one of the first to be implemented (early nineties). This, and the fact that there is now some experience in its use, explains why it is so often quoted. Let us follow Miller (1993) and Coelho (1993) for a brief summary.

Privatisation of the water industry in England and Wales during 1989 led to the creation of 10 **holdings**, the Water Holding Companies, each one including a Water and Sewerage Company (WaSC); besides these, 23 more companies were appointed for water supply, the so called **Water Only Companies** (WoC). These Appointments run for a minimum period of 25 years.

A clear distinction was introduced between the role of service supplier and those of customer protection and quality regulation, which were attributed to:

- The Drinking Water Inspectorate (DWI) - Drinking Water Quality,
- The National Rivers Authority (NRA) - Environmental Regulation,
- The Office of Water Services (OFWAT) - Economic Regulation and Customer Protection; OFWAT is also in charge of facilitating competition and promoting efficiency.

One of the most innovative aspects of this structure is the importance given to the consumer, and for his protection the 1989 Water Act establishes that:

- Consumers must have access to documents where their rights are presented in a clear and simple way,
- Consumers have the right to examine the records of water distribution,
- Companies must make public their performance, that is, their behaviour in terms of minimum levels of service legally established,
- A Guaranteed Standards Scheme assures to the consumer a compensation of 5 pounds per day or per event, in the case that the minimum levels of service are not achieved.

Consumers are directly represented by the Customer Service Committees, created by OFWAT for each WaSC operating area. The Director General of Water Services, head of OFWAT and responsible in the last analysis for the defense of the consumers' rights, has an obligation to act when levels of service stay below the legal limits.

The Appointment stipulates that companies must provide an annual report to the Director General, where, among other information, they must compare their performance with certain levels of reference and forecast their future evolution. Designated by the Director General (DG), the following indicators for levels of service have been set out in the 1991 version:

- *DG1 - Raw Water Availability*
  Indicates the population supplied by a company from water sources where the reliable output in abnormally prolonged dry conditions is less than the peak demand of the customers and there is a risk of water shortage.

- *DG2 - Pressure of Mains Water*
  Monitors the number of properties assessed as being at risk of experiencing a water pressure below the reference level of service for some period during the year.

- *DG3 - Interruptions to Supply*
  Monitors the number of properties that have experienced an interruption to the water supply for more than 12 hours without prior notice.

- *DG4 - Hosepipe restrictions*
  Shows the population actually subject to water restrictions in a reporting year.

- *DG5 - Flooding Incidents from Sewers*
  Identifies the number of buildings at risk of flooding from sewage more than twice in ten years because of inadequate sewers.

- *DG6 - Response to Billing Queries*
  A measure of the speed with which a company responds to both written and telephone billing queries.

- *DG7 - Replies to Written Complaints*
  Deals with responses to written complaints.

Due to its subjectivity, DG8 was retired. Meanwhile, further levels of service have been proposed, mainly connected with quality and environmental improvement, rather than customer service:

- DG9, drinking water quality compliance measured at the customers tap,
- DG10 (A), the volume of water entering distribution from treatment works affected by Undertaking and Relaxations,
- DG10 (B), the number of properties affected by section 19 Undertakings for the distribution system,
- DG11, the number of sewage treatment works whose performance is unsatisfactory,
- DG12, the quantity of sludge disposed of in an unsatisfactory manner.

Information also requires to be supplied on combined sewer overflows, sea outfalls and leakage control.

Miller (1993) emphasises that the Director General may use these cross company comparisons to stimulate efficiency improvements in those companies exhibiting low levels of performance when compared to best practice elsewhere. The importance to a company of achieving a "best practice" designation and the influence this may have upon their subsequent negotiations with the Director General with respect to charging policy, is therefore considerable.

3.3.2. Possible Criteria for Measuring Levels of Service

The system described is an example of something that already exists but, of course, it could not be totally transferred to other countries, even if it could be considered as an excellent model. It is not important to discuss now if a universal system is possible. However, a set of desirable characteristics may be suggested (Neves, 1995), even if the system is confined to internal use (Alegre 1993):

(i) **Globality**

The rules must cover all the main aspects that may affect the users.

(ii) **Simplicity**

Restrict the rules to those aspects that really have practical importance.

(iii) **Independence**

The system must be independent of the kind of institution, public, private, or mixed.

(iv) **Objectivity**

The system must assure only one interpretation of the results. Therefore, it should be based on indicators that can be objectively measured.

A complete formulation of criteria can be seen in Alegre (1993). Here, only the main aspects will be described, as well as some application examples. The analysis must be based on periods of one year, and the results of each indicator classified in four levels:

- Level 4: very good service with respect to that indicator,

- Level 3: good service, but some improvements must be planned,

- Level 2: the service has some deficiencies which must be rapidly solved,

- Level 1: the service has too many deficiencies.

With respect to indicators, one group is proposed for technical aspects (e.g. availability of water, quality of the water, pressure in the network) and another group for management style (e.g. response to complaints). The total number of indicators is 20.

Evaluation criteria

Two examples will be shown in order to illustrate how the quality of the service can be objectively measured.

Example 1

Indicator: Consumption restrictions.

Definition: Number of customers that, in the period of one year, were subject to consumption restrictions, divided by the product (number of customers × 365).

Classification: - Level 4: less that 5%

- Level 3: from 5% to 10%
- Level 2: from 10% to 15%
- Level 1: more than 15%.

Example 2

Indicator: Number of failures in the network.

Definition: Number of reported failures in the network, divided by its length.

Classification: - Level 4: less that 10/km

- Level 3: from 10/km to 50/km
- Level 2: from 50/km to 150/km
- Level 1: more than 150/km.

As a final observation, the central idea is to fix values, so that anyone can understand what is a high standard of service, and what is a low service standard. This is important at national level, but also at the international level, in order that citizens and suppliers become objectively informed about the interpretation given to quality standards in other regions. This may even constitute a stimulus to improve service towards the best practice.

3.3.3. Companies Policy on Levels Service

The importance of the levels of service assessment in terms of public image and even under the financial point of view has already been emphasised. Therefore, it is natural to expect that some companies will develop their own evaluation systems, as happened in England, where more exhaustive systems have been implemented in some cases, for self-defense. An inquiry carried out by Coelho (1993) has shown that companies had a tendency to consider DG levels as minimum standards, below which the performances should be considered worrying.

The responses have shown that the most relevant points for the companies were the continuity of the supply, followed by the pressure in the network and the quality of the water, probably the more sensitive for the consumers. For monitoring and data acquisition, model simulation combined with occasional measurement campaigns and/or district metering (macro flow measurements in zones of average size, usually for leakage control) and telemetering, were the most used information sources, complemented by the consumers complaints.

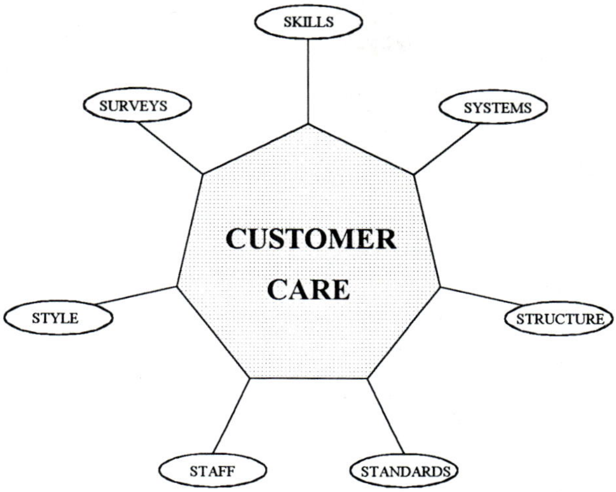

Figure 1. Customer care strategy

Computerised systems to treat the complaints are very popular, and sophisticated solutions have been implemented in some cases, based on the consumers perception about some variables connected to the level of the service, such as the pressure, interruptions to supply, taste and odour of the water.

Customer Care Strategy (CCS) is a very sensitive point for the companies. Figure 1. shows the so-called "Seven S" as an example adopted by one of them (Miller 1993).

Let us follow the explanations of the company:

Standards

Besides external standards, internal ones must be refined to support internal customer relationships. The skill courses reinforce personal standards.

Skills

From time to time staff receive 2 to 3 days training to develop face to face, telephone and letter writing skills.

Systems

Implementation of integrated computerised systems eliminates less efficient non-integrated computer and manual systems. This leads to fewer delays, mistakes and reworking of information.

Reward systems promote the linking of individual performance and achievement of personal objectives to corporate objectives.

Structure

A customer centre providing access to all services via a single telephone number for all customers is important. Customers will speak immediately to customer service staff who have both the technology and the authority to resolve many queries immediately.

Style

Corporate image expressed through its logo, letterheads, uniforms, vehicles and accommodation is also relevant.

Surveys

Customer surveys should be repeated at intervals to register the impact of changes.

Staff

An open management style is recommended to improve two way communication and encourage initiative and motivation at all levels. Introduction of team briefing and an employee suggestion schemes are two examples..

Finally, the company recognises that new methods of reinforcing customer care services have to be continually found out.

3.4. REFERENCES

Alegre H. (1993) Instrumentos de Apoio à Gestão Técnica de Sistemas de Distribuição de Água. Teses e Programas de Investigação, Laboratório Nacional de Engenharia Civil (LNEC), Lisbon, Portugal, ISBN 972-49-1608-1.

Coelho S. (1993) Avaliação de Níveis de Serviço: Introdução ao Modelo Britânico. LNEC Seminar, Lisbon, Portugal.

Hueb J. (1993) Aspectos Operacionais Ligados com a Gestão dos Serviços. LNEC Seminar, Lisbon, Portugal.

Matos R. (1993) Regulamentação, Normalização e Certificação do Sector da Indústria da Água no Quadro do Mercado Europeu. LNEC Seminar, Lisbon, Portugal.

Miller A. (1993) Levels of Services and Customer Care Strategy in a Major UK Water Service Company. LNEC Seminar, Lisbon, Portugal.

Neves M.J. (1994) Levels of Service in Water Supply Systems. Proceedings of the NATO Advanced Study Institute New Technologies for Large Water Supply Systems, Albena, Bulgaria.

Neves A. (1995) Níveis de Serviço em Saneamento Básico. Relatório Interno da Comissão de Coordenação da Região do Norte, Porto, Portugal.

4. Examples of Existing Management Systems in Europe

Paolo Veltri
Dipartimento di Difesa del Suolo
Università della Calabria
87040, Montalto Uffugo (Cs)
ITALY

In most cases the existing services appear to be complex models, because more than one form of water distribution service is present at the same time. The current arrangements have been influenced by many cultural, economical, political, historical, social factors, so that no system could be easily exported to other situations, but it can represent a useful tool of comparison, in order to bring on new water management systems in the countries which now adopt obsolete and uneconomic systems.

4.1. FRANCE

In France water resource management has been divided since 1964 into Six River Basin Agencies, recently renamed Agences d'Eau, based on river catchment areas. The tasks of the Agencies are now related to many aspects of water management, including water pollution control, within an administrative framework in which final responsibilities are the concern of the Prefect of the Department. The Agencies collect two kinds of charges, for drinking and polluted waters, which are then partially transferred to communes in order to construct and upgrade treatment facilities. Each Agency is formed by a parliament. In 1992 a new law decreed the creation of the Local Water Commission, whose tasks are to plan water management and collect information on the ongoing water politics. The State is always present, together with other partners, namely local municipalities or private enterprises. The body responsible for water provision service is the commune, which can choose the management system best in keeping with municipal needs. Communes themselves (through "regies"), intermunicipal water associations, inter-municipal multi-purpose associations directly

NATO ASI Series, Partnership Sub-Series, 2. Environment – Vol. 15
Water Supply Systems – New Technologies
Edited by Č. Maksimović, F. Calomino, and J. Snoxell

manage about 30% of communes mainly in rural areas. Mixed or private companies manage by delegation the remaining 70% in a sort of concession or leasing contract. The same body is often charged with the water supply as well as sewage disposal services.

In the case of public or semi-public managements the investments are financed through subsidies. Water charges have to take into account both operating and investment costs, they are fixed by the commune (or intermunicipal association), which may also provide for extra-subsidies when prices are not balanced. Many institutional subjects, such as central government and regional administrations, control legal, sanitary and financial management.

Also in the case of private company management, the commune is always responsible for the service; it also owns facilities and decides on the investment program on the proposals of the contractor. The commune and the company freely contract duration and water charges, including an automatic price revision formula as well as the possibility to renegotiate prices if relevant changes take place. These control mechanisms also regulate private management.

There is a very great number of communes in France, and drinking water prices vary greatly, ranging from 0.42 FF/m^3 to 10.92 FF/m^3.

4.2. ENGLAND

Drinking water and sewage disposal services were centrally organized in England by a 1974 law, which created 10 Water Authorities, corresponding to the main river catchment areas. They collected charges for the services from the consumers, but also received subsidies from the government to promote investments. So in 1989 the English water system was again re-shaped, and the above 10 Authorities became privately-owned companies, which presently manage both drinking and waste water services. In addition, there are now 23 statutory water only companies, which manage drinking water services, while waste water disposal is carried out by the main companies. The entity responsible for water provision are the companies themselves, which obviously own all the facilities. Some of the companies are listed on the Stock Exchange, and the most important French private companies have recently purchased shares. As for charges, the English system is based on the well known price-cap method. Prices are up-dated through formula mentioned previously. According to their nature many regulatory constraints are provided by law in terms of economics, quality and environmental aspects. In particular the management is subjected to the control of national regulatory organizations, such as OFWAT. Since water meter installation does not cover all apartments, customers do not pay charges for the volume of water actually delivered, but bills based on the value of the houses. The annual average charges for the drinking water service vary in England (Zabel and others, 1994) from 73 £/house to 115 £/house. Because the number of management companies is limited and charges are generally the same within the same company no great price differences are noticed.

A very similar criteria is common in the US water supply companies.

4.3. ITALY

Many legislative provisions have to be taken into account when dealing with water resources in Italy. First of all Italy is divided in 20 Regions and, according to a 1989 law on Soil Conservation, River Basins have been created in the whole territory which are based on river catchments areas. Moreover other laws concern sewage disposal and treatment plants. When dealing only with drinking water supply, the distribution services are at present rapidly changing, because of two recent laws passed by Parliament: the first one dates back to 1990 and is a general re-shaping of local government authorities; the second one dates to 1994 and deals with provisions in water resources matters. It is necessary to focus on the situation existing before these new laws in order to look at the possible changes and the problems to face. The new legislative framework inherited from the past different management systems can be summarized as follows:

- in most great communes the service management is performed by Communal or Intercommunal Association Companies ("Aziende municipalizzate") which are also often in charge of other municipal services, such as sewage disposal, gas, electricity,
- in small communes the direct management ("by economy") is nearly always adopted,
- in some cases concessions to private companies have been implemented or public commercial companies created;
- in three cases (Sicilia, Sardegna and Puglia) Regional State Bodies have been set up, which are responsible for both water provision and management service.

In the case of management "by economy" the commune itself manages the distribution service, being the water service included into the general municipal structure. Even if the water charges have to be fixed by trying to balance costs and profits, it is very difficult to estimate accurately expenses; frequently the water price does not cover the costs, so that the balance has to be filled through subsidies. The prices fixed by the local authority have also to be approved by both Provincial and National Price Committees, while local public health agencies are responsible for health standards.

Among communal or intercommunal Companies, the Azienda Municipalizzata is a financially independent structure of the municipality, which manages the service on behalf of the commune itself, and has a separate budget. The investments are financed according to their nature, but only the local authority can allocate funds for the company. Direct management and management by Azienda Municipalizzata share the same rules to fix charges and the same control entities.

Also in the case of Regional State Bodies price setting and controls are the same as the other management systems, except for those made by Provincial Price Committee.

The Public Commercial Companies are share holding companies, created by the commune, whose participants are public entities, namely departments and companies. The managing company performs the service on the basis of a contract, it finances investments and owns the facilities. Again, price setting and controls are the same as the other management forms. The economic management is at the risk of the company.

Finally, according to private company management, the commune delegates the service on the basis of a contract, which can last up to 30 years; investments can be financed according to their nature; facilities are municipally-owned, except those directly financed by the company, which, in any case, will return to the municipality when the contract expires, after assessing their depreciated value. Prices are proposed by the company, but they must be approved by both Provincial and National Price Committees. Controls are the same as mentioned above. Also in this case the economic management is at the risk of the company.

As a result of the above management systems, water has not yet been administered as an economic good. Prices, which have been determined by political and social elements, are often unbalanced, the service quality is outside of any control, investments are often delayed and they are allocated by external funds without influencing charge policy, the conditions of facilities and, in particular, distribution networks are often obsolete. Users are now accustomed to think that low tariffs mean low quality of service. High domestic consumptions and corresponding low prices clearly reflect the way water is supplied in Italy.

The last law which recently passed the Italian Parliament has re-shaped water service management policy. First of all, this law promotes the integrated management of all water public services, from water supply to sewage disposal and wastewater treatment; secondly water services must have a territorial basis, according to some optimum territorial areas ("Ambiti territoriali ottimali"), while now several hundreds of municipal services are spread even at a regional scale (at a national level there are now about 6.000 drinking water entities and about 4.000 wastewater treatment ones). Moreover, the Price Committees have been abolished; the tariff, which has to be intended as a compensation for the service, must in any case cover total investment and running costs. Once again, the commune is responsible for the water service, and it can choose the most suitable management system among the following forms:

- direct management (this form, however, even if based on economic criteria, will disappear),
- delegated management by concession,
- management by a special company ("Azienda Speciale"), which will likely take the place of the former "Azienda Municipalizzata", with a new legal form,
- management by joint-stock companies, with public and private capitals.

The relationships between the local authority and the management entity, if separate, will be regulated through a formal contract, which will establish the duration, the criteria of economic and financial planning, control rules, warrants, penalties, and the kind of tariff. In any event tariffs will be fixed by the local authority.

4.4. GERMANY

Germany is divided in 16 individual Länder, 5 of which are new Länder created after the fall of the Berlin wall and 1990 German unification.

Provisions which allow municipal self-government date back to 1808, when municipal structures for water supplies and sewerage were established. Over this period the municipalities of the Federal Republic of Germany developed a wide range of provisions in order to handle all the municipal services, holding in any case the service responsibility or a large degree of control. On the contrary, in the former Eastern Democratic Republic the municipal water management was abolished in favour of centralized planning, based on several river authorities as well as on the so-called People-owned water utilities, one for each district. As for the service organization, another important difference must be highlighted (Kraemer, 1994): Western municipalities separated the water supply service from sewage disposal and wastewater treatment, because water supply had traditionally been considered as a commercial activity, including also eventually other public services, such as gas, electricity, urban heating; on the contrary, the former Eastern Germany set up integrated models of water supply and sewage disposal. Though the integrated management of both services, which anyhow are under direct municipal control, is profitable, separate management organizations in Western Germany have a historical origin, because sewage disposal has always been looked at as an essential public service, including those citizens who can not pay service charges.

Following the fall of the former Eastern Germany, which accepted the West German law and administrative regulations, a very fast process of water service re-organization is now operating in Germany, whose main aims are (Kraemer, 1994):

- the restoration of municipal control and administration in the cities of the new Länder, by assimilating West German management models,
- the creation of integrated systems.

Regarding the existing management models, the local government is responsible for water supply and it can freely choose the best means to reach its aim, according to city size, historical reasons, customs and so on.

Each municipality can choose its water management structure among many models, such as direct municipal management, joint management with other local communities, management with a third public or private body. A simplified classification includes:

- direct municipal management (Regiebetriebe),
- management by "Stadtwerke", which includes both forms of municipal enterprises, the so-called "Eigenbetriebe", namely a semi-autonomous municipal agency, and other commercial forms, such as the so-called "Eigengesellschaften", or mixed companies, or private companies,
- management by intermunicipal associations (Zweckverbände), whose members can be either local authorities or private persons,
- management by public companies with other public sector organizations.

The Stadtwerke model prevails in the biggest cities, whereas in rural areas and in smaller communes the Zweckverbände are common.

When a third body is responsible for the management a delegation contract between the municipality and the company rules the relationship. The delegation can be carried out by contract or statute.

From an economic point of view, water charges must enable the service to break even, whatever the adopted management form, being in any case the Cost Recovery method adopted. The charges can be either prices, if the management system is under private law, or fees when public law is applicable in the management system. Depending on the company features, charge control is carried out by municipal/intermunicipal authority or a Cartel Office.

Investment strategies as well as overall control are up to local authority.

According to their nature, other controls are finally carried out by several bodies.

4.5. REFERENCES

Kraemer R.A. (1994) Restructuring Water Institutions in Germany. Hydrotop 94, 12-15 avril, Marseille.

Zabel T.F., Rees Y., Correia F.N., Neves E.B. (1994) Institutional Arrangements for Water Management in Selected EC Countries. Hydrotop 94, 12-15 avril, Marseille.

5. Institutional Framework for Training of Specialists for Water Supply Sysytems
*The Status and Problems to be Solved *

Čedo Maksimović
IRTCUD, Institute of Hydraulic Engineering
Faculty of Civil Engineering
University of Belgrade
P.O. Box 895, 11 000 Belgrade
YUGOSLAVIA

5.1. INTRODUCTION

In broad terms the institutional framework requires the analysis of a wide list of topics such as economic and social conditions, legal framework and enforcement policies, hierarchy of institutions at various levels, decision making procedures and, interaction between public and private sector, technology transfer and training incentives, public awareness, etc. These topics have been discussed at various conferences and symposia (such as Lisboa Conference, 1995, for the group of several EU countries and USA). All of the above factors affect the way and means in which the introduction of new technological achievements find its way into Water Supply Systems companies. Thus in this Section, an emphasis will be placed only on the educational aspects of institutional framework.

To achieve high standards or high levels of service in water supply systems (WSS) it is essential that all specialists and other employees who take part in planning (design), production, distribution, recycling (reuse) have an opportunity to learn how they can contribute to the improvement of either water quality, safety of supply, efficiency in production, reduction in consumption, environmental protection, etc. These tasks are not easy to achieve unless there is an organized system of continuous training and/or occasional update of knowledge, improvement of skills and mastering of new technologies.

Since the present book is targeted at the international market one should bear in mind the big differences not only between developed and developing countries, but also in

NATO ASI Series, Partnership Sub-Series, 2. Environment – Vol. 15
Water Supply Systems – New Technologies
Edited by Č. Maksimović, F. Calomino, and J. Snoxell

technological levels and in institutional frameworks in various regions of developed countries (North America, i.e. USA and Canada, Western Europe, etc.).

Presentation of the current situation would require more space then is possible here. However, there is no doubt that water supply projects have a high priority among the other infrastructure and environmental protection projects. This should normally imply that for successful development of this essential urban infrastructure system, an appropriate organizational, social and human resources infrastructure should be built in order to support the system. Unfortunately this is not often the case. Several levels of institutional framework will be tackled here (Figure 1.).

Level	Major shortcomings (region and country dependent)
Government and ministries	• lack of coordination and interference of competence • improper priorities in training • conflicting regulations • lack of updated information
Education at universities	• outdated curricula • lack of holistic approach neglected • poor coordination between various specialists
Professional associations	• too much local expertise • lack of international flavor • low influence on companies politics
Water companies and water supply enterprises	• lack of funding in developing countries • mismanagement • improper training of staff • outdated technologies
Operational management units	• outdated technologies • poor maintenance • low reliability of data
Consulting companies	• lack of holistic approach • low responsibility for the system's long term performances • slow response to new technologies

Figure 1. Major problems determing needs for institutional framework improvements in WSS

5.2. COORDINATION AT GOVERNMENTAL LEVEL

Coordination between ministries responsible for different aspects of decision making affecting WS projects is generally low. It is the case in many countries that there is a lack of coordination and cooperation between ministries of water resources, environment, public health, city planning and construction, forestry and agriculture and ministry of finance. This lack of cooperation causes inefficient planning. The result is often that water supply projects are badly conceived, delayed, poorly managed, systems are badly maintained and technologically outdated.

5.3. DISTRIBUTION OF TEACHING AT UNIVERSITY LEVEL

Water supply systems are complex and require either multidisciplinary teaching or a well balanced curricula within a particular department. The departments of sanitary engineering (in most countries) or as separate departments or even universities (Eastern Europe) have usually been overloaded by traditional "design school" programs, which lack integration covering resources management, treatment, environmental protection, modern operational management and supported by a high level of information processing.

Traditionally those engineers were able to design and construct systems which conveyed the required (predefined) amount of water. There was not much concern about the amount of water being measured, stored and conveyed in real world water balance, operational control, diagnosis and energy efficiency. Environmental concern and long term adaptability is the weak point of a system designed in this way. In some countries they are being replaced by departments or institutes of environmental science and engineering (Figure 2. - Denmark, Sweden).

Until recently, the talented young engineers from developing countries (especially from Central and Eastern Europe) have had a very good background in fundamental disciplines (physics, mathematics, mechanics, hydraulics, etc.) and, because of the lack of access to modern computer and information technologies, were rather unskilled in coping with advanced modeling and design techniques. On the other hand, one often hears young engineers and managers from developed countries complaining about being overwhelmed by computers and powerful modeling and graphics facilities but being unable to sufficiently understand the physical background of the processes involved, the phenomena being simulated and the complexity of the water-environment interaction. The term DRIPS (data rich information poor system) complements to the situation in which a lot of data is being collected and stored poorly or improperly processed with very little proper information being generated.

It is not only the low accuracy and reliability of data that contribute to this kind of situation, but also the lack of proper conceptualization and right interaction between the monitoring and data acquisition programs and data processing and information generation.

The author of this Section has often experienced the situation where complex urban water projects fail or malfunction because of the poor coordination of various specialists (hydraulic, mechanical, chemical, automation, electrical, software, structural engineers, etc.) in the design phase and their counterparts in the operational management phase.

A large number of water supply systems exist in which the elementary information such as annual or daily water balance is known only by a fair guess, with uncertainties of the order of at least 50%, are rather the rule than the exception.

This is proof of the low awareness of the importance of the basic quantities which are essential in several tasks such as:

- energy consumption management,
- rehabilitation strategies,
- resources management,
- human resources,
- water balance.

In this respect, the future of the education and training of urban water infrastructure specialists has to be reexamined at both undergraduate (university) and on-the-job level.

The proper coordination in modernizing the existing curricula for undergraduate engineering schools and organization of the proper programs and training tools for continuous education in the field is a strikingly important task that will need to be fulfilled (CALWARE, 1992-1994). Recent trials, such as for water supply (Maksimović, 1991) and for urban drainage are just small efforts that have to be complemented by a number of other activities.

Target group	Topic to be taught
Top level politicians and design makers, bankers	• water = money (in real terms) • new technologies • integrated (holistic) approach • human resources development • improve before investing
University (departments of sanitary, environmental engineering and science)	• introduction of innovative methodology • informatic impact • integrated approach (multidisciplinar) • new curricula
Top level managers	• data ≠ reliable data • new technologies (need to be introduced) • impact of informatics • human resources development
Chef engineers and technology managers	• innovative technologies • water balance • energy efficiency • human resources development • maintenance and reliability of data
Professional associations	• appropriate technology to be adjusted • new technologies ≠ appropriate ones • some do it better • training of members • training sites need support
Engineers in consulting and water supply companies	• new technologies • informatic tools can solve part of my problem • efficiency • reliability of data and improvement of system performances • what was wrong with my project
Software developers	• how to prevent "garbage in garbage out" result with my package • reliability assessment • development of educational versions • test cases • colorful ≠ reliable
Equipment manufacturers	• operational reliability

Figure 2. Needs for training

5.4. PROFESSIONAL ASSOCIATIONS

The professional associations at national levels are supposed to play a major role in the organization of professional meetings (seminars, symposia, training courses, etc.) for innovation of knowledge and for transfer of new technologies. However, it is often the case that these events tend to be much of a localized nature and that there is no sufficient international interference and transfer of the appropriate problem solving tools.

The thematic meeting of international associations such as IWSA, although often attended by a few hundred participants do not have much influence on the work of operational managers of engineers working in treatment plants (the "end users" of new technologies), especially in developing countries.

Restructuring and reorganization of the educational and training units within the professional organization is needed such that it enables sufficient information on modern and appropriate innovative technologies to hit the right target groups of specialists:

- top level managers for strategic planning and decision making,
- chief engineers in consulting companies for implementation at design levels,
- chiefs of operational management units for implementation of the new and appropriate technology in the real world,
- information specialists for improvement in the reliability of data on the existing status and characteristics of the system and for merging tools from other disciplines such as GIS, CAD, etc. with water related projects.

The appropriate training tools have to be developed and implemented for training of all of the above groups jointly or separately.

5.5. CONSULTING COMPANIES

The major shortcomings of the infrastructure projects being effected by the consulting companies seems to be:

(i) lack of holistic approach. Project design and management is affected by ignoring the integrated approach at the level of the city, region or river basin, and

(ii) poor quality or lack of reliable data on the characteristics and capacity of the existing systems - followed by low responsibility for the long term performance of the systems being designed.

The latter fact combined with poor maintenance of the WSS especially in developing countries causes fast deterioration and outdating of systems.

One possible way of coping with this problem could be an improvement in regulations by introducing the obligation of water and consulting companies to perform full scale tests on the existing systems before undergoing reconstruction, rehabilitation or upgrade.

An example of successful policy of this nature is the procedure for upgrade of urban drainage systems in the UK.

5.6. WATER SUPPLY COMPANIES

There is an apparent difference in the companies of developed and developing countries.

The characteristics of the transition in the companies undergoing privatization in the EU (UK, France, Italy) is that the companies with relatively high levels of technological management systems (UK) are being forced to shrink these systems as a direct consequence of the request for higher profit. There is not enough experience in the long term consequences and possible worsening of the operational safety of the system. On the other hand the WSS in Italy are just being privatized and there is not enough experience on the effect of the change of ownership on the change in the level of technological modernization and reliability of data collected from the systems and in real time.

The situation in developing countries is much different because of the fact that "cheap labor" replaces technology. Generally, the levels of reliability of data on the systems' characteristics is very low. Outdated, poorly maintained and malfunctioning measuring, regulation and operation equipment is almost a rule.

The instrumentation in control rooms of many companies is similar to that shown in Figure 3.

There is generally a low level of awareness of the importance of data (as simple as water balance). The system can often run with a high level of leakage (losses) without the management being aware of it. From the point of view of long term sustainability these systems are often badly managed.

An important problem is a psychological barrier being placed by senior managers which prevents the introduction of new technology even where the financial resources are "almost' available.

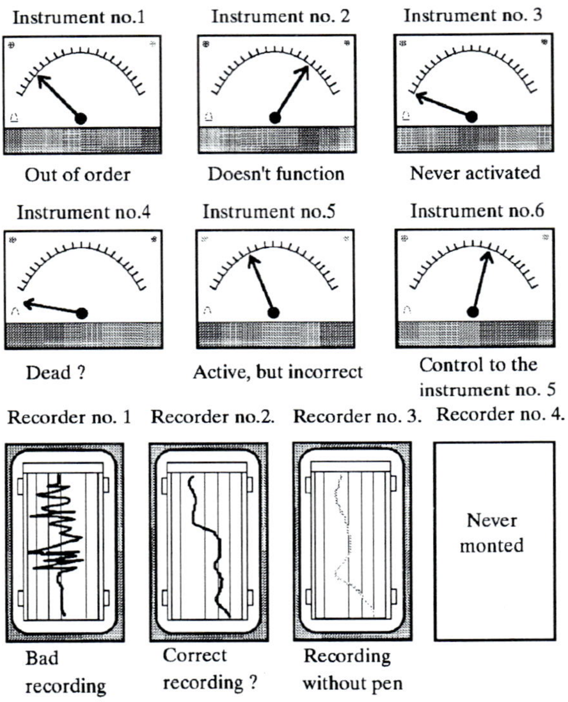

Figure 3. Typical status of instrumentation on a control room of an 'average' WSS

The decisions about big investments on structures, pipelines, dams, etc. are made much more easily than in modernization of equipment, transmission, introduction of informatic support, etc.

There is a major need and room for training of specialists and for the introduction of advanced management at all levels in those countries. However one should be cautions in planning. Gradual, realistically achievable steps are recommended.

5.7. PUBLIC AWARENESS

It is becoming more important to water companies to establish and maintain a permanent contact and cooperation with the water consumers and public in general. As the environmental and ecological awareness and concern of the population increases, the importance of providing the right information to the consumers increases as well.

The projects for new water intakes (especially reservoirs and dams) are becoming more and more difficult to justify and defend. The public has been mobilized and motivated to oppose them. Thus the environmental impact assessment has not only to be done correctly, but all possible measurements have to be well documented and prepared for presentation in an "easy to understand" and popular fashion so that the good projects can get a "green light". Otherwise, good projects can easily be compromised and difficulties in approval experienced. For the reason of preserving the cooperation of the public in rational and prudent use of a vital resource such as water, it is important to devote some intellectual and technological resources for the development of appropriate tools (posters, ride, TV spots, booklets, flyers, software demos, etc.) that can serve this purpose.

Although the local audience has to be addressed in the correct way, taking into account social, cultural, religious, economic and all other aspects, a lot of fact and convincing material is universal and can be shared between various countries.

5.8. REFERENCES

CALWARE (1992-1994) Computer Aided Learning in Water Resources - EU project under the TECHWARE framework.

Lisboa Conference (1995) Proc. of the International Conference on Institutional Dimensions of Water Resources Management. Lisboa, Portugal.

Maksimović Č. (1991) Hypothetical Catchment - An Educational Tool for Training in Sensitivity Runoff Models. Proc. Urban Drainage and New Technologies, UDT '91, Dubrovnik.

Index

- A -

- B -

- C -

- D -

- E -

- F -

- G -

- H -

- I -

- J -

- K -

- L -

- M -

- N -

- O -

- P-

- Q -

- R -

- S -

- T -

- U -

- V -

- W -

- Z -

The Partnership Sub-Series incorporates activities undertaken in collaboration with NATO's Cooperation Partners, the countries of the CIS and Central and Eastern Europe, in Priority Areas of concern to those countries.

The volumes published as a result of these activities are:

Vol. 1: **Clean-up of Former Soviet Military Installations.** Edited by R. C. Herndon, P. I. Richter, J. E. Moerlins, J. M. Kuperberg, and I. L. Biczó. 1995

Vol. 2: **Cleaner Technologies and Cleaner Products for Sustainable Development.** Edited by H. M. Freeman, Z. Puskas, and R. Olbina. 1995

Vol. 3: **Remediation and Management of Degraded River Basins.** Edited by V. Novotny and L. Somlyódy. 1995

Vol. 4: **Earthquakes Induced by Underground Nuclear Explosions.** Edited by R. Console and A. Nikolaev. 1995

Vol. 5: **Transportation Infrastructure.** Edited by R. M. Gutkowski and J. Kmita. 1996

Vol. 6: **Sustainable Development of the Lake Baikal Region.** Edited by V. A. Koptyug and M. Uppenbrink. 1996

Vol. 7: **Transboundary Water Resources Management.** Edited by J. Ganoulis, L. Duckstein, P. Literathy, and I. Bogardi. 1996

Vol. 8: **Urban Air Pollution.** Edited by I. Allegrini and F. De Santis. 1996

Vol. 9: **Advances in Groundwater Pollution Control and Remediation.** Edited by M. M. Aral (Kluwer Academic Publishers). 1996

Vol. 10: **Assessing the Risks of Nuclear and Chemical Contamination in the Former Soviet Union.** Edited by E. S. Kirk (Kluwer Academic Publishers). 1996

Vol. 11: **Ventilation and Indoor Air Quality in Hospitals.** Edited by M. Maroni (Kluwer Academic Publishers). 1996

Vol. 12: **The Aral Sea Basin.** Edited by P. P. Micklin and W. D. Williams. 1996

Vol. 13: **Radioecology and the Restoration of Radioactive-Contaminated Sites.** Edited by F. F. Luykx and M. J. Frissel (Kluwer Academic Publishers). 1996

Vol. 14: **Economics of Atmospheric Pollution.** Edited by E. C. van Ierland and K. Górka. 1996

Vol. 15: **Water Supply Systems.** Edited by C. Maksimović, F. Calomino, and J. Snoxell. 1996

Vol. 16: **Bioindicator Systems for Soil Pollution.** Edited by N. M. van Straalen and D. A. Krivolutsky (Kluwer Academic Publishers). 1996

Vol. 17: **Remediation of Soil and Groundwater – Opportunities in Eastern Europe.** Edited by E. A. McBean, J. Balek, and B. Clegg (Kluwer Academic Publishers). 1996

Vol. 18: **Environmental Engineering and Pollution Prevention – European Network on Excellence and Partnership.** Edited by J. Wotte, W. A. Halang, and B. J. Kraemer (Kluwer Academic Publishers). 1996

Vol. 19: **East-West Life Expectancy Gap in Europe – Environmental and Non-Environmental Determinants.** Edited by C. Hertzman, S. Kelly, and M. Bobak (Kluwer Academic Publishers). 1996